Lecture Notes in Mechanical Engineering

For further volumes:
http://www.springer.com/series/11236

Himanshu C. Patel · Gunamani Deheri
Harshvadan S. Patel · Shreya M. Mehta
Editors

Proceedings of International Conference on Advances in Tribology and Engineering Systems

ICATES 2013

Springer

Editors

Himanshu C. Patel
Department of Mathematics
Lalbhai Dalpatbhai College of Engineering
Ahmedabad, Gujarat
India

Harshvadan S. Patel
University Area
Lalbhai Dalpatbhai College of Engineering
Ahmedabad, Gujarat
India

Gunamani Deheri
Department of Mathematics
Sardar Patel University
Vallabh Vidyanagar, Gujarat
India

Shreya M. Mehta
Department of Mechanical Engineering, University Area
Lalbhai Dalpatbhai College of Engineering
Ahmedabad, Gujarat
India

ISSN 2195-4356 ISSN 2195-4364 (electronic)
ISBN 978-81-322-1655-1 ISBN 978-81-322-1656-8 (eBook)
DOI 10.1007/978-81-322-1656-8
Springer New Delhi Heidelberg New York Dordrecht London

Library of Congress Control Number: 2013948869

Printed on acid-free paper

Springer is part of Springer Science+Business Media (www.springer.com)

Foreword I

In the ancient world, mechanical engineering was developed to satisfy the needs of the military. Archimedes used differential gears in his machines and catapults were used widely by armies. So, mechanical engineering was known as military engineering. During the eighteenth century, the French were the first to begin university-level education in civil engineering. All those engineers who practiced engineering for non-military applications were Civil Engineers. With the development of the steam engine, when military engineers started to provide technology for non-military applications, the British initiated the first professional society for mechanical engineering. As engineers developed mathematical analysis and measurement methods, university education in mechanical engineering came about.

Every machine required movement of metal on metal and the issues of friction, wear, and corrosion became of interest to mechanical engineers. In 1966, when H. Peter Jost pointed out that losses due to friction may be reduced by a systematic study of materials and lubricants, several national centers for tribology were set up in the UK. In India, tribology is studied at a number of engineering institutes. IIT Delhi, for example, has a dedicated Center for Research in Industrial Tribology.

Gujarat Technological University (GTU) is a relatively new University and it is encouraging its affiliate colleges to develop research programs in different areas of engineering. Research in tribology today includes a study of the way in which energy is dissipated at the interface between moving solids, the atomic- and molecular-scale mechanisms of interfacial wear, and the effect of the environment on these processes. Tribology requires an understanding of the property of materials, a study of the effects of fluid film lubrication and simulations of the processes of deformation of bearings. The performance of a magnetic fluid-based journal bearing, the effects of velocity-slip and viscosity variation in spherical bearings, the power law fluid film lubrication of journal bearings with squeezing and with variation of temperature effects—including the performance under cryogenic conditions, are being studied by researchers. The use of ferro-fluids and transient analysis of plain circular bearing with micro-polar fluid are also active areas of research. These studies have led to better efficiencies in multi-cylinder I.C. engines and in development of high-density hard disk drives. Experimental investigation on life cycle analysis of the moly (Mo) coated piston ring in

C.I. engines, the work on tribological properties of CuO Nano-lubricants at elevated temperatures and the work on new composite materials are included in this publication. Research in tribology needs inter-disciplinary work in mathematics, physical sciences, chemical sciences, material sciences, mechanical engineering, and almost all of the engineering disciplines. The work of eminent researchers from all these fields is included in this volume.

An era of good governance in Gujarat is creating an environment, where our businesses and industries have started working closely with the researchers at GTU. By maintaining sustained and continuous interaction with the researchers at GTU, researchers from other national and international institutes can benefit from these close relationships formed by GTU. The close linkages, which have been established by GTU with the industries, can help make our research directly useful and relevant to the needs of the industries. The International Conference on Advances in Tribology and Engineering Systems (ICATES-2013) is being organized by Gujarat Technological University, which is the largest university in the state of Gujarat. GTU is pleased to welcome professors and researchers, working at premier institutes within India and abroad. Let us hope this conference serves to create active research links between the visiting researchers and professors and researchers at GTU.

Gujarat Dr. Akshai Aggarwal

Foreword II

Tribology is an essential aspect of engineering systems. Traditionally, tribology has held an important place in mechanical and aerospace engineering research. However, of late, nano-tribology has gained significance in almost all walks of engineering, thus making the topic of tribology truly multidisciplinary in nature. It is heartening to know that the Department of Mathematics at L. D. College of Engineering, Ahmedabad is organizing the International Conference on Advances in Tribology and Engineering Systems in October 2013, under the aegis of Gujarat Technological University. Researchers from both India and abroad are contributing wholeheartedly. It is my sincere hope that this conference serves to strengthen the partnerships between academic tribologists and industry practitioners. I would like to congratulate both of the organizing institutions, viz., Gujarat Technological University and L. D. College of Engineering. In particular, I felicitate Dr. M. N. Patel and Dr. H. C. Patel for organizing the event, and offer my best wishes for a resounding success of this conference.

Gujarat — Dr. Jayanti S. Ravi

Preface

Tribology is the science and technology of friction, lubrication, and wear. Primarily, mechanical engineers and mathematicians are concerned with this discipline. Bio-tribology is also a fast-emerging field. Tribology of orthopaedic biomaterials and artificial joints is the field of active interest to many. Nano-tribology which studies friction phenomenon at the nanometer scale is also a promising new development in tribology.

The organizers of the *International Conference on Advances in Tribology and Engineering Systems* (ICATES) wish to provide a platform for deliberations on theoretical calculations and experimental results in different areas of tribology. The papers are so selected that, as far as possible, equal emphasis is laid on both theoretical and experimental research. The response to the conference was overwhelming on both national and international fronts. The submitted papers were by renowned experts in the field.

Broadly, the contents of this set of proceedings can be classified into two aspects, namely: analytical methods and experimental validation. There are several papers incorporating the contribution of magnetic fluid (micro polar fluid and power law fluid) toward the reduction of adverse effects of roughness and slip. Nano-tribology in the context of rheology, also figures in some of the studies. A good number of investigations deal with the evaluation of friction and wear from the point of view of design.

There are studies to indicate that tribological properties can be enhanced by mainly using polymers and nanocomposites; and by considering fiber surface modifications. The discussions contained in these proceedings also underline the importance of CuO nano-lubricants at elevated temperatures.

Organized by

Gujarat Technical University

Gujarat Technological University, Ahmedabad

L. D. College of Engineering, Ahmedabad

Organizing Committees

Steering Committee

International Advisory Committee

Review Committee

Dr. Zulfiqar Khan	Associate Dean, R&E, Bournemouth University, Dorset, UK
Dr. P. L. Wong	Associate Professor, Mechanical and Biomedical Engineering Department, City University of Hong Kong, P. R. China
Dr. Amaya Igartua	Head Tribology Unit, Tekniker, Spain
Dr. B. N. J. Persson	Research Scientist, IFF, Forschungszentrum, Julich, Germany
Dr. Guido Paolicelli	Technologist, CNR, Institute of Nano Science, Modena, Italy
Dr. P. Menezes	Post-Doc, University of Wisconsin Milwaukee, Industrial Engineering
Dr. Barun Chakrabarti	Head (R&D), Larsen & Toubro, Mumbai, India
Dr. Sujeet Kumar Sinha	Associate Professor, Mechanical Engineering Department, IIT Kanpur, India
Dr. Harish Hirani	Associate Professor, Mechanical engineering Department, IIT Delhi, India
Dr. Jayashree Bijwe	Professor, Industrial Tribology Machine Dynamics and Maintenance Engineering Centre (ITMMEC), IIT Delhi, India
Dr. Satish V. Kailash	Associate Professor, Mechanical Engineering Department, IISc Bangalore, India
Dr. G. M. Deheri	Associate Professor, Department of Mathematics, Sardar Patel University, Vallabh Vidhyanagar, Gujarat, India
Dr. D. P. Vakharia	Professor, Mechanical Engineering Department, SVNIT, Surat, Gujarat, India
Dr. G. H. Upadhyay	HOD, Metallurgy Engineering Faculties, GEC, Gandhinagar, Gujarat, India
Dr. R. N. Patel	Professor, Mechanical Engineering, Institute of Technology, Nirma University, Ahmedabad, Gujarat, India
Dr. H. C. Patel	Associate Professor, L. D. College of Engineering, Ahmedabad, Gujarat, India
Dr. D. V. Bhatt	Assistance Professor, Mechanical Engineering Department, SVNIT, Surat, Gujarat, India
Dr. K. N. Mistry	Principal, Sarvajanik College of Engineering and Technology, Surat, Gujarat, India

State Advisory Committee

Dr. G. M. Deheri	Associate Professor, Department of Mathematics, Sardar Patel University, Vallabh Vidhyanagar, Gujarat, India
Dr. D. P. Vakharia	Professor, Mechanical Engineering Department, SVNIT, Surat, Gujarat, India

Website Committee

Editors

[illegible] Committee

[illegible]

Website Committee

Dr. H. B. Raval [illegible] Associate Professor, L.D. College of Engineering, Ahmedabad, Gujarat, India

Mr. H. K. [illegible] [illegible] Student, L.D. College of Engineering, Ahmedabad, Gujarat, India

[illegible] Research Student, L.D. College of Engineering, Ahmedabad, Gujarat, India

Editors

Dr. H. S. Patel [illegible] Associate Professor, L.D. College of Engineering, Ahmedabad, Gujarat, India

Dr. G. M. Deheri [illegible] Associate Professor, Department of Mathematics, S. P. University, Vallabh Vidyanagar, Gujarat, India

Dr. [illegible] Patel [illegible] Professor, Applied Mechanics [illegible] Engineering [illegible] [illegible], Gujarat, India

[illegible] Associate Professor, L.D. College of Engineering, Ahmedabad

Contents

Contributors

Nikhilkumar D. Abhangi Faculty of Engineering, Department of Mathematics, Marwadi Education Foundation's Group of Institution, Rajkot, India, e-mail: nikhil.abhangi@gmail.com

Dr. S. K. Acharya Department of Mechanical Engineering, National Institute of Technology, Rourkela 769008, India

P. I. Andharia Department of Mathematics, Maharaja Krishnakumarsinhji Bhavnagar University, Bhavnagar, Gujarat 364002, India, e-mail: pareshandharia@yahoo.com

P. K. Arya Tribology and Combustion Division, CSIR, IIP, Dehradun 248005, India

Dr. D. V. Bhatt Mechanical Engineering Department, Sardar Vallabhbhai National Institute of Technology, Surat, Gujarat 395007, India

Dr. M. K. Bhatt Department of Mechanical Engineering, Sardar Vallabhbhai National Institute of Technology, Surat, Gujarat 395007, India

Jayashree Bijwe Industrial Tribology Machine Dynamics and Maintenance Engineering Centre, Indian Institute of Technology, Delhi, India, e-mail: jbijwe@gmail.com

M. S. Bobji Department of Mechanical Engineering, Indian Institute of Science, Bangalore 560012, India

P. C. Chhotani Department of Mechanical Engineering, Chhotubhai Gopalbhai Patel Institute of Technology, Bardoli, Gujarat 394350, India, e-mail: paresh.c.chhotani@gmail.com

R. S. Damor Department of Applied Mathematics and Humanities, Sardar Vallabhbhai National Institute of Technology, Surat 395007, India, e-mail: rameshsvnit2010@gmail.com

Gunamani Deheri Department of Mathematics, Sardar Patel University, Vallabh Vidyanagar, Anand, Gujarat 388120, India

Nidhi Dureja Industrial Tribology Machine Dynamics and Maintenance Engineering Centre, Indian Institute of Technology, Hauz Khas, New Delhi 110016, India

Z. M. Fang Department of Mechanical Engineering and Automation, Shanghai University, Shanghai 200072, China

Piyush P. Gohil Department of Mechanical Engineering, CSPIT, CHARUSAT, Anand, Gujarat, 388421, India, e-mail: piyushgohil.me@charusat.ac.in

K. Gouthami Department of Mathematics, K. L. University, Vaddeswaram, Guntur, Andhra Pradesh 522502, India

S. P. Harsha Vibration and Noise Control Laboratory, Mechanical and Industrial Engineering Department, Indian Institute of Technology, Roorkee, Uttarakhand 247667, India; Mechanical and Industrial Engineering Department, Indian Institute of Technology, Roorkee 247667, India

S. B. Jadeja Department of Mechanical Engineering, B. H. Gardi College of Engineering and Technology, Rajkot, Gujarat 360005, India

T. Moasunep. Jamir Department of Mechanical Engineering, Indian Institute of Technology, Guwahati 781039, India, e-mail: t.jamir@iitg.ernet.in

R. Jayaganthan Department of Metallurgical and Materials Engineering, Indian Institute of Technology, Roorkee, Uttrakhand 247667, India

R. B. Jotania Department of Physics, University School of Sciences, Gujarat University, Ahmedabad, Gujarat 380009, India

Ajay Kumar Kadiyala Industrial Tribology Machine Dynamics and Maintenance Engineering Centre, Indian Institute of Technology, Hauz Khas, New Delhi 110016, India

S. K. Kakoty Department of Mechanical Engineering, Indian Institute of Technology, Guwahati, Guwahati 781039, India, e-mail: sashin@iitg.ernet.in

Sanjeev Katoch Center for Material Science and Engineering, National Institute of Technology, Hamirpur 177005, India, e-mail: katochsanjeev@gmail.com

Neelima Khare Refueling Technology Division, Bhabha Atomic Research Centre, Mumbai 400085, India, e-mail: neelima.khare@gmail.com

Chacko Preno Koshy Department of Mechanical Engineering, National Institute of Technology, Calicut, Kerala 673601, India, e-mail: chackopreno@yahoo.co.in

J. Vijaya Kumar Department of Mathematics, Vasireddy Venkatadri Institute of Technology, Nambur, Guntur, Andhra Pradesh 522508, India

Sushil Kumar Department of Applied Mathematics and Humanities, Sardar Vallabhbhai National Institute of Technology, Surat 395007, India, e-mail: sushilk@ashd.svnit.ac.in

Z. Li Department of Mechanical Engineering and Automation, Shanghai University, Shanghai 200072, China

P. K. Limaye Refueling Technology Division, Bhabha Atomic Research Centre, Mumbai 400085, India

Nirav P. Maniar Department of Mechanical Engineering, Dharmsinh Desai University, Nadiad, Gujarat 387001, India, e-mail: niravpmaniar@rediffmail.com

Puneet Mathur Department of Mechanical Engineering, National Institute of Technology, Kurukshetra, Haryana 136119, India, e-mail: puneet1802@gmail.com

Shruti S. Mehta Department of Mathematics, Sardar Patel University, Vallabh Vidyanagar, Gujarat, India

R. Mittal Department of Metallurgical and Materials Engineering, Indian Institute of Technology, Roorkee 247667, India, e-mail: rashmimittal3@gmail.com

Niharika Mohanta Department of Mechanical Engineering, National Institute of Technology, Rourkela 769008, India, e-mail: mohanta.niharika@gmail.com

N. R. Panchal Department of Physics, M. G. Science Institute, Ahmedabad, Gujarat 380009, India, e-mail: Nital_panchal@yahoo.co.in

S. S. Panda Regency Institute of Technology, Yanam 533464, India

D. H. Pandya Vibration and Noise Control Laboratory, Mechanical and Industrial Engineering Department, Indian Institute of Technology, Roorkee, Uttarakhand 247667, India, e-mail: veddhrumi@gmail.com

G. Paolicelli Dipartimento di Scienze Fisiche Informatiche e Matematiche (FIM), Università di Modena e Reggio Emilia, Modena, Italy

Hiral H. Parikh Department of Mechanical Engineering, Dr. Jivraj Mehta Institute of Technology, Mogar, Gujarat 388340, India

H. A. Patel Department of Mathematics, Gujarat Power Engineering and Research Institute, Mehsana, Gujarat 382710, India, e-mail: himeshpatel87@yahoo.com

H. C. Patel Department of Mathematics, Lalbhai Dalpatbhai College of Engineering, Ahmedabad, Gujarat 380015, India

Chirag N. Patel Faculty of Engineering, Pacific Academy of Higher Education and Research University, Udaipur, Rajasthan 313024, India, e-mail: cnpatel.693@gmail.com

Dr. H. S. Patel Applied Mechanics Department, Lalbhai Dalpatbhai College of Engineering, Ahmedabad, Gujarat 380015, India

Jimit R. Patel Department of Mathematics, Sardar Patel University, Vallabh Vidyanagar, Anand, Gujarat 388120, India, e-mail: patel.jimitphdmarch2013@gmail.com

M. P. Patel Department of Mathematics, Gujarat Power Engineering and Research Institute, Mehsana, Gujarat 382710, India

N. D. Patel Department of Mathematics, Sardar Patel University, Vallabh Vidyanagar, Gujarat 388120, India, e-mail: ndpatel2002@gmail.com

N. S. Patel Faculty of Technology, Mechanical Engineering Department, Dharmsinh Desai University, Nadiad, Gujarat 387001, India, e-mail: nimeshsp@yahoo.co.in

P. D. Patel Department of Mechanical Engineering, Nirma University, Ahmedabad, Gujarat 380015, India, e-mail: pdpatel.auto@gmail.com

Pradip M. Patel Department of Mechanical Engineering, Lalbhai Dalpatbhai College of Engineering, Ahmedabad, Gujarat 380015, India

R. J. Patel Refueling Technology Division, Bhabha Atomic Research Centre, Mumbai 400085, India

R. M. Patel Department of Mathematics, Gujarat Arts and Science College, Ahmedabad, Gujarat 380006, India, e-mail: rmpatel2711@gmail.com

R. N. Patel Department of Mechanical Engineering, Nirma University, Ahmedabad, Gujarat 380015, India

U. A. Patel Department of Mechanical Engineering, Lalbhai Dalpatbhai College of Engineering, Ahmadabad, Gujarat 380015, India, e-mail: utkarsh12345@rediffmail.com

Vimal B. Patel Department of Mechanical Engineering, CSPIT, CHARUSAT, Anand, Gujarat 388421, India

Prashant Pendyala Department of Mechanical Engineering, Indian Institute of Science, Bangalore 560012, India

Dhaneshwar Prasad Dr. S. R. K. Government Arts College, Yanam, Pondicherry 533464, India, e-mail: rpdhaneshwar@gmail.com

P. K. Rajendrakumar Department of Mechanical Engineering, National Institute of Technology, Calicut, Kerala 673601, India

R. Raghavendra Rao Department of Mathematics, K. L. University, Vaddeswaram, Guntur, Andhra Pradesh 522502, India, e-mail: rrrsvu@sify.com

Dr. V. J. Savsani Department of Mechanical Engineering, Pandit Deendayal Petroleum University, Gandhinagar, Gujarat 380015, India

Rakesh Sehgal Department of Mechanical Engineering, National Institute of Technology, Hamirpur 177005, India, e-mail: rakeshsehgal.nitham@ gmail.com

Atul S. Shah Department of Mechanical Engineering, Government Polytechnic Waghai, Dang 394730, India, e-mail: asshah97@rediffmail.com

Hetal N. Shah Department of Mechanical Engineering, Venus International College of Engineering, Gandhinagar, Gujarat 382422, India, e-mail: amihetnadiad@gmail.com

M. E. Shimpi Department of Mathematics, Birla Vishvakarma Mahavidyalaya Engineering College, Vallabh Vidyanagar, Anand, Gujarat 388120, India, e-mail: mukesh.shimpi@gmail.com

A. K. Shukla Department of Applied Mathematics and Humanities, Sardar Vallabhbhai National Institute of Technology, Surat 395007, India, e-mail: ajayshukla2@rediffmail.com

B. M. Shukla Tribology and Combustion Division, CSIR, IIP, Dehradun 248005, India

S. D. Shukla Department of Mathematics, Shri R. K. Parikh Arts and Science College, Petlad, Gujarat 388450, India, e-mail: snehaldshukla@gmail.com

D. Singh Department of Metallurgical and Materials Engineering, Indian Institute of Technology, Roorkee 247667, India

Eshan Singh PGRPE, AAT, CSIR, IIP, Dehradun 248005, India, e-mail: eshan@iip.res.in

Vishal Singh Center for Material Science and Engineering, National Institute of Technology, Hamirpur 177005, India

Sujeet K. Sinha Department of Mechanical Engineering, Indian Institute of Technology, Kanpur 208016, India, e-mail: sujeet@iitk.ac.in

N. L. Soni Refueling Technology Division, Bhabha Atomic Research Centre, Mumbai 400085, India

B. K. Sonigra Department of Mechanical Engineering, Sardar Vallabhbhai National Institute of Technology, Surat, Gujarat 395007, India

S. V. Subrahmanyam K. L. University, Vaddeswaram, Guntur, Andhra Pradesh, India, e-mail: sudamshekhar@gmail.com

Dr. B. M. Sutaria Department of Mechanical Engineering, Sardar Vallabhbhai National Institute of Technology, Surat, Gujarat 395007, India, e-mail: bms@med.svnit.ac.in

G. D. Thakre Tribology and Combustion Division, CSIR, IIP, Dehradun 248005, India

Manu V. Thottackkad Department of Mechanical Engineering, National Institute of Technology, Calicut, Kerala 673601, India

M. Tripathi Dipartimento di Scienze Fisiche Informatiche e Matematiche (FIM), Università di Modena e Reggio Emilia, Modena, Italy, e-mail: manoj.triapthi@unimore.it

S. H. Upadhyay Vibration and Noise Control Laboratory, Mechanical and Industrial Engineering Department, Indian Institute of Technology, Roorkee, Uttarakhand 247667, India

P. A. Vadher Department of Physics, Government Science College, Gandhinagar, Gujarat 382016, India

M. B. Vaghela Department of Mechanical Engineering, B. H. Gardi College of Engineering and Technology, Rajkot, Gujarat 360005, India, e-mail: manoj0085@gmail.com

Dr. D. P. Vakharia Department of Mechanical Engineering, Sardar Vallabhbhai National Institute of Technology, Surat, Gujarat 395007, India

S. Valeri Dipartimento di Scienze Fisiche Informatiche e Matematiche (FIM), Istituto Nanoscienze CNR, Modena, Italy

D. J. Varia Department of Applied Mechanics, Government Engineering College, Patan, Gujarat 384265, India, e-mail: djvaria@gmail.com

Rajiv Verma Department of Mechanical Engineering, National Institute of Technology, Kurukshetra, Haryana 136119, India

J. Vijaya Kumar Department of Mathematics, K. L. University, Vaddeswaram, Guntur, Andhra Pradesh, 522502, India

W. Wang Department of Mechanical Engineering and Automation, Shanghai University, Shanghai 200072, China, e-mail: mewwang@shu.edu.cn

P. L. Wong Department of Mechanical and Biomedical Engineering, City University of Hong Kong, Hong Kong, China

Arti Yadav Department of Mechanical Engineering, Indian Institute of Science, Bangalore 560012, India, e-mail: arti@mecheng.iisc.ernet.in

H. K. Yadav Mechanical and Industrial Engineering Department, Indian Institute of Technology, Roorkee 247667, India, e-mail: himanshukyadav2@gmail.com

Z. M. Zhang School of Mechatronics Engineering and Automation, Shanghai University, Shanghai, China

Part I
Fluid Film Lubrication

Experimental Thermal Analysis of Bronze Elliptical and Offset-Halves Journal Bearing Profiles

Rakesh Sehgal

Abstract The present research study is aimed at experimental and theoretical investigation of the thermal behavior of metallic (Phosphor Bronze) elliptical and offset-halves journal bearing configurations of same geometrical size under low to high operating conditions (loads ranging from 800 to 1,900 N and speeds ranging from 2,000 to 5,000 rpm) using three different commercially available grades of oils (Hydrol 68, 2T and Mak Multigrade oil). Experimentally, it is established that offset-halves journal bearing runs cooler than elliptical bearing for all the three grades of oils under all operating conditions thus making it suitable for all operating loads and speeds. The lowest operating temperatures are obtained for 2T oil (oil 2) thus making it most suitable for use under all operating conditions.

Keywords Elliptical journal bearing · Offset-halves journal bearing · Temperature profile · Load capacity

1 Introduction

In the past few decades industries world over have gone through tremendous changes. This is due to ever increasing demand for high speed machinery, be it manufacturing sector, power generation sector, transportation sector or cybernetics. Bearings, especially hydrodynamic journal bearings are extensively used in high speed rotating machines because of their low friction, high load capacity, and good damping characteristics. With the development of technology, hydrodynamic bearings are expected to work with higher load capacity having small sizes, more stability and less degradation of the lubricating oil due to rise in temperature while

R. Sehgal (✉)
Department of Mechanical Engineering, NIT Hamirpur, Hamirpur,
Himachal Pradesh 177007, India
e-mail: rakeshsehgal.nitham@gmail.com

H. C. Patel et al. (eds.), *Proceedings of International Conference on Advances in Tribology and Engineering Systems*, Lecture Notes in Mechanical Engineering,
DOI: 10.1007/978-81-322-1656-8_1,

machinery is running. One problem associated with conventional (circular) journal bearings is that these become highly unstable under high operating speeds. Researchers are continuously striving to overcome this problem and have suggested many design modifications in the circular journal bearings to not only increase stability under high speed operation but also enhance the load capacity and minimize the thermal degradation of the lubricating oil used. One such design modification is to go for multilobe/non-circular configurations such as two lobe bearings (elliptical, offset-halves/orthogonally displaced, lemon bore), three lobe and four lobe bearings etc. These bearing profiles operate with more than one active oil film, the number of active oil films being a function of number of lobes. This feature reduces the cavitation zone and hence accounts for the superior stiffness, enhanced shaft stability, and reduced temperature in the oil film as compared to the circular journal bearings.

Though different researchers have attempted to investigate the performance of non-circular journal bearings, most of these studies have been theoretical and focused on stability analysis. A very few studies (especially experimental) have been reported on the thermal behavior of non-circular journal bearings which is important to establish optimum operating parameters for these bearings.

Pinkus and Lynn [1], Singh and Gupta [2], Tayal et al. [3], Booker and Govindachar [4], Singh et al. [5], Mehta and Singh [6], Read and Flack [7], Hussain et al. [8], Banwait and Chandrawat [9], Sehgal et al. [10–12], Mishra et al. [13], have investigated various non-circular journal bearing profiles. Literature reveals that thermal studies on metallic non-circular journal bearing profiles especially elliptical and offset-halves are very scarce. Chauhan et al. [14–16] have recently attempted few investigations on non-circular journal bearing profiles made from acrylic (Methyl Methacrylite). Further, these investigations are limited to very low range of loads up to 600 N and operating speeds up to 4,000 rpm. However, these bearings find wide applications in high speed machinery because of their better stability and cool running characteristics. Moreover, the design procedures for these non-circular bearing profiles have not been standardized and no design curves/monograms for these bearings are available in design hand books. Hence there is a compelling need to investigate these bearing profiles made from common bearing materials like bronze, aluminium etc. over a wide range of operating parameters such as load, speed using commonly available commercial lubricants. This will help to produce relevant design data to analyse and standardize the design procedures for these non-circular journal bearing profiles for the required operating conditions.

In the present work, experimental evaluation of circumferential bush-oil film interface temperatures at the central plane of elliptical and offset-halves journal bearings made from phosphor bronze under various load, speed, and constant flow rate conditions for three commercial grade oils namely Hydrol 68, Mak 2T and Mak Multigrade has been attempted with an aim to optimize the set of operating parameters as well as to explore the suitability of an oil for the operation of a particular bearing under these operating conditions. In the subsequent sections of this paper, Hydrol 68, Mak 2T and Mak Multigrade oils are referred to as oil 1, oil 2 and oil 3 respectively.

2 Methodology

The objectives of this research study have been achieved by conducting experiments for specifically designed elliptical (Figs. 1, 2) and offset-halves (Figs. 3, 4) journal bearing profiles on an existing journal bearing test rig modified (Fig. 5) suitably to accommodate loads up to 2,000 N and speeds up to 5,500 rpm. The elliptical test bearing is not truly elliptical but has the centres of its upper and

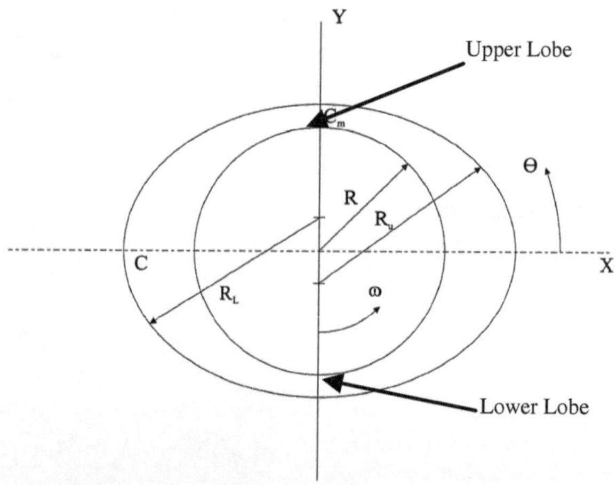

Fig. 1 Schematic diagram of elliptical journal bearing

Fig. 2 Actual elliptical journal test bearing

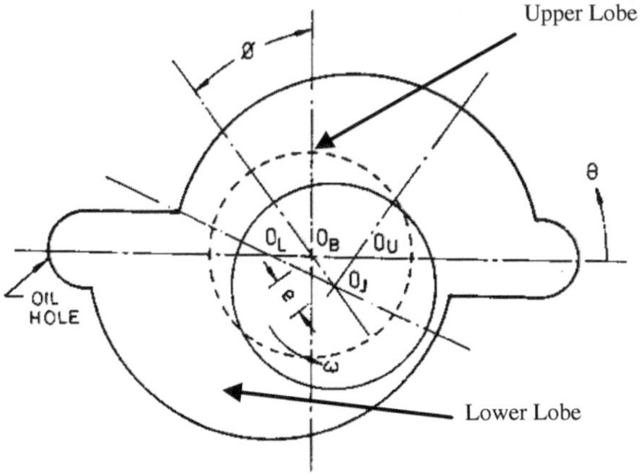

Fig. 3 Schematic diagram of offset-halves journal bearing

Fig. 4 Actual offset-halves journal test bearing

Fig. 5 Modified journal bearing test rig

lower lobes shifted on a vertical line passing through its geometric centre. The offset-halves journal bearing has been designed by orthogonally displacing the two halves (upper and lower) of a cylindrical bearing. The bearings are provided with temperature sensors at 12 circumferential locations (30° angular position).

The journal (Fig. 6) is made of C45 steel material and is mounted horizontally on two pedestal bearings. A chrome plated journal sleeve is tightened at middle portion of journal with lock nuts and bearing slides over it. Dimensions of the elliptical test bearing are given in Table 1. Table 2 gives test operating conditions.

Fig. 6 Actual C45 steel journal

Table 1 Dimensions of elliptical test journal bearing

Parameter	Dimension	Tolerance	Roughness (μm)
Outer dia. of brg, OD	85 mm	±0.2 mm	10
Max. inner dia. of brg., D_{Imax}	65.4 mm	±0.2 mm	10
Min. inner dia. of brg., D_{Imin}	65.2 mm	±0.2 mm	10
Length, L	65 mm	±0.2 mm	10
Radial clearance, C	300 μm	±50 μm	
Min. clearance, C_m	200 μm	±50 μm	
Oil hole	6.35 mm	±0.15 mm	
Rel. sensor position	30°	±1°	

Table 2 Test operating conditions

Oil inlet temperature, T_0	40 °C
Lubricants	Oil 1, oil 2 and oil 3
Rotational speeds, n	2,000, 3,000, 4,000 and 5,000 rpm
Loads, W	800, 1,000, 1,200, 1,500 and 1,900 N
Oil inlet pressure, P	0.1–0.2 MPa

Table 3 Properties of oils and bush material

	Oil 1 (Hydrol 68)	Oil 2 (Mak 2T)	Oil 3 (Mak Multigrade)
Viscosity, μ (at $T_o = 33$ °C)	0.075 Pas	0.065 Pas	0.200 Pas
Viscosity, μ (at $T_o = 100$ °C)	0.00771 Pas	0.004861 Pas	0.01239 Pas
Density, ρ	880 kg/m^3	868 kg/m^3	885 kg/m^3
Thermal conductivity, K_{oil}	0.126 W/m °C	0.126 W/m °C	0.126 W/m °C
Viscosity index	98	135	110
Flash point, °C	230	94	200
Pour point, °C	−9	−24	−21
Barus viscosity—pressure index, α	2.3 e^{-8} Pa^{-1}		
Temp. viscosity—coefficient, γ	0.03 K^{-1}		
Thermal cond. of bush, K_{bush}	54 W/m °C		
Coeff. of thermal expn. of bush, h_{bush}	17 e^{-6} K^{-1}		

Various properties of oils and bush material are given in Table 3. Elliptical and offset-halves journal bearings have been tested for load range (800–1,900 N), speed range (2,000–5,000 rpm) and constant flow rate of 6.8 l/min. The test data for these bearings has been presented and analysed for 800, 1,200 and 1,900 N loads and 2,000, 3,000, 4,000 and 5,000 rpm journal speeds.

3 Results and Discussion

Experimental results obtained for elliptical and offset-halves journal bearings are presented and discussed in the following sub-sections.

3.1 Elliptical Journal Bearing

The variation of circumferential temperatures at mid plane of elliptical journal bearing at 800, 1,200 and 1,900 N load, speed (2,000−5,000 rpm) and constant flow rate of 6.8 l/min for three grade oils are plotted in Figs. 7a−c, 8a−c and 9a−c respectively. It is observed that for all oils, the circumferential temperature at bush-oil interface increases with increase in journal speed. Maximum temperature was observed at lower lobe for all oils and speed conditions. For oil 1, maximum temperature rise (25.5 °C) was found for maximum load (1,900 N) and speed (5,000 rpm) conditions whereas the minimum temperature rise (9.3 °C) was found for minimum load (800 N) and speed (2,000 rpm). For oil 2, the maximum temperature rise (24.7 °C) was observed for maximum load (1,900 N) and speed (5,000 rpm) conditions whereas the minimum temperature rise (9.3 °C) was found for minimum load (800 N) and speed (2,000 rpm). For oil 3, the maximum temperature rise was found to be 27.9 °C for maximum load

Fig. 7 Circumferential variation of mid plane temperatures in elliptical journal bearing at load 800, and speeds (2,000–5,000 rpm) for **a** Hydrol 68, **b** 2T and **c** Mak Multigrade oil

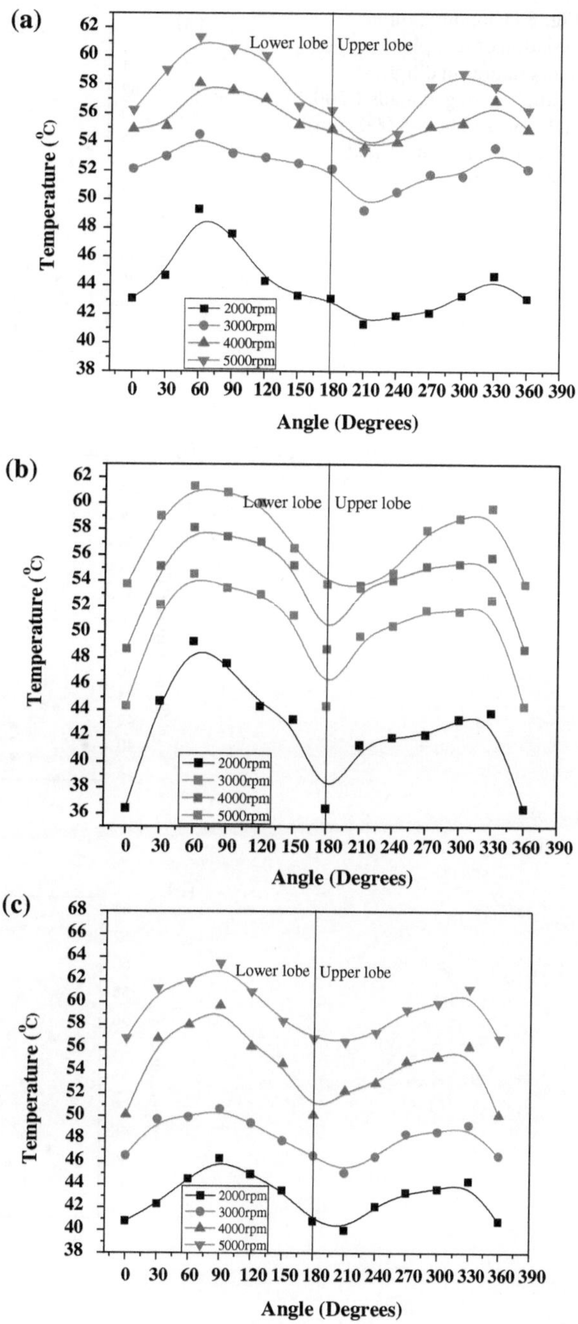

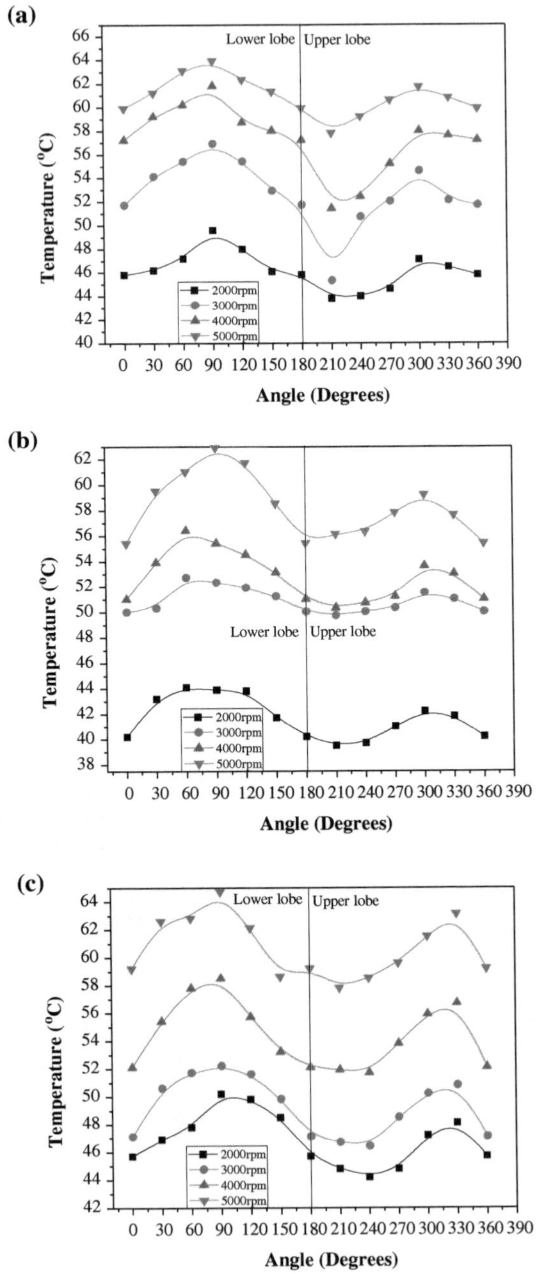

Fig. 8 Circumferential variation of mid plane temperatures in elliptical journal bearing at loads 1,200, and speeds (2,000–5,000 rpm) for **a** Hydrol 68, **b** 2T and **c** Mak Multigrade oil

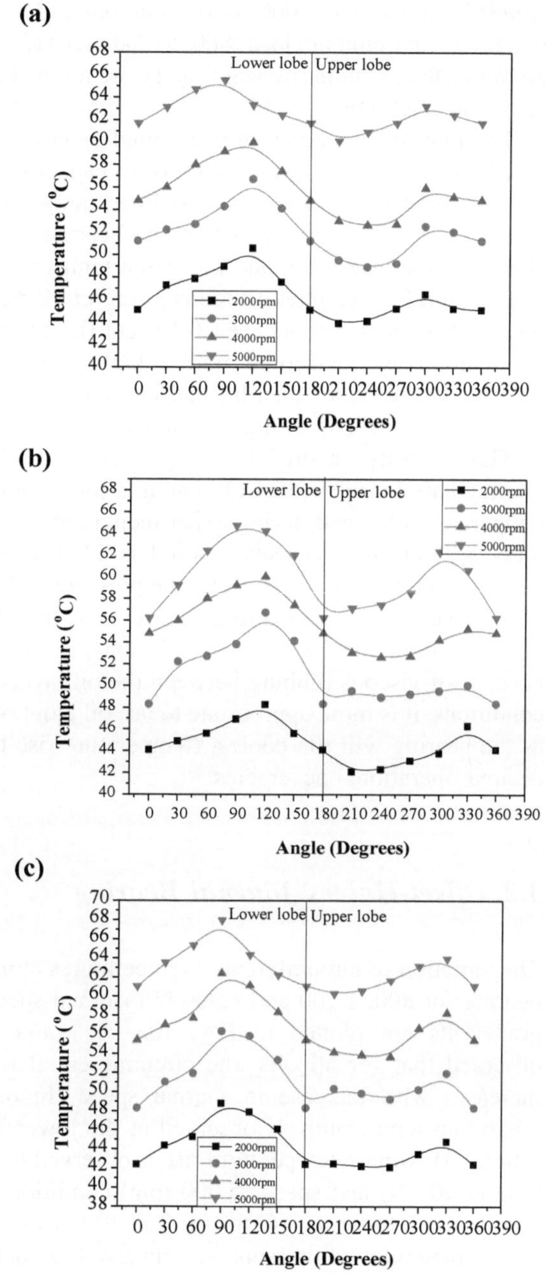

Fig. 9 Circumferential variation of mid plane temperatures in elliptical journal bearing at loads 1,900, and speeds (2,000–5,000 rpm) for **a** Hydrol 68, **b** 2T and **c** Mak multigrade oil

(1,900 N) and speed (5,000 rpm) conditions and minimum temperature rise (6.3 °C) was found for minimum load (800 N) and speed (2,000 rpm). Thus the maximum and minimum temperature rise was found to be for oil 3 and oil 2 respectively under same operating conditions.

The inlet oil temperature was found to gradually increase at higher load and speed conditions. This was due to lower capacity of oil tank in journal bearing test rig, non-availability of external cooling arrangement for exit oil from bearing and mixing of hot oil coming out of the bearing with cold oil present in the tank. However, the maximum and minimum temperature rise has been calculated with respect to reference inlet oil temperature of 40 °C. Average inlet oil pressure was observed to fluctuate between 0.15 and 0.2 MPa. This tendency of oil pressure fluctuation was found to be more at higher operating speeds and loads leading to shearing of layers of oil film and consequent rise in oil temperature and hence lowering of viscosity at higher temperatures.

The viscosity, at oil inlet temperature (40 °C) is highest for oil 3 (121 cP), intermediate for oil 1 (55.4 cP) and minimum for oil 2 (43.3 cP). At the maximum temperature obtained during experimentation using oil 3 (61 °C), the viscosity of oil 3 is 25 cP, the viscosity of oil 1 is 22 cP and the viscosity of oil 2 is 18 cP. Because of decrease in viscosity, the pressure and hence the load carrying capacity of the bearing is expected to decrease. It is, therefore, not advisable to use highly viscous oil in the bearings as this may result in high temperature in bearing because of viscous rubbing between the oil layers. Therefore, under the operating conditions, it is more appropriate to use oil 2 out of the three grade oils considered, as the bearing will run coolest (temperature rise being lowest) thus providing the desired operational accuracies.

3.2 Offset-Halves Journal Bearing

The variation of circumferential temperatures at mid plane of offset-halves journal bearing for 800, 1,200 and 1,900 N load and speeds (2,000–5,000 rpm) for three grade oils are plotted in Figs. 10a–c, 11a–c and 12a–c respectively. It is observed that for all oils, the circumferential temperature at bush-oil interface increases with increase in journal speed. In offset-halves journal bearing the maximum temperature is observed at the lower lobe for all load and speed conditions. Maximum temperature rise is observed to be 28.6 °C for oil 3 at maximum load (1,900 N) and speed (5,000 rpm) conditions. The minimum oil temperature rise is found to be 3.4 °C for oil 2 at 800 N load 2,000 rpm speed.

Average inlet oil pressure was observed to fluctuate between 0.13 and 0.2 MPa. Under the operating conditions, it is more appropriate to use oil 2 out of the three grade oils considered, as the bearing will run coolest. Comparison of circumferential temperatures in elliptical and offset-halves journal bearings for 2T oil (oil 2) at maximum speed (5,000 rpm); maximum load (1,900 N) and minimum speed (2,000 N) and minimum load (800 N) is given in Fig. 13. It is observed that offset-

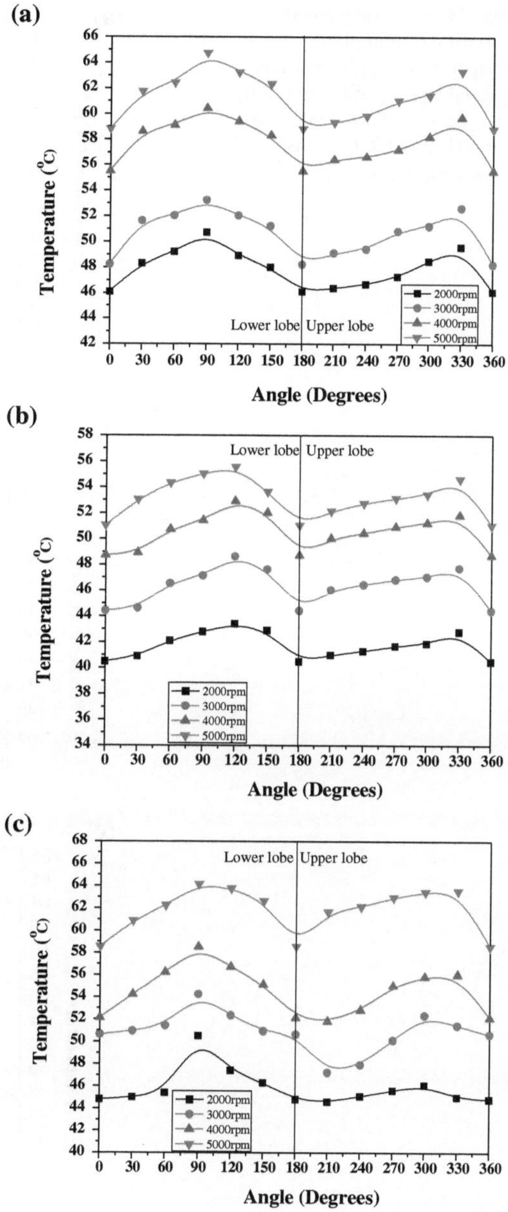

Fig. 10 Circumferential variation of mid plane temperatures in offset-halves journal bearing at loads 800, and speeds (2,000–5,000 rpm) for **a** Hydrol 68, **b** 2T and **c** Mak multigrade oil

halves journal bearing runs cooler and provides thermally stable operation in comparison to elliptical journal bearing under all operating conditions. This is true for oil 1 and oil 2 as well.

Fig. 11 Circumferential variation of mid plane temperatures in offset-halves journal bearing at loads 1,200, and speeds (2,000–5,000 rpm) for **a** Hydrol 68, **b** 2T and **c** Mak multigrade oil

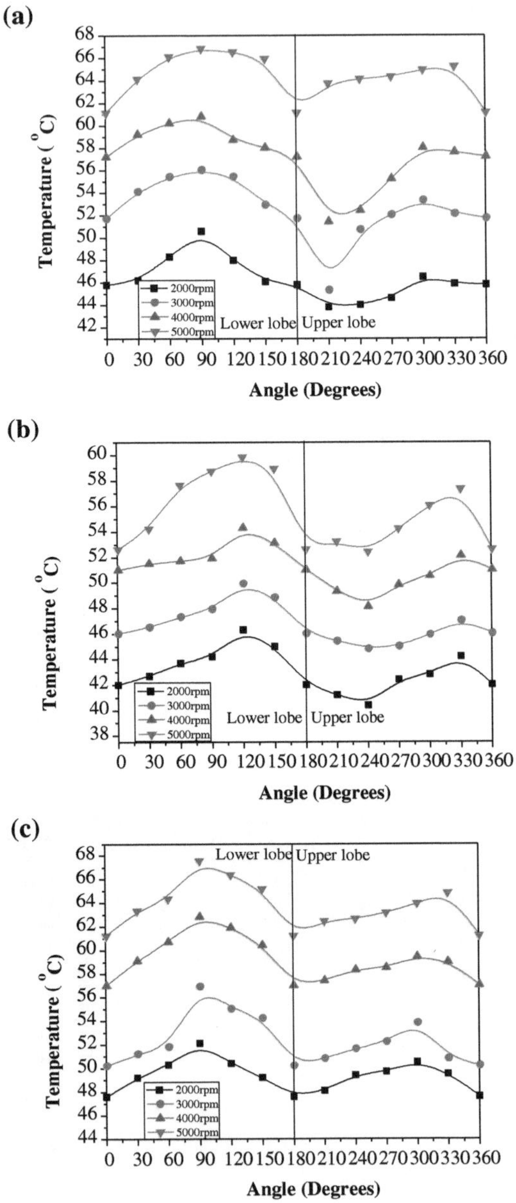

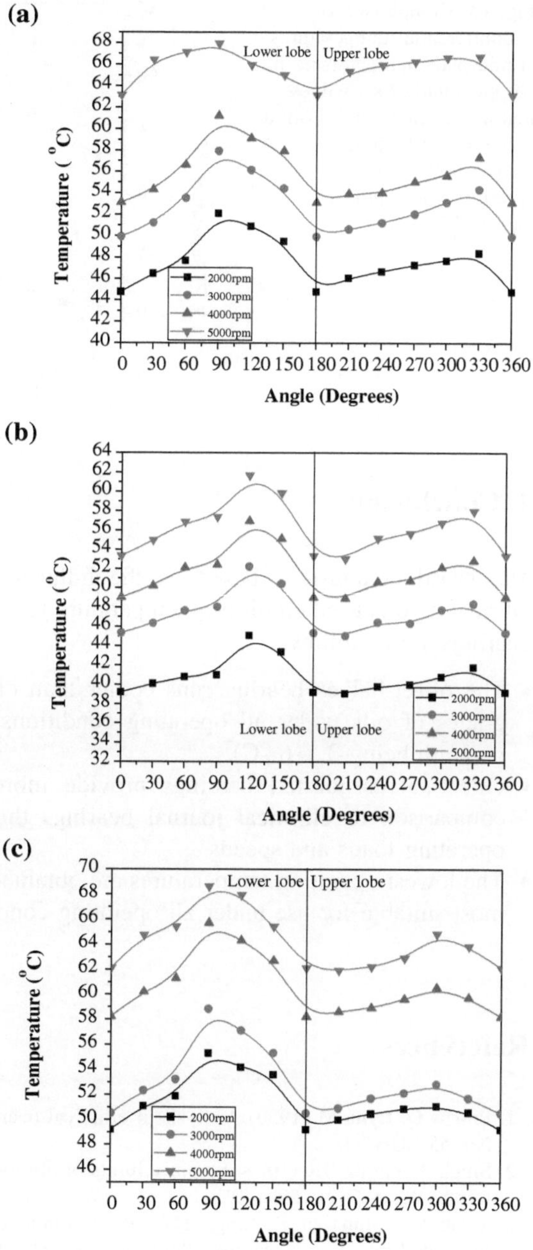

Fig. 12 Circumferential variation of mid plane temperatures in offset-halves journal bearing at loads 1,900, and speeds (2,000–5,000 rpm) for **a** Hydrol 68, **b** 2T and **c** Mak multigrade oil

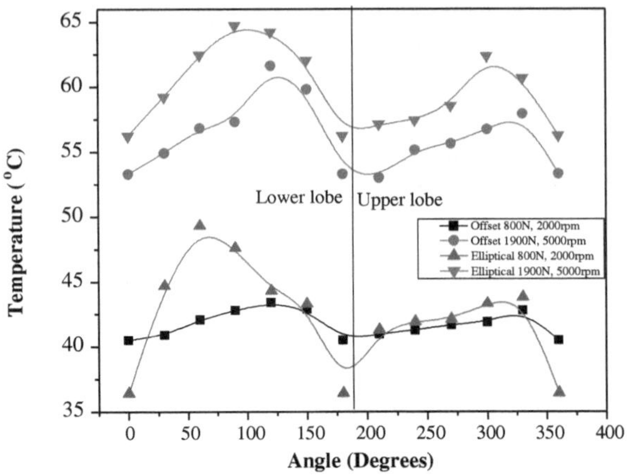

Fig. 13 Comparison of circumferential temperatures at mid-plane temperatures in elliptical and offset-halves journal bearings for 2T oil at speed = 2,000 rpm; load = 800 N and speed = 5,000 rpm; load = 1,900 N

4 Conclusion

The broad conclusions based on the influence of load, operating speed, and lubricating oils used on oil film temperature for elliptical and offset-halves journal bearings are as follows:

- The offset-halves bearing runs cooler than elliptical bearing for all the three grades of oils under all operating conditions (the difference in oil film temperature being 3.5–6 °C).
- Offset-halves journal bearings provide more thermally stable operation in comparison to elliptical journal bearings thus making them suitable for all operating loads and speeds.
- The lowest operating temperatures are obtained for 2T oil (oil 2) thus making it most suitable for use under all operating conditions.

References

1. Pinkus O, Lynn M (1956) Analysis of elliptical bearings. Trans ASME 78:965–973 (Paper No. 55-LUB-22)
2. Singh A, Gupta BK (1982) Stability limits of elliptical journal bearings supporting flexible rotors. Wear 77(2):159–170
3. Tayal SP, Sinhasan R, Singh DV (1982) Finite element analysis of elliptical bearings lubricated by a non-Newtonian fluid. Wear 80:71–81
4. Booker JF, Govindachar S (1984) Stability of offset journal bearing systems. In: Proceedings of IMechE, C283/84, pp 269–275
5. Singh DV, Sinhasan R, Nair KP (1989) Elastothermohydrodynamic effects in elliptical bearings. Tribol Int 22:43–49
6. Mehta NP, Singh A (1986) Stability analysis of finite offset-halves pressure dam bearing. Trans ASME J Tribol 108:270–274

7. Read LJ, Flack RD (1987) Temperature, pressure and film thickness measurements for an offset half bearing. Wear 117:197–210
8. Hussain A, Mistry K, Biswas S, Athre K (1996) Thermal analysis of non-circular bearing. Trans ASME J Tribol 118:246–254
9. Banwait SS, Chandrawat HN (2000) Effect of misalignment on thermohydrodynamic analysis of elliptical journal bearings. IE (I) J 81:93–101
10. Sehgal R, Swamy KNS, Athre K, Biswas S (2000) A comparative study of the thermal behaviour of circular and non-circular journal bearings. Lubr Sci 12:329–344
11. Sehgal R (2010) Experimental measurement of oil film temperatures of elliptical journal bearing profile using different grade oils. Tribol Online 5(6):291–299
12. Sehgal R, Chauhan A, Sharma RK (2013) An experimental investigation of oil film temperatures in elliptical profile journal bearing. Tribol Online 8(1):1–6
13. Mishra PC, Pandey RK, Athre K (2007) Temperature profile of an elliptic bore journal bearing. Tribol Int 40:453–458
14. Chauhan A, Sehgal R, Sharma RK (2010) Thermohydrodynamic analysis of elliptical journal bearing with different grade oils. Tribol Int 43:1970–1977
15. Chauhan A, Sehgal R, Sharma RK (2011) A study of thermal effects in offset-halves journal bearing profile using different oils. Lubr Sci 23(5):233–248
16. Chauhan A, Sehgal R, Sharma RK (2011) Thermohydrodynamic studies based on different oils in offset-halves journal bearings. Lubr Sci 23:375–392

Effect of Bearing Deformation on the Performance of a Magnetic Fluid-Based Infinitely Rough Short Porous Journal Bearing

M. E. Shimpi and G. M. Deheri

Abstract An attempt has been made to investigate theoretically the performance of a transversely rough porous infinitely short journal bearing considering bearing deformation under the presence of a magnetic fluid lubricant. Christensen and Tonder's stochastic model has been used to develop the stochastic Reynolds' type equation. This associated equation is solved to obtain the pressure distribution paving the way for the calculation of the load carrying capacity. The results indicate that the roughness pattern and the height of the roughness have significant effects on the performance characteristics. It is noticed that the bearing deformation further influences the adverse effect rendered by transverse roughness. However, for a suitable choice of eccentricity, the magnetization relatively improves the performance of the bearing system for a long range of deformation, at least in the case of negatively skewed roughness. Lastly, it is revealed that higher bearing deformation hampers the positive effect of eccentricity ratio.

Keywords Deformation · Short bearing · Magnetic fluid · Squeeze film · Surface roughness

1 Introduction

Lin [1] theoretically analyzed the squeeze film behaviour in a finite journal bearing in the presence of a couple stress fluid. It was found that the couple stress effects not only increased the load carrying capacity significantly but also lengthened the response time of the squeeze film.

M. E. Shimpi (✉)
Department of Mathematics, BVM Engineering College, Anand, Vallabh Vidyanagar, Gujarat 388120, India
e-mail: mukesh.shimpi@gmail.com

G. M. Deheri
Department of Mathematics, Sardar Patel University, Anand, Vallabh Vidyanagar, Gujarat 388120, India

H. C. Patel et al. (eds.), *Proceedings of International Conference on Advances in Tribology and Engineering Systems*, Lecture Notes in Mechanical Engineering, DOI: 10.1007/978-81-322-1656-8_2,

Turaga et al. [2] adopted the model of Christensen and Tonder [3–5]; to study the influence of roughness patterns on the steady state and dynamic characteristics of hydrodynamic journal bearing with rough surfaces. It was concluded that the transverse surface roughness induced a significant effect on the performance of the bearing system.

In order to analyze the effect of using current carrying wear model in the design of a hydrodynamic journal bearing lubricated with Ferrofluid et al. [6] developed the modified Reynolds type equation for a Ferrofluid under an applied magnetic field. The results established that the magnetic lubrication resulted in higher load carrying capacity and reduced friction coefficient as compared to that of a conventional fluid based lubrication.

Gururajan and Prakash [7] extended the investigation carried out by Gururajan and Prakash [8] by incorporating the velocity slip in a thin walled infinitely short rough porous journal bearing operating under the steady conditions in a hydrodynamic regime. It was found that a strong interaction between roughness and slip effects was in place. However, for an effective performance the slip parameter deserved to be minimized.

The numerical analyses of very narrow journal bearings were presented by Stahil [9]. A number of relevant dimensionless design quantities were computed for each combination of different width, diameter ratio and eccentricities.

Hsu et al. [10] theoretically studied the combined influence of couple stress and surface roughness on the lubrication performance of a short journal bearing. It was established that the couple stress effect and the longitudinal roughness improved the load carrying capacity and decreased the friction parameters.

Deheri et al. [11] considered the performance of a longitudinally rough slider bearing with squeeze film formed by a magnetic fluid adopting the Christensen and Tonder's stochastic modeling of roughness. It was concluded that the longitudinal surface roughness resulted in an improved load bearing capacity.

Haque and Guha [12] theoretically investigated the performance characteristics isotropically rough porous hydrodynamic journal bearing of finite width with the effect of slip flow on the basis of Beavers-Joseph criterion. The results obtained in this article confirmed that the influence of roughness on the steady state performance of the journal bearing could not be neglected.

Chen et al. [13] derived a modified Reynolds equation in order to study the effects of non Newtonian lubricants. A comparison showed that surface roughness had less effect on performance results as compared to lubricants with non Newtonian property.

Nada et al. [14] investigated the lubrication of a finite hydrodynamic journal bearing under the presence of magnetic fluid taking couple stress effect into consideration. It was observed that the combined influence of couple stress and magnetic effects on the bearing performance characteristics were significant. The results indicated that fluids with couple stress were better than Newtonian fluids and the performance characteristic enhanced owing to magnetic effect.

Urreta et al. [15] carried out a discussion on the solution of the Reynolds' Equation to obtain the pressure distribution in a hydrodynamic journal bearing, based on viscosity modulation for Ferrofluid. It was conclusively established that magnetic fluid could be used to manufacture active journal bearings.

Shimpi and Deheri [16] analyzed the performance of a magnetic fluid based transversely rough short bearing. The graphical representation made it clear that the negative effect of standard deviation could be compensated largely due to the magnetization parameter in the case of negatively skewed roughness resorting to suitable values of eccentricities ratio.

Recently, Patel et al. [17] theoretically investigated the behaviour of a magnetic fluid based squeeze film for a hydrodynamic short bearing. It was found that the load carrying capacity increased nominally due to magnetic fluid lubricant while the coefficient of the friction decreased significantly. Besides, this study confirmed the importance of forms of the magnitude of the magnetic field.

1.1 Analysis

The geometry and configuration of the bearing system is in Fig. 1.

The assumptions of usual hydrodynamics lubrication theory are taken into consideration in the development of the analysis. The following Christenson and Tonder [3–5] bearing surfaces are assumed to be transversely rough. The expression for film thickness is considered as

$$h(x) = \overline{h(x)} + h_s$$

where $\overline{h}$ is the mean film thickness, h_s is assumed to have the probability density function

Fig. 1 The configuration of the bearing system

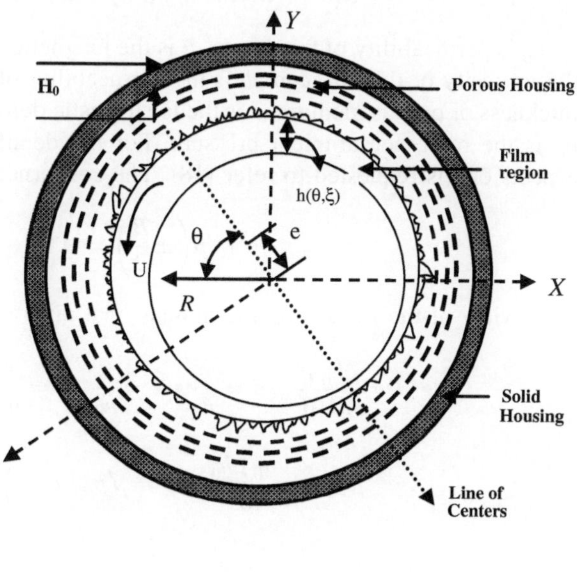

$$F(h_s) = \begin{cases} \frac{32}{35b}\left[1 - \left(\frac{h_s}{b}\right)^2\right]^3, & -b \le h_s \le b \\ 0, & otherwise \end{cases}$$

where b is the maximum deviation from the mean film thickness. The mean α, the standard deviation σ and the parameter ε, which is the measure of symmetry of the random variableh_s are defined by the relationships

$$\alpha = E(h_s),\ \sigma^2 = E\left[(h_s - \alpha)^2\right],\ \varepsilon = E\left[(h_s - \alpha)^3\right],$$

where E denotes the expected value defined as

$$E(R) = \int_{-b}^{b} RF(h_s)dh_s$$

Stochastically averaging and adopting the properties of magnetic fluid lubrication [16, 18], the associated generilized Reynolds equation is derived as

$$\frac{\partial}{\partial z}\left[g(h)\frac{\partial}{\partial z}\{p - 0.5\mu_0\overline{\mu}H^2\}\right] = \frac{6\mu U}{R}\frac{\partial}{\partial\theta}(h + p_a p'\delta) \tag{1}$$

where,

$$h = h_0(1 + e\cos\theta); \quad H^2 = k\left(z - \frac{B}{2}\right)\left(z + \frac{B}{2}\right),$$

$$g(h) = (h + p_a p'\delta)^3 + 3\alpha(h + p_a p'\delta)^2 + 3(\sigma^2 + \alpha^2)(h + p_a p'\delta) + \alpha^3 + 3\sigma^2\alpha + \varepsilon + 12\phi H_0,$$

μ_0 is permeability of free space, $\overline{\mu}$ is the magnetic susceptibility of particles,μ is the viscosity of the lubricant,ϕ is the permeability of porous facing and H_0 is the thickness of porous medium, δ is the local elastic deformation of the porous facing, p_a is the reference ambient pressure. For the details regarding the deformation aspects one is requested to refer [18]. The concerned boundary conditions are

$$p\left(\pm\frac{B}{2}\right) = 0. \tag{2}$$

In view of the following non-dimensional quantities,

$$P = \frac{h_0^3 p}{\mu UB^2},\ \overline{\sigma} = \frac{\sigma}{h_0},\ \overline{\alpha} = \frac{\alpha}{h_0},\ \overline{\varepsilon} = \frac{\varepsilon}{h_0^3},\ \psi = \frac{\phi H_0}{h_0^3},$$

$$\mu^* = -\frac{\mu_0\overline{\mu}h_0^3}{\mu U},\ \overline{p} = p'p_a,\ \overline{\delta} = \frac{\delta}{h},\ Z = \frac{z}{B}.$$

integrating the stochastically averaged Reynolds Eq. (1) under the boundary conditions (2) one obtains the expression for the non-dimensional pressure distribution as

$$P = \left[-0.5\mu^* - \frac{3h_0^3\, e}{RG(h)}\left(1 + \overline{p\delta}\right)\sin\theta\right]\left(Z^2 - 0.25\right) \tag{3}$$

where,

$$\begin{aligned}
G(h) &= h_0^3\left[A_1 + A_2\cos\theta + A_3\cos^2\theta + A_4\cos^3\theta\right];\\
A_1 &= \left(1+\overline{p\delta}\right)^3 + 3\overline{\alpha}\left(1+\overline{p\delta}\right)^2 + 3\left(\overline{\sigma}^2 + \overline{\alpha}^2\right)\left(1+\overline{p\delta}\right)\\
&\quad + 3\overline{\sigma}^2\overline{\alpha} + \overline{\alpha}^3 + 12\psi + \overline{\varepsilon}\\
A_2 &= 3e\left[\left(1+\overline{p\delta}\right)^3 + 2\overline{\alpha}\left(1+\overline{p\delta}\right)^2 + \left(\overline{\sigma}^2+\overline{\alpha}^2\right)\left(1+\overline{p\delta}\right)\right]\\
A_3 &= 3e^2\left[\left(1+\overline{p\delta}\right)^3 + \overline{\alpha}\left(1+\overline{p\delta}\right)^2\right];\ A_4 = e^3\left(1+\overline{p\delta}\right)^3.
\end{aligned}$$

The load carrying capacity in non-dimensional form then, is calculated as

$$W = \int_0^{\pi}\int_{-0.5}^{0.5} P\, dZ\, d\theta$$

which leads to

$$\begin{aligned}
W &= \frac{\mu^*\pi}{12}\\
&\quad + \frac{2\pi h_0^3\, e\overline{\alpha}\left(1+\overline{p\delta}\right)}{9Rh_0^2}\left[\frac{8(4A_1+3A_3)}{4(4A_1+3A_3)^2 - 3(4A_2+3A_4)^2}\right]\\
&\quad + \frac{2\pi h_0^3\, e\overline{\alpha}\left(1+\overline{p\delta}\right)}{9Rh_0^2}\left[\frac{1}{A_1} + \frac{16\sqrt{3}(4A_1+A_3)}{4(4A_1+A_3)^2 - (4A_2+A_4)^2}\right].
\end{aligned} \tag{4}$$

1.2 Results and Discussion

It is clearly seen that the load carrying capacity enhances by $0.261\mu^*$ due to magnetization as compared to that of a conventional lubricant. As the expression for non-dimensional load carrying capacity is linear in μ^*, it is easily observed that increasing values of magnetization cause increased load carrying capacity.

It is noticed that in the absence of magnetization this investigation gives the deformation effect on the behaviour of a rough porous infinitely short journal bearing. Further, for a smooth surface this reduces to the performance of a infinitely short bearing in the absence of deformation. In addition, if there is no deformation this essentially turns to the effect of porosity on infinitely short journal bearing. It is revealed from the Eq. (4) that the bearing can support a load even in the absence of flow.

A comparison of this study with the investigation of Patel et al. [17] tends to indicate that the load carrying capacity is relatively reduced here due to transverse roughness and deformation. Probably, this may be due to the fact that the roughness retards the motion of the lubricant thereby reducing the pressure distribution. However, this situation is slightly improved in the case of negatively skewed roughness.

It is easily observed from Figs. 2, 3, 4 and 5 that the parabolic profile of the pressure is significantly affected by roughness and deformation.

The variation of load carrying capacity with respect to the magnetization presented in Figs. 6, 7, 8, 9, 10 and 11 makes it clear that the load carrying capacity rises sharply due to the magnetic fluid lubricant. However, the effect of variance on the load carrying capacity with respect to magnetization is marginal.

Figures 12, 13, 14, 15 and 16 deal with the variation of load carrying capacity with respect to the standard deviation associated with the roughness. It is clear that

Fig. 2 Distribution of pressure with respect to Z and μ^*

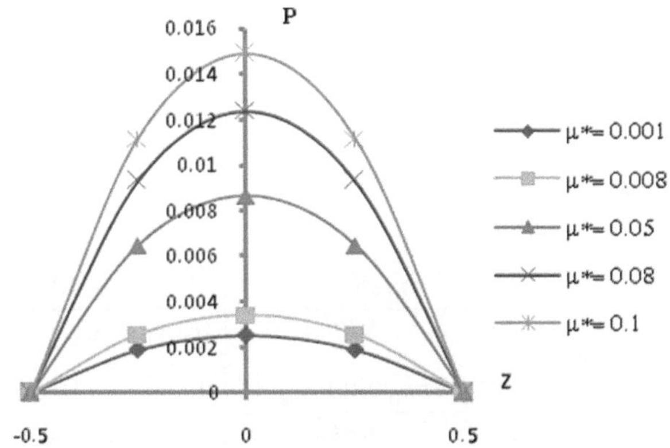

Fig. 3 Distribution of pressure with respect to Z and $\overline{\alpha}$

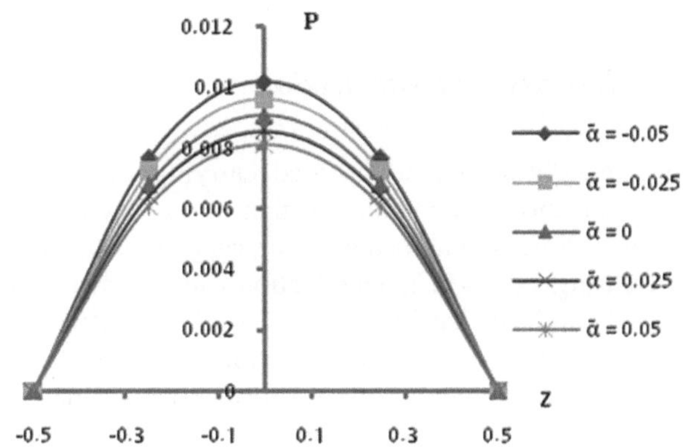

Fig. 4 Distribution of pressure with respect to Z and $\overline{\varepsilon}$

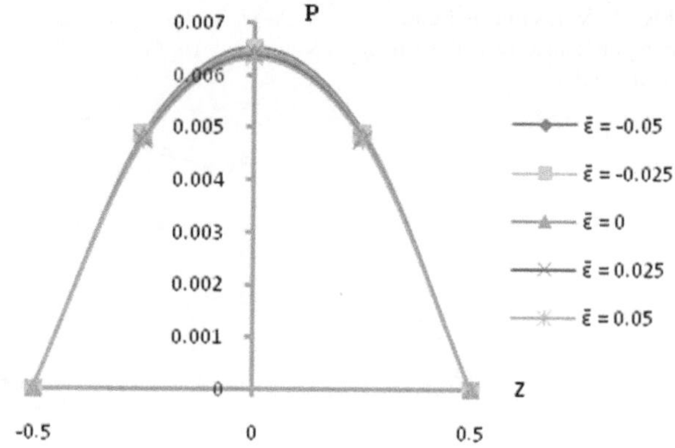

Fig. 5 Distribution of pressure with respect to Z and $\overline{\delta}$

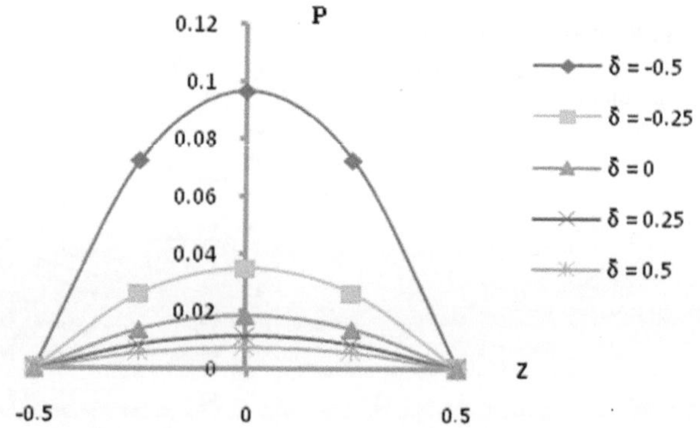

Fig. 6 Variation of Load carrying capacity with respect to μ^* and $\overline{\sigma}$

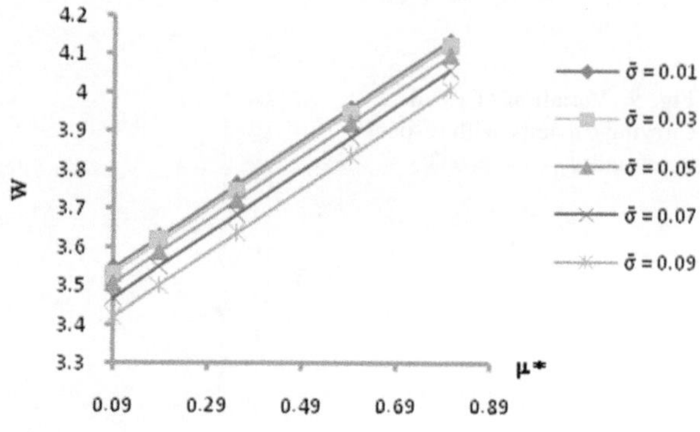

Fig. 7 Variation of Load carrying capacity with respect to μ^* and $\overline{\alpha}$

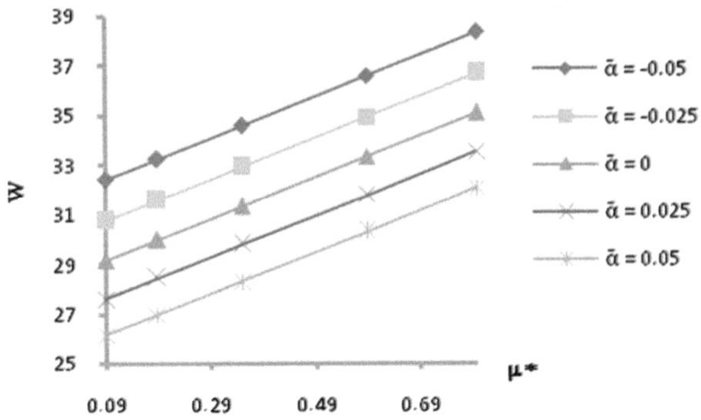

Fig. 8 Variation of Load carrying capacity with respect to μ^* and $\overline{\varepsilon}$

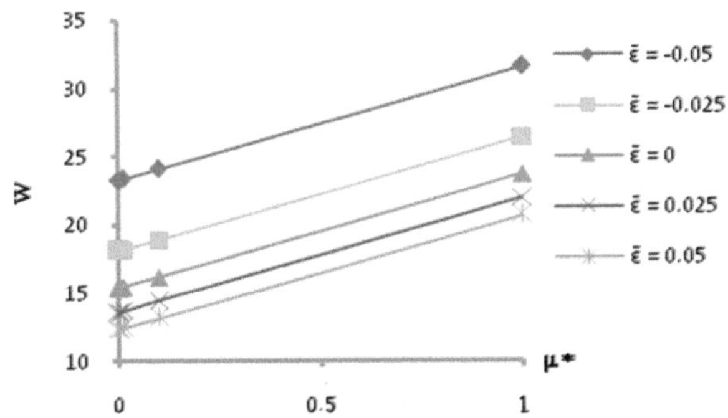

Fig. 9 Variation of Load carrying capacity with respect to μ^* and ψ

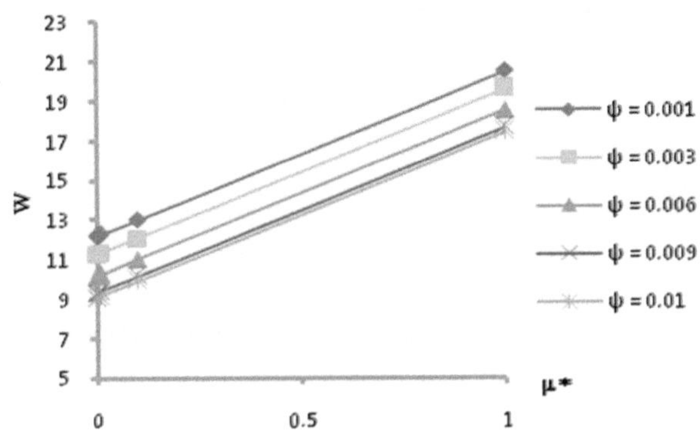

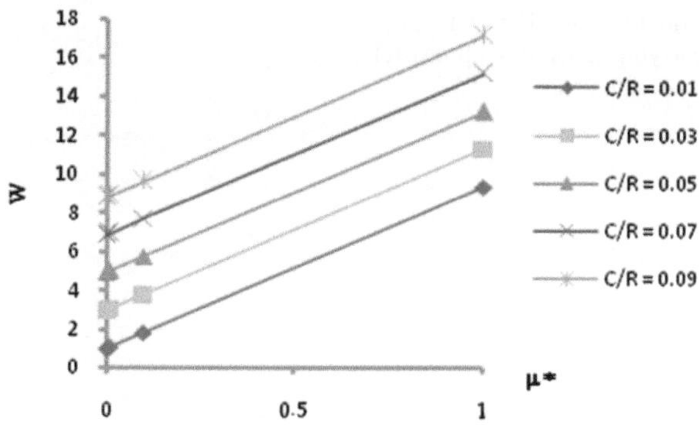

Fig. 10 Variation of Load carrying capacity with respect to μ^* and C/R

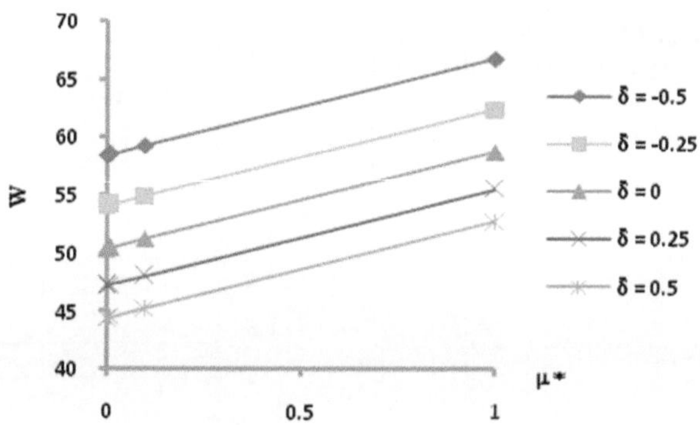

Fig. 11 Variation of Load carrying capacity with respect to μ^* and $\overline{\delta}$

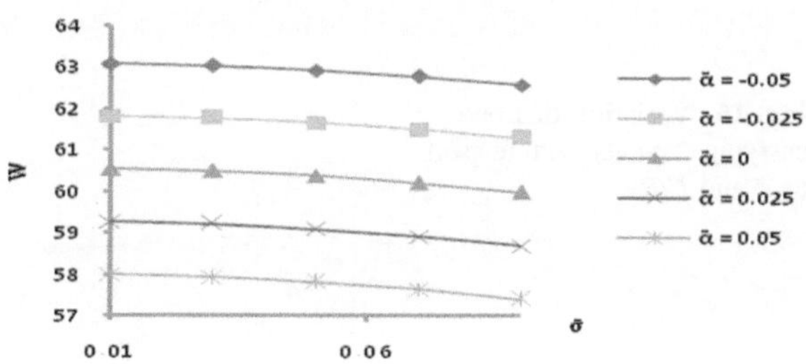

Fig. 12 Variation of Load carrying capacity with respect to $\overline{\sigma}$ and $\overline{\alpha}$

standard deviation has a considerable adverse effect on the performance of the bearing system in the sense that it decreases the load carrying capacity considerably. Interestingly, however it is noticed that the rate of reduction in the load carrying capacity with respect to standard deviation gets decreased for higher values of porosity.

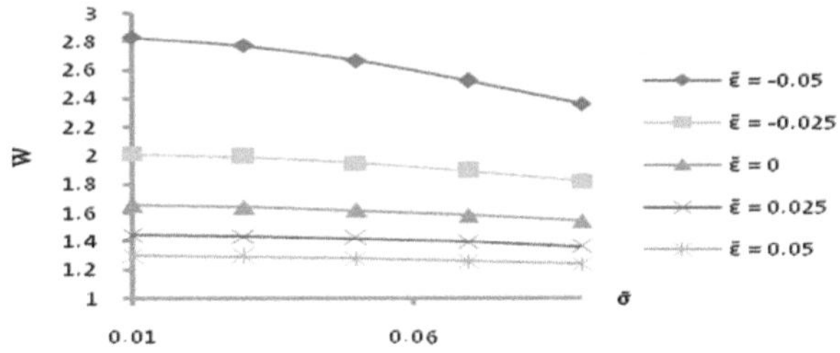

Fig. 13 Variation of Load carrying capacity with respect to $\overline{\sigma}$ and $\overline{\varepsilon}$

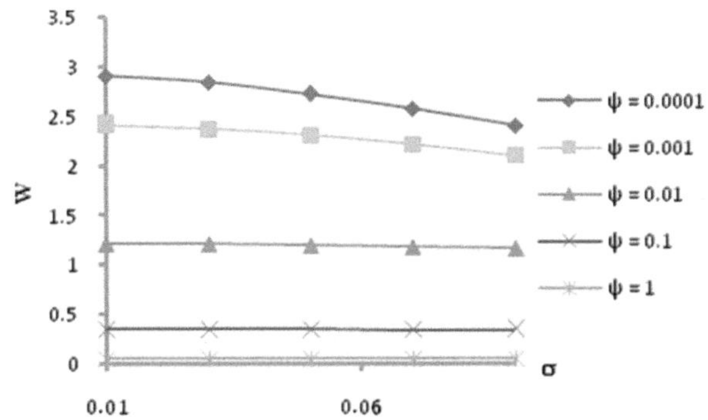

Fig. 14 Variation of Load carrying capacity with respect to $\overline{\sigma}$ and ψ

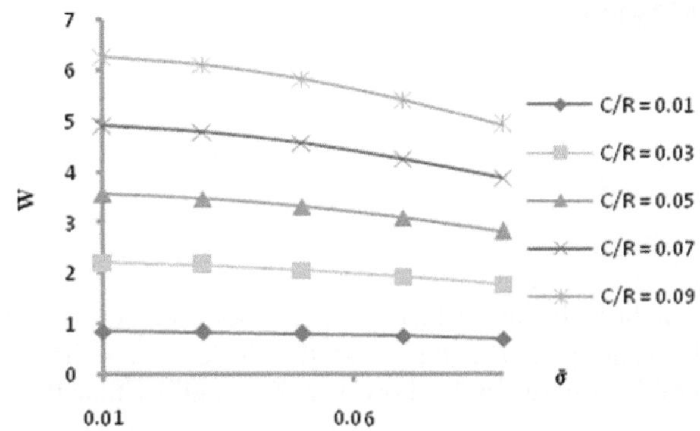

Fig. 15 Variation of Load carrying capacity with respect to $\overline{\sigma}$ and C/R

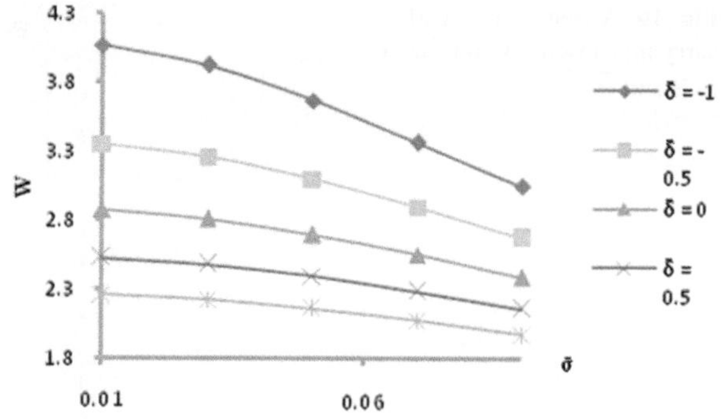

Fig. 16 Variation of Load carrying capacity with respect to $\overline{\sigma}$ and $\overline{\delta}$

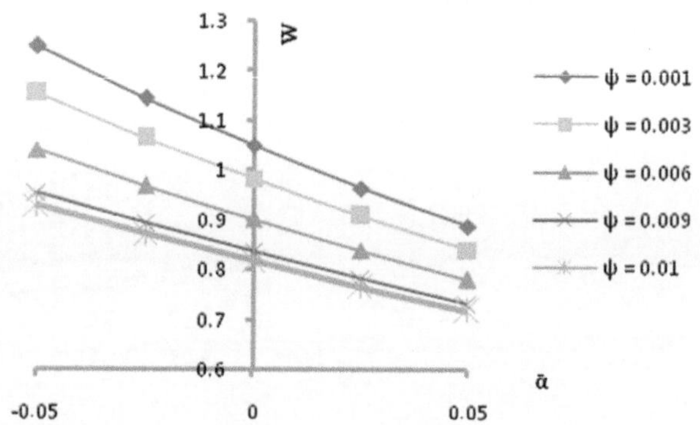

Fig. 17 Variation of Load carrying capacity with respect to $\overline{\alpha}$ and ψ

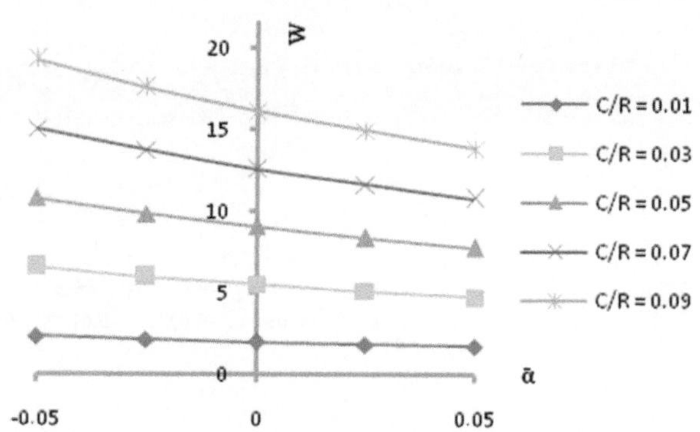

Fig. 18 Variation of Load carrying capacity with respect to $\overline{\alpha}$ and C/R

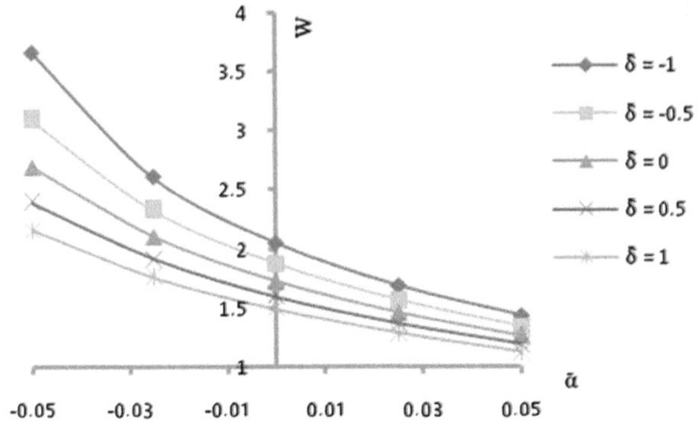

Fig. 19 Variation of Load carrying capacity with respect to $\overline{\alpha}$ and $\overline{\delta}$

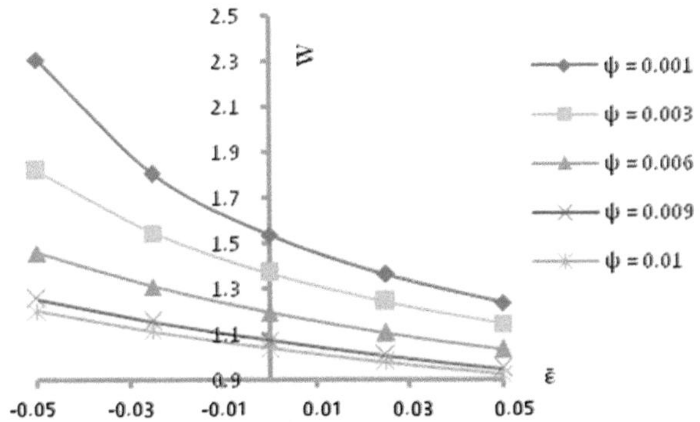

Fig. 20 Variation of Load carrying capacity with respect to $\overline{\varepsilon}$ and ψ

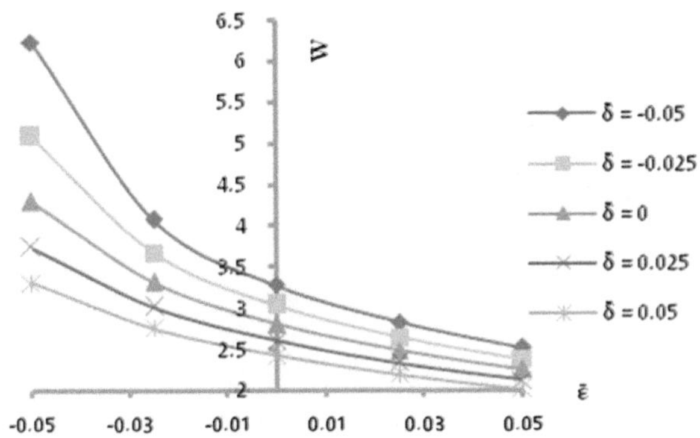

Fig. 21 Variation of Load carrying capacity with respect to $\overline{\varepsilon}$ and $\overline{\delta}$

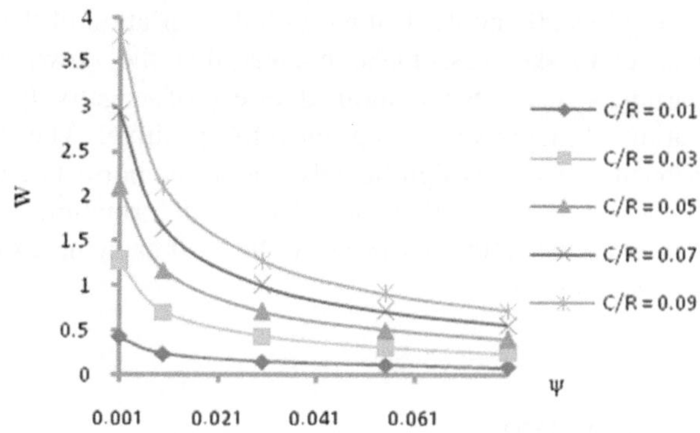

Fig. 22 Variation of Load carrying capacity with respect to ψ and C/R

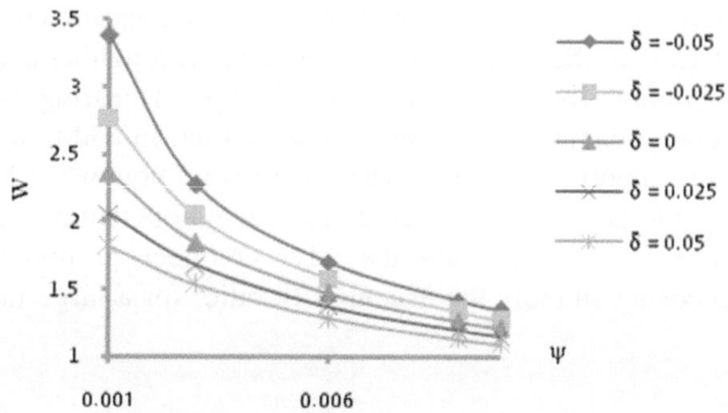

Fig. 23 Variation of Load carrying capacity with respect to ψ and $\overline{\delta}$

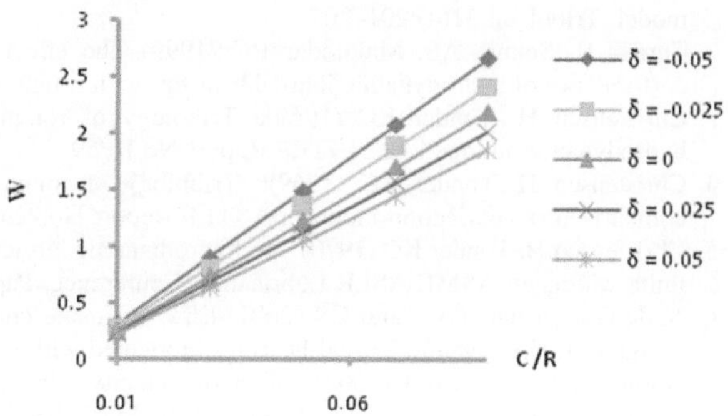

Fig. 24 Variation of Load carrying capacity with respect to C/R and $\overline{\delta}$

The effect of variance on the distribution of load carrying capacity is presented in Figs. 17, 18 and 19. These figures suggest that the load carrying capacity decreases when variance (positive) increases while variance (negative) increases the load carrying capacity. The effect of variance is relatively sharp with the increase in the eccentricity.

Figures 20 and 21 dealing with the variation of the load carrying capacity with respect to skewness make it clear that the skewness follow the trends of the variance. Thus, the combined effect of negatively skewed roughness and the variance (negative) is significantly positive. The fact that the load carrying capacity decreases significantly due to the porosity can be seen from Figs. 22 and 23. Figure 24 says that the effect of deformation cannot be disregarded for all values of the ratio C/R because the load carrying capacity increases sharply with respect to C/R.

2 Conclusion

Although, the effect of transverse surface roughness and deformation is relatively adverse, this investigation provides some measures to mitigate this negative effect at least in the case of negatively skewed roughness. Thus, the roughness must be accounted for while designing this type of bearing system, even if suitable magnetic strength is taken into consideration. In addition, this type of bearing system can support a load even when there is no flow while fails to happen in the case of traditional lubricants. It is interesting to note that the effect of magnetization goes a long way in reducing the adverse effect of porosity and standard deviation choosing suitably the eccentricity ratio, for a large range of deformation.

References

1. Lin JR (1998) Squeeze film characteristics of finite journal bearings: couple stress fluid model. Tribol Int 31(4):201–207
2. Turaga R, Sekhar AS, Majumdar BC (1999) The effect of roughness parameter on the performance of hydrodynamic journal bearings with rough surfaces. Tribol Int 32(5):231–236
3. Christensen H, Tonder KC (1969)a Tribology of rough surfaces: stochastic models of hydrodynamic lubrication. SINTEF Report No.10/69
4. Christensen H, Tonder KC (1969)b Tribology of rough surfaces: parametric study and comparison of lubrication models. SINTEF Report No.22/69, 1969b
5. Christensen H, Tonder KC (1970) The hydrodynamic lubrication of rough bearing surfaces of finite width. In: ASME-ASLE Lubrication Conference, Paper no.70
6. Nada GS, Osman TA, Safar ZS (2001) Effect of using current-carrying-wire models in the design of hydrodynamic journal bearings lubricated with Ferrofluid. Tribol Lett 11(1):61–70
7. Gururajan K, Prakash J (2003) Effect of velocity slip in a narrow rough porous journal bearing. In: Proceedings of the Institution of Mechanical Engineers, Part J: J Eng Tribol, January 1, 217(1):59–70
8. Gururajan K, Prakash J (2000) Effect of surface roughness in a narrow porous journal bearing. J Tribol 122(2):472–476
9. Stahil J (2002) Narrow journal bearings. In: Proceedings of the Institution of Mechanical Engineers, Part J: J Eng Tribol May 1, 216(5):343–346
10. Hsu C-H, Lin J-R, Chiang H-L (2003) Combined effects of couple stresses and surface roughness on the lubrication of short journal bearings. Indus Lubri Tribol 55(5):233–243

11. Deheri GM, Andharia PI, Patel RM (2004) Longitudinally rough slider bearings with squeeze film formed by a magnetic fluid. Indus Lubri Tribol 56(3):177–187
12. Haque R, Guha SK (2005) On the steady-state performance of isotropically rough porous hydrodynamic journal bearings of finite width with slip-flow effect. In: Proceedings of the Institution of Mechanical Engineers, Part C: Journal of Mechanical Engineering Science November 1, 219(11):1249–1267
13. Chen H, Chen D, Wang J, Li Y-J (2006) Calculated journal bearing lubrication of non-Newtonian medium with surface roughness effects. Front Mech Eng China 3:270–275
14. Nada GS, Osman TA (2007) Static performance of finite hydrodynamic journal bearings lubricated by magnetic fluids with couple stresses. Tribol Lett 27(3):261–268
15. Urreta H, Leicht Z, Sanchez A, Agirre A, Kuzhir P, Magnac G (2009) Hydrodynamics bearings lubricated with magnetic fluids. J Phy: conference Series, 1499(1), Article ID 012113
16. Shimpi ME, Deheri GM (2012) Effect of deformation in magnetic fluid based transversely rough short bearing. J Tribol Materia Lubr Interface 6(2):1–5
17. Patel NS, Vakharia DP, Deheri GM (2012) A study on the performance of a magnetic fluid-based hydrodynamic short journal bearing. p 7
18. Prajapati BL (1995) On certain theoretical studies in hydrodynamic and electro-magneto hydrodynamic lubrication. Sardar Patel University, Dissertation

Effects of Velocity-Slip and Viscosity Variation in Squeeze Film Lubrication of Spherical Bearings

R. Raghavendra Rao, K. Gouthami and J. Vijaya Kumar

Abstract In this paper, a theoretical study of effects of velocity-slip and viscosity variation on squeeze film lubrication of spherical bearings is analyzed. The modified Reynolds accounting for velocity-slip and viscosity variation using deterministic theory is mathematically derived. It is applied to study the effects of velocity-slip and viscosity variation in squeeze film lubrication of spherical bearings. In order to get the expression for pressure the modified Reynolds equation is solved. Then by making use of this expression, we obtain the expressions for load carrying capacity which in tern is used to find the expression for response time. These expressions are numerically computed and the results are presented graphically. From the numerical computations of the results, it is found that the beneficial result for hydrodynamic lubrication due to the presence of increased viscosity near the bearing surface was indicated. Although the effects of velocity-slip at the bearing is to decrease both the load capacity, frictional force and the coefficient of friction increases which leads to an un-favorable results. For a gas-lubricated hydrostatic bearing, the gas film pressure and the load decrease with increasing molecular mean free path.

Keywords Reynolds equation · Velocity-slip · Viscosity variation · Squeeze film lubrication · Load capacity · Squeeze time

Nomenclature

C Radial clearance
e Eccentricity

R. Raghavendra Rao (✉) · K. Gouthami
Department of Mathematics, K L University, Vaddeswaram, Guntur 522502, Andhra Pradesh, India
e-mail: rrrsvu@sify.com

J. Vijaya Kumar
Department of Mathematics, Vasireddy Venkatadri Institute of Technology, Nambur, Guntur 522508 Andhra Pradesh, India

H. C. Patel et al. (eds.), *Proceedings of International Conference on Advances in Tribology and Engineering Systems*, Lecture Notes in Mechanical Engineering, DOI: 10.1007/978-81-322-1656-8_3,

F	Frictional drag
b	Width of the bearing
h	Film thickness
k	Ratio of the viscosities in different layers
l	Length of the bearing
P	Hydrodynamic pressure
r, θ	Cylindrical coordinates
r_1	Radius of the journal (or) spheres
μ	Coefficient of viscosity
$\overline{\beta}$	Velocity-Slip
$\overline{a}$	Thickness of the peripheral layer
V	Squeeze velocity
W	Load capacity
$\overline{W}$	Non dimensional load
T	Squeezing time
$\overline{T}$	Non dimensional squeezing time
Q	Flow flux in spherical bearing

1 Introduction

In general, most of the lubricated systems can be considered to consist of moving (stationary) surfaces (plane/curve, loaded/unloaded) with a thin film of an external material (lubricant) between them. The presence of such a thin film between these surfaces not only helps to support considerable load but also minimizes friction. The characteristics such as pressure in the film, frictional force at the surface, flow rate of the lubricant etc. of the system depend upon the nature of the surfaces, the nature of the lubricant film boundary conditions etc.

The equation governing the pressure generated in the lubricant film can be obtained by coupling the equations of motion with the equation of continuity and was first derived by Reynolds [1] in 1866 and is known as "Reynolds Equation". In deriving this equation, the thermal, compressibility, viscosity variation, slip at the surfaces, inertia and surface roughness effects were ignored. Later this Reynolds equation is modified in 1949 by Cope [2] including viscosity and density variation along the fluid film. In 1957–1958 the viscosity variation across the film thickness has been considered by Zienkiewicz and Cameron [3, 4] who also pointed out that temperature gradient and viscosity variation across the film may not be ignored. In the year 1962, Dowson [5] unified the various attempts in generalizing the Reynolds Equation by considering the variation of fluid properties across as well as along the fluid film thickness by neglecting the slip effects at the bearing surfaces. Since then many workers including myself have studied the effects of velocity-slip and viscosity variation in lubricated systems by considering Reynolds Equation with energy equation [6–13]. Hydrodynamically lubricated spherical bearings are widely used in a variety of applications involving severe operating conditions of speeds, loads etc. It is observed fact that the fluids containing additives or contaminants enhance the lubrication process. These additives have a desirable effect of enhancement of viscosity and a consequent rise in load capacity. Usually these additives are in the form of long chain organic compounds. In recent years experimental studies [14, 15] have shown clear evidence of load enhancement and friction reduction effects due to the pressure additives. Therefore the effects of additives on the fluid reheology of a lubricant received great research attention. Since the flow behavior of Newtonian lubricant blended with various additives could not be described accurately by the classical continuum theory many micro continuum theories have been proposed [16–18]. Among these theories the Stokes theory [18] is the simplest theory that accounts for the effects of couple stresses, body couple and asymmetric tensors. An inverse solution for finite journal bearing lubricated with couple stress fluids by Elshakawy and Guedouar [19] and a theoretical study of squeeze film behavior for a finite journal bearing lubricated with couple stress by [20] are investigated.

1.1 Basic Equations

The one-dimensional Reynolds equation [21] can be written as follows

$$\frac{d}{dx}\left[F_4\frac{dP}{dx}\right] = -V \tag{1}$$

where $F_4 = \dfrac{(h-a)^3}{12\eta_2} + \dfrac{a^3 + 3a^2(h-a) + 3a(h-a)^2}{12\eta_1} + \dfrac{h^2}{2\beta}$ taking $\beta = \frac{\eta_1}{\lambda}$ as the slip parameter.

Consider the squeeze film in two eccentric spherical surfaces of radii r_1 and r_2 which are approaching each other with a velocity 'V' as shown in the Fig. 1.

The film thickness 'h' is given by $h = C(1 - \varepsilon\ Cos\ \theta)$.
where $C = (R_1 - r_1)$which is the clearance width and $\varepsilon = \frac{e}{c}$ is the eccentricity ratio.

The flow flux in this case is obtained by using the equation of continuity as follows

$$Q = 2\pi V r_1 Sin^2\,\theta \tag{2}$$

The flow flux in this case is giving by

$$Q = \left(\frac{F_4}{r}\frac{dP}{d\theta}2\pi\, r_1 Sin\,\theta\right) \tag{3}$$

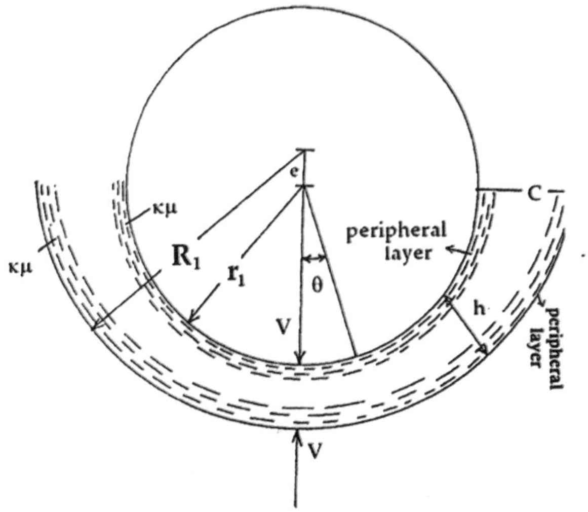

Fig. 1 Squeeze film lubrication in Spherical bearing

$$F_4 = \frac{1}{12\,\mu k}\left[\{C\,(1-\varepsilon\,Cos\;\theta)-a\}^3(k-1)+\{C\,(1-\varepsilon\,Cos\;\theta)\}^3\right] + \left[\frac{\{C\,(1-\varepsilon\,Cos\;\theta)\}^2}{2\beta}\right] \tag{4}$$

The boundary condition for Eq. (3) is

$$P = 0 \quad at \quad \theta = \frac{\pi}{2}$$

Integrating the Eq. (3) and incorporating the above condition, we get

$$P(\theta) = \int_0^{\frac{\pi}{2}} \frac{r_1^2\;V\;Sin\;\theta}{F_4}\,d\theta \tag{5}$$

The load capacity W is given by

$$W = 2\pi\,r_1^2 \int_0^{\frac{\pi}{2}} P\;Sin\,\theta\,Cos\,\theta\;\,d\theta \tag{6}$$

Using Eq. (5) and (6), the load capacity W can be obtained as

$$W = \pi\,r_1^4\,V \int_0^{\frac{\pi}{2}} \frac{Sin^3\,\theta}{F_4}\,d\theta \tag{7}$$

where

$$F_4 = \frac{1}{12\,\mu k}\left[\left\{C^3(1-\varepsilon Cos\,\theta) - \frac{a}{C}\right\}^3(k-1) + C^3(1-\varepsilon Cos\,\theta)^3\right]$$
$$+\,C^2\frac{(1-\varepsilon Cos\,\theta)^2}{2\beta}$$
$$= \frac{C^3}{12\,\mu k}\left[\left\{\frac{1}{k}(1-\varepsilon Cos\,\theta) - \overline{a}\right\}^3(k-1)\right.$$
$$\left. + \frac{1}{k}(1-\varepsilon Cos\,\theta)^3 + \frac{(1-\varepsilon Cos\,\theta)^2}{\overline{\beta}}\right]$$

The load capacity

$$\overline{W} = 12\,\pi \int_0^{\frac{\pi}{2}} \frac{Sin^3\,\theta}{\overline{F_4}} d\,\theta \tag{8}$$

$$\overline{F}_4 = \left[\left\{\frac{1}{k}(1-\varepsilon Cos\,\theta) - \overline{a}\right\}^3(k-1)\right.$$
$$\left. + \frac{1}{k}(1-\varepsilon Cos\,\theta)^3 + \frac{(1-\varepsilon Cos\,\theta)^2}{\overline{\beta}}\right]$$

Squeezing time T for the surfaces to approach from the initial eccentric position ($\varepsilon = 0$) to a final eccentric position ($\varepsilon = \varepsilon_1$) is given by

$$T = \frac{2\pi\, r_1^4}{W} \int_0^{\varepsilon_1} I_2 d\varepsilon \tag{9}$$

where $I_2 = \int_0^{\frac{\pi}{2}} \frac{Sin^3\theta}{F_4} d\theta$ and F_4 is given by the Eq. (4).

$$\overline{T} = \frac{2\pi\, r_1^4}{\frac{\overline{W}}{l}} \int_0^{\varepsilon_1} \int_0^{\frac{\pi}{2}} \frac{Sin^3\theta}{F_4} d\theta\, d\varepsilon$$

Squeezing time

$$\overline{T} = 2\pi \int_0^{\varepsilon_1} \int_0^{\frac{\pi}{2}} \frac{Sin^3\theta}{F_4} d\theta\, d\varepsilon \tag{10}$$

2 Results and Discussions

The load capacity $\overline{W}$ as represented in Eq. (8) is integrated numerically. The squeezing time $\overline{T}$ represented as double integral in Eq. (10) is evaluated numerically and graphs have been plotted in Figs. 2, 3, 4, 5, 6, 7, 8, 9, 10, 11, and 12. It is seen from these figures that the load capacity increases as the peripheral layer thickness increases for $k > 1$and it decreases for $k < 1$. These parameters increase as the slip decreases.

In Figs. 2 and 3, the load capacity $\overline{W}$ is plotted with $\overline{a}$ (peripheral layer viscosity) for various values of k ($k = 0.5$, 1, 1.5, 2 etc.). $k = 0.5$ means peripheral layer viscosity is lower than middle layer viscosity. $k = 1.0$ means there is no change in viscosity. $k > 1.0$means peripheral layer viscosity is greater than middle layer viscosity.

It is seen from the graph that when $k = 1$, there is no change in the load capacity even when there is a change in $\overline{a}$. When $k = 0.5$, $\overline{W}$ decreases as $\overline{a}$ increases and it is lower than $k = 1$. It means that when the peripheral layer viscosity is less than the middle layer viscosity the load capacity is less and it further decreases as the peripheral layer thickness increases when

$k = 1.5$ or $k = 2.0$ etc., $\overline{W}$ increases as $\overline{a}$ increases and it is more than $k = 1.0$. It means that when the peripheral layer viscosity is more than the middle layer viscosity the load capacity increases and increase is more as the peripheral layer thickness increases. The effect is more reduced up to $k = 1.5$.

In Figs. 4 and 5, $\overline{W}$ is plotted with k for various $\overline{a}$. It is found that as k increases the load capacity increases and the load capacity is more as $\overline{a}$

Fig. 2 Variation of $\overline{W}$ with $\overline{a}$ for various values of k

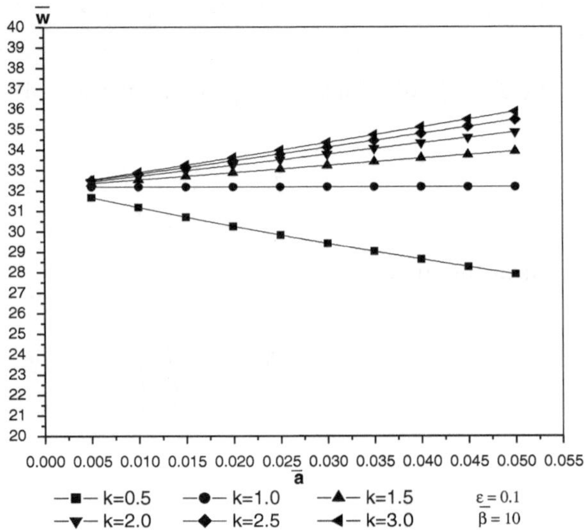

Fig. 3 Variation of $\overline{W}$ with $\overline{a}$ for various values of k

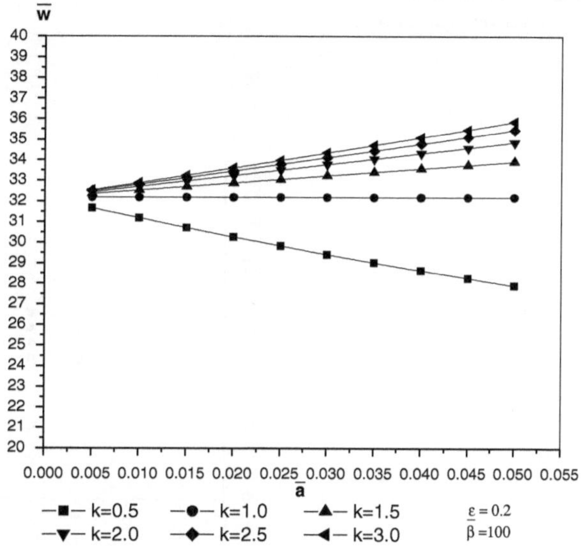

Fig. 4 Variation of $\overline{W}$ with k for various values of $\overline{a}$

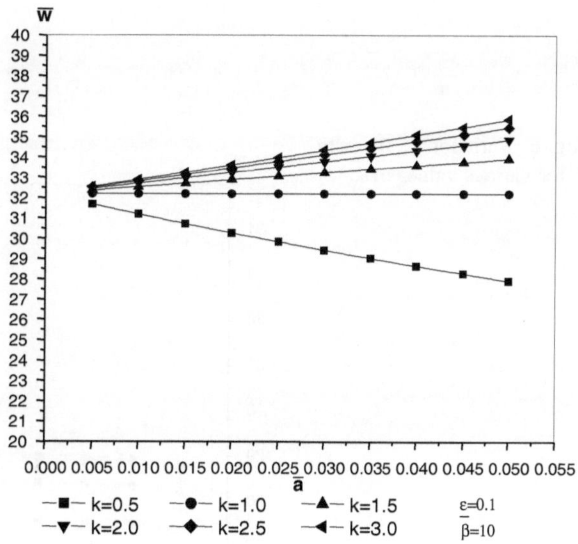

increases. The above graphs are drawn considering $k > 1$ and all the lines meet when $k = 1$.

In Figs. 6,7, $\overline{W}$ is plotted with$\overline{\beta}$ for various $\overline{a}$ and k. It is seen from Fig. 6, that $\overline{W}$ increases as $\overline{\beta}$ increases (slip parameter). But it is known that higher values of $\overline{\beta}$

Fig. 5 Variation of $\overline{W}$ with k for various values of $\overline{a}$

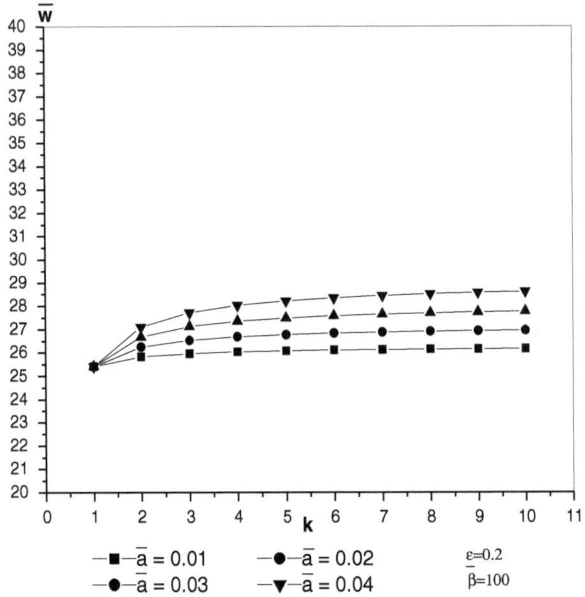

Fig. 6 Variation of $\overline{W}$ with $\overline{\beta}$ for various values of k

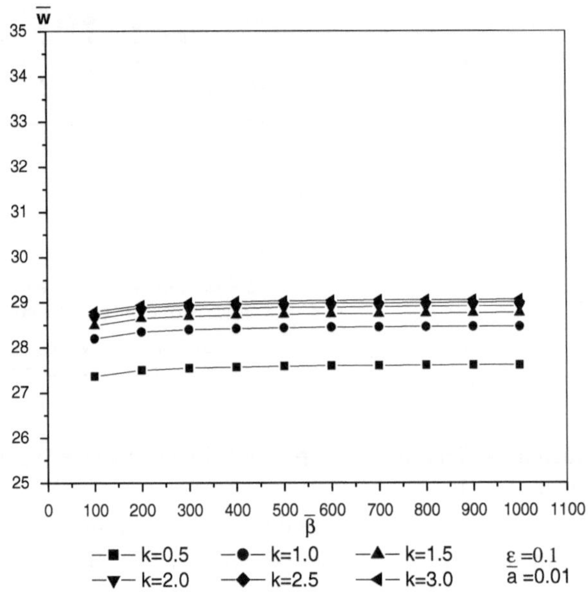

Fig. 7 Variation of $\overline{W}$ with $\overline{\beta}$ for various values of k

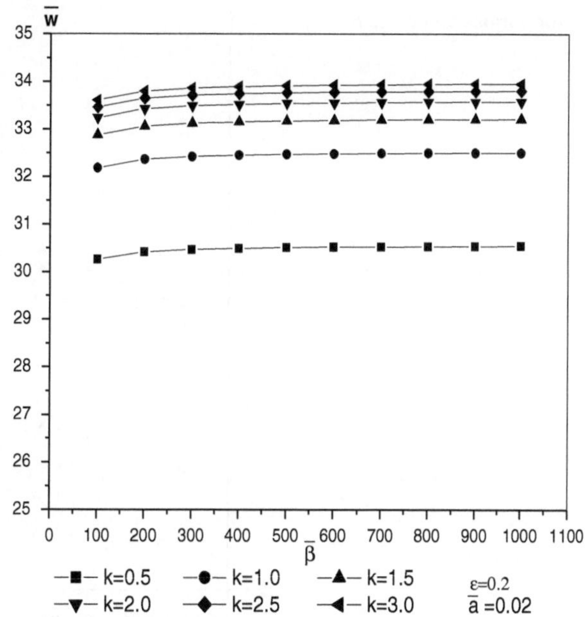

Fig. 8 Variation of $\overline{T}$ with $\overline{a}$ for various values of $\overline{\beta}$

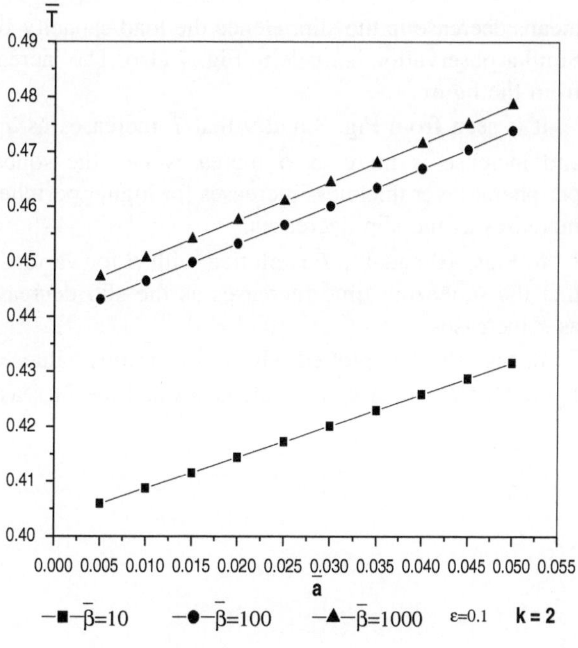

Fig. 9 Variation of $\overline{T}$ with $\overline{a}$ for various values of $\overline{\beta}$

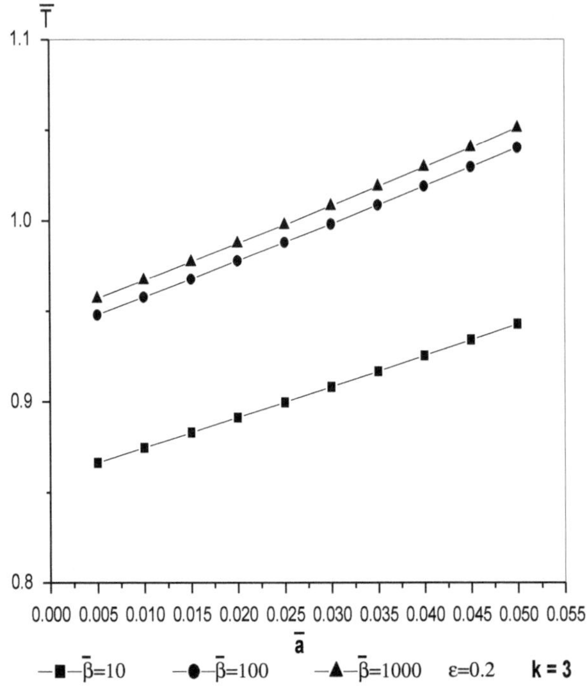

means decrease in the slip. Hence the load capacity $\overline{W}$ increases as slip decreases. Similar observation is made in Fig. 7 also. The increase is also very small as seen from the figure.

It is seen from Fig. 8 and 9 that $\overline{T}$ increases as $\overline{a}$ increases for $k = 2$ $(k > 1)$ and increase is more as $\overline{\beta}$ increases i.e., the squeezing time increases as the peripheral layer thickness increases for higher peripheral layer viscosity and it also increases as the slip decreases.

In Figs. 10 and 11, $\overline{T}$ is plotted with $\overline{\beta}$ for various values of k and $\overline{a}$. It is seen that the squeezing time increases as the slip decreases and the increase is more as k increases.

In Fig. 12, $\overline{T}$ is plotted with k for various values of $\overline{a}$. It is also seen from the figure that $\overline{T}$ increases as k increases and the increase is more as $\overline{a}$ increases.

Fig. 10 Variation of $\overline{T}$ with $\overline{\beta}$ for various values of k

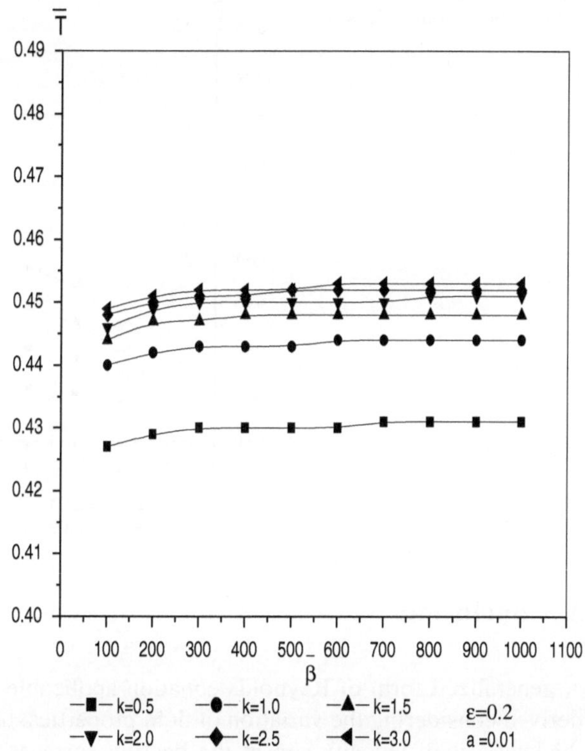

Fig. 11 Variation of $\overline{T}$ with $\overline{\beta}$ for various values of $\overline{a}$

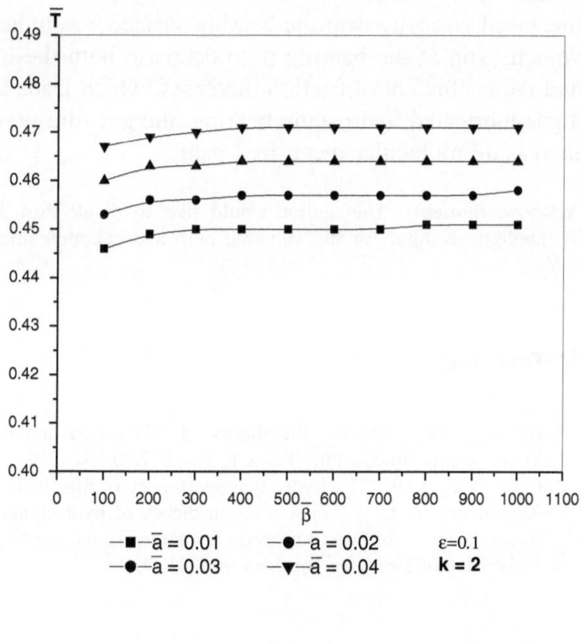

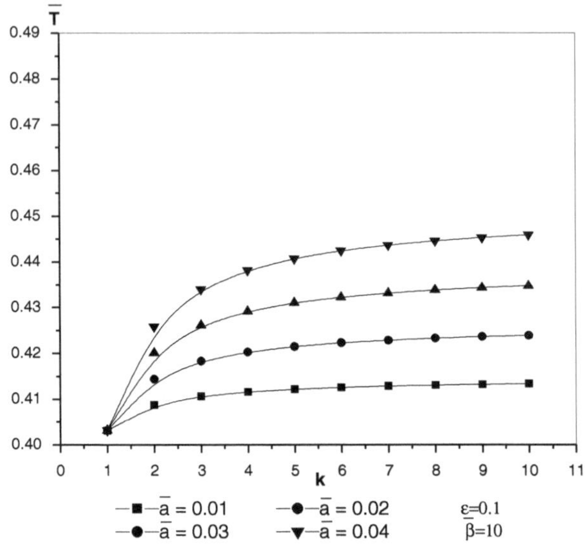

Fig. 12 Variation of $\overline{T}$ with k for various values of $\bar{a}$

3 Conclusion

A generalized form of Reynolds equation applicable to fluid film lubrication was derived considering the variation of fluid properties, both across and along the film thickness, with velocity-slip at the bearing surfaces. The effects of velocity-slip and viscosity variation in squeeze film lubrication of spherical bearings have been studied. The beneficial result for hydrodynamic lubrication due to the presence of increased viscosity near the bearing surface was indicated. Although the effects of velocity-slip at the bearing is to decrease both the load capacity, frictional force and the coefficient of friction increases, which leads to an unfavorable results. For a gas-lubricated hydrostatic bearing, the gas film pressure and load decrease with increasing molecular mean free path.

Acknowledgments The author would like to thank Prof J. B. Shukla, Indian Institute of Technology, Kanpur for his valuable help and encouragement during the completion of this study.

References

1. Reynolds O (1886) On the theory of lubrication and its application to Mr. Beauchamp Tower's experiment. Phil Trans R Soc 177(1):157–234
2. Cope W F (1949) The hydrodynamic theory of film lubrication. Proc R Soc 197:201–217
3. Zienkiewiez O C (1957) A note on theory of hydrodynamic lubrication of parallel surface thrust bearings. In:Proceedings on 9th international conference on applied mechanics, Vol 4. University of Brussels, Brussels, pp 251–258

4. Cameron A (1958) The viscous wedge. Trans ASME 1:248
5. Dowson D (1962) A generalized Reynolds equation for fluid film lubrication. Inst J Mech Sci 4:159–170
6. Shukla J B (1964) Theory for the squeeze film for Power law lubricants. Trans ASME 86:441–444
7. Shukla J B et al (1982) Effects of consistency variation of power law lubricants in squeeze films. Wear 76:299–319
8. Sinha P et al (1983) Viscosity variation considering cavitation in a journal bearing lubricant containing additives. Wear 86:43–56
9. Prakash J, Sinha P (1977) Theoritical effects of solid particles on the lubrication of journal bearings considering cavitation. Wear 41:233–249
10. Raghavendra Rao R, Prasad K R (2001) Effects of Velocity-slip on the elastohydrodynamic lubrication of heavily loaded rollers, Bull Pure Appl Sci 20(2):277–295
11. Raghavendra Rao, R, Prasad, K. R (2002) Effects of velocity-slip and viscosity variation in rolling and normal motion. J Aeronaut Soc India 54(4):399–407
12. Raghavendra Rao R, Prasad K. R (2003) Effects of velocity- slip and viscosity variation for lubrication of Roller Bearings. J Def Sci 53(4):431–442
13. Raghavendra Rao R, Prasad K R Effects of velocity-slip and viscosity variation on Journal Bearings. J ANZIAM 46:143–155
14. Oliver DR (1988) Load enhancement effects due to polymer thickening in a short model journal bearing. J Non-Newtonian Fluid Mech 30:185–196
15. Spikes HA (1994) The behaviour of lubricants in contacts: current understanding and future possibilities. J Porc Instn Mech Engrs 28:3–15
16. Ramanaiah G (1979) Squeeze films between finite plates lubricated by fluids with couple stress. Wear 54:315–320
17. Airman T, Turk MA, Sylvester ND (1973) Micro continuum fluid mechanics. A rev Int J Eng Sci 11:905–930
18. Stokes VK (1966) Couple stresses in fluids. Phys Fluids 9:1709–1715
19. Elsharkawy AA, Guedouar LH (2001) An inverse solution for finite journal bearings lubricated with couple stress fluids. Tribol Int 34:107–108
20. Lin JR, Hsu CH, Lai C (2002) Surface roughness effects on the oscillating squeeze-film behavior of long partial journal bearing. Comp Struct 80:297–303
21. Raghavendra Rao R et al (2013) Effects of Velocity-Slip and Viscosity Variation in squeeze film lubrication of two circular plates J Tribol Ind 35(1):51–60

Magnetic Fluid-Based Squeeze Film Between Rotating Curved Rough Circular Plates

Nikhilkumar D. Abhangi, G. M. Deheri and Shruti S. Mehta

Abstract Efforts have been made to study an analyze the behaviour of a magnetic fluid based squeeze film between curved rough rotating circular plates when the rotating curved upper plate lying along a surface determined by an exponential function approaches the rotating curved lower plate along the surface governed by a hyperbolic function. The lubricant used here is a magnetic fluid in the presence of an external magnetic field oblique to the radial axis. The concern Reynolds equation is averaged with respect to the random roughness parameter and solved with appropriate boundary conditions. This is then applied to find out the expression for load carrying capacity. It is seen that the effects of surface roughness on the squeeze film characteristic depend on the pattern of roughness structure. The results show that the bearing system registers a considerably enhanced performance as compared to that of bearing system working with a conventional lubricant. This study tends to suggest that although, the bearing suffers owing to transverse roughness in general, there are ample scopes for obtaining a comparatively better performance in the case of negatively skewed roughness by properly choosing curvature parameters and rotation ratio especially when negative variance is involved.

Keywords Circular plates · Rotation · Roughness · Magnetic fluid · Load carrying capacity

N. D. Abhangi (✉)
Department of Mathematics, Faculty of Engineering, Marwadi Education Foundation's Group of Institution, Rajkot, India
e-mail: nikhil.abhangi@gmail.com

G. M. Deheri · S. S. Mehta
Department of Mathematics, Sardar Patel University, Vallabh Vidyanagar, Gujarat, India

H. C. Patel et al. (eds.), *Proceedings of International Conference on Advances in Tribology and Engineering Systems*, Lecture Notes in Mechanical Engineering,
DOI: 10.1007/978-81-322-1656-8_4,

1 Introduction

Archibald [1] discussed the squeeze film behaviour between various geometrical configurations of flat surfaces. Hays [2] studied the squeeze film phenomena between curved plates considering curvature of sine form and keeping minimum film thickness as constant. Murti [3] investigated the squeeze film performance between curved circular plates describing the film thickness by an exponential expression. This analysis was based on the assumption that the central film thickness, instead of minimum film thickness as assumed by Hays [2]; was kept constant. It was shown here that the load carrying capacity increase sharply with the curvature in the case of concave pads. In the above mentioned studies the lower plate was taken to be flat. Patel and Deheri [4] modified the approach of Murti [3] to analyze the squeeze film behaviour between curved circular plates lying along the surfaces determined by hyperbolic function. Prakash and Vij [5] simplified the analysis of Murti [3] by incorporating the well known Morgan Cameron approximation when the porous facing thickness was assumed small. In fact, they presented several bearing configurations such as circular, annular, elliptical, rectangular and conical. Circular geometry registered the highest transient load carrying capacity.

Bhat and Hingu [6] investigated the behaviour of squeeze film trapped between curved porous circular plates approaching normally flat non-porous circular plates. Vora and Bhat [7] dealt with the same problem when the plates were rotating. Wu [8] conducted a study on the squeeze film behaviour between rotating porous annular disk and showed that rotation caused reduced load carrying capacity and response time. Ting [9] used an analogous method who studied the squeeze film between rotating porous annular disk. Gupta, Vora and Bhat [10] analyzed the problem of squeeze film between rotating porous curved annular plates. Prajapati [11] considered the squeeze film behaviour between rotating porous circular plates by incorporating elastic deformation effects.

All the above mention studies dealt with conventional lubricant Verma [12] and Agarwal [13] analyzed the application of a magnetic fluid as lubricant. Later on, Bhat and Deheri [14] presented a study on the squeeze film behaviour between porous annular disks using a magnetic fluid lubricant with the external magnetic field oblique to the lower disk. Modifying this approach Bhat and Deheri [15] discussed the magnetic fluid based squeeze film behaviour in curved porous circular disks. In the above two studies it was concluded that the application of magnetic fluid lubricant enhanced the performance of the squeeze film behaviour. The magnetic fluid based squeeze film behaviour between curved plates lying along the surfaces determined by exponential and hyperbolic function was discussed by Patel and Deheri [16, 17]. Here also it was noticed that the application of magnetic fluid lubricant improved the performance of the bearing system.

The random character of the surface roughness was duly recognized by several investigators, who used a stochastic approach to mathematically model the roughness of the bearing surfaces (Tzeng and Seibel [18], Christensen and Tonder

[19–21]), Tonder [22], theoretically analyzed the transition between surface distributed waviness and random roughness. Christen and Tonder [19–21] modified the approach of Tzeng and Seibel to proposed a comprehensive general analysis both for transverse as well as longitudinal surface roughness based on a general probability density function. This method of Christen and Tonder [19–21] was adopted in several investigations to study the effect of surface roughness on the performance of the bearing system (Ting [9], Prakash and Tiwari [23], Prajapati [24], Guha [25], Gupta and Deheri [26]). Andharia, Gupta and Deheri [27] discussed the effect of transverse surface roughness on the performance of a hydrodynamic squeeze film in a spherical bearing using a general stochastic analysis. It was observed that the effect of transverse surface roughness on the performance of the bearing system was mostly adverse. Recently, Deheri, Patel and Abhangi [28] discussed the behaviour of squeeze film performance between magnetic fluid based curved annular plates.

Here it has been sought to study and analyze a magnetic fluid based squeeze film between curved transversely rough rotating circular plates where in, the upper plate lies along a surface determined by an exponential expression while the lower plate lies along a surface governed by a hyperbolic function.

2 Analysis

The geometrical configuration of the rotating bearing system is displayed in Fig. 1.

The bearing surfaces are assumed to be transversely rough. Following Christensen and Tonder [19–21] the thickness $h(x)$ of the lubricant film is considered as

$$h(x) = \overline{h}(x) + h_s \tag{1}$$

where $\overline{h}(x)$ is the mean film thickness while h_s is the deviation from the mean film thickness characterizing the random roughness of the bearing surfaces. The deviation h_s is assumed to be stochastic in nature and described by the probability density function (Abhangi and Deheri [29])

$$f(h_s), -c \leq h_s \leq c \tag{2}$$

where c is the maximum deviation from the mean film thickness. The mean α, the standard deviation σ and the parameter ε which is the measure of symmetry associated with the random variable h_s, are determined by the relations

$$\alpha = E(h_s) \tag{3}$$

$$\sigma^2 = E\left[(h_s - \alpha)^2\right] \tag{4}$$

and

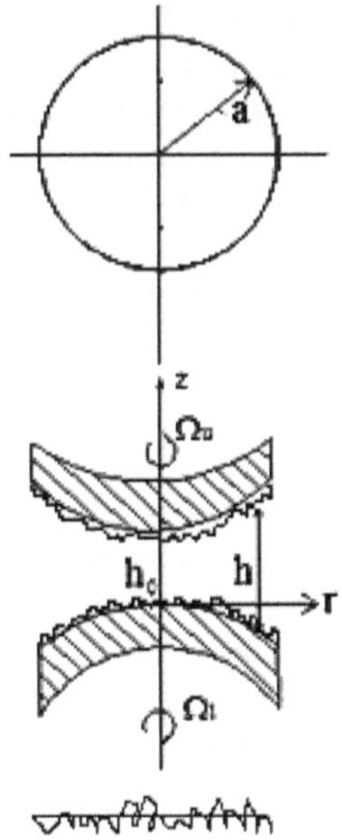

Fig. 1 Configuration of bearing system

$$\varepsilon = E\left[(h_s - \alpha)^3\right] \tag{5}$$

where E is the expectancy operator,

$$E(R) = \int_{-c}^{c} Rf(h_s)dh_s \tag{6}$$

It is assumed that the upper plate lying along the surface determined by

$$Z_u = h_0\left[\exp(-Br^2)\right]\ 0 \le r \le a \tag{7}$$

approaches with normal velocity $\dot{h}_0 = \frac{dh_0}{dt}$, to the rotating lower plate lying along the surface

$$Z_l = h_0\left[\frac{1}{1+Cr} - 1\right];\ 0 \le r \le a \tag{8}$$

where h_0 is central distance between the plates, B and C are the curvature parameters of the corresponding plates. The central film thickness $h(r)$ then, is defined by

$$h(r) = h_0\left[\exp(-Br^2) - \frac{1}{1+Cr} + 1\right] \tag{9}$$

The lower and upper plate rotate with angular velocities is Ω_l and Ω_u.

The magnetic fluid taken to be the lubricant is a suspension of solid magnetic particles of sub domain size in a liquid carrier. The density of the particle is of the order 10^{23} per cubic meter. Depending upon the ferromagnetic material and the method of preparation, the mean diameter of a particle varies from 3 to 15 nm.

These fluids, which are properly stabilized, undergo practically no aging or separation. They remain liquid in a magnetic field and after removal of field recover their characteristics. Particles within the liquid experience a force due to the field gradient and move through the liquid imparting drag to it causing it to flow. Thus, the magnetic fluids can be made to move with the help of a magnetic field gradient, even in the regions where there is no gravity. This property makes these fluids useful in space ships which often go in zero gravity regions. Further details concerning the magnetic fluid and application aspects one can have a glance at Bhat [30], Verma [12] and Kaloni and Venkatasubramanian [31].

The advantage of magnetic fluid lubricants over the conventional ones is that the former can be retained at the desired location by an external magnetic field. In sealed systems as in food processing machines, contamination due to conventional lubricants can be prevented by using magnetic fluid lubricants.

Axially symmetric flow of the magnetic fluid between the plates is taken into consideration under an oblique magnetic field

$$\overline{H} = (H(r)\cos\phi(r,z), 0, H(r)\sin\phi(r,z)) \tag{10}$$

whose magnitude H vanishes at $r = a$; for instance; $H^2 = ka(a - r)$; $0 \leq r \leq a$, where k is a suitably chosen constant so as to have a magnetic field of required strength, which suits the dimensions. The direction of the magnetic field plays a significant role since $\overline{H}$ has to satisfy the relations (Bhat and Deheri [15])

$$\nabla \cdot \overline{H} = 0,\ \nabla \times \overline{H} = 0 \tag{11}$$

Therefore, $\overline{H}$ arises out of a potential function and the inclination angle ϕ of the magnetic field $\overline{H}$ with the lower plate is determined by

$$\cot\phi \frac{\partial \phi}{\partial r} + \frac{\partial \phi}{\partial z} = \frac{1}{2(a - r)} \tag{12}$$

whose solution is obtained from the equation

$$c_1^2 \cos ec^2\phi = a - r,\ z = -2c_1\sqrt{(a - c_1^2 - r)} \tag{13}$$

where c_1 is a constant of integration.

With the usual assumptions of hydromagnetic lubrication the modified Reynolds equation governing the film pressure p can be obtained as (Prajapati [11], Gupta and Deheri [26])

$$\frac{1}{r}\frac{d}{dr}\left\{rg(h)\frac{d}{dr}\left(p-\frac{1}{2}\mu_0\overline{\mu}H^2\right)\right\}=12\mu\overset{\bullet}{h_0}$$
$$+\rho\left(\frac{3}{10}\Omega_r^2+\Omega_r\Omega_l+\Omega_l^2\right)\frac{1}{r}\frac{d}{dr}\left(r^2g(h)\right) \tag{14}$$

where μ_0 is free space permeability, $\overline{\mu}$ is magnetic susceptibility of particles

$$g(h)=h^3+3h^2\left(\alpha^2+\sigma^2\right)+3h\sigma^2+3\sigma^2\alpha+\alpha^3+\varepsilon$$

and

$$\Omega_r=\Omega_u-\Omega_l$$

Introducing the non-dimensional quantities

$$\overline{h}=\frac{h}{h_0},\ R=\frac{r}{a},\ \mu^*=-\frac{\mu_0\overline{\mu}kh^3}{\mu\overset{\bullet}{h_0}},\ P=-\frac{h_0^3p}{\mu a^2\overset{\bullet}{h_0}},$$
$$\Omega_f=\frac{\Omega_l}{\Omega_u},\ S=-\frac{\rho\Omega_u^2h_0^3}{\mu\overset{\bullet}{h_0}},\ \sigma^*=\frac{\sigma}{h_0},\ \varepsilon^*=\frac{\varepsilon}{h_0^3},\ \alpha^*=\frac{\alpha}{h_0}, \tag{15}$$
$$B=\mathrm{Ba}^2,\ C=\mathrm{Ca}$$

and solving the associated Reynolds equation with the concerned boundary conditions

$$P(1)=0,\ \frac{dP}{dR}=-\frac{\mu^*}{2},\ \text{at } R=0 \tag{16}$$

one gets the expression for nondimensional pressure distribution as

$$P=\frac{1}{2}\mu^*R^2(1-R)$$
$$-6\int_1^R\frac{R}{G(h)}dR-\frac{S}{20}\left(3\Omega_f^2+4\Omega_f+3\right)\left(1-R^2\right) \tag{17}$$

where

$$G(h)=\overline{h}^3+3\overline{h}^2\alpha^*+3\overline{h}\left(\alpha^{*^2}+\sigma^{*^2}\right)+3\sigma^{*^2}\alpha^*+\varepsilon^*+\alpha^{*^3}$$

The dimensionless load carrying capacity then is given by

$$\overline{W}=-\left[Wh_0^3\Big/2\pi\mu a^4\overset{\bullet}{h_0}\right]$$
$$=\frac{\mu^*}{40}+3\int_0^1\frac{R^3}{G(h)}dR-\frac{S}{80}\left(3\Omega_f^2+4\Omega_f+3\right) \tag{18}$$

where the load carrying capacity W is obtained from the relation

$$W = 2\pi \int_0^a r p(r) dr \tag{19}$$

3 Results and Discussions

Equation (17) suggests that the nondimensional pressure P increases by $\frac{1}{2}\mu^* R^2(1 - R)$ while the load carrying capacity gets increased by $\frac{\mu^*}{40}$ (Eqn. (18)). As the expression of W is linear in μ^*, it is natural that increasing values of μ^* will lead to increased in load carrying capacity.

Figure 2 shows the effect of the standard deviation associated with roughness on the variation of load carrying capacity. The standard deviation has a considerable adverse effect on the performance of the bearing system as the load carrying capacity decreases.

Figure 3 depict the effect of variance on the distribution of load carrying capacity. From these figures it is noticed that α^*(+ve) decreases the load carrying capacity while the load carrying capacity gets increased due to α^*(−ve). Besides, it is revealed that the combined effect of upper plate's curvature parameter and the negative variance is significantly positive.

The trends of load carrying capacity with respect to ε^* is akin to that of variance.

It is observed that the upper plate curvature parameter has a significantly positive effect when combined with the effect of rotation parameters and lower plate's curvature parameters. The trends of lower plate's curvature parameter are opposite to that of the upper plate's curvature parameter when considered with rotation.

Lastly, the combined effect of Ω_f and S has been presented in Fig. 4. Here it is seen that the load carrying capacity increases considerably with respect to S while the load carrying capacity decreases with respect to Ω_f up to the value −0.667 and then the load carrying capacity increases.

Fig. 2 Load carrying capacity with respect to μ^* and σ^*

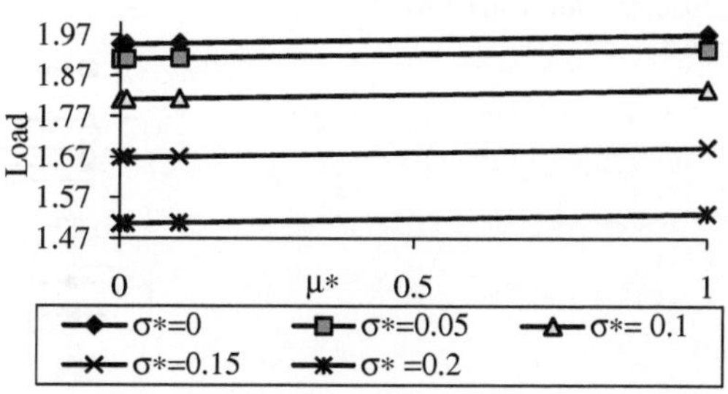

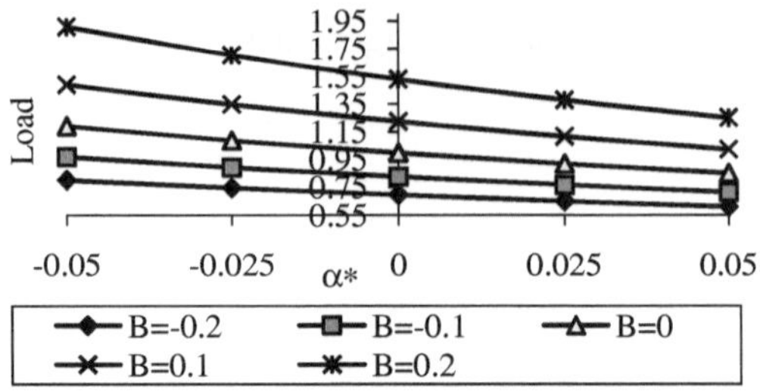

Fig. 3 Load carrying capacity with respect to α^* and **B**

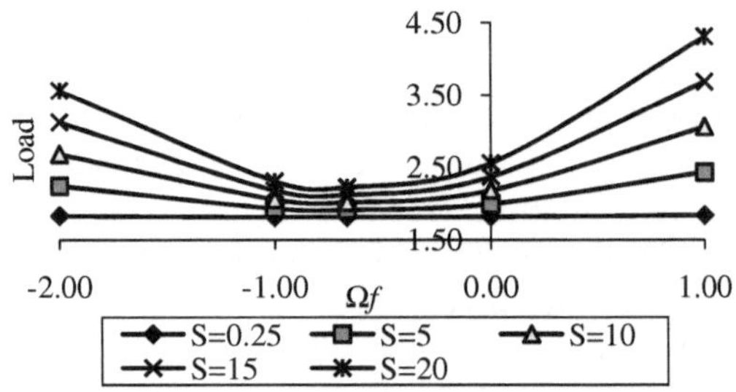

Fig. 4 Load carrying capacity with respect to Ω_f and S

Furthermore the combined effect positive effect of ε^*(+ve) and the lower plate curvature parameter **C** is relatively more as compared to the positive effect of ε^*(+ve) and the upper plate's curvature parameter **B**. Interestingly it is observed that the rate of increase in load carrying capacity with respect to the magnetization parameter is a little more with respect to the lower plate's curvature parameter in comparison with the upper plates curvature parameter (Fig. 5).

It is revealed from the study that the adverse effect of roughness can be reduced by the positive effect of magnetization parameter in the case of negatively skewed roughness by choosing suitable values of lower plate curvature parameter and upper plate curvature parameters (Fig. 6).

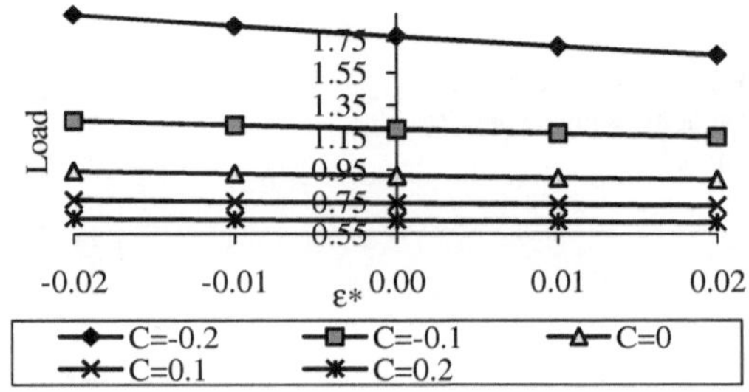

Fig. 5 Load carrying capacity with respect to ε^* and **C**

Fig. 6 Load carrying capacity with respect to ε* and **B**

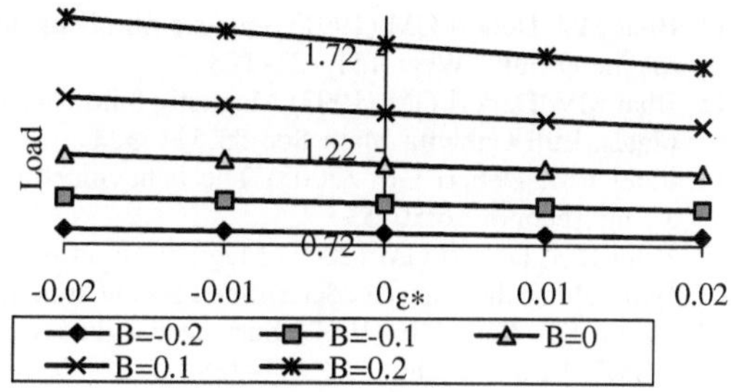

4 Conclusion

This investigation makes it clear that, although, the magnetization has a strong positive effect, the rotation plays a key role in enhancing the performance of the bearing system in the case of negatively skewed roughness especially, when negative variance is involved. So, the roughness must be given due consideration while designing the bearing system even if, suitable values of magnetization parameter and curvature parameter are chosen from bearing's life period point of view.

Acknowledgments The authors acknowledge with thanks the suggestions of reviewers.

References

1. Archibald FR (1956) Load capacity and time relations for squeeze films. ASME 78(D):A231–245
2. Hays DF (1963) Squeeze films for rectangular plates. ASME 58(D):243–248
3. Murti PRK (1975) Squeeze films in curved circular plates. ASME J Lubr Technol 97:650–654
4. Patel RM, Deheri GM (2002) Analysis of the behaviour of the squeeze film between curved plates. J Indian Acad Math 24(2):333–338
5. Prakash J, Vij SK (1973) Load capacity and time height relations for squeeze films between porous plates. Wear 24:309–322
6. Bhat MV, Hingu JV (1979) Squeeze films in a curved porous circular plates. Wear 52:193–196
7. Vora KH, Bhat MV (1980) The load capacity of a squeeze film between curved porous rotating circular plates. Wear 65:39–46
8. Wu H (1971) The squeeze film between rotating porous annular disks. Wear 18:461–470
9. Ting LL (1975) Engagement behaviour of lubricated porous annular disks. Part 1: squeeze film phase, surface roughness and elastic deformation effects. Wear 34:159–182
10. Gupta JL, Vora KH, Bhat MV (1982) The effect of rotational inertia on the squeeze film load between porous annular curved plates. Wear 79:235–240
11. Prajapati BL (1991) Behavior of squeeze film between rotating porous circular plates: surface roughness and elastic deformation effects. Pure Appl Mathematika Sci 33(1–2):27–33
12. Verma PDS (1986) Magnetic fluid based squeeze film. Int J Eng Sci 24(3):395–401
13. Agrawal VK (1986) Magnetic fluid based porous inclined slider bearing. Wear 107:133–139

14. Bhat MV, Deheri GM (1991) Squeeze film behaviour in porous annular disks lubricated with magnetic fluid. Wear 151:123–128
15. Bhat MV, Deheri GM (1993) Magnetic fluid based squeeze film between two curved circular plates. Bull Calcutta Math Soc 85:521–524
16. Patel RM, Deheri GM (2003) The behaviour of the squeeze film between annular plates. J Eng Technol 16:50–53
17. Patel RM, Deheri GM (2002) Magnetic fluid based squeeze film between two curved plates lying along the surfaces determined secant functions. Indian J Eng Mater Sci 9:43–48
18. Tzeng ST, Saibel E (1967) Surface roughness effect on slider bearing lubrication. Trans ASME J Lubr Technol 10:334–338
19. Christensen H, Tonder KC (1969a) Tribology of rough surfaces: stochastic models of hydrodynamic lubrication. SINTEF report no.10/69-18
20. Christensen H, Tonder KC (1970) The hydrodynamic lubrication of rough bearing surfaces of finite width. ASME-ASLE lubrication conference, Cincinnati, Ohio paper no. 70-Lub-7
21. Christensen H, Tonder KC (1969b) Tribology of rough surfaces: parametric study and comparison of lubrication models. SINTEF report no. 22/69-18
22. Tonder KC (1972) Surface distributed waviness and roughness. In: Proceedings of first world conference in industrial tribology, vol 3A, pp. 1–8
23. Prajapati BL (1992) Squeeze film behaviour between rotating porous circular plates with a concentric circular pocket: surface roughness and elastic deformation effects. Wear 152:301–307
24. Prakash J, Tiwari K (1983) Roughness effects in porous circular squeeze-plates with arbitrary wall thickness. J Lubr Technol 105:90–95
25. Guha SK (1993) Analysis of dynamic characteristics of hydrodynamic journal bearings with isotropic roughness effects. Wear 167:173–179
26. Gupta JL, Deheri GM (1996) Effect of roughness on the behaviour of squeeze film in a spherical bearing. Tribol Trans 39:99–102
27. Andharia PI, Gupta JL, Deheri GM (1999) Effect of transverse surface roughness on the behavior of squeeze film in a spherical bearing. J Appl Mech Eng 4:19–24
28. Deheri GM, Patel RM, Abhangi ND (2011) Magnetic fluid-based squeeze film behavior between transversely rough curved annular plates: a comparative study. Ind Lubr Tribol 63(4):254–270
29. Abhangi ND, Deheri GM (2012) Numerical modelling of squeeze film performance between rotating transversely rough curved circular plates under the presence of a magnetic fluid lubricant. ISRN mechanical engineering. doi: 10.5402/2012/873481
30. Bhat MV (2003) Lubrication with a magnetic fluid. Team Spirit Pvt. Ltd, India
31. Kaloni PN, Venkatasubramanian S (2008) Flow of a magnetic fluid between eccentric rotating disks. J Magn Magn Mater 320:142–149

Performance of a Magnetic Fluid-Based Squeeze Film Between Infinitely Long Rough Porous Rectangular Plates

R. M. Patel, G. M. Deheri and P. A. Vadher

Abstract An attempt has been made to discuss the performance of a magnetic fluid based squeeze film between porous infinitely long transversely rough rectangular plates. A magnetic fluid is used as a lubricant and the external magnetic field is oblique to the lower plate. The roughness of the bearing surfaces is characterized by a stochastic random variable with non-zero mean, variance and skewness. The associated Reynolds' equation is solved with suitable boundary conditions to obtain the pressure distribution in turn, which is used to get the expression for load carrying capacity resulting in the calculation of response time. The graphical representation tends to suggest that the bearing system registers an enhanced performance as compared to that of a bearing system dealing with a conventional lubricant. In fact, the pressure, load carrying capacity and response time increase with increasing magnetization parameter. It is noticed that the bearing suffers in general due to transverse roughness. However, the present study reveals that the negatively skewed roughness improves the performance of the bearing system. In addition, it is observed that the aspect ratio significantly affects the performance of the bearing system. Besides, the present study indicates that although the combined negative influence of the porosity, standard deviation and the aspect ratio dominates the positive effect of the magnetization parameter, there exists some scopes for improving the performance of the bearing system by choosing a suitable combination of the magnetization parameter and the aspect ratio in the case of negatively skewed roughness because even the load carrying

R. M. Patel (✉)
Department of Mathematics, Gujarat Arts and Science College, Gujarat 380006, Ahmedabad, India
e-mail: rmpatel2711@gmail.com

G. M. Deheri
Department of Mathematics, Sardar Patel University, Gujarat 388120, Vallabh Vidyanagar, India

P. A. Vadher
Department of Physics, Government Science College, Gujarat 382016, Gandhinagar, India

H. C. Patel et al. (eds.), *Proceedings of International Conference on Advances in Tribology and Engineering Systems*, Lecture Notes in Mechanical Engineering, DOI: 10.1007/978-81-322-1656-8_5, © Springer India 2014

capacity increases due to negative variance. This article reveals that the roughness must be given due consideration while designing the bearing system even though there is the presence of a strong magnetic field.

Keywords Rectangular plates · Surface roughness · Reynolds' equation · Magnetic fluid · Load carrying capacity

1 Introduction

The classical theory of squeeze film between plane parallel surfaces was presented by Archibald [1]. Wu [2–4] investigated the performance of squeeze film behavior for a porous bearing considering mainly annular and rectangular geometries. Later on, the squeeze film behavior between porous plates of various shapes was analyzed by Prakash and Vij [5] resorting to Morgan-Cameron approximation. It was made clear that the aspect ratio played a crucial role on the performance of this type of bearing systems. In fact, a comparison was made between the squeeze film behavior of various geometries of equivalent surface area which established that the circular plates recorded the highest transient load carrying capacity, other parameters remaining same.

The above studies considered the bearing surfaces to be smooth. However, bearing surfaces particularly after having some run-in and wear develop roughness. It has now been well established that the roughness of the bearing surfaces significantly affects the bearing performance, especially in bearing working in the boundary lubrication regime. The roughness appears to be random in character which was recognized by several investigators who adopted a stochastic approach to mathematically model the roughness of the bearing surfaces [6–9]. Christensen and Tonder's work [7–9] which was based upon the concept of viewing the film thickness as a stochastic process, resulted in a averaged Reynolds' type equation. Modifying and developing the approach of Tzeng and Seibel [6], Christensen and Tonder [7–9] presented a comprehensive general analysis for both version of surface roughness namely, transverse and longitudinal. Later on, this approach of Christensen and Tonder [7–9] formed the basis of the analysis to analyze the effect of surface roughness on the performance of the bearing system in a number of investigations [10–14]. Andharia, Gupta and Deheri [15, 16] dealt with the analysis of the effect of surface roughness on the performance of a squeeze film bearing using the general stochastic analysis (without making use of a specific probability distribution) for describing the random roughness.

All these above studies were based on conventional lubricants. Oil based or other lubricant fluid based magnetic fluid can be used as a lubricant. The advantage of magnetic fluid as a lubricant over the conventional ones is that the former can be retained at the desired location by an external magnetic field. The magnetic fluid is prepared by suspending fine grains coated with surfactants and dispersing it in the magnetically passive solvent such as kerosene, hydrocarbons and fluorocarbons. Each particle experiences a force when magnetic field is applied to the magnetic fluid. Use of magnetic fluid as a lubricant modifying the performance of the bearing has been very well recognized. Verma [17] and Agrawal [18] investigated the application of magnetic fluid as a lubricant. While Verma [17] considered tangential slip velocity at the porous matrix lubricant interface, Agrawal [18] discussed no slip condition. Bhat and Deheri [19, 20] analyzed the performance of a magnetic fluid based squeeze film behavior between curved annular disks and curved circular plates lubricated with a magnetic fluid and found that its performance with the magnetic fluid as a lubricant was better than with a conventional lubricant. The study of Bhat and Deheri [20] was further extended by Patel and Deheri [21] to study the behavior of squeeze film formed by a magnetic fluid between curved annular plates. Patel and Deheri [22] discussed the performance of a magnetic fluid based squeeze film between transversely rough annular plates. It was shown that the effect of transverse roughness was adverse in general.

All these above studies establish that the performance of the bearing system gets enhanced by the presence of a magnetic fluid lubricant. Hence, it was deemed proper to study and analyze the behavior of a magnetic fluid based squeeze film between transversely rough infinitely long porous rectangular plates.

2 Analysis

The configuration of the bearing is shown in Fig. 1.

It consists of the geometrical structure of rectangular plates. The upper plate moves normally towards the lower plate with uniform velocity $\dot{h} = {}^{dh}/_{dt}$. The bearing surfaces are assumed to be transversely rough. The geometry of the local film thickness can be thought of as consisting of two parts:

$$h(x) = \bar{h}(x) + h_s(x)$$

where $\bar{h}(x)$ is the mean film thickness and $h_s(x)$ is the deviation from the mean film thickness characterizing the random roughness of the bearing surfaces. $h_s(x)$ is considered to be stochastic in nature and governed by the probability density

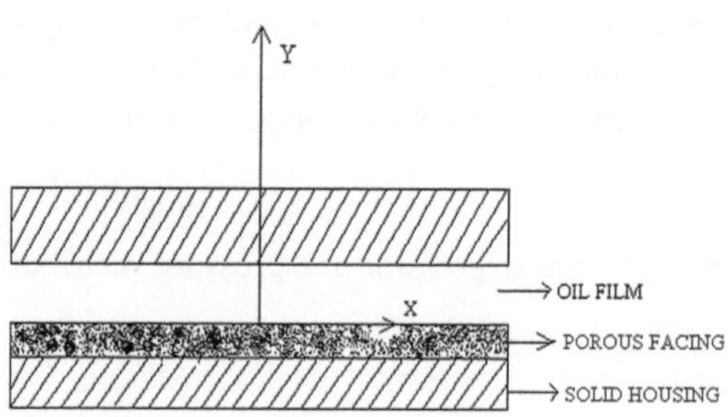

Fig: 1 Configuration the bearing

function f (h_s), $-c \le h_s \le c$ where c is the maximum deviation from the mean film thickness. The mean α, the standard deviation σ and the parameter ε which is the measure of symmetry of random variable h_s, are defined by relationships.

$$\alpha = E\,(h_s)$$
$$\sigma^2 = E\left[(h_s - \alpha)^2\right]$$

and

$$\varepsilon = E\left[(h_s - \alpha)^3\right]$$

where the expectancy operator E is defined by

$$E\,(R) = \int_{-c}^{c} Rf(h_s)dh_s$$

while

$$f(h_s) = \begin{cases} \dfrac{35}{32c^7}\left(c^2 - h_s^2\right)^3, & \text{if } -c \le h_s \le c \\ 0, & \text{elsewhere} \end{cases}$$

Assuming axially symmetric flow of the magnetic fluid between the rectangular plates under an oblique magnetic field $\bar{H}$ whose magnitude H is a function of z vanishing at $\pm$ b/2, the modified Reynolds' equation governing the film pressure p is obtained as [5, 20].

$$\frac{d^2}{dz^2}\left[p - 0.5\mu_0\bar{\mu}H^2\right] = \frac{12\mu\dot{h}}{g(h) + 12\varphi H_0} \tag{1}$$

where

$$g(h) = h^3 + 3\sigma^2 h + 3h^2\alpha + 3h\alpha^2 + 3\sigma^2\alpha + \alpha^3 + \varepsilon,$$
$$H^2 = kb^2\cos\left(\frac{\pi z}{b}\right),$$

k being suitably chosen constant from dimensionless point of view, μ is the viscosity, μ_0 is the permeability of the free space, $\bar{\mu}$ is magnetic susceptibility, ϕ is the permeability of the porous matrix and H_0 is the thickness of the porous facing.

Integrating the above equation with respect to the boundary conditions

$$p(\pm b/2) = 0 \tag{2}$$

one gets the expression for pressure distribution as

$$p = 0.5\mu_0\bar{\mu}H^2 + \frac{6\mu\dot{h}}{g(h) + 12\varphi H_0}\left[z^2 - \frac{b^2}{4}\right]$$

Introducing the non dimensional quantities

$$\bar{z} = z/b \quad \mu^* = -\frac{\mu_0\bar{\mu}h^3}{\mu\dot{h}} \quad k_1 = \frac{a}{b}$$

$$\psi = \frac{\varphi H_0}{h^3} \quad \bar{h} = \frac{h}{h_0} \quad \bar{h}_1 = \frac{h_1}{h_0}$$

$$\bar{h}_2 = \frac{h_2}{h_0} \quad \sigma^* = \frac{\sigma}{\bar{h}} \quad \alpha^* = \frac{\alpha}{\bar{h}}$$

$$\varepsilon^* = \frac{\varepsilon}{h^3}$$

$$g(\bar{h}) = 1 + 3\sigma^{*2} + 3\alpha^* + 3\alpha^{*2} + 3\sigma^{*2}\alpha^* + \alpha^{*3} + \varepsilon^*,$$

the dimensionless pressure is obtained as

$$\bar{p} = \frac{\mu^*\cos(\pi\bar{z})}{2k_1} + \frac{6\left[\frac{1}{4} - z^{-2}\right]}{k_1\left(g(\bar{h}) + 12\psi\right)} \tag{3}$$

The load carrying capacity per unit width

$$w = a\int_{-b/2}^{b/2} p(z)dz$$

$$= ab^3\left(\frac{0.5k\mu_0\bar{\mu}}{\pi} - \frac{\mu\dot{h}}{g(h) + 12\varphi H_0}\right)$$

of the bearing in non dimensional form can be expressed as

$$\bar{w} = \frac{1}{k_1}\left[\mu^* + \frac{\pi}{\left(g(\bar{h}) + 12\psi\right)}\right] \tag{4}$$

The response time Δt taken by the upper plate to reach a film thickness h_2 from an initial film thickness h_1 can be determined from the equation,

$$\Delta t = -\frac{\mu\bar{w}a^2b^2h_0^2}{w}\int_{h_1}^{h_2}\frac{dh}{g(h) + 12\varphi H_0}$$

leading to the non dimensional form

$$\Delta\bar{t} = -\overline{w} \int_{\bar{h}_1}^{\bar{h}_2} \frac{d\bar{h}}{g(\bar{h}) + 12\psi} \tag{5}$$

where

$$\overline{h_1} = \frac{h_1}{h_0},\ \bar{h}_2 = \frac{h_2}{h_0} \text{ and } \bar{h} = \frac{h}{h_0}$$

3 Results and Discussions

Equations (3, 4) present dimensionless pressure and load carrying capacity respectively. The corresponding non-magnetic case concerned with a smooth bearing [5] can be obtained by taking magnetization and roughness parameters to be zero in these expressions. It is easily observed from the Eq. (3) that the pressure increases by

$$\frac{\mu^* \cos(\pi\bar{z})}{2k_1}$$

while increase in load carrying capacity is

$$\frac{\mu^*}{k_1}$$

as indicated by Eq. (4). This alone tells that a suitable combination of magnetization parameter and the aspect ratio may result in a enhanced performance. The method adopted here tends to suggest that the performance of a rough bearing system can be obtained by considering only μ^* to be zero.

Figures 2, 3, 4, 5 and 6 give the variation of the non-dimensional load carrying capacity with respect to magnetization parameter for different values of porosity, roughness parameters and aspect ratio respectively.

From these figures it is observed that while the presence of magnetic fluid enhances the performance of the bearing system, the transverse roughness affects the system adversely. Further, it is noticed that the negative influence induced by the standard deviation is relatively more as regards to the adverse effect introduced by other two roughness parameters. Besides, porosity results in decreased load carrying capacity. However, it is interesting to note that negatively skewed roughness increases the load carrying capacity while even negative variance leads to an increase in load carrying capacity Figs. 4, 5. In addition, the effect of porosity with respect to the magnetization parameter is negligible up to certain extent ($\psi = 0.0001$) Fig. 3 while the effect of the standard deviation with respect

Fig: 2 Variation of load carrying capacity with respect to μ^* and ψ

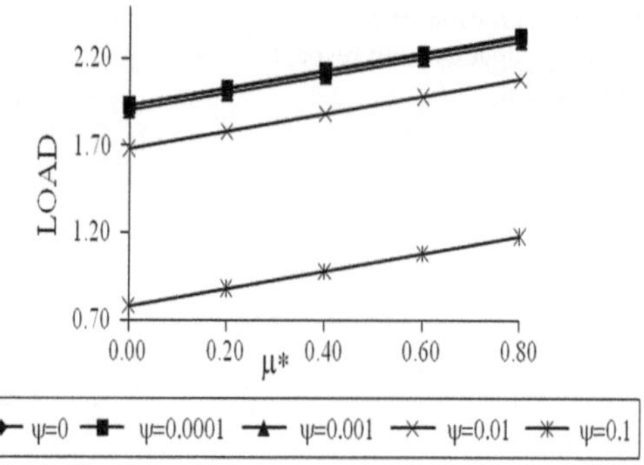

Fig: 3 Variation of load carrying capacity with respect to μ^* and σ^*

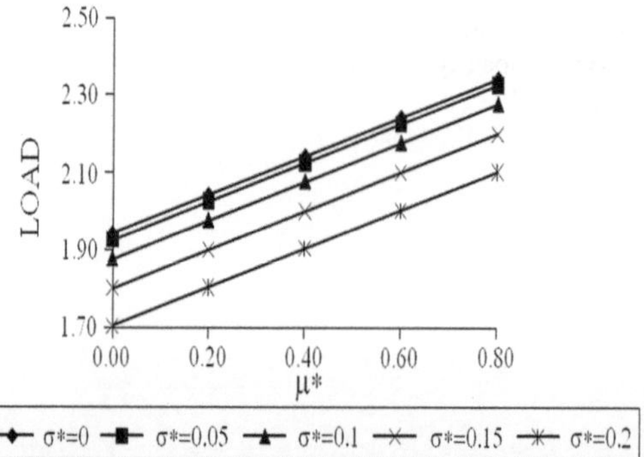

Fig: 4 Variation of load carrying capacity with respect to μ^* and α^*

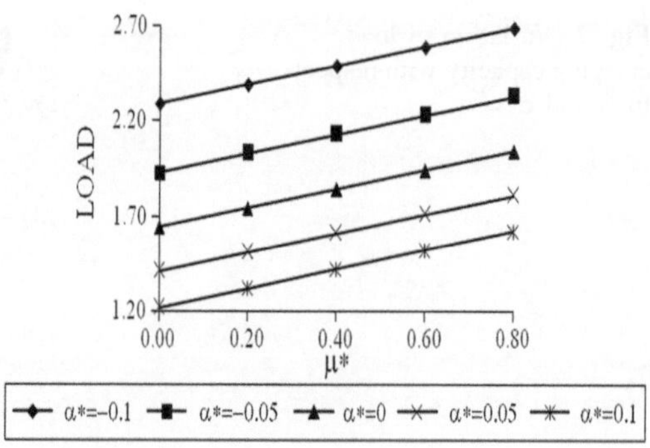

to μ^* is almost negligible up to some extent ($\sigma^* = 0.05$) Fig. 4. However, the effect of aspect ratio is adverse as indicated by Fig. 5. It is noticed that the positive effect of the negatively skewed roughness is comparatively sharp.

The distribution of load carrying capacity with respect to porosity for different values of the roughness parameters and aspect ratio is given in Figs. 7, 8, 9 and 10 respectively.

Fig: 5 Variation of load carrying capacity with respect to μ^* and ε^*

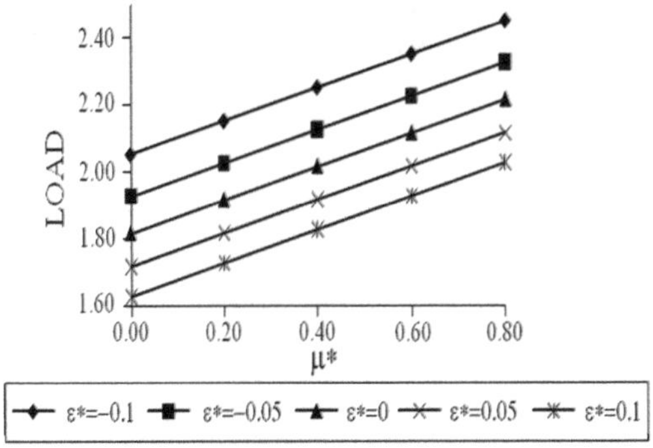

Fig: 6 Variation of load carrying capacity with respect to μ^* and k_1

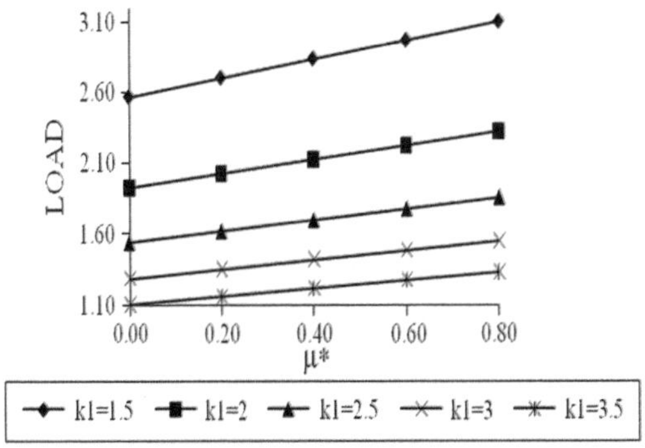

Fig: 7 Variation of load carrying capacity with respect to ψ and σ^*

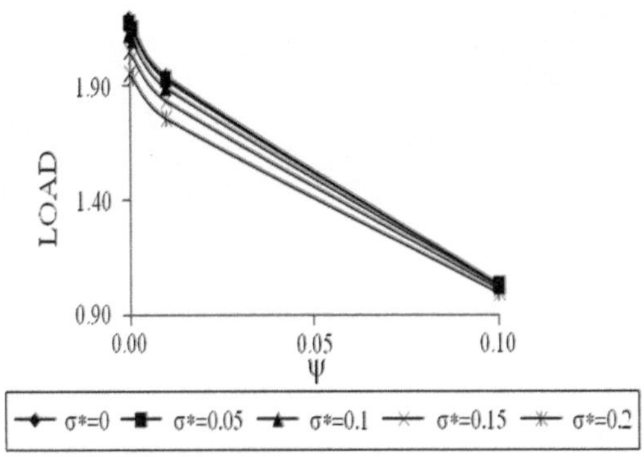

These figures suggest that the combined effect of porosity parameter, roughness parameters and aspect ratio is considerably adverse. Here it appears that the decreased load carrying capacity due to porosity gets further decreased owing to the standard deviation. A comparison reveals that the effect of aspect ratio is sharp as compared to the porosity in decreasing the load carrying capacity.

Fig: 8 Variation of load carrying capacity with respect to ψ and α*

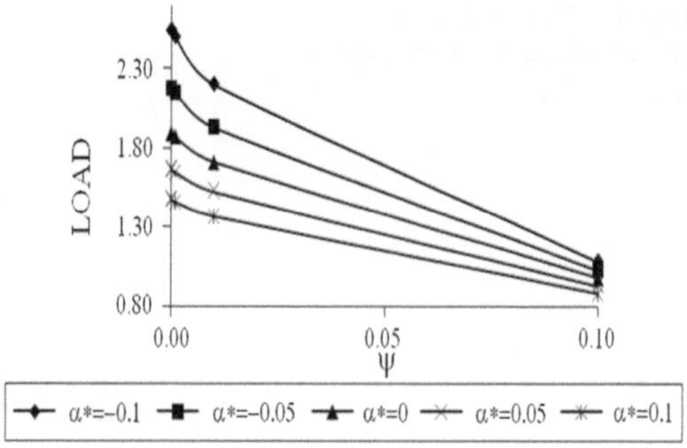

Fig: 9 Variation of load carrying capacity with respect to ψ and ε*

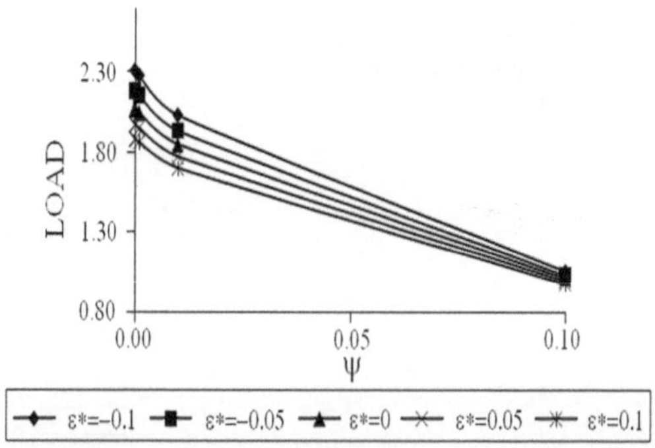

Fig: 10 Variation of load carrying capacity with respect to ψ and k_1

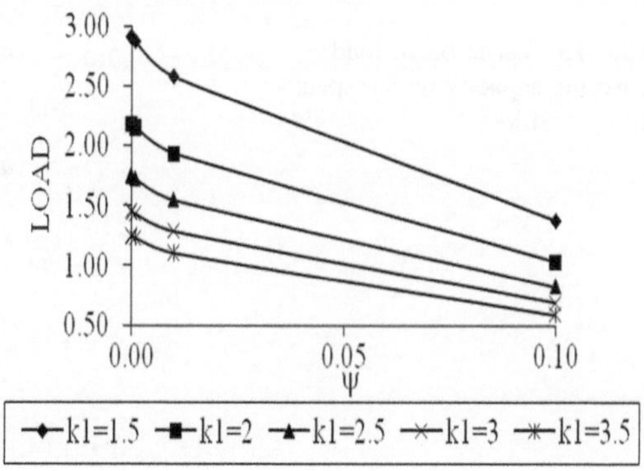

The effect of standard deviation on the distribution of load carrying capacity is depicted in Figs. 11, 12 and 13.

As can be seen the standard deviation induces a sharp decrease in the load carrying capacity. Figures 14, 15 depict the variation of load carrying capacity with respect to the variance.

It is clear that skewness positive and variance positive decrease the load carrying capacity while the negatively skewed roughness and negative variance

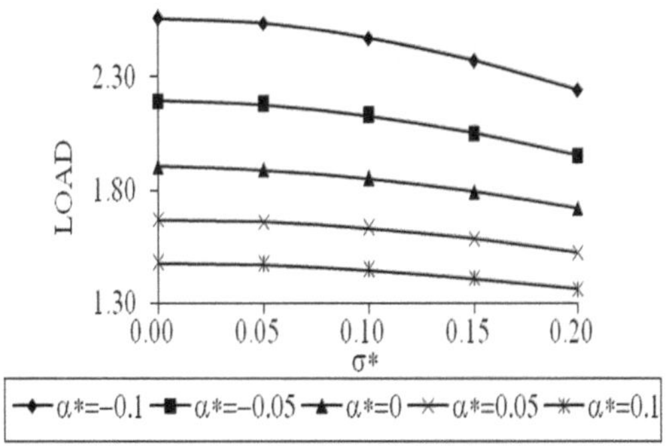

Fig: 11 Variation of load carrying capacity with respect to σ^* *and* α^*

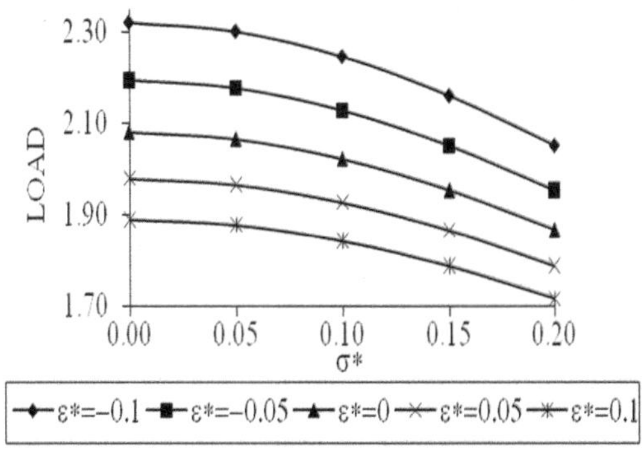

Fig: 12 Variation of load carrying capacity with respect to σ^* and ε^*

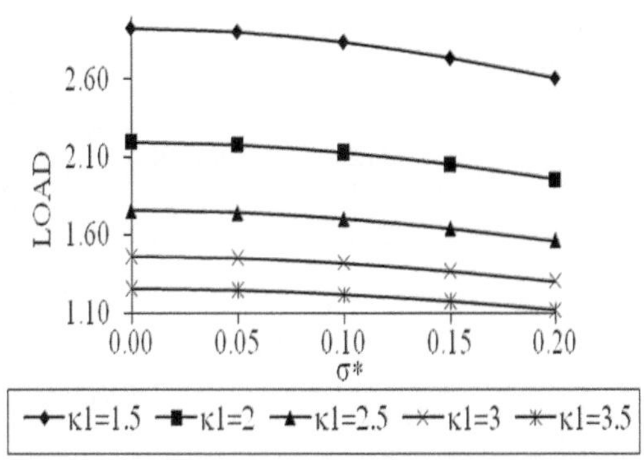

Fig: 13 Variation of load carrying capacity with respect to σ^* and k_1

increase the load carrying capacity. These figures indicate that the combined effect of the negatively skewed roughness and negative variance is significantly positive. Finally, Fig. 16 tells that the effect of negatively skewed roughness is more pronounced than that of negative variance when considered with the aspect ratio.

A close scrutiny of these figures tends to reveal that the porosity effect may be neglected up to certain extent while there may be a possibility of the fluid being nearly squeezed out. A comparison with the study of Prakash and Vij [5] informs

Fig: 14 Variation of load carrying capacity with respect to α^* and ε^*

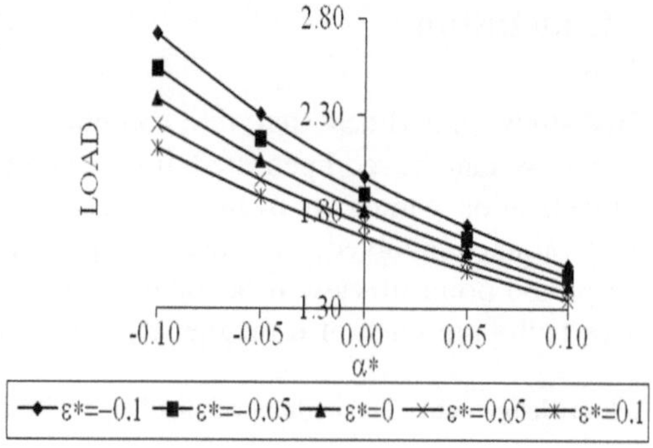

Fig: 15 Variation of load carrying capacity with respect to α^* and k_1

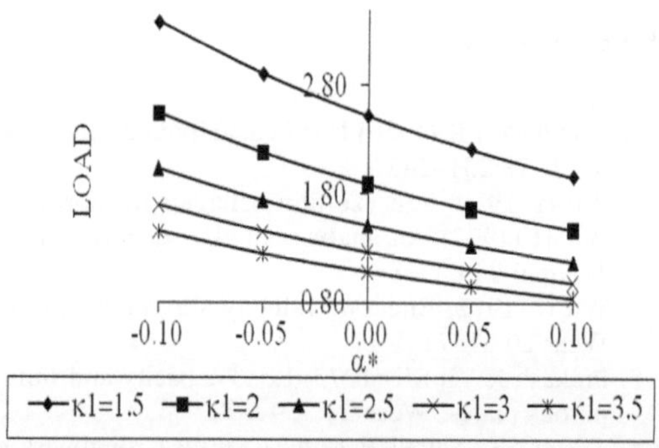

Fig. 16 Variation of load carrying capacity with respect to ε^* and k_1

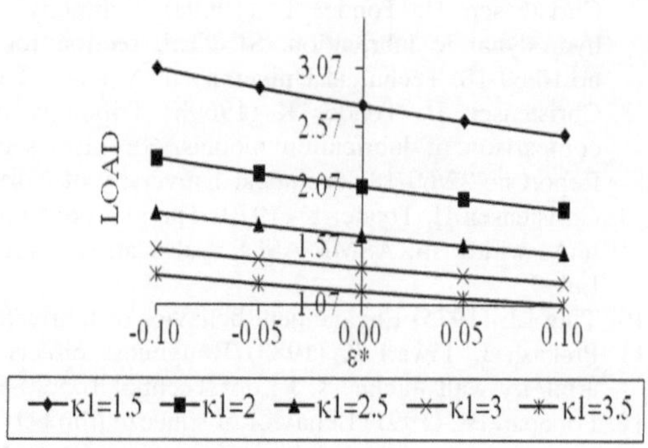

that the performance of the bearing system is a little bit improved here. It is more appealing to note that this bearing with magnetic fluid lubricant can support a good amount of load even when there is no flow.

4 Conclusion

This study underlines that the combined adverse effect of the porosity and roughness can be compensated up to certain extent by the positive effect of magnetization parameter in the case of negatively skewed roughness when negative variance involved, by choosing a proper value of the aspect ratio. Thus, from life period point of view it becomes mandatory to account for roughness even if there is the presence of a magnetic fluid, while designing the bearing system.

Acknowledgement The authors acknowledge with thanks the suggestions of reviewer/editor.

References

1. Archibald FR (1956) Load capacity and time height relations for squeeze films. J Basic Eng Ser D 78:231–245
2. Wu H (1970) Squeeze film behavior for porous annular disks. J Lubr Technol 92:593–596
3. Wu H (1972) An analysis of the squeeze film between porous rectangular plates. J Lubr Technol 94:64–68
4. Wu H (1972) Effect of velocity slip on the squeeze film between porous rectangular plates. Wear 20:67–71
5. Prakash J, Vij SK (1973) Load capacity and time height relations for squeeze films between porous plates. Wear 24:309–322
6. Tzeng ST, Saibel E (1967) Surface roughness effect on slider bearing lubrication. Trans ASLE 10:334–340
7. Christensen H, Tonder K (1969a) Tribology of rough surfaces, stochastic models of hydrodynamic lubrication. SINTEF, section for machine dynamics in tribology, Report no.10/69-18. Technical University of Norway, Trondheinm, Norway
8. Christensen H, Tonder K (1969b) Tribology of rough surfaces, parametric study and comparison of lubrication models. SINTEF, section for machine dynamics in tribology, Report no.22/69-18. Technical University of Norway, Trondheinm, Norway
9. Christensen H, Tonder K (1970) The hydrodynamic lubrication of rough bearing surfaces of finite width. In: ASME-ASLE Lubrication conference Cincinnati, Ohio, 12–15 Oct 1970, Lub-7
10. Ting LL (1975) Engagement behavior of lubricated porous annular discs. Wear 34:159–182
11. Prakash J, Tiwari K (1983) Roughness effects in porous circular squeeze –plates with arbitrary wall thickness. J Lubr Technol 105:90–95
12. Prajapati BL (1991) Behavior of squeeze film between rotating porous circular plates, surface roughness and elastic deformation effects. Pure Appl Mathematica Sci 33(1–2):27–36
13. Guha SK (1993) Analysis of dynamic characteristics of hydrodynamic journal bearings with isotropic roughness effects. Wear 167(1):173–180
14. Gupta JL, Deheri GM (1996) Effect of roughness on the behavior of squeeze film in a spherical bearing. Tribol Trans 39:99–102
15. Andharia PI, Gupta JL, Deheri GM (1999) Effect of transverse roughness on the behavior of squeeze film in a spherical bearings. Int J Appl Mech Eng 4(1):19–24
16. Andharia PI, Gupta JL, Deheri GM (2001) Effect of surface roughness and hydrodynamic lubrication of slider bearings. Tribol Trans 44(2):291–297
17. Verma PDS (1986) Magnetic fluid based squeeze films. Int J Eng Sci 24(3):395–401
18. Agrawal VK (1986) Magnetic fluid based porous inclined slider bearing. Wear 107:133–139

19. Bhat MV, Deheri GM (1991) Squeeze film behavior in porous annular disks lubricated with magnetic fluid. Wear 151:123–128
20. Bhat MV, Deheri GM (1993) Magnetic fluid based squeeze film in curved porous circular disks. J Magn Magn Mater 127:159–162
21. Patel RM, Deheri GM (2002) On the behavior of squeeze film formed by a magnetic fluid between curved annular plates. Indian J Math 44:353–359
22. Patel RM, Deheri GM (2004) Magnetic fluid based squeeze film behavior between annular plates and surface roughness effect. In: Proceedings of AIMETA international tribology conference, pp 631–638

Power Law Fluid Film Lubrication of Journal Bearing with Squeezing and Temperature Effects

Dhaneshwar Prasad, S. S. Panda and S. V. Subrahmanyam

Abstract This paper presents theoretical investigations of the rheological effects of lubricant on the performance of the Journal bearing system under steady state condition including squeezing. Runga Kutta Fehlberg method is employed to solve the Reynolds and the energy equations governing the flow of power law fluids simultaneously. Those equations are coupled due to the consistency which is a function of pressure and temperature both. The results show that this simple innovative model can reasonably calculate delta profile and hence the pressure and the temperature. The obtained results that the pressure and the temperature both increase with the power law flow index n and decrease with the increase of the squeezing parameter q. These results are found to be similar to the results available in the literature.

Keywords Hydrodynamic lubrication · Journal bearing · Squeezing · Consistency variation of power law · Thermal effect

Nomenclature

c	Radial clearance
h	Oil film thickness
m	Consistency index
n	Flow behaviour index
p	Hydrodynamic pressure
Q	Flow flux

D. Prasad (✉)
S.R.K. Government Arts College, Yanam, India
e-mail: rpdhaneshwar@gmail.com

S. S. Panda
Regency Institute of Technology, Yanam 533464, India

S. V. Subrahmanyam
KL University, Vijayawada, India
e-mail: sudamshekhar@gmail.com

H. C. Patel et al. (eds.), *Proceedings of International Conference on Advances in Tribology and Engineering Systems*, Lecture Notes in Mechanical Engineering,
DOI: 10.1007/978-81-322-1656-8_6,

R	Radious of the journal
t	Time of approach
T	Temperature
u,v	Velocity components
V	Squeeze velocity
W	Load capacity
W_R	$W_\pi/W_{\pi/2}$ load ratio
ε	Eccentricity ratio
θ	Angular co-ordinate
$\overline{m}$	$m\left(\frac{U}{c}\right)^n\alpha$
U	$c\frac{d\varepsilon}{dt}$
p_e	$\frac{\rho C_p C U}{k}$
$\overline{D}$	$D\sin\theta\frac{2n+1}{n}$
$\overline{E}$	$\frac{E}{D}\left(\frac{2n+1}{n}\right)^n\sin^n\theta$
$\overline{\gamma}$	$\frac{\beta}{\rho Cp\,\alpha}$
$\overline{R}$	$\frac{R}{C}$
$\overline{h}$	$\frac{h}{C}$
B	$\left(\frac{2n+1}{2n}\right)^n\sin^n\theta$

1 Introduction

The squeeze is an important issue from many tribological aspects. This feature significantly affects minimum film thickness, pressure and temperature distributions in concentrated contact film mechanism. During squeezing, the surfaces come closer to each other so it can also increase the risk of wear, scuffing and pitting if the surfaces are not enough smooth. Further, it is commonly observed in the bearings of automotive engines, aircraft engines, machine tools, turbo machinery, and skeletal joints. There are several investigations done on squeezing and some of them are Lin, Oliver and Scot [1–3].

Conventionally, the prediction of squeeze film motion assumes that the lubricant behaves as a Newtonian viscous fluid. However, experimental results show that the addition of small amounts of long-chained additives to a Newtonian fluid minimizes the sensitivity of the lubricant to change in shear rate and provides beneficial effects on the load-carrying and frictional characteristics [2, 3]. Moreover, a base oil blended with additives can stabilize the behavior of lubricants in elasto hydrodynamic contacts and reduce friction and surface damage, which describe the rheological behavior of non-Newtonian lubricants [4].

In addition, high pressure in concentrated contact can influence the temperature rise there, and hence it introduces the field of thermo hydrodynamic lubrication which measures the performance of journal bearings with thermal effects in the lubrication process. Since it can naturally yield the peak bearing temperature, then the bearing failure can be predicted at the design stage when the maximum temperature exceeds a certain limit. The research into THD lubrication has drawn research effort. For example, Ferron [5] solved Dowson's [6] generalized Reynolds equation simultaneously with the energy equation and reported excellent results. Khonsari and Beaman [7] obtained THD solutions under severe boundary conditions considering the mixing of the recirculating fluid and the supply oil.

On the line of non-Newtonian fluid model, Power law lubricant model has got attention in the recent years because of its simplicity and potential to describe many lubricants such as silicon fluids, polymer solutions Chu et al. [8], Suneetha et al. [9]. In fact, this power law model characterizes two different types of non-Newtonian fluids i.e. Viscoelastic and dilatants plus Newtonian as well when index n of the power law model is unity [10]. Dein and Elrod [11] examined the analysis of lubrication of journal bearing with the same non-Newtonian fluid model and developed a new numerical technique based on perturbation expansion for velocity under Couette dominated flow condition.

Xiong and Wang [12] investigated a steady state problem of smooth surface hydrodynamic lubrication of a pocketed pad plain journal bearing based on Payvar—Salant mass conservation model leaving the importance of thermal effect. Balasoiu et al. [13] presented a 3D analysis of cylindrical porous journal bearing characterized by a self circulating lubricating system that eliminates the necessity of external pump. However, the thermal effect was also ignored. Further, in contrast to the heavy loaded system it has been believed that in conformal contact such as in journal bearings and thrust bearing elastic deformation can be ignored under low load because the fluid pressure is insufficient to cause large deformation of the surfaces Yagi and Sugimura [14].

Hence in this paper, hydrodynamic lubrication of rigid journal bearing with the power law is studied including thermal and squeezing effects. The consistency of the lubricant is assumed to vary with pressure and fluid film mean temperature. The last assumption and the condition of rigidity of the bearing surfaces in fact provide us the solution in almost closed form.

2 Mathematical Formulation

2.1 Fluid Flow Governing Equations

The fluid flow governing equations of the hydrodynamic lubrication with some usual assumptions are [15] (Fig. 1).

$$\frac{\mathrm{d}p}{\mathrm{d}x} = \frac{\partial}{\partial y}\left[m\left|\frac{\partial u}{\partial y}\right|^{n-1}\frac{\partial u}{\partial y}\right] \tag{1}$$

Fig. 1 Journal bearing

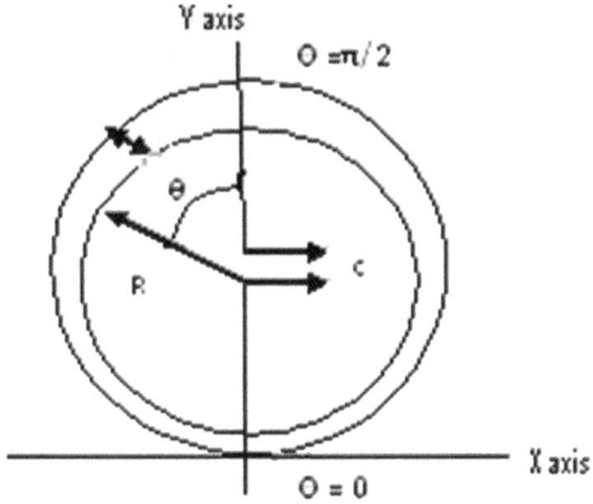

$$\frac{\partial u}{\partial x}+\frac{\partial v}{\partial y}=0 \tag{2}$$

where

$$m=m_0 e^{\alpha p-\beta(T_m-T_0)} \tag{3}$$

with

$$T_m=\frac{1}{h}\int_0^h T\,dy \tag{4}$$

and

$$h=c\,(1-\varepsilon\cos\theta) \tag{5}$$

Let y = h_1 be the height where the velocity gradient $\frac{\partial u}{\partial y}=0$
The boundary conditions for the above governing equations are:

$$u=0,\ at\ y=0;\ \text{and}\ u=0\ at\ y=h \tag{6}$$

The velocity boundary conditions for the geometry under consideration are:

$$\frac{\partial u_1}{\partial y}\geq 0,\ \ 0\leq y\leq h_1 \tag{7}$$

$$\frac{\partial u_2}{dy}\leq 0\ ,\ \ h_1\leq y\leq h, \tag{8}$$

Using the sign of the velocity gradient and integrating Eq. (1) twice for $0\leq y\leq h_1$; one may obtain

$$u_1 = \left(\frac{n}{n+1}\right)\left(-\frac{1}{m}\frac{dp}{dx}\right)^{1/n}\left[(h_1)^{\frac{n+1}{n}} - (h_1 - y)^{\frac{n+1}{n}}\right], \tag{9}$$

And similarly for $h_1 \leq y \leq h$

$$u_2 = \left(\frac{n}{n+1}\right)\left(-\frac{1}{m_1}\frac{dp}{dx}\right)^{1/n}\left[(h - h_1)^{\frac{n+1}{n}} - (y - h_1)^{\frac{n+1}{n}}\right], \tag{10}$$

At $y = h_1$ the velocities are continuous so from Eqs. (9) and (10) one can get $(h - h_1)^{\frac{n+1}{n}} = (h_1)^{\frac{n+1}{n}}$ which gives $h_1 = h/2$, Thus the flow is symmetrical about the middle point of the film thickness.

Hence we can write the equation (9) and (10) as

$$u_1 = \left(\frac{n}{n+1}\right)\left(-\frac{1}{m}\frac{dp}{dx}\right)^{1/n}\left[\left(\frac{h}{2}\right)^{\frac{n+1}{n}} - \left(\frac{h}{2} - y\right)^{\frac{n+1}{n}}\right], \tag{11}$$

$$u_2 = \left(\frac{n}{n+1}\right)\left(-\frac{1}{m}\frac{dp}{dx}\right)^{1/n}\left[\left(\frac{h}{2}\right)^{\frac{n+1}{n}} - \left(y - \frac{h}{2}\right)^{\frac{n+1}{n}}\right], \tag{12}$$

Now the volume flux Q of the fluid is defined as

$$Q = \int_0^h u \; dy \tag{13}$$

or

$$Q = \int_0^{h/2} u_1 \, dy + \int_{h/2}^{h} u_2 \, dy \tag{14}$$

or

$$Q = \left(\frac{2n}{2n+1}\right)\left(-\frac{1}{m}\frac{dp}{dx}\right)^{1/n}\left(\frac{h}{2}\right)^{\frac{2n+1}{n}} \tag{15}$$

Now the equation of continuity $\frac{\partial u}{\partial x} + \frac{\partial v}{\partial y} = 0$ can be solved with respect to the boundary conditions for v which are $v = -V$ at $y = 0$ and $v = 0$ at $y = h$ we get

$$\frac{\partial Q}{\partial x} = -V \tag{16}$$

Using Eqs. (15) into (16) it comes over to be

$$\frac{\partial}{\partial x}\left[\left(\frac{2n}{2n+1}\right)\left(-\frac{1}{m}\frac{dp}{dx}\right)^{1/n}\left(\frac{h}{2}\right)^{\frac{2n+1}{n}}\right] = -V \tag{17}$$

Assuming $x = R\theta$ and $dx = R\,d\theta$ Eq. (17) can be written as

$$\frac{1}{R}\frac{\partial}{\partial\theta}\left\{\frac{2n}{2n+1}\left(\frac{-1}{mR}\frac{dp}{d\theta}\right)^{1/n}\left(\frac{h}{2}\right)^{\frac{2n+1}{n}}\right\} = -V \tag{18}$$

where $h = c\,(1 - \varepsilon\cos\theta)$ and $V = -c\,\dfrac{d\varepsilon}{dt}\cos\theta$

Substituting for V in Eq. (18) we get

$$\frac{1}{R}\frac{\partial}{\partial\theta}\left\{\frac{2n}{2n+1}\left(\frac{-1}{mR}\frac{dp}{d\theta}\right)^{1/n}\left(\frac{h}{2}\right)^{\frac{2n+1}{n}}\right\} = c\,\frac{d\varepsilon}{dt}\cos\theta \tag{19}$$

Integrating Eq. (19) we get

$$\frac{dp}{d\theta} = -mR\left(\frac{2n+1}{2n}\,Rc\,\frac{d\varepsilon}{dt}\right)^{n}\left(\frac{2}{h}\right)^{2n+1}\sin^{n}\theta \tag{20}$$

2.2 Heat Fluid Flow Equation

The heat energy equation with usual assumptions is taken as [15]

$$\rho\,c_p\left(u\frac{\partial T}{dx} + v\frac{\partial T}{dy}\right) = k\frac{\partial^2 T}{\partial y^2} + m\left|\frac{\partial u}{\partial y}\right|^{n-1}\left(\frac{\partial u}{\partial y}\right)^2$$

This is modified for the problem under consideration is [19]:

$$\rho\,c_p\,u\left(\frac{dT_m}{dx}\right) = k\frac{\partial^2 T}{\partial y^2} + m\left|\frac{\partial u}{\partial y}\right|^{n-1}\left(\frac{\partial u}{\partial y}\right)^2 \tag{21}$$

The boundary conditions for the above equation are:

$$T = T_{11} \text{ at } y = 0,\; T = T_{12} \text{ at } y = h \tag{22}$$

Now the values of T_1 and T_2 are calculated in the region $0 \le y \le h/2$ and $h/2 \le y \le h$ respectively and are obtained as

$$\begin{aligned} T_1 = A&\left[\left(\frac{h}{2}\right)^{\frac{n+1}{n}}\frac{y^2}{2} - \frac{n^2}{(2n+1)(3n+1)}\left(\frac{h}{2}-y\right)^{\frac{3n+1}{n}}\right] \\ &- B\,\frac{n^2}{(2n+1)(3n+1)}\left(\frac{h}{2}-y\right)^{\frac{3n+1}{n}} + c_1 y + d_1 \end{aligned} \tag{23}$$

$$T_2 = A\left[\left(\frac{h}{2}\right)^{\frac{n+1}{n}}\frac{y^2}{2} - \frac{n^2}{(2n+1)(3n+1)}\left(y-\frac{h}{2}\right)^{\frac{3n+1}{n}}\right] - B\,\frac{n^2}{(2n+1)(3n+1)}\left(y-\frac{h}{2}\right)^{\frac{3n+1}{n}} + c_2 y + d_2 \tag{24}$$

Use of the temperature matching condition $T_1 = T_2$ at $y = h/2$ and the matching heat flux condition

$$\text{one may get}\, c_1 = c_2 = c(\text{say})\text{and}\; d_1 = d_2 = d(\text{say}) \tag{25}$$

Using the last temperature boundary conditions (22) and (25), in (23) and (24) one can get

$$c = \frac{1}{h}\left[T_{12} - 2A\left(\frac{h}{2}\right)^{\frac{3n+1}{n}} + (A+B)\frac{n^2}{(2n+1)(3n+1)}\left(\frac{h}{2}\right)^{\frac{3n+1}{n}} - d\right] \tag{26}$$

and

$$cd = T_{11} + (A+B)\frac{n^2}{(2n+1)(3n+1)}\left(\frac{h}{2}\right)^{\frac{3n+1}{n}} \tag{27}$$

where

$$A = \frac{\rho\, c_p}{k}\,\frac{dT_m}{dx}\left(-\frac{1}{m}\frac{dp}{dx}\right)^{\frac{1}{n}}\frac{n}{n+1} \quad \text{and} \quad B = \left(\frac{m}{k}\right)\left(-\frac{1}{m}\frac{\mathrm{d}p}{\mathrm{d}x}\right)^{\frac{n+1}{n}} \tag{28}$$

Thus, T_{11}, T_{12} are explicitly known functions of x and y analytically. Finally, the mean temperature T_m defined as in (4) is obtained as

$$T_m = \frac{1}{h}\left[\int_0^{h/2} T_1\, dy + \int_{h/2}^{h} T_2\, dy\right] \tag{29}$$

or

$$T_m = \frac{T_{11}+T_{12}}{2} + \left(\frac{h}{2}\right)^{\frac{3n+1}{n}}\left\{\frac{\rho\, c_p D}{k}\,\frac{dT_m}{dx}\left(-\frac{1}{m}\frac{\mathrm{d}p}{\mathrm{d}x}\right)^{\frac{1}{n}} + \left(\frac{mE}{k}\right)\left(-\frac{1}{m}\frac{\mathrm{d}p}{\mathrm{d}x}\right)^{\frac{n+1}{n}}\right\} \tag{30}$$

or

$$\frac{\mathrm{d}T_m}{\mathrm{d}x} = \frac{\left(T_m - \frac{T_{11}+T_{12}}{2}\right)\left(\frac{2}{h}\right)^{\frac{3n+1}{n}} - \left(\frac{mE}{k}\right)\left(-\frac{1}{m}\frac{dp}{dx}\right)^{\frac{n+1}{n}}}{\frac{\rho\, c_p D}{k}\left(-\frac{1}{m}\frac{dp}{dx}\right)^{\frac{1}{n}}} \tag{31}$$

We can write Eq. (30) using x = Rθ as

$$\frac{dT_m}{d\theta} = \frac{\left(T_m - \frac{T_{11}+T_{12}}{2}\right)\left(\frac{2}{h}\right)^{\frac{3n+1}{n}} - \left(\frac{mE}{k}\right)\left(\frac{2}{h}\right)^{\left(\frac{(2n+1)(n+1)}{n}\right)}\left(\frac{2n+1}{2n} RC \sin\theta\right)^{n+1}\left(\frac{d\varepsilon}{dt}\right)^{n+1}}{\frac{\rho c_p D}{k}\left(\frac{2}{h}\right)^{\frac{2n+1}{n}}\left(\frac{2n+1}{2n} RC \sin\theta \frac{d\varepsilon}{dt}\right)} \tag{32}$$

where $D = \dfrac{n^3}{(n+1)(2n+1)(4n+1)} - \dfrac{n}{3(n+1)}$ and $E = \dfrac{n^2}{(2n+1)(4n+1)}$

2.3 Dimensionless Schemes

$$\overline{m} = m\left(\frac{U}{c}\right)^n \alpha\, U = c\frac{d\varepsilon}{dt}\; p_e = \frac{\rho\, C_p\, C\, U}{k}\; \overline{D} = D\sin\theta\frac{2n+1}{n}$$

$$\overline{E} = \frac{E}{D}\left(\frac{2n+1}{n}\right)^n \sin^n\theta$$

$$\overline{\gamma} = \frac{\beta}{\rho Cp\, \alpha} \quad \overline{R} = \frac{R}{C}\; \overline{h} = \frac{h}{C} \quad B = \left(\frac{2n+1}{2n}\right)^n \sin^n\theta$$

The dimension less equations

$$\frac{d\overline{T_m}}{d\theta} = \frac{2}{P_e\overline{h}\overline{D}}\left(\overline{T}_m - \frac{\overline{T}_{11}+\overline{T}_{12}}{2}\right) - \overline{E}\,\overline{m}\,\left(\overline{R}\right)^{n+1}\overline{\gamma}\left(\frac{2}{\overline{h}}\right)^{2n+1} \tag{33}$$

$$\frac{d\overline{p}}{d\theta} = -\overline{m}\,\left(\overline{R}\right)^{n+1}\left(\frac{2}{\overline{\overline{h}}}\right)^{2n+1} B \tag{34}$$

2.4 Load

$$\begin{aligned} W &= 2\int_0^{\pi} p\,\cos\theta\, R\, d\theta = -2R\int_0^{\pi} \sin\theta\,\frac{dp}{d\theta}\,d\theta \\ &= -2R\int_0^{\pi}\left(-\overline{m}\,\left(\overline{R}\right)^{n+1}\left(\frac{2}{\overline{\overline{h}}}\right)^{2n+1}\left(\frac{2n+1}{n}\right)^n \sin^{n+1}\theta\right)d\theta \end{aligned} \tag{35}$$

2.5 Result and Discussion

Theoretical aspects of numerically computed results for various bearing characteristics are elaborated through figures and tables which follow. These characteristics are functions of the flow behavior index n. Results are calculated by the following behavior of n i.e., in between 0.4 and 1.15. For numerical calculation following sets of values are used:

$$\bar{R} = 14.96, \alpha = 1.6 \times 10^{-6}\, dyne^{-1}\, m^2,\ \bar{\gamma} = 0.4,$$
$$\varepsilon = 0.252,\ d\varepsilon/dt = 1.0, Pe = 2.0$$

In order to study the qualitative behavior of consistency variations of the incompressible lubricant, pressure and temperature must be computed first. This is achieved by solving the simultaneous Eqs. (33) and (34) numerically for dimensionless pressure $\bar{P}$ and temperature $\bar{T}$ by Runge–Kutta fourth order method. The variation of $\bar{P}$ and $\bar{T}$ are shown in Figs. 2 and 3 respectively. All these cases have one feature in common that the variation in n does not change the general shape of the profile.

2.6 Pressure Distribution

The pressure distribution $\bar{P}$ versus θ for various values of n have been depicted in Fig. 2. $\bar{P}$ decreases continuously when θ from 0 to π increases. The pressure profile $\bar{P}$ against θ for each n is similar to that of Peng and Khonsari [16] and Singh et al. [17]. Xiong and Wang [12], Chen et al. [18]. The pressure imposition in this Fig. 2 i.e. instead of $0 \le \theta \le \pi$, if we take $-\pi \le \theta \le \pi$ profile becomes very similar to [13, 19, 20].

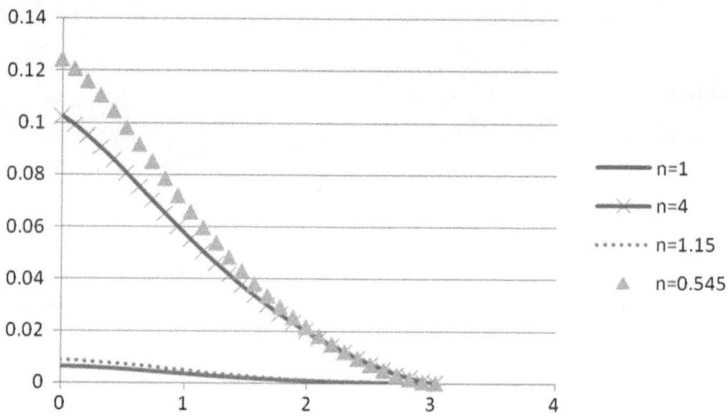

Fig. 2 Pressure profile

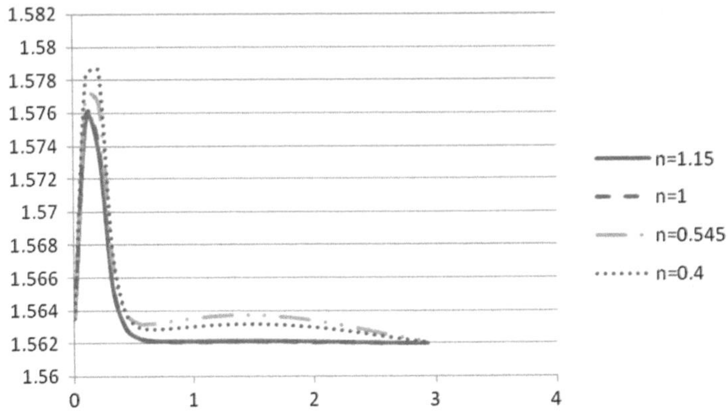

Fig. 3 Temperature profile

2.7 Temperature Distribution

The temperature distribution for various value of n is presented in Fig. 3. It is interesting to note that $\bar{T}$ decreases with respect to θ increase except in the vicinity of zero where the trend is reversed. The similar trend has been found by Liu et al. [21]. The difference is however not very significant both for Newtonian as well as non-Newtonian fluid.

2.8 Load Ratio

A ratio $W_R = W_{\pi}/W_{\pi/2}$ of load capacities is used to study the variation in the load with n. Fig. 4 shows the load ratio W_R is a function of n for various values of ε. This indicates that for the low value of ε (epsilon) the load ratio is higher than the

Fig. 4 Load ratio

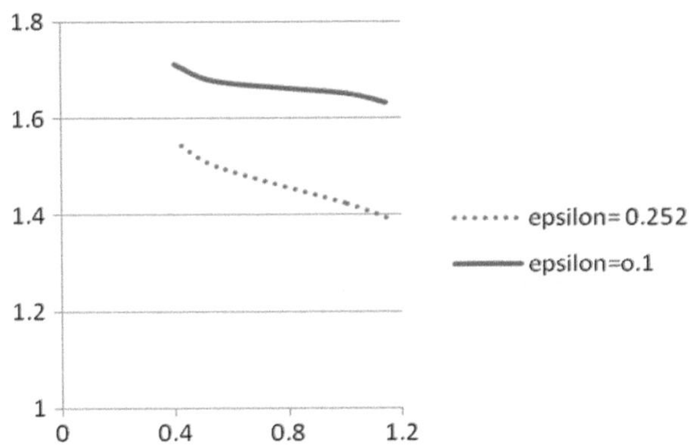

load capacity ratio at higher value of ε.; this indicates that for low values of ε the load in a full journal bearing is much greater than the load in a half journal bearing. Figure 4 shown that the load ratio decreases as n increases; this indicates that a decrease in actual load capacity of both the full and half journal bearing. The same has been reported by Singh and Sinha [22].

3 Conclusion

As it is difficult to non dimensional zed pressure and temperature in case of power law lubricant Singh and Sinha [22], an attempt has been made to complete this assignment further it is concluded that the load ratio decreases with increase in n. It also shown that there is a significant change in pressure and temperature for non Newtonian lubricants.

References

1. Lin JR, Chou TL, Liang LJ, Hung TC (2012) Non- Newtonian dynamic characteristics of parabolic film slider bearing: micropolar fluid model. Tribol Int 48:226–231
2. Oliver DR (1988) Load enhancement effects due to polymer thickening in a short model journal bearing. J Non-Newtonian Fluid Mech 30:185–196
3. Scott W, Suntiwattana P (1995) Effect of oil additives on the performance of a wet friction clutch material. Wear 181(183):850–855
4. Spikes HA (1994) The behaviour of lubricants in contacts: current understanding and future possibilities. J Proc Inst Mech Eng 28:3–15
5. Ferron J, Frene J, Boncompain R (1983) A study of the thermohydrodynamic performance of a plain journal bearing comparison between theory and experiments. ASME J Lubr Tech 105:422–428
6. Dowson D (1962) A generalized Reynolds equation for fluid-film lubrication. Int J Mech Sci 4:159–170
7. Khonsari MM, Beaman JJ (1987) Thermohydrodynamic analysis of laminar incompressible journal bearings. ASLE Trans 29:141–150
8. Chu HM, Li WL, Chang YP (2006) Thin film EHD lubrication—a power law fluid. Tribol Int 39:1474–1481
9. Sunitha P, Kumar VB, Prasad KR (2013) Generalised Reynolds equation for power law fluid application to parallel plates and spherical bearings squeezing considering thermal variation. Int J Adv Eng Technol 4(1):72–77
10. Chu H-M, Li W-L, Chang Y-P (2006) Thin film elastohydrodynamic lubrication—a power law fluid model. Tribol Int 39:1474–1481
11. Dien IK, Elrod HG (1983) A generalised study state Reynolds equation for non—Newtonian fluids with application to journal bearings. ASME J Lubr Technol 105:385
12. Xiong S, Wang QJ (2012) A steady state hydrodynamic lubrication model with the Payvar-Salant Mass conservation model. J Tribol 134:031703-1, 031703-16
13. Balasoiu AM, Braun Mj, Moldovan SI (2013) A parametric study of porous self circulating hydrodynamic bearing. Tribol Int 61:176–193
14. Yagi K, Sagimura J (2013) Elastic deformation in thin film hydrodynamic lubrication. Tribol Int 59:170–180

15. Safar ZS (1978) Dynamically loaded bearing operating with non Newtonian lubricant films. Wear 55:295–304
16. Peng ZC, Khonsari MM (2006) A thermohydrodynamic analysis of foil journal bearing. ASME Trans 128:534–541
17. Sing U, Roy I, Sahu M (2008) Steady state THD analysis of cylindrical fluid film journal bearing with an axial groove. Tribol Int 41:1135–1144
18. Chen CY, Chen QD, Li WL (2013) Characteristics of journal bearings with anisotropic slip. Tribol Int 61:144–155
19. Wang XL, Zhu K, Wen S (2001) THD analysis of journal bearing lubricated with couple stress fluid. Tribol Int 34:335–343
20. Thomsen K, Klit P (2011) A study on complaints layers and its influence on dynamic response of a hydrodynamic journal bearing. Tribol Int 44:1872–1877
21. Liu D, Zjang W, Zhang T (2008) A simplified one dimensional thermal model for journal bearing. J Tribol 130
22. Sing C, Sinha P (1981) Non-Newtonian squeeze film and journal bearing. Wear 70:311–319

Rough Porous Circular Convex Pad Slider Bearing Lubricated with a Magnetic Fluid

S. D. Shukla and G. M. Deheri

Abstract An endeavor has been made to discuss the performance of a transversely rough porous circular convex pad slider bearing in the presence of a magnetic fluid lubricant. Jenkins model for the flow of magnetic fluid has been adopted. The bearing surfaces are assumed to be transversely rough and the transverse surface roughness of the bearing surfaces is characterized by a stochastic random variable with non-zero mean, variance and skewness. With the aid of suitable boundary conditions the associated stochastically averaged Reynolds' equation is solved to obtain the expression for pressure distribution resulting in the calculation of load carrying capacity. The computed values of dimensionless load carrying capacity are displayed in graphical forms. The results make it clear that the bearing working with magnetic fluid as a lubricant records a better performance than that of an identical bearing working with a conventional lubricant. The negatively skewed roughness induces increased load carrying capacity which goes a long way in mitigating the adverse effect of the standard deviation and porosity, taking recourse to suitable values of magnetization parameter. The bearing can support a load even when there is no flow which is not true in the case of conventional lubricant. This article suggests some measures for extending the bearing's life period. It is interestingly note that the film thickness ratio turns in a marginally better performance as compared to most of recent studies. In spite of the fact that the roughness affects the system adversely, a companison of this investigation with few earlier ones indicates that the overall performance is fairly improved here.

Keywords Slider bearing · Roughness · Porosity · Magnetic fluid

S. D. Shukla (✉)
Department of Mathematics, Shri R. Parikh Arts and Science College,
Petlad, Gujarat 388450, India
e-mail: snehaldshukla@gmail.com

G. M. Deheri
Department of Mathematics, S.P. University, Vallabha Vidyanagar, Gujarat 388120, India

H. C. Patel et al. (eds.), *Proceedings of International Conference on Advances in Tribology and Engineering Systems*, Lecture Notes in Mechanical Engineering,
DOI: 10.1007/978-81-322-1656-8_7, © Springer India 2014

Nomenclature

h	Fluid film thickness at any point
B	Bearing length
H	Magnitude of magnetic field
P	Lubricant pressure
U	Shaft surface speed
W	Total load carrying capacity
h_1	Minimum film thickness
h_2	Maximum film thickness
P^*	Dimensionless pressure
W^*	Dimensionless load carrying capacity
η	Dynamic viscosity of fluid
ρ	Fluid density
λ^2	Material constant of Jenkins model
φ	Porosity
σ	Standard deviation
ε	Skewness
α	Variance
β^*	Material parameter
σ^*	Non-dimensional standard deviation
φ^*	Non-dimensional porosity
ε^*	Non-dimensional skewness
α^*	Non-dimensional variance
μ^*	Magnetization parameter
$\overline{\mu}$	Magnetic susceptibility
μ_0	Permeability of the free space

1 Introduction

Slider bearing is by all means, the most interesting machine part in tribological point of view. Slider bearings have great use in heavy machinery, turbines, mills, forging machinery and gears. Basic advantage of slider bearings is their load carrying capacity and their service life. Gupta and Kavita [9] developed a mathematical model to investigate the behavior of a plane inclined slider bearing under the effect of uniform small rotation. Porous bearings do not need continuous lubrication, therefore; their structures are simple and reduce costs. For this reason they are more advantageous. Prakash and Vij [15] discussed the load carrying capacity and time height relations for squeeze film performance between porous plates. In that study various geometries such as circular, annular, elliptical, rectangular, conical and truncated conical were considered. It was established that the circular geometry registered the highest load carrying capacity.

Magnetic fluids are stable suspensions of colloidal magnetic particles of the order 10 nm in suitable non-magnetic carrier liquids. Because of the industrial applications of magnetic fluids, the investigation on them fascinated the researchers and engineers vigorously since five decades. Magnetic fluid is widely used in sealing of hard disc drives, rotating X-ray tubes under engineering application, contrast medium in X-ray examinations and for positioning tamponade for retinal detachment repair in eye surgery. Jenkins [10] used a simple continuum model for a paramagnetic fluid to analyze a simple shearing flow and parallel flow through a pipe and examined the possibility of maintaining a steady circular flow in a circular cylinder by rotating a magnetic fluid. Rosenweig [18], in his monographs, has given an authoritative introduction to the research on magnetic liquids and revealed interesting information about the effect of magnetization. Sunil et al. [20] studied the effect of rotation on thermosolutal convection in a ferromagnetic fluid considering a horizontal layer of an incompressible ferromagnetic fluid. Ram et al. [17] solved the non-linear differential equation under Neuringer-Rosensweig model by using power series approximations and discussed the effect of magnetic field-dependent viscosity and velocity components and pressure profile. Agrwal [1] studied the squeeze film performance by taking a magnetic fluid as a lubricant. Lin et al. [11] investigated the parallel circular squeeze film disks with a non-Newtonian ferrofluid in the presence of a transverse magnetic field. Using the Shliomis Ferro hydrodynamic model, the modified Reynolds' equation was derived. Here, they compared the results with Newtonian and non-Newtonian ferrofluid cases and found that non-Newtonian ferrofluid lubricated squeeze films provided a higher load carrying capacity. Oladeinde and Akpobi [13] presented a mathematical model for the hydrodynamic lubrication of finite slider bearing with velocity slip and couple stress lubricants. At the juncture film thickness ratio was obtained for which load capacity was maximized with or without slip.

When the gap between two mating surfaces becomes smaller, the effects of roughness become more important. In most of the applications, the smooth bearing surfaces would not be valid for the accurate prediction of the performance and life of the bearings. Thus, surface roughness has been studied with much interest in the recent years because all bearing surfaces are rough to some extent. Also, to increase the performance of hydrodynamic lubrication in different bearings, it is essential to study the influence of surface roughness. The random character of the roughness was recognized by Tzeng and Seibel [21] who employed a stochastic approach to study the roughness. To investigate the effects of surface roughness, many theories have been proposed and implemented in the context of surface lubrication. The efforts were made by Christensen [7], Christensen and Tonder [8] and Chow and Cheng [6] within the frame work of the stochastic theory. Bujurke and Naduvinamani [4] have studied the effect of roughness on squeeze film characteristics between two rectangular plates of which the upper plate has a roughness structure and lower plate has a porous material. Here, it was shown that the effect of longitudinal roughness increased load capacity of the bearing compared to smooth case. Litwin [12] has experimentally investigated the effect of roughness on the water-lubricated polymer bearings and showed that the effect of roughness to the sliding direction of the bearing could increase hydrodynamic load capacity.

Patel et al. [14] investigated the performance of a magnetic fluid based squeeze film between rough circular plates. Here, it was shown that with a proper selection of the thickness ratio parameter a magnetic fluid based squeeze film bearing with variable porous matrix thickness, in the case of negatively skewed roughness could be made to perform considerably better than that of a bearing working with a conventional lubricant. Chiang et al. [5] considered the analysis of performance of a magnetic hydrodynamic tilted bearing under the effect of surface roughness. Here, it was found that the tilted bearing lubricated with ferro-fluid had the higher built-up pressure distribution and load carrying capacity.

Here, it has been sought to analyze the performance of a magnetic fluid based rough porous circular convex pad slider bearing.

2 Analysis

The geometry and orientation of the bearing structure is given in Fig. 1.

The bearing consists of a stator with a circular convex pad surface with crown height δ and a slider moving with a uniform velocity U in the X-direction. The film thickness h is taken as (Shah and Bhat [19]).

$$h = 4\delta\left(\frac{x^2}{B^2} - \frac{x}{B}\right) + h_2 - (h_2 - h_1)\frac{x}{B}$$

Fig. 1 The configuration of the bearing system

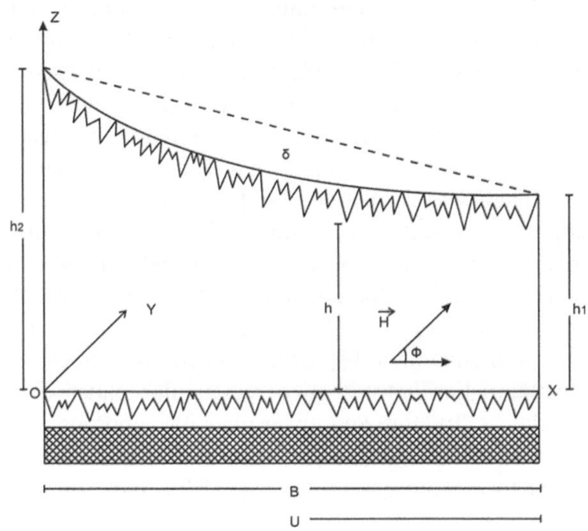

Following the expression for film thickness is considered in the form of (Christensen [7] and Christensen and Tonder [8]).

$$h(x) = \overline{h}(x) + h_s$$

where $\overline{h}$ is the mean film thickness, while h_s is a randomly varying portion measured from the mean level characterizing the random roughness. Here h_s is assumed to have the probability density function $F(h_s)$ over the domain $[-b, b]$.

$$F(h_s) = \begin{cases} \frac{32}{35b}\left(1 - \frac{h_s^2}{b^2}\right)^3 & -b \le h_s \le b \\ 0, & \text{elsewhere} \end{cases}$$

The details regarding the mean α, standard deviation σ, and skewness ε associated with the characterization of roughness can be seen from (Christensen [7] and Christensen and Tonder [8]).
where

$$\alpha = E(h_s),$$

$$\sigma^2 = E\left[(h_s - \alpha)^2\right],$$

$$\varepsilon = E\left[(h_s - \alpha)^3\right],$$

while E is the expectancy operator defined by

$$E(R) = \int_{-b}^{b} RF(h_s)dh_s$$

The magnetic field is oblique to the stator and its magnitude is given by

$$H^2 = kx(B - x)$$

where $k = 10^{14}A^2m^{-4}$ chosen so as to have a magnetic field of strength over 10^5 (Bhat and Deheri [3] and Verma [22]). Taking into account the usual assumptions of the hydromagnetic lubrication with no slip condition at the boundaries, no end effect and no side effects with the aid of the modeling proposed by Andharia et al. [2] and Ram and Verma [16], one gets

$$\frac{\mathrm{d}}{\mathrm{d}x}\left[\frac{a(h)}{1 - \frac{\rho\lambda^2\overline{\mu}H}{2\eta}}\frac{\mathrm{d}}{\mathrm{d}x}\left(P - \frac{\mu_0\overline{\mu}H^2}{2}\right)\right] = 6U\eta\frac{\mathrm{d}h}{\mathrm{d}x} \tag{1}$$

where

$$a(h) = h^3 + 3\alpha h^2 + 3(\sigma^2 + \alpha^2)h + \varepsilon + 3\sigma^2\alpha + \alpha^3 + 12\varphi H$$

The concerned boundary conditions are

$$P = 0, \quad x = 0 \quad \text{and} \quad 1$$

Using the following dimensionless scheme

$$x^* = \frac{x}{B},\ P^* = \frac{h_1^2 P}{U\eta B},\ \mu^* = \frac{k\mu_0\overline{\mu}Bh_1^2}{U\eta},\ \beta^* = \frac{\beta^2}{\sqrt{\mu^*}}$$
$$\varepsilon^* = \frac{\varepsilon}{h_1^3},\ \sigma^* = \frac{\sigma}{h_1},\ \alpha^* = \frac{\alpha}{h_1},\ n = \frac{h_2}{h_1},\ \delta^* = \frac{\delta}{h_1} \tag{2}$$
$$\beta^2 = \frac{\rho\lambda^2\overline{\mu}\sqrt{k}B}{2\eta},\ h^* = \frac{h}{h_1},\ A(h) = \frac{a(h)}{h_1^3}$$

and fixing the following symbol,

$$S = \frac{(2x^* - 1)}{4}\sqrt{(x^* - 1)x^*} + \frac{1}{8}\sin^{-1}(2x^* - 1)$$

one obtains the pressure distribution in dimensionless form as

$$P^* = \frac{\mu^*}{2}(x^* - 1)x^* + \frac{6h^*}{A(h)}(x^* - \beta^2 S) \tag{3}$$

Substitution of the pressure field above leads to the expression for dimensionless load carrying capacity, given by

$$\begin{aligned} W^* &= \int_0^1 P^* dx^* \\ &= \frac{\mu^*}{12} + \int_0^1 \frac{6h^*}{A(h)}(x^* - \beta^2 S)dx^* \end{aligned} \tag{4}$$

3 Results and Discussion

It is clearly seen that Eq. (3) determines the dimensionless pressure distribution while the distribution of load carrying capacity in non-dimensional form is given by Eq. (4). It is observed from Eq. (3) that the non-dimensional pressure gets increased by $(\mu^*/2)(x^* - 1)x^*$ while the dimensionless load carrying capacity enhances by $\mu^*/12$ as compared to the case of conventional lubricants. It also becomes clear that as the expression for load is linear with respect to the magnetization parameter, the load carrying capacity increases with the increasing magnetization parameter. Setting μ^* to be zero one gets the performance of the corresponding conventional rough porous bearing system. Further, taking non-porous medium in the absence of roughness this study reduces to the performance of a circular convex pad slider bearing.

However, for smaller values of the standard deviation the effect on the load carrying capacity with respect to the magnetization parameter can be nominal in the Fig. 2. The effect of standard deviation is negligible up to 0.05.

The variation of load carrying capacity with respect to the standard deviation presented in Figs. 3, 4 and 5 suggests that the load carrying capacity decreases as the standard deviation increases, thereby affecting the system adversely.

The effect of skewness on the load carrying capacity is presented in Figs. 6, 7 and 8. It is clearly seen that the negatively skewed roughness increase the load carrying capacity while load capacity decreases due to positively skewed roughness.

The positive variance induces a sharp decrease in load carrying capacity which can be seen in Fig. 9. It can be easily seen that the load falls rapidly when α^* is beyond 0.05 approximately for all values of porosity.

Lastly, the effect of material parameter given in the Figs. 10, 11 and 12 on the load carrying capacity establishes that the effect of material parameter is decreased

Fig. 2 Variation of L.C.C. with respect to μ^* and σ^*

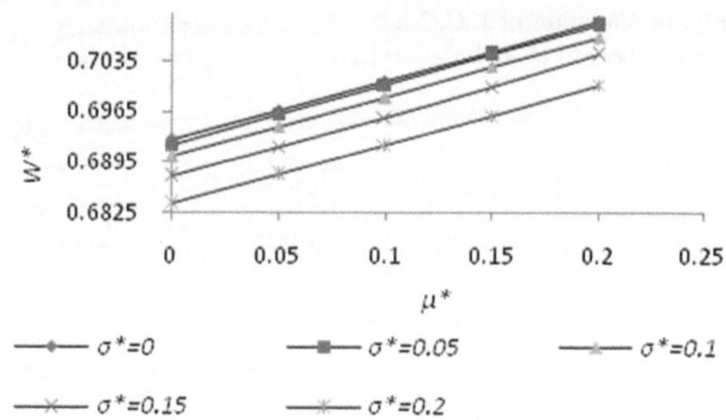

Fig. 3 Variation of L.C.C. with respect to σ^* and α^*

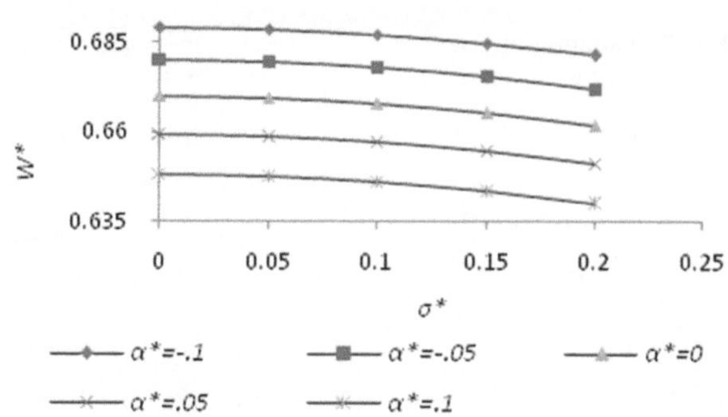

Fig. 4 Variation of L.C.C. with respect to σ^* and φ^*

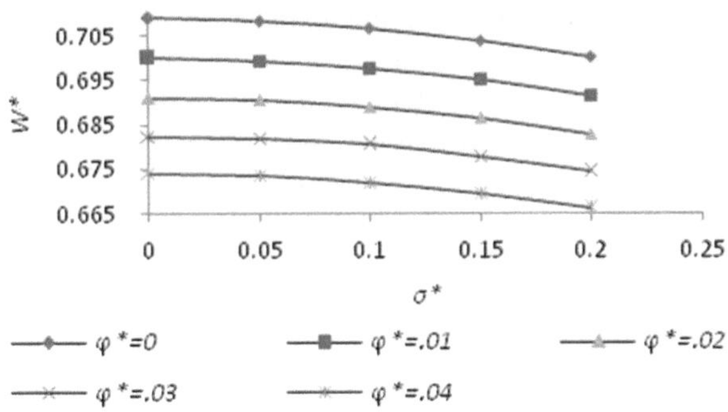

Fig. 5 Variation of L.C.C. with respect to σ^* and n

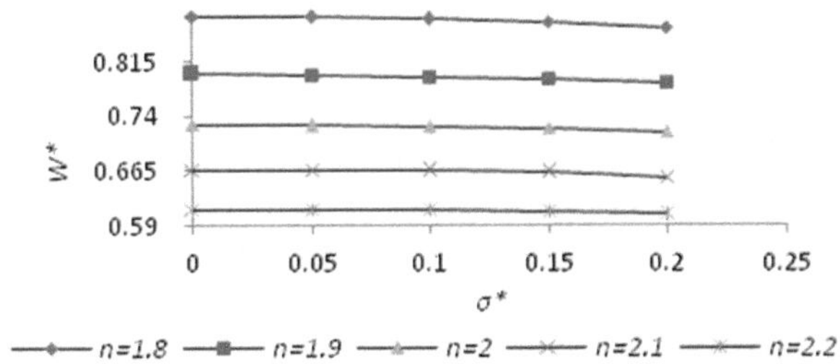

Fig. 6 Variation of L.C.C. with respect to ε^* and α^*

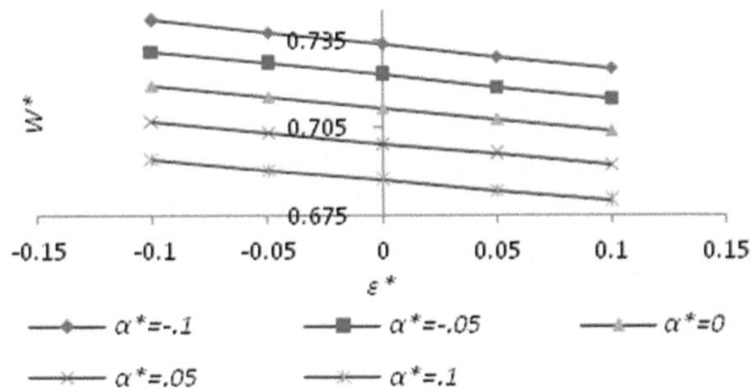

Fig. 7 Variation of L.C.C. with respect to ε^* and φ^*

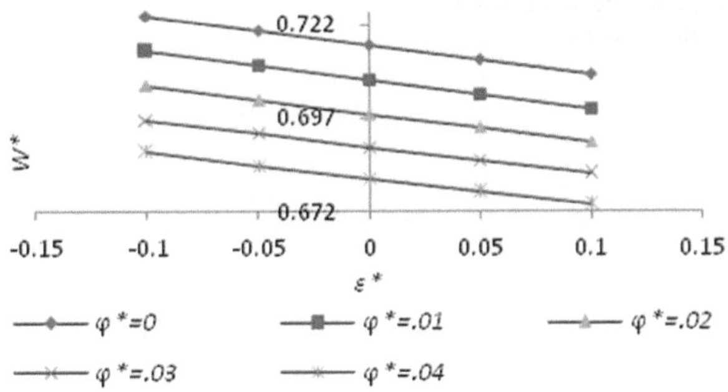

Fig. 8 Variation of L.C.C. with respect to ε^* and n

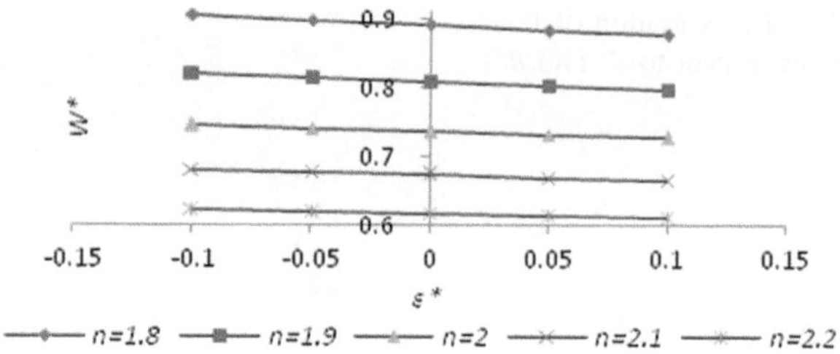

Fig. 9 Variation of L.C.C. with respect to α^* and φ^*

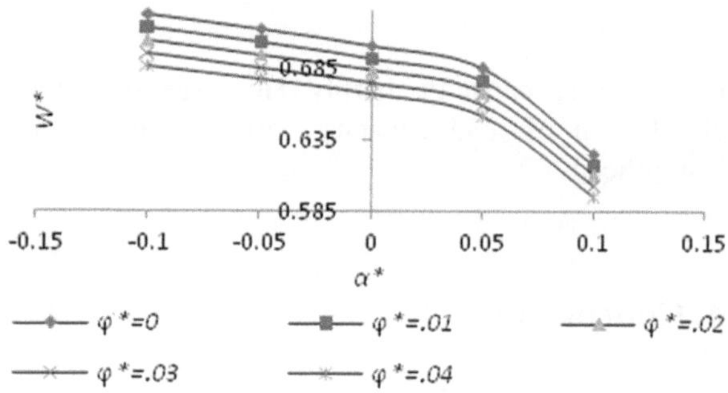

Fig. 10 Variation of L.C.C. with respect to β^* and σ^*

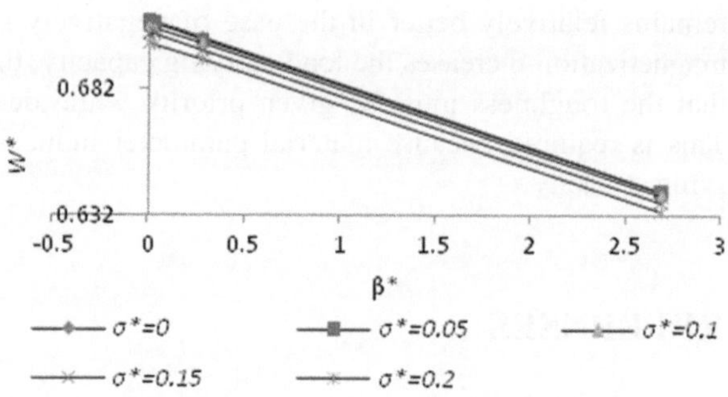

Fig. 11 Variation of L.C.C. with respect to β^* and α^*

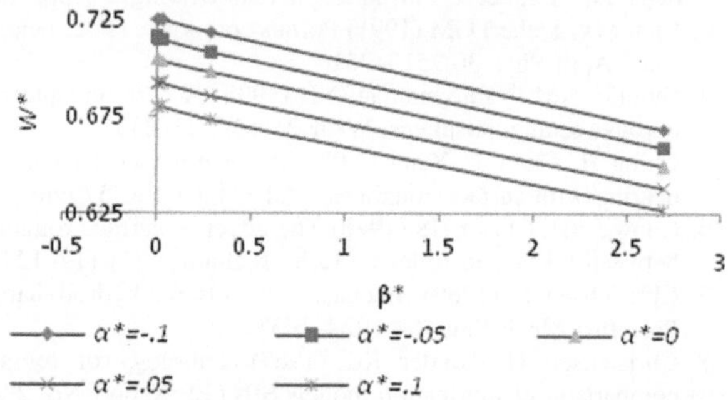

Fig. 12 Variation of L.C.C. with respect to β^* and n

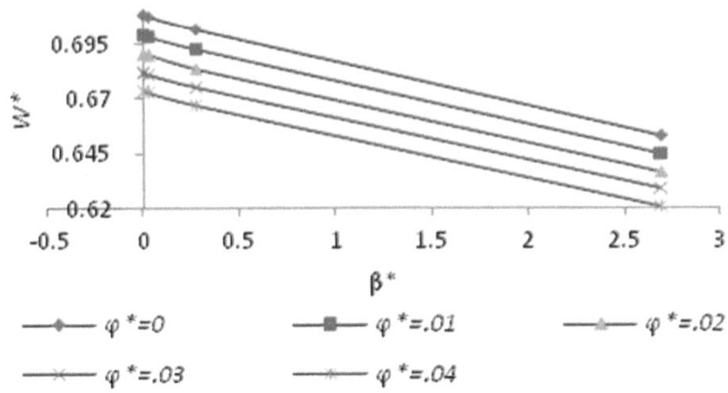

load carrying capacity sharply. However, the effect of standard deviation on the distribution of load capacity with respect to the material parameter remains insignificant.

4 Conclusion

There exists a limited scope of the magnetization for compensating the adverse effect of porosity, roughness and material parameter. However, the situation remains relatively better in the case of negatively skewed roughness. Although, magnetization increases the load carrying capacity, this investigation makes it sure that the roughness must be given priority while designing such bearing system. This is required because material parameter induces sharp decrease in load carrying capacity.

REFERNSES

1. Agrwal VK (1986) Magnetic fluid based porous inclined slider bearing. Wear 107:133–139
2. Andharia PI, Gupta JL, Deheri GM (1999) Effects of transverse surface roughness on the behavior of squeeze film in a spherical bearing. J Appl Mech Eng 4:19–24
3. Bhat MV, Deheri GM (1991) Porous composite slider bearing lubricated with magnetic fluid. Jpn J Appl Phys 30:2513–2514
4. Bujurke NM, Naduvinamani NB (1998) A note on squeeze film between rough anisotropic porous rectangular plates. Wear 217(2):225–230
5. Ching H, Chou T, Yang C (2012) Performance analysis of magnetic hydrodynamic tilted bearing with surface roughness. Adv Mater Res 579:407–415
6. Chow LSH, Cheng HS (1976) The effect of surface roughness on the average film thickness between lubricated rollers. J Lubr Technol 98(1):117–124
7. Christensen H (1969) Stochastic models for hydrodynamic lubrication of rough surfaces. Proc Inst Mech Eng 184:1022–1033
8. Christensen H, Tonder KC (1969) Tribology of rough surface: parametric study and comparison of lubrication models SINTEF Report No. 22/69-18

9. Gupta RS, Kavita P (1986) Analysis of rotation in the lubrication of porous slider bearing: small rotation. Wear 111(3):245–258
10. Jenkins JT (1971) Some simple flows of a para-magnetic fluid. J Phys 32:931–939
11. Lin JR, Lu RF, Lin MC, Wang PY (2013) Squeeze film characteristics of parallel circular disks lubricated by ferrofluid with non-Newtonian couple stresses. Tribol Int 61:56–61
12. Litwin W (2011) Influence of surface roughness topography on properties of water-lubricated polymer bearings: experimental research. Tribol Trans 54(3):351–361
13. Oladeinde MH, Akpobi JA (2010) A study of load carrying capacity of finite slider bearings with slip surfaces and stokesian couple-stress. Int J Eng Res Africa 1:57–66
14. Patel RM, Deheri GM, Patel HC (2011) Effect of surface roughness on the behavior of a magnetic fluid-based squeeze film between circular plates with porous matrix of variable thickness. Acta Polytech Hung 8(5):171–190
15. Prakash J, Vij SK (1973) Load capacity and time-height relation for squeeze films between porous plates. Wear 24(3):309–322
16. Ram P, Verma PDS (1999) Ferro fluid lubrication in porous incline slider bearing. Indian j Pure Appl Math 30(12):1273–1281
17. Ram P, Bhandari A, Sharma K (2010) Effect of magnetic field-dependent viscosity on revolving ferrofluid. J Magn Magn Mater 322(21):3476–3480
18. Rosenweig RE (1985) Ferro hydrodynamics. Cambridge University press, Cambridge
19. Shah RC, Bhat MV (2010) Ferro fluid based convex pad slider bearing. Natl J Srilankan Sci 32:55–69
20. Sunil PK, Bharti D, Sharma RC (2005) The Effects of rotation on thermosolutal convection in a ferromagnetic fluid. Int J Appl Mech Eng 10(4):713–730
21. Tzeng ST, Saibel E (1967) Surface roughness effect on slider bearing lubrication. Trans ASME J Lubr Tech 10:334–338
22. Verma PDS (1986) Magnetic fluid based squeeze film. Int J Eng Sci 24(3):395–401

Slip Velocity and Roughness Effect on Magnetic Fluid-Based Infinitely Long Bearings

Jimit R. Patel and Gunamani Deheri

Abstract Efforts have been made to analyze the performance of a magnetic fluid based infinitely long rough bearing with slip velocity. The magnitude of the magnetic field is taken to be an unusual one. It is known that in this type of bearing system the standard deviation affects the system most as compared to the other two roughness parameters. Therefore the effect of standard deviation remains in focus. Christensen and Tonder's model has been adopted for stochastically averaging the Reynolds' type equation. The solution of this second order non linear partial differential equation gives rise to the expression for pressure distribution leading to the calculation of load carrying capacity. Further, friction is calculated at both the plates. The computed results presented in graphical forms reveal that while the magnetization has a strong positive effect, the combined effect of the slip parameter and the standard deviation is significantly adverse. However, this investigation says that keeping the slip coefficient at minimum, the magnetization may compensate the adverse effect of the standard deviation. In addition, this article offers an additional degree of freedom through the form of the magnitude of the magnetic field for designing the bearing system.

Keywords Long bearing · Roughness · Slip velocity · Magnetic fluid · Load carrying capacity

1 Introduction

Now a day investigation of the properties of the flow and the application possibilities of suspensions of magnetic nano particles are a fertile ground for new researches. A magnetic fluid is a liquid dispersion of magnetic nanoparticles

J. R. Patel (✉) · G. Deheri
Department of Mathematics, Sardar Patel University, Vallabh, Vidyanagar, Anand, Gujarat 388120, India
e-mail: patel.jimitphdmarch2013@gmail.com

H. C. Patel et al. (eds.), *Proceedings of International Conference on Advances in Tribology and Engineering Systems*, Lecture Notes in Mechanical Engineering, DOI: 10.1007/978-81-322-1656-8_8, © Springer India 2014

coated by surfactants for stabilization. The biomedical application and the investigation of the rheological properties of magnetic fluids attained considerable importance during the last decade. The peculiarity of magnetic fluids is the conversion of normal liquid behavior with a magnetic control of their flow and properties. Another distinct feature of a magnetic fluid is that the magnetic fluid can be retained and controlled at a desired location by an external magnetic field. Some developments have been documented in Odenbach [17] for magnetic fluid research.

Ochonski [16] mooted some designs of magnetic fluid exclusion seals for rolling bearing and explored the possibility of using them in industrial sealing application. Ahmad and Singh [2] adopted a theoretical model for the magnetic fluid lubrication of porous pivoted slider bearing considering slip velocity. For the minimization of slip parameter and permeability parameter to increase the load carrying capacity was discussed. Hsu et al. [13] presented the investigation of squeeze film characteristics between rotating circular disks with an electrically conducting lubricant in the presence of a transverse magnetic field. The influence of rotational inertia resulted in lowered MHD load carrying capacity. However, the externally applied magnetic field provided an increase in the load carrying capacity.

Montazeri [15] numerically analyzed the hydrodynamic journal bearings lubricated with ferrofluid. The use of ferrofluid as the lubricant improved the hydrodynamic performance characteristics of journal bearing and reduced the friction coefficient. Lin et al. [14] launched a numerical investigation into the magneto hydrodynamic steady and dynamic characteristics of one dimensional slider bearings lubricated with an electrically conducting fluid in the presence of a transverse magnetic field. The MHD steady load carrying capacity increased significantly. Ruckenstein [21] derived an equation for the slip velocity when a non equilibrated drop spread over a solid surface. Gibelli [12] presented a kinetic theory for the variation of first and second order velocity slip coefficients.

Tzeng and Seibel [22] recognized the random character of roughness of the bearing surfaces and employed a stochastic approach to understand the effect of roughness. Later on this approach was modified and developed by Christenson and Tonder [7, 9, 10] to study the effect of surface roughness (both transverse as well as longitudinal) on the performance of a bearing system in general.

Patel and Deheri [18] considered the ferrofluid lubrication of a rough porous inclined slider bearing with slip velocity. For an overall improved performance of the bearing system the slip parameter was required to be kept at minimum. Because of the adverse effect of transverse roughness, the bearing system registered a slightly enhanced performance as compared to that of a bearing system dealing with a conventional lubricant. Recently, Patel and Deheri [19] evaluated the effect of surface roughness on the performance of a magnetic fluid based parallel plate porous slider bearing considering slip velocity. All though, porosity, slip velocity and standard deviation decreased the load carrying capacity; this negative effect could be minimized by the magnetic fluid lubricant in the case of negatively skewed roughness. In the current investigation it has been sought to study and analyze the effect of surface roughness and slip velocity on the performance of a magnetic fluid based infinitely long bearing.

2 Analysis

Figure 1 shows the geometry and configuration of present study. The bearing system is infinite in Z direction. The slider moves with uniform velocity u in X direction. The length of the bearing is L and the breadth B is in Z direction.

The bearing surfaces are assumed to be transversely rough. Following Christensen and Tonder [7, 9, 10], the thickness $h(x)$ of the lubricant film is considered as

$$h(x) = \overline{h}(x) + h_s$$

where $\overline{h}(x)$ is the mean film thickness and h_s is the deviation from the mean film thickness characterizing the random roughness of the bearing surfaces. h_s is assumed to be stochastic in nature and governed by the probability density function

$$f(h_s) = \begin{cases} \frac{35}{32c}\left(1 - \frac{h_s^2}{c^2}\right)^3, & -c \leq h_s \leq c \\ 0, & \text{elsewhere} \end{cases}$$

where c is the maximum deviation from the mean film thickness. The mean α, the standard deviation σ and the parameter ε, which is the measure of symmetry of the random variable h_s, are defined by the relationships illustrated in Christensen and Tonder [8]. The magnetic field is oblique to the stator as considered in Agrawal [1]. The detailed discussion of the effect of various forms of magnitude of the magnetic field has been incorporated in Prajapati [20] and Bhat [5]. According to their discussions, here the magnitude of the magnetic field is expressed as

$$M^2 = kL^2 \frac{x}{L} \sin\left(1 - \frac{x}{L}\right)$$

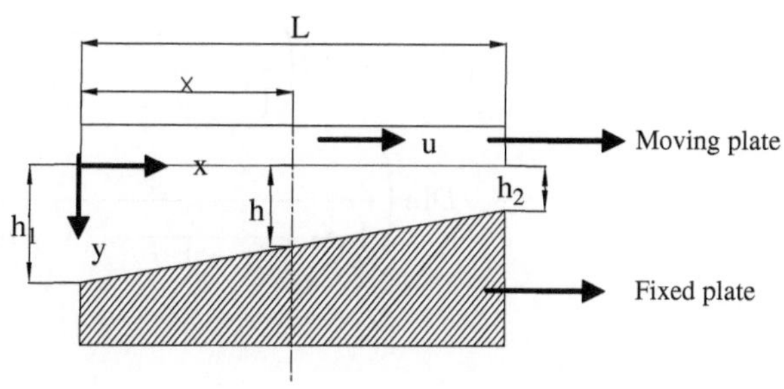

Fig. 1 Geometry and configuration of the bearing system

where k is a suitably chosen constant from dimensionless point of view (Bhat and Deheri [6]).

It is known that in this pattern of roughness in long bearing system the standard deviation plays a central role as compared to the other two parameters. Hence under the usual assumptions of hydrodynamic lubrication neglecting the effect of variance and skewness the associated Reynolds equation [3, 5, 11, 20] turns out to be

$$\frac{\mathrm{d}}{\mathrm{d}x}\left(p-\frac{\mu_0\overline{\mu}M^2}{2}\right)=6\mu u\,\frac{h-\lambda h_2}{g(h)} \tag{1}$$

where

$$m=\frac{h_1-h_2}{h_2},h=h_2\left\{1+m\left(1-\frac{x}{L}\right)\right\}$$

and

$$g(h)=\left(h^3+3\sigma^2h\right)\left(\frac{4+sh}{2+sh}\right)$$

The associated boundary conditions are

$$p=0\,\mathrm{at}\,x=0\,\mathrm{and}\,x=L$$

Introducing the non-dimensional quantities,

$$P=\frac{h_2^3}{\mu uL^2}p,\ \mu^*=\frac{h_2^3k\mu_0\overline{\mu}}{\mu u},\ X=\frac{x}{L},\ \overline{\sigma}=\frac{\sigma}{h_2},\ \overline{L}=\frac{L}{h_2},\ \overline{s}=s.h_2$$

the pressure distribution in non dimensional form is obtained as

$$\begin{aligned} P=&\frac{\mu^*}{2}X\sin(1-X)+\frac{6}{m}\frac{1}{\overline{\overline{L}}}\\ &\left[(A+\lambda G)\tan^{-1}\left(\frac{m(1-X)/\sqrt{3}\overline{\sigma}}{1+\frac{1}{3\overline{\sigma}^2}\{1+m(1-x)\}}\right)\right.\\ &+(B+\lambda E)\ln\left(\frac{4+\overline{S}\{1+m(1-x)\}}{4+\overline{S}}\right)\\ &+(C-\lambda F)\ln\left(\frac{1+3\overline{\sigma}^2}{\{1+m(1-x)\}^2+3\overline{\sigma}^2}\right)\\ &\left.+\lambda D\ln\left(\frac{\frac{1}{3\overline{\sigma}^2}+1}{\left(\frac{1}{3\overline{\sigma}^2}+\frac{1}{\{1+m(1-x)\}^2}\right)\{1+m(1-x)\}}\right)\right] \end{aligned} \tag{2}$$

where in

$$\lambda = Aa + Bb + Cc/Dd - Eb + Fc - Ga$$

$$A = \left(1/\sqrt{3\overline{\sigma}}\right)\left(\left(8/\left(\overline{S}\left(\left(16/\overline{S}^2\right) + (3\overline{\sigma}^2)\right)\right)\right) - 1\right)$$

$$B = (2/\overline{S})\left(1/\left(\left(16/\overline{S}^2\right) + (3\overline{\sigma}^2)\right)\right)$$

$$C = (1/\overline{S})\left(1/\left(\left(16/\overline{S}^2\right) + (3\overline{\sigma}^2)\right)\right)$$

$$D = (1/6\overline{\sigma}^2)$$

$$E = (1/2)\left(1/\left(\left(16/\overline{S}^2\right) + (3\overline{\sigma}^2)\right)\right)$$

$$F = (1/4)\left((1/3\overline{\sigma}^2) - \left(1/\left(\left(16/\overline{S}^2\right) + (3\overline{\sigma}^2)\right)\right)\right)$$

$$G = (2/\overline{S})\left(1/\sqrt{3\overline{\sigma}}\right)\left(1/\left(\left(16/\overline{S}^2\right) + (3\overline{\sigma}^2)\right)\right)$$

$$a = \tan^{-1}\left(\frac{m/\sqrt{3\overline{\sigma}}}{1 + \frac{1}{3\overline{\sigma}^2}\{1+m\}}\right),\ b = \ln\left(\frac{4 + \overline{S}\{1+m\}}{4+\overline{S}}\right)$$

and

$$c = \ln\left(\frac{1 + 3\overline{\sigma}^2}{\{1+m\}^2 + 3\overline{\sigma}^2}\right),\ d = \ln\left(\frac{\left(\frac{1}{3\overline{\sigma}^2} + \frac{1}{\{1+m\}^2}\right)\{1+m\}}{\frac{1}{3\overline{\sigma}^2} + 1}\right)$$

The non dimensional load carrying capacity of the bearing is given by

$$W = \frac{h_2^3}{\mu u L^4} w = \int_0^1 P \mathrm{d}X = \frac{\mu^*}{2}(1 - \sin(1)) + \frac{6}{m}\frac{1}{\overline{L}} I \tag{3}$$

where

$$I = \int_0^1 \left[(A + \lambda G)\tan^{-1}\left(\frac{m(1-X)/\sqrt{3\sigma}}{1 + \frac{1}{3\overline{\sigma}^2}\{1 + m(1-x)\}}\right) + (B + \lambda E)\ln\left(\frac{4 + \overline{S}\{1 + m(1-x)\}}{4 + \overline{S}}\right) + (C - \lambda F)\ln\left(\frac{1 + 3\overline{\sigma}^2}{\{1 + m(1-x)\}^2 + 3\overline{\sigma}^2}\right) + \lambda D \ln\left(\frac{\frac{1}{3\overline{\sigma}^2} + 1}{\left(\frac{1}{3\overline{\sigma}^2} + \frac{1}{\{1+m(1-x)\}^2}\right)\{1 + m(1-x)\}}\right)\right] \mathrm{d}X$$

The frictional force $\overline{F}$ per unit width on the lower plane of moving plate is obtained as

$$\overline{F} = \int_0^1 \overline{\tau} \mathrm{d}X \quad \text{where } \overline{\tau} = \left(\frac{h_2}{\mu u}\right)\tau$$

$\overline{\tau}$ is the non dimension shearing stress while
$\tau = \frac{\mathrm{d}p}{\mathrm{d}x}\left(Y - \frac{h}{2}\right) + \frac{\mu u}{h}$ (Basu et al. [4])
The expression for non dimensional frictional force is found to be

$$\overline{\tau} = \frac{\mathrm{d}p}{\mathrm{d}X}\overline{L}\{1 + m(1 - X)\}\left(Y - \frac{1}{2}\right) + \frac{1}{\{1 + m(1 - X)\}}$$

At $Y = 0$ (moving plate) one observes that the non dimensional frictional force assumes the form,

$$\begin{aligned}
\overline{F_0} = {} & \frac{\overline{L}\mu^* m}{4}(\sin(1) - 1) + \frac{\ln(1 + m)}{m} \\
& + \left(3\sqrt{3\overline{\sigma}} \big/ 2m\right)(A + \lambda G)\ln\left((1 + m)^2 + 3\overline{\sigma}^2 \big/ 1 + 3\overline{\sigma}^2\right) \\
& + 3(B + \lambda E)\left[1 + \frac{4}{m}\ln\left(\frac{4 + \overline{S}}{4 + \overline{S}\{1 + m\}}\right)\right] \\
& - 6(C - \lambda F)\left[1 + \frac{\sqrt{3\overline{\sigma}}}{m}\tan^{-1}\left(\frac{-m\sqrt{3\overline{\sigma}}}{3\overline{\sigma}^2 + \{1 + m\}}\right)\right] \\
& - 3\lambda D\left[1 + \frac{2\sqrt{3\overline{\sigma}}}{m}\tan^{-1}\left(\frac{-m\sqrt{3\overline{\sigma}}}{3\overline{\sigma}^2 + \{1 + m\}}\right)\right]
\end{aligned} \tag{4}$$

Further at $Y = 1$ (fixed plate) the frictional force in non dimensional form turns out to be

$$\begin{aligned}
\overline{F_1} = {} & \frac{\overline{L}\mu^* m}{4}(1 - \sin(1)) + \frac{\ln(1 + m)}{m} \\
& - \left(3\sqrt{3\overline{\sigma}} \big/ 2m\right)(A + \lambda G)\ln\left((1 + m)^2 + 3\overline{\sigma}^2 \big/ 1 + 3\overline{\sigma}^2\right) \\
& - 3(B + \lambda E)\left[1 + \frac{4}{m}\ln\left(\frac{4 + \overline{S}}{4 + \overline{S}\{1 + m\}}\right)\right] \\
& + 6(C - \lambda F)\left[1 + \frac{\sqrt{3\overline{\sigma}}}{m}\tan^{-1}\left(\frac{-m\sqrt{3\overline{\sigma}}}{3\overline{\sigma}^2 + \{1 + m\}}\right)\right] \\
& + 3\lambda D\left[1 + \frac{2\sqrt{3\overline{\sigma}}}{m}\tan^{-1}\left(\frac{-m\sqrt{3\overline{\sigma}}}{3\overline{\sigma}^2 + \{1 + m\}}\right)\right]
\end{aligned} \tag{5}$$

3 Result and Discussions

It is seen clearly that the non dimensional pressure distribution is determined from Eq. (2) while Eq. (3) gives the distribution of non-dimensional load carrying capacity. Equations (4) and (5) present the friction at both the plates. The performance characteristics depend on several parameters such as $\mu^*, \overline{\sigma}, m$ and $\overline{S}$ as can be seen from Eqs. (2) and (3) respectively. These parameters describe the effect of magnetization, standard deviation, aspect ratio and slip velocity. Setting the roughness parameters to be zero, this investigation reduces to the study of the performance of a magnetic fluid based infinitely long bearing with smooth surfaces in the absence of slip. Further, taking the magnetization parameter to be zero this study reduces to the performance analysis of a long bearing as discussed in Basu et al. [4] when slip velocity is non existing.

It is observed from Eq. (2) that the pressure increases by $(\mu^*/2)X\sin(1-X)$ while the increase in the dimensionless load carrying capacity is $(\mu^*/2)(1-\sin(1))$ as indicated by Eq. (3) as compared to traditional lubricants.

In Figs. 2, 3, 4, 5 variation of load carrying capacity with respect to the magnetization parameter is presented for various value of $\overline{L}, m, \overline{\sigma}$ and $\overline{S}$. It is clear that the load carrying capacity increases sharply due to the magnetization. It is noticed that the effect of aspect ratio m is not that significant for higher values. Equation (3) suggests that the expression for non-dimensional load carrying capacity is linear with respect to the magnetization parameter μ^*. Accordingly, increasing values of μ^* will lead to increased load carrying capacity which can be

Fig. 2 Variation of load carrying capacity with respect to μ^* and $\overline{L}$

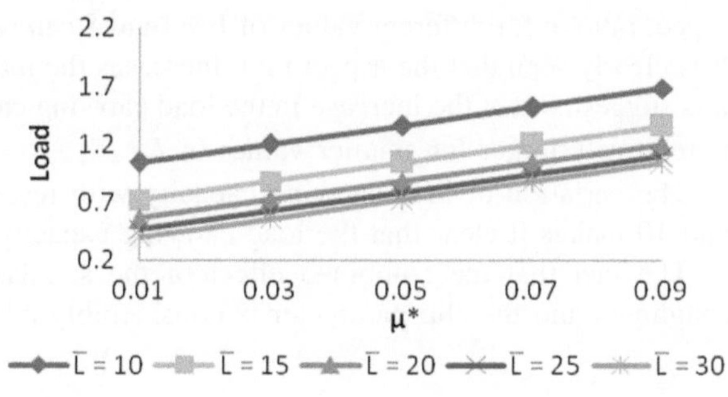

Fig. 3 Variation of load carrying capacity with respect to μ^* *and* m

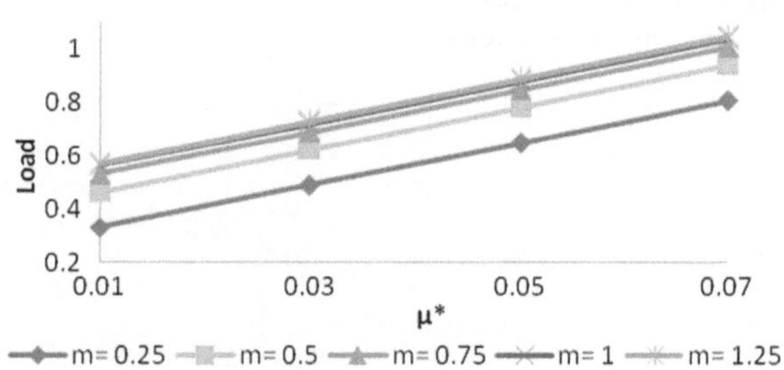

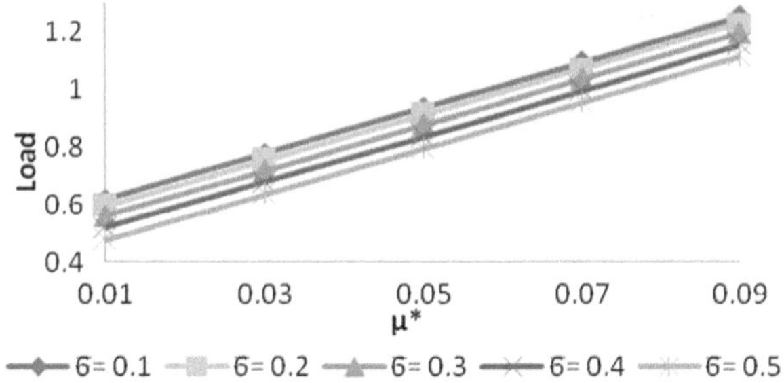

Fig. 4 Variation of load carrying capacity with respect to μ^* and $\overline{\sigma}$

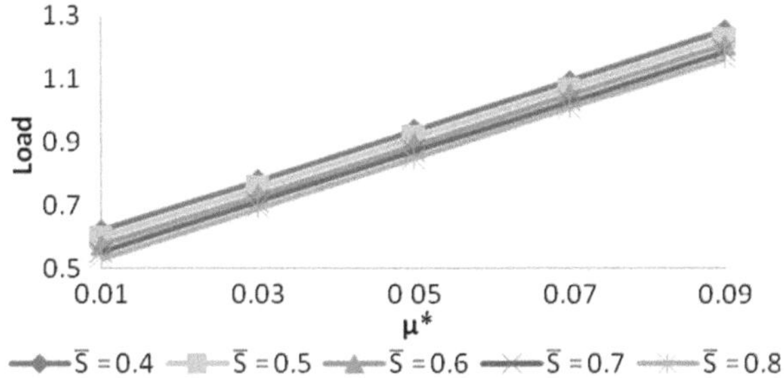

Fig. 5 Variation of load carrying capacity with respect to μ^* and $\overline{S}$

seen from Figs. 2, 3, 4, 5. From Fig. 5 it can be inferred that for an effective performance, the slip parameter deserves to be kept at minimum.

The distribution of non dimensional load carrying capacity with respect to the aspect ratio m for different values of $\overline{L}, \overline{\sigma}$, and $\overline{S}$ can be obtained from Figs. 6, 7, 8. It is clearly seen that the aspect ratio increases the load carrying capacity. Further, it is suggested that the increase in the load carrying capacity due to the aspect ratio is relatively larger for smaller values of $\overline{L}$.

The variation of load carrying capacity with respect to $\overline{L}$ presented in Figs. 9 and 10 makes it clear that the load carrying capacity decreases steadily.

The fact that the combined effect of the standard deviation associated with roughness and the slip parameter is considerably adverse, is manifest in Fig. 11.

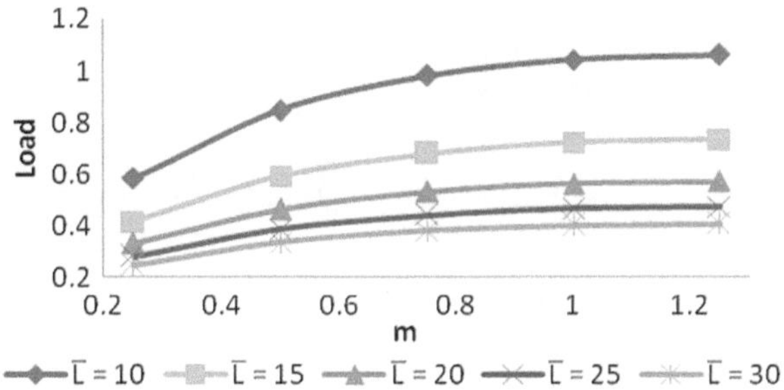

Fig. 6 Variation of load carrying capacity with respect to m and $\overline{L}$

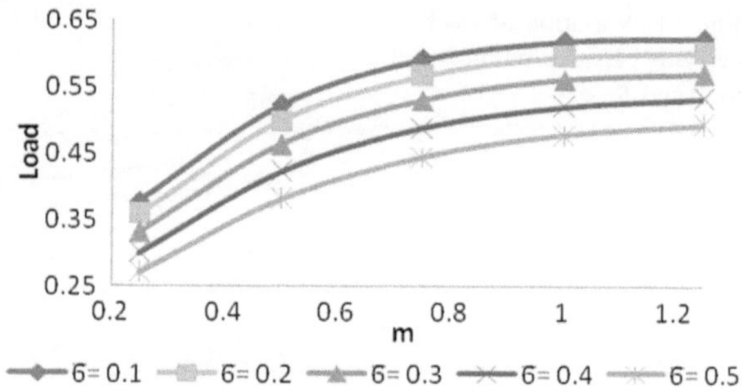

Fig. 7 Variation of load carrying capacity with respect to m and $\overline{\sigma}$

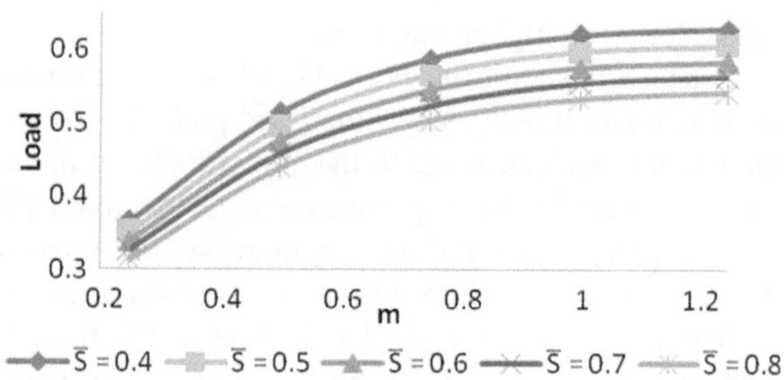

Fig. 8 Variation of load carrying capacity with respect to m and $\overline{S}$

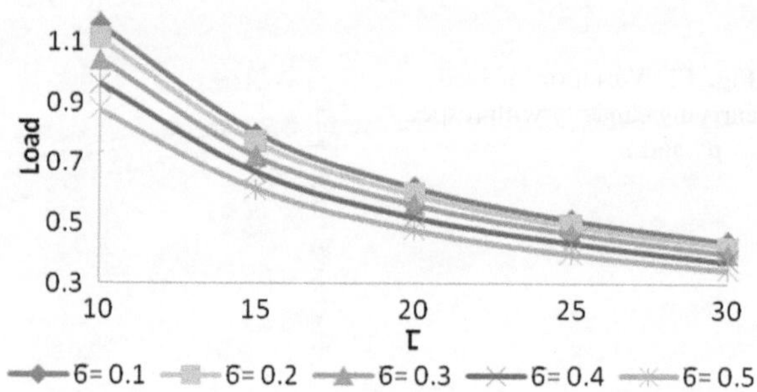

Fig. 9 Variation of load carrying capacity with respect to $\overline{L}$ and $\overline{\sigma}$

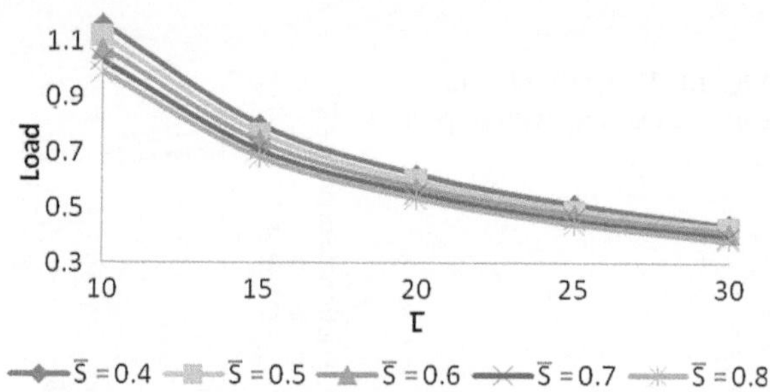

Fig. 10 Variation of load carrying capacity with respect to $\overline{L}$ and $\overline{S}$

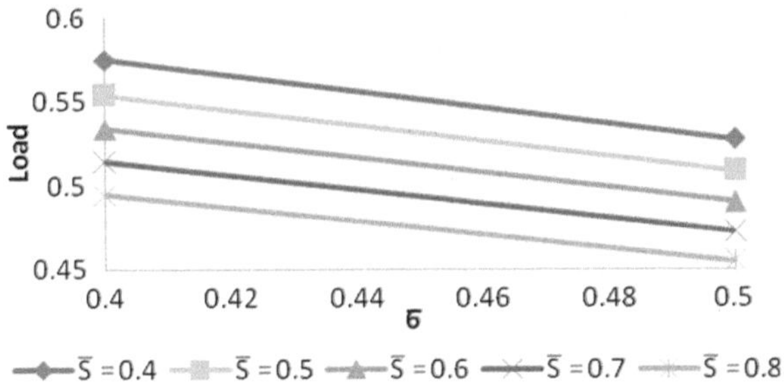

Fig. 11 Variation of load carrying capacity with respect to $\overline{\sigma}$ and $\overline{S}$

This also goes on to suggest that the adverse effect of the above combination gets quite sharp for higher values of $\overline{L}$.

It is clearly seen from Figs. 12, 13, 14, 15 that while the friction decreases at the moving plate, it increases at the fixed plate. It is revealed from Figs. 13, 14, 15 that the friction decreases due to the slip velocity at the moving plate while the friction increases due to the slip parameter at the fixed plate. It is also observed from Figs. 12, 13, 14 that the friction increases with respect to the standard deviation at the moving plates while this trends gets reversed at the stationary plate.

It is noticed from Figs. 16, 17 that at the runner plate the combined effect of standard deviation and the slip velocity on the friction with respect to the aspect ratio are just negligible for a large range of the slip parameter. This is far from being true at the stationary plate which can be seen from Fig. 18, 19.

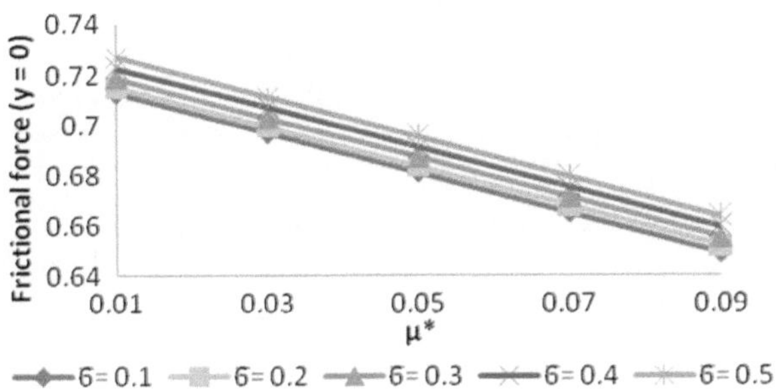

Fig. 12 Variation of load carrying capacity with respect to μ^* and $\overline{\sigma}$

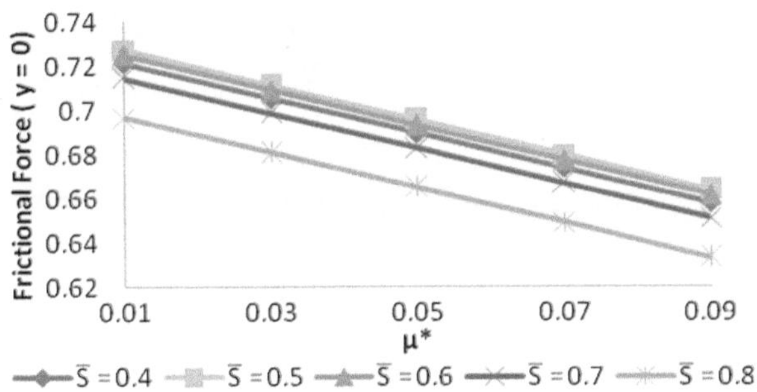

Fig. 13 Variation of load carrying capacity with respect to μ^* and $\overline{S}$

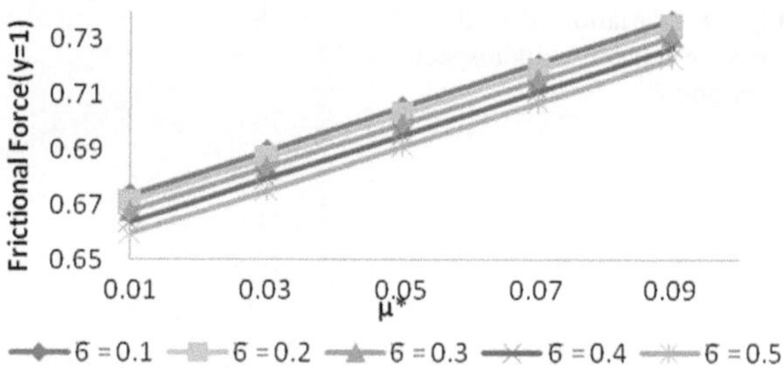

Fig. 14 Variation of load carrying capacity with respect to μ^* and $\overline{\sigma}$

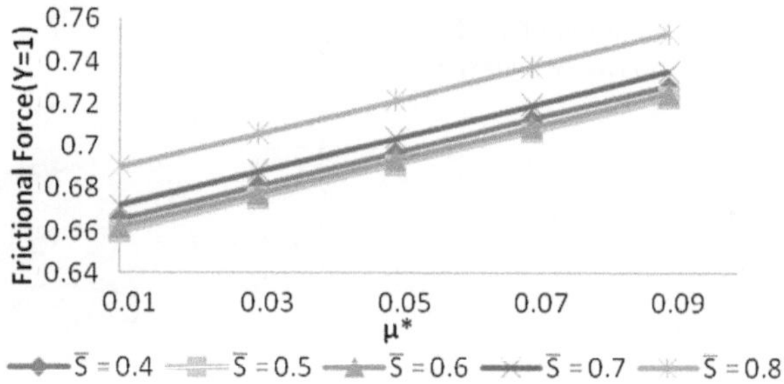

Fig. 15 Variation of load carrying capacity with respect to μ^* and $\overline{S}$

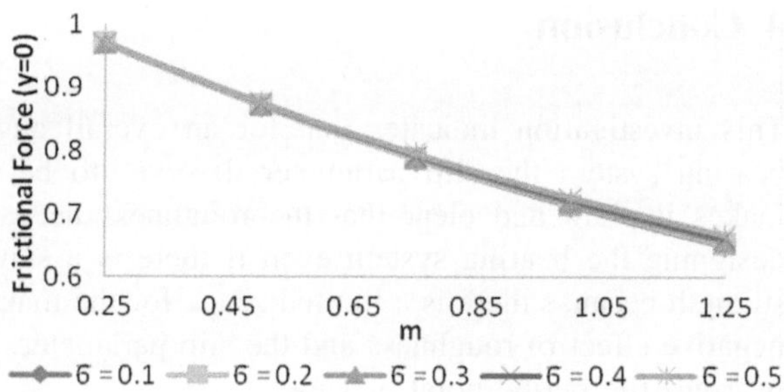

Fig. 16 Variation of load carrying capacity with respect to m and $\overline{\sigma}$

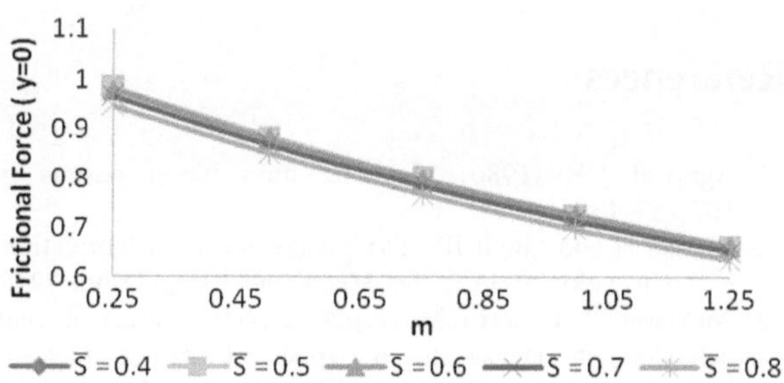

Fig. 17 Variation of Load carrying capacity with respect to m and $\overline{S}$

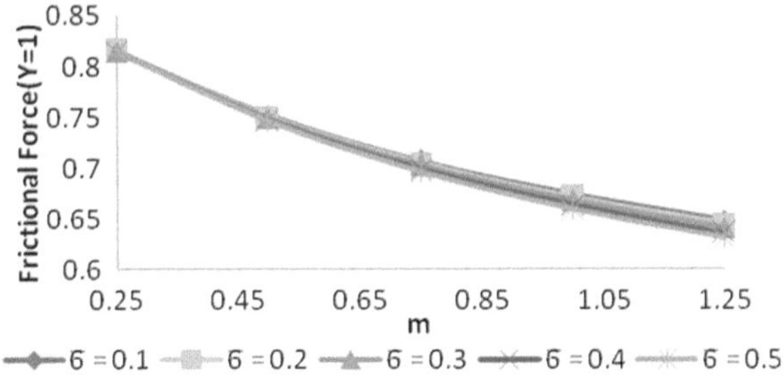

Fig. 18 Variation of load carrying capacity with respect to m and $\overline{\sigma}$

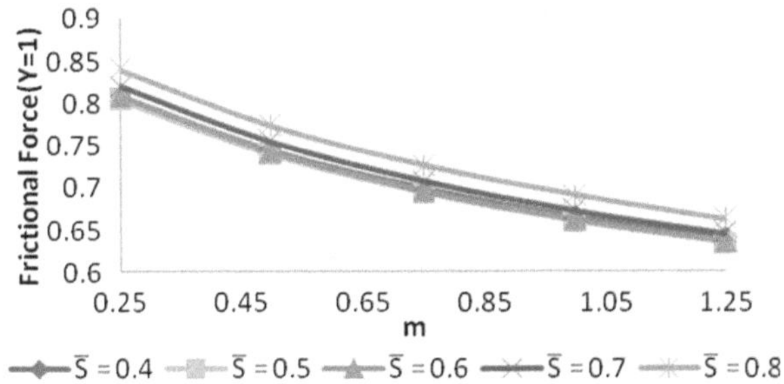

Fig. 19 Variation of load carrying capacity with respect to m and $\overline{S}$

4 Conclusion

This investigation indicates that for an overall augmented performance of the bearing system the slip parameter deserves to be reduced. Further, this study makes it plain and clear that the roughness needs to be accounted for while designing the bearing system even if there is a suitable choice of the magnetic strength because there is a limited scope for the magnetization to compensate the negative effect of roughness and the slip parameter. This is necessary because of bearings life period point of view.

References

1. Agrawal VK (1986) Magnetic fluid based porous inclined slider bearing. WEARS 107:133–139
2. Ahmad N and Singh JP (2007) Magnetic fluid lubrication of porous-pivoted slider bearings with slip velocity. Proc Inst Mech Eng J Eng Tribol 221(5):609–613
3. Andharia PI, Deheri GM, Gupta JL (2001) Effect of longitudinal surface roughness on the behaviour of squeeze film in a spherical bearing. Int J Appl Mech Eng 6:885–897
4. Basu SK, Sengupta SN, Ahuja BB (2005) Fundamentals of Tribology. Prentice-Hall of India private limited, New Delhi
5. Bhat MV (2003) Lubrication with a magnetic fluid. Team Spirit (India) Pvt., Ltd, New Delhi
6. Bhat MV, Deheri GM (1995) Porous slider bearing with squeeze film formed by a magnetic fluid. Pure Appl Math Sci 39(1–2):39–43

7. Christensen H, Tonder K C (1970) The hydrodynamic lubrication of rough bearing surfaces of finite width. In: Paper no.70-lub-7, ASME-ASLE Lubrication conference, Cincinnati, pp 12–15
8. Christensen H, Tonder KC (1971) The hydrodynamic lubrication of rough bearing surfaces of finite width. Trans ASME J Lubr Technol 93:324–330
9. Christensen H, Tonder KC (1969a) Tribology of rough surfaces: stochastic models of hydrodynamic lubrication. SINTEF, report no. 10/69-18
10. Christensen H, Tonder KC (1969b) Tribology of rough surfaces: parametric study and comparison of lubrication models. SINTEF, report no. 22/69-18
11. Deheri GM, Andharia PI, Patel RM (2005) Transversely rough slider bearings with squeeze film formed by a magnetic fluid. Int J Appl Mech Eng 10(1):53–76
12. Gibelli L (2012) Response to comment on velocity slip coefficients based on the hard-sphere boltzmann equation. Phys Fluids, 24:079101
13. Hsu C-H, Lai C, Hung C-R, Lin J-R (2008) Magneto-hydrodynamic squeeze film characteristics between circular discs including rotational inertial effects. Proc Inst Mech Eng J Eng Tribol, 222(2):157–164
14. Lin J-R, Hung C-R, Hsu C-H, Lai C (2009) Dynamic stiffness and damping characteristics of one dimensional magneto-hydrodynamic inclined-plane slider bearings. Proc Inst Mech Eng J Eng Tribol, 223(2):211–219
15. Montazeri H (2008) Numerical analysis of hydrodynamic journal bearings lubricated with ferrofluid. Proc Inst Mech Eng J Eng Tribol 222(1):51–60
16. Wlodzimierz Ochonski (2005) New designs of magnetic fluid exclusion seals for rolling bearings. Ind Lubr Tribol 57(3):107–115
17. Odenbach S (2004) Recent progress in magnetic fluid research. J Phys Condens Matter 16(32):R1135–R1150
18. Patel ND, Deheri GM (2011) A ferrofluid Lubrication of a rough porous inclined slider bearing with slip velocity. J Mech Eng Technol 4(1):15–34
19. Patel ND, Deheri GM (2011) Effect of surface roughness on the performance of a magnetic fluid based parallel plate porous slider bearing with slip velocity. J Serb Soc Comput Mech 5(1):104–118
20. Prajapati, BL (1995) On certain theoretical studies in hydrodynamic and electro-magneto hydrodynamic lubrication. Ph.D. Thesis, S.P. University, Vallabh Vidya-Nagar
21. Ruckenstein E (2011) Slip velocity during the flow of a liquid over a solid surface. Eur Phys J Spec Top 197:203–209
22. Tzeng ST, Saibel E (1967) Surface roughness effect on slider bearing lubrication. Trans ASLE 10:334–340

Squeeze Film Performance in Parallel Rough Circular Disks Lubricated by Ferrofluid with Non-newtonian Couple Stress Effect

H. A. Patel, M. P. Patel, H. C. Patel and G. M. Deheri

Abstract This paper aims to analyze the effect of transverse surface roughness on the performance of squeeze film in parallel circular disks with non-Newtonian ferrofluid under the presence of transverse magnetic field. The modified Reynolds' equation obtained by applying the Shliomis ferrohydrodynamic model incorporating the Stokes micro-continuum theory, has been stochastically averaged taking recourse to Christensen and Tonder's model for transverse roughness. With the aid of suitable boundary conditions the equation has been solved to derive the expression for the pressure distribution, thus resulting in the calculation of load carrying capacity. The results establish that the transverse surface roughness significantly affects the squeeze film performance. This article offers some measures to compensate the adverse effect of roughness under suitable conditions due to the positive effect of non-Newtonian ferrofluid.

Keywords Squeeze film · Circular disks · Roughness · Ferrofluid · Couple stress

H. A. Patel (✉) · M. P. Patel
Department of Mathematics, Gujarat Power Engineering and Research Institute, Mehsana 382710, Gujarat, India
e-mail: himeshpatel87@yahoo.com

H. C. Patel
Department of Mathematics, L. D. College of Engineering, Ahmedabad 380015, Gujarat, India

G. M. Deheri
Department of Mathematics, Sardar Patel University, Vallabh Vidhynagar, Anand 388120, Gujarat, India

H. C. Patel et al. (eds.), *Proceedings of International Conference on Advances in Tribology and Engineering Systems*, Lecture Notes in Mechanical Engineering, DOI: 10.1007/978-81-322-1656-8_9, © Springer India 2014

1 Introduction

It is well established that ferrofluids can be retained at a desired location by the application of an external magnetic field. Because of this advantage ferrofluids are widely applicable. In some of the investigations one comes across the application of ferrofluid as a lubricant to analyze the squeeze film performance characteristics of thin film lubricating mechanism [1–5]. Because of the fact that the viscosity of ferrofluid increases when subjected to external magnetic fields, the magnetic fluid lubricated squeeze film results in a higher load carrying capacity as compared to the conventional lubricants. Patel et al. [6] studied the performance of magnetic fluid based rotating rough circular step bearing. Here it was shown that the magnetic fluid went a long way in mitigating the negative effect of transverse surface roughness. Recently, Lin et al. [7] considered the squeeze film performance in parallel circular disks lubricated by ferrofluids with non-Newtonian couple stresses. Here, it has been sought to investigate the effect of transverse roughness on the configuration of Lin et al. [7].

1.1 Analysis

The stochastically averaged [8–10] modified Reynolds' equation governing the pressure distribution for the performance of ferrofluid lubricated squeeze film in circular disk with non-Newtonian couple stress is found to be [7]

$$f(h, l_c, \phi, \tau)\frac{1}{r}\frac{\mathrm{d}}{\mathrm{d}r}\left\{r\frac{\mathrm{d}p}{\mathrm{d}r}\right\} = 12\eta_0(1+\tau)(1+2.5\phi)\frac{\mathrm{d}h}{\mathrm{d}t}$$

where

$$\begin{aligned} f(h, l_c, \phi, \tau) = {} & h^3 - 12\frac{l_c^2}{(1+\tau)(1+2.5\phi)}h\left(3\overline{\alpha}^2 + 3\overline{\sigma}^2\right) \\ & + \left(24\frac{l_c^3}{(1+\tau)^{3/2}(1+2.5\phi)^{3/2}}\right) \\ & \cdot\left(\tanh\left(\frac{\sqrt{(1+\tau)(1+2.5\phi)}}{2l_c}h\left(3\overline{\alpha}^2 + 3\overline{\sigma}^2\right)\right)\right) \\ & + 3h^2\overline{\alpha} + 3\overline{\sigma}^2\overline{\alpha} + \overline{\alpha}^3 + \overline{\varepsilon} \end{aligned}.$$

Solving this equation with the boundary conditions

(1) $r = 0, \dfrac{dp}{dr} = 0$ and (2) $r = R, p = 0$ the expression for non dimensional pressure distribution is obtained as

$$P = \frac{3\,(1+\tau)\,(1+2.5\phi)}{f^{*}(h^{*},c,\phi,\tau)}\left(1-r^{*2}\right)$$

Integrating the film pressure over the film region, yields the load carrying capacity

$$W = \frac{3\,\pi\,(1+\tau)(1+2.5\,\phi)}{f^{*}(h^{*},c,\phi,\tau)}$$

It is clearly seen that the results of Lin et al. [7] can be obtained from this study by setting the roughness parameters to be zero. For the quantitative analysis the Figs. 1, 2, 3, 4, 5 and 6 are taken into consideration.

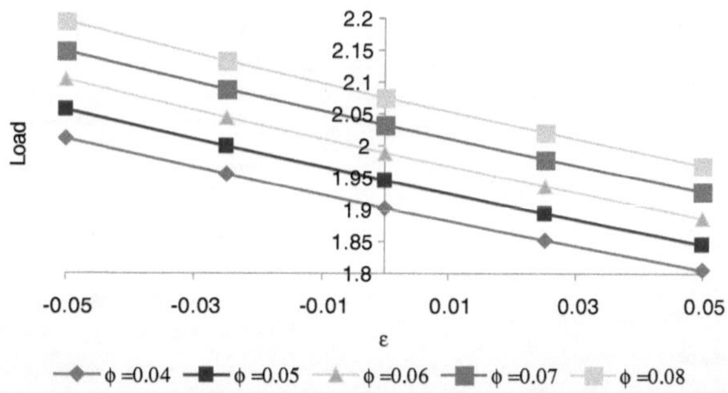

Fig. 1 Variation of load carrying capacity w.r.t ε & ϕ

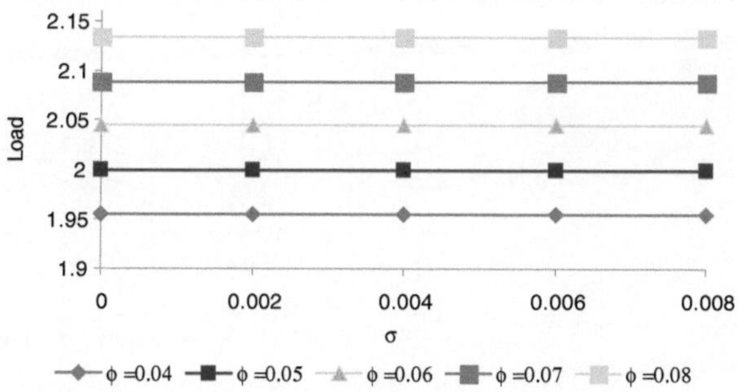

Fig. 2 Variation of load carrying capacity w.r.t σ & ϕ

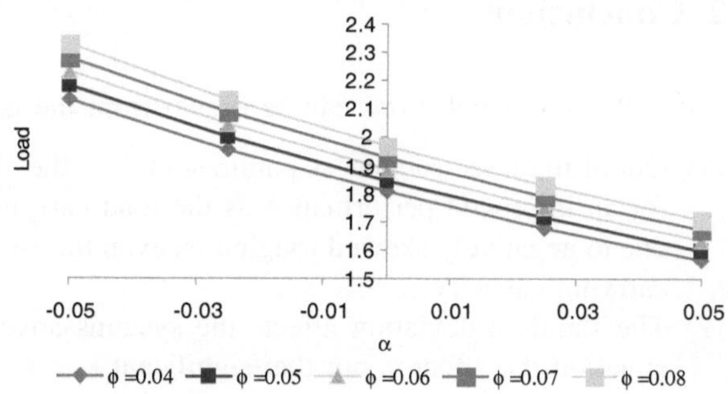

Fig. 3 Variation of load carrying capacity w.r.t α & ϕ

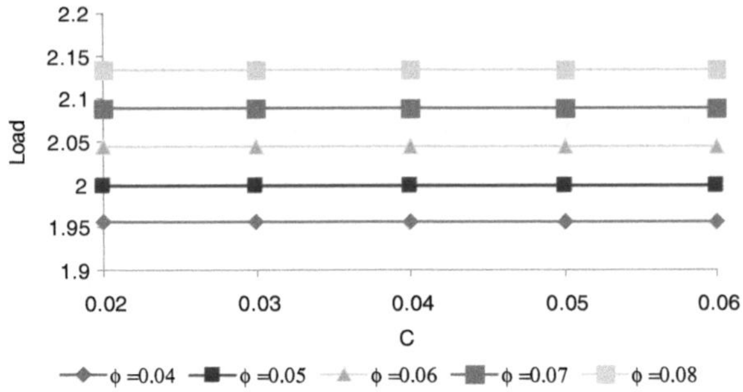

Fig. 4 Variation of load carrying capacity w.r.t c & ϕ

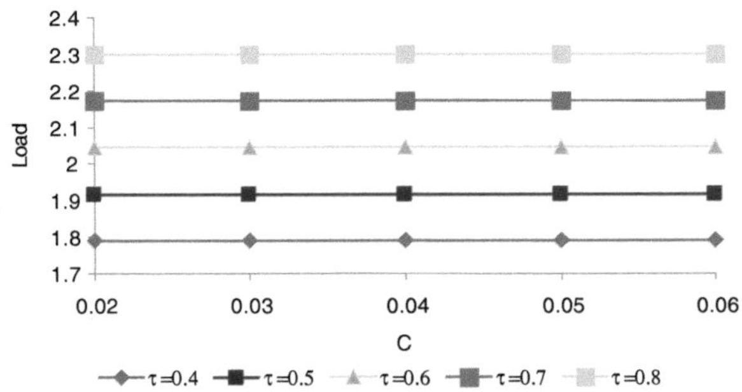

Fig. 5 Variation of load carrying capacity w.r.t c & τ

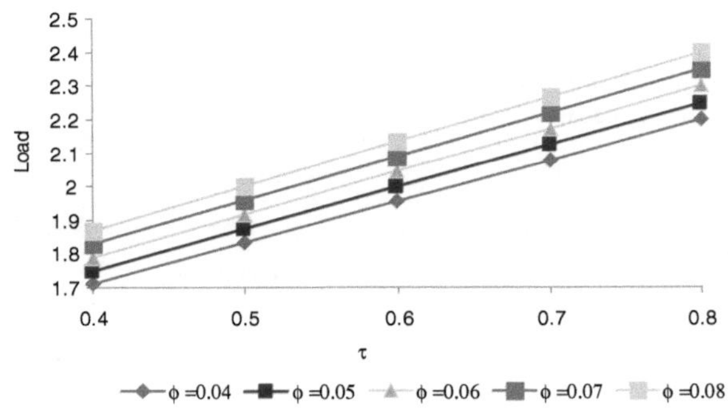

Fig. 6 Variation of load carrying capacity w.r.t τ & ϕ

2 Conclusions

The following conclusions can be drawn from the graphical representation.

(1) Out of the three roughness parameters, it is the skewness which holds the key for an improved performance as the load carrying capacity increases sharply due to negatively skewed roughness, even the variance (-ve) increases the load carrying capacity.

(2) The standard deviation affects the systems adversely, though in most of the situation the effect is not that significant.

(3) The positive effect induced by negative skewness further enhances when variance (-ve) occurs.
(4) In some circumstances, Shliomis model scores over the Newringer Rosenwicz model.
(5) This investigation strongly suggests that the roughness aspects must be accorded priority while designing the bearing system, even if suitable magnetic strength is in place.

References

1. Tipei N (1982) Theory of lubrication with ferrofluids: application of short bearings. J Lub Tech—Trans ASME 104:510–515
2. Zhang Y (1991) Static characteristics of magnetized journal bearing lubricated with ferrofluid. J Tribol—Trans ASME 113:533–538
3. Ahmed N, Singh JP (2007) Magnetic fluid lubrication of porous-pivoted slider bearing with slip velocity. Proc Inst Mech Eng J—J Eng Tribol 221:609–613
4. Deheri GM, Patel HC, Patel RM (2006) Performance of magnetic fluid based circular step bearing. Mechanika 1:22–27
5. Shimpi ME, Deheri GM (2012) A study on the performance of a magnetic fluid based squeeze film in curved porous rotating rough annular plates and deformation effect. Tribol Int 47:90–99
6. Patel HC, Deheri GM, Patel RM (2008) Performance of magnetic fluid based rotating rough circular step bearing. Int J Appl Mech Eng 13(2):441–455
7. Lin JR, Lu RF, Lin MC, Wang PY (2013) Squeeze film characteristics of parallel circular disks lubricated by ferrofluids with non-Newtonian couple stresses. Tribol Int 61:56–61
8. Christensen H, Tonder KC (1969a) Tribology of rough surfaces: stochastic models of hydrodynamic lubrication. SINTEF Report No.10/69-18
9. Christensen H, Tonder KC (1969b) Tribology of rough surfaces: parametric study and comparison of lubrication models. SINTEF Report No.22/69-18
10. Christensen H, Tonder KC (1970) The hydrodynamic lubrication of rough bearing surfaces of finite width. In: ASME-ASLE lubrication conference. Paper No.70-Lub-7

The Performance Analysis of a Magnetic Fluid-Based Hydrodynamic Long Journal Bearing

N. S. Patel, D. P. Vakharia, G. M. Deheri and H. C. Patel

Abstract An attempt has been made to investigate the performance characteristics of a hydrodynamic long journal bearing taking recourse to a magnetic fluid lubricant. With the usual assumptions of hydromagnetic concern Reynold's equation for the fluid film pressure is solved with Sommerfeld boundary conditions. This leads to the calculation of load carrying capacity. The results are presented graphically. It is found that the pressure is marginally affected by clearance ratio. It is noticed that the bearing system records an enhanced performance owing to magnetic fluid lubricant. Undoubtedly, the eccentricity ratio plays a performance role in the augmented performance. In addition, there appears to be less possibility of cavitation due to magnetic fluid lubricant.

Keywords Hydrodynamic · Magnetic fluid · Journal bearing

N. S. Patel (✉)
Mechanical Engineering Department, Faculty of Technology, Dharmsinh Desai University, Nadiad 387001 Gujarat, India
e-mail: nimeshsp@yahoo.co.in

D. P. Vakharia
Mechanical Engineering Department, Sardar Vallabhbhai National Institute of Technology, Surat 395007 Gujarat, India

G. M. Deheri
Department Mathematics Department, Sardar Patel University, Vallabh Vidhynagar, Anand 388120 Gujarat, India

H. C. Patel
Department of Mathematics, L. D. College of Engineering, Ahmedabad 380015 Gujarat, India

H. C. Patel et al. (eds.), *Proceedings of International Conference on Advances in Tribology and Engineering Systems*, Lecture Notes in Mechanical Engineering,
DOI: 10.1007/978-81-322-1656-8_10,

1 Introduction

It is well known that journal bearings are widely used in industry. The classical theory of journal bearing is well investigated one as can be seen from Cameron [1], Majumdar [2], Hamrock [3]. Prakash and Gururajan [4] studied the effect of velocity slip in an infinitely long rough porous journal bearing operating in the hydrodynamic regime. It was shown that the slip-roughness interaction considerably affected the bearing performance. Wang and Khonsari [5] presented an analytical expression for the dynamic pressure and dynamic fluid force in axially grooved long journal bearings considering oil inlet pressure and inlet position. In this paper, Reynold's-Floberg-Jakobsson boundary conditions are assumed to account for the appropriate starting position of the cavitation, the reformation of oil film at the end of cavitation and the effect of oil inlet pressure and inlet position. Shah and Bhat [6] dealt with the ferrofluids lubrication of a squeeze film in a long journal bearing. They made a comparison of squeeze film behavior using the ferrofluids flow models of Neuringer-Rosenswaig, Jenkins and Shliomis with uniform and non-uniform magnetic fields. Patel et al. [7] analyzed the performance of a fsherrofluid based short journal bearing. It was established that the load carrying capacity increased nominally while the coefficient of friction decreases significantly because of magnetic fluid lubrication. Patel et al. [8] discussed the magnetic fluid lubrication of a hydrodynamic short porous journal bearing. The results indicated that the magnetic fluid turned in a better performance of the bearing system as compared to a conventional lubricant. Interestingly, the coefficient of friction decreased significantly. Recently, Agrawal et al. [9] investigated the hydrodynamic lubrication in infinitely long journal bearing considering oscillating velocity. This was achieved, formatting the Reynold's equation by the finite difference method.

Here, it has been proposed to study and analyzed the magnetic fluid lubrication of an infinitely long journal bearing.

1.1 Analysis

The configuration of the bearing which is infinitely long in Z-direction is presented in Fig. 1. The journal has radius R rotates inside a bearing and the space between journal and bearing is filled with a magnetic fluid. If the journal is infinitely long, the pressure gradient $\frac{\partial p}{\partial x}$ is much larger than the pressure gradient $\frac{\partial p}{\partial z}$, as a result of which the later can be neglected. The magnetic field is oblique to the stator as in Agrawal [10] and its magnitude is given by (Fig. 2)

$$H^2 = kR^2\theta(2\pi - \theta)$$

where, k is a constant to suit the dimensions and the strength of the magnetic field.

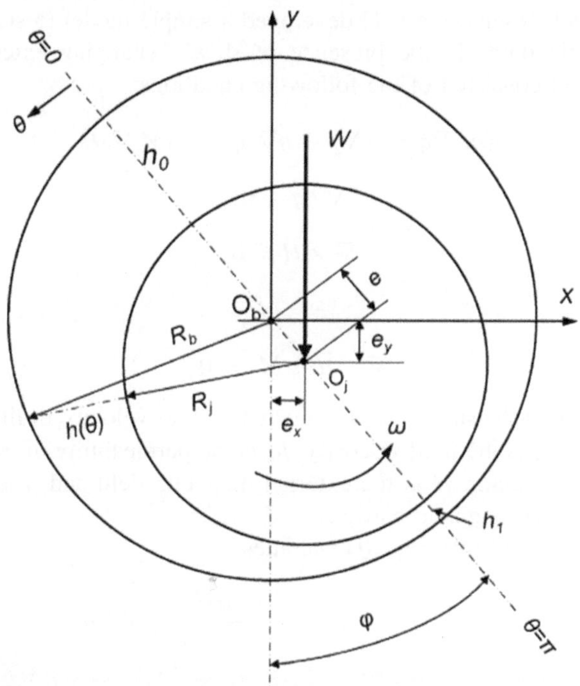

Fig. 1 Journal bearing

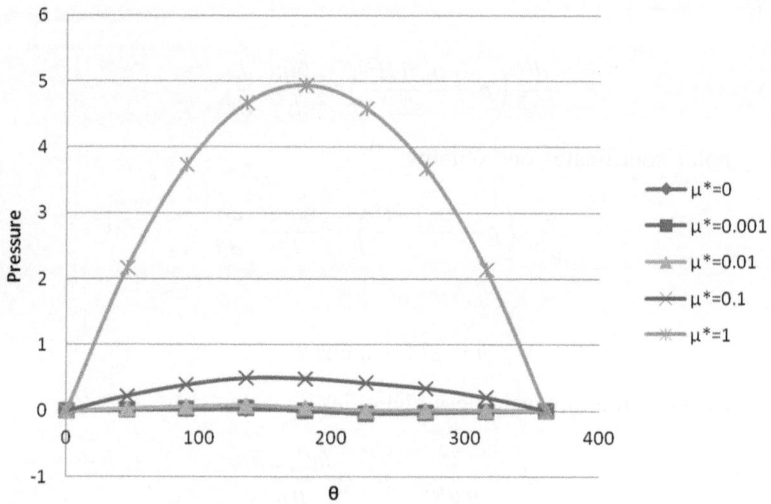

Fig. 2 Non-dimensional pressure distribution P versus θ for different values of μ^*

Neuringer and Rosensweig [11] developed a simple model to study the steady flow of magnetic fluids in the presence of slowly changing external magnetic fields. The model consisted of the following equations:

$$\rho(\bar{q}.\nabla)\bar{q} = -\nabla p + \eta\nabla^2\bar{q} + \mu_o(\bar{M}.\nabla)\bar{H} \tag{i}$$

$$\nabla \cdot \bar{q} = 0 \tag{ii}$$

$$\nabla \times \bar{H} = 0 \tag{iii}$$

$$\bar{M} = \bar{\mu}\bar{H} \tag{iv}$$

$$\nabla \cdot (\bar{H} + \bar{M}) = 0 \tag{v}$$

where ρ is the fluid density, $\bar{q} = (u, v, w)$ is the fluid velocity in film region, p is the film pressure, η is the fluid viscosity, μ_o is the permeability of free space, $\bar{M}$ is the magnetization vector, $\bar{H}$ is the external magnetic field and $\bar{\mu}$ is the magnetic susceptibility of the magnetic particles.

Using Eqs. (iii) and (iv), Eq. (i) becomes

$$\rho(\bar{q}.\nabla)\bar{q} = -\nabla\left(p - \frac{\mu_0\bar{\mu}H^2}{2}\right) + \eta\nabla^2\bar{q} \tag{vi}$$

This shows that an extra pressure $\frac{\mu_0\bar{\mu}H^2}{2}$ is introduced into the Navier–Stokes equation when magnetic fluid is used as a lubricant. Thus, the modified Reynolds equation for magnetohydrodynamic long journal bearing under the usual assumptions of hydrodynamic lubricantion (Bhat and Deheri [12], Bhat [13]) turns out to be

$$\frac{d^2}{dx^2}\left(p - \frac{\mu_o\,\overline{\mu}\,H^2}{2}\right) = \frac{6\eta u}{h^3}\cdot\frac{dh}{dx} \tag{1}$$

Using polar coordinates one obtains,

$$\frac{d^2}{d\theta^2}\left(p - \frac{\mu_o\,\overline{\mu}\,H^2}{2}\right) = \frac{6\eta uR}{h^3}\cdot\frac{dh}{d\theta}$$

where,

$$h = c(1 + \varepsilon\cos\theta)$$

Introduction of the dimensionless quantities

$$P = \frac{c^3}{\eta\,uR^2}, \quad \mu^* = \frac{kc^3\mu_o\,\overline{\mu}}{\eta\,u},$$

leads to the expression for the pressure distribution in dimensionless form

$$P = \left[\frac{\mu * \theta(2\pi - \theta)}{2} + 6\left(\frac{c}{R}\right)\int\limits_0^\theta \left(\frac{1}{(1+\varepsilon\cos\theta)^2}\right)d\theta + c_1\int\limits_0^\theta \left(\frac{1}{(1+\varepsilon\cos\theta)^3}\right)d\theta + c_2\right]$$

To evaluate integration part, the following Sommerfeld substitutions have been made,

$$\cos\gamma = \frac{\varepsilon + \cos\theta}{1+\varepsilon\cos\theta},\quad \sin\theta = \frac{(1+\varepsilon^2)^{1/2}\sin\gamma}{1+\varepsilon\, os\gamma},\quad \cos\theta = \frac{\cos\gamma - \varepsilon}{1-\varepsilon\cos\gamma},\quad d\theta = \frac{(1-\varepsilon^2)^{1/2}d\gamma}{1-\varepsilon\cos\gamma}$$

The associated Sommerfeld boundary conditions are

$$p = 0\,\text{at}\,\theta = 0$$

$$\text{and,}\quad p(0) = p(2\pi)$$

Using above substitutions and boundary conditions and reverting equation from γ to θ, one gets

$$P = \left[\frac{\mu * \theta(2\pi - \theta)}{2} + 6\left(\frac{c}{R}\right)\varepsilon\frac{(2+\varepsilon\cos\theta).\sin\theta}{(2+\varepsilon^2)(1+\varepsilon\cos\theta)^2}\right] \tag{2}$$

The load carrying capacity in x direction is given by

$$w_x = w\cos\varphi = \int\limits_0^{2\pi} P\cos\theta d\theta$$

The load carrying capacity in z direction is obtained from

$$w_z = w\sin\varphi = \int\limits_0^{2\pi} P\sin\theta d\theta$$

on substitution for P from Eq. (2), one obtains

$$w_x = w\cos\varphi = 0,$$

that is,

$$\varphi = \pi/2$$

Similarly, one finds that

$$w_z = w\sin\varphi = w$$

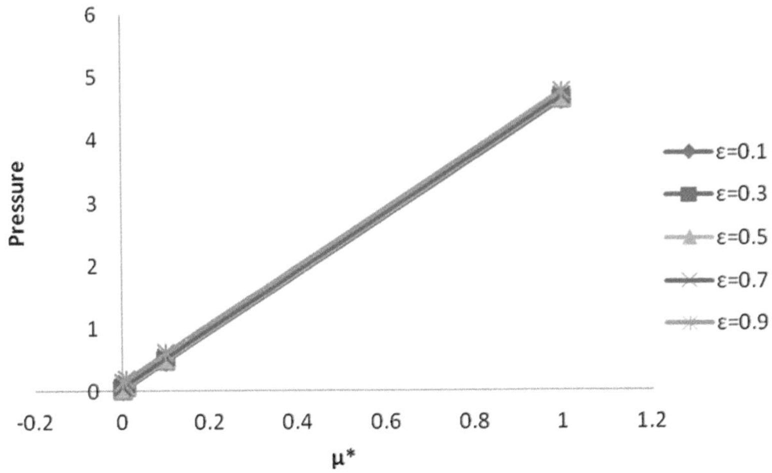

Fig. 3 Non-dimensional pressure distribution P versus ε for different values of μ^*

Thus, the dimensionless load carrying capacity in x direction is obtained from

$$W = \frac{c^2}{\eta\, u\, B^3}\, w$$

as

$$W = \left(\frac{2}{3}\pi^3\mu * + \frac{12\pi\varepsilon}{(2+\varepsilon^2)(1-\varepsilon^2)^{1/2}}\right) \tag{3}$$

1.2 Results and Discussion

It is seen from the Eq. (2) that the non-dimensional pressure increases by $\mu^*\theta(2\pi-\theta)/2$, while from the Eq. (3) one can easily observe that the load carrying capacity enhances by $2\pi^2\ \mu^*/3$ as compared to the case of conventional lubricants. Figures 3, 5 and 6 indicate that the pressure sharply increases with increasing magnetization no matter what the values of ε and c/R. Further, it can be noticed that the expression involved in Eq. (3) is linear with respect to the magnetization parameter μ^*. Accordingly, the load carrying capacity would increase with the increase in the magnetization irrespective of the choice of ε. It is appealing to note from Figs. 2 and 4 that the magnetization has a very strong positive effect on this type of bearing system in the sense that it overcomes the cavitation effect due to negative pressure as the pressure turns out to be positive even in the divergent region. Probably this is due to the fact that the magnetization increases the overall

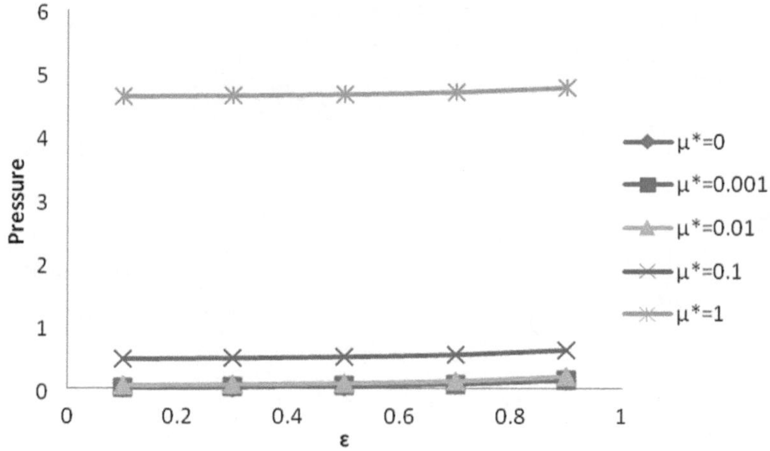

Fig. 4 Dimensional pressure distribution P versus μ^* for different values of θ

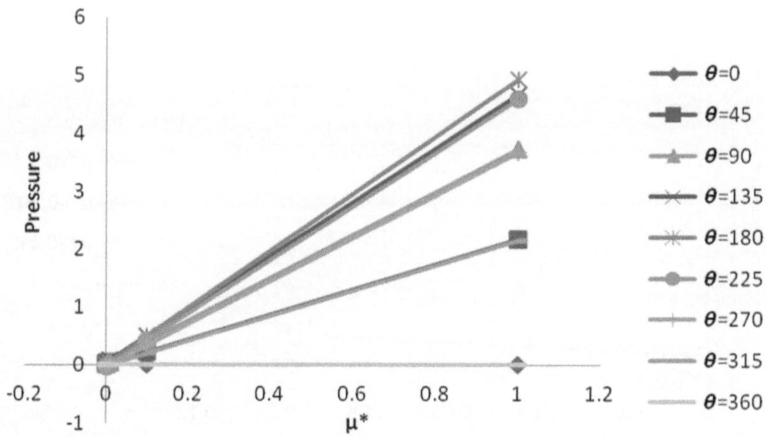

Fig. 5 Non-dimensional pressure distribution P versus c/R for different values of μ^*

effective viscosity of the lubricant. The effect of c/R on pressure distribution and ε on the load carrying capacity are at the best nominal (Figs. 4, 7). Figure 8 shows that the load carrying capacity increases significantly with increase in magnetization.

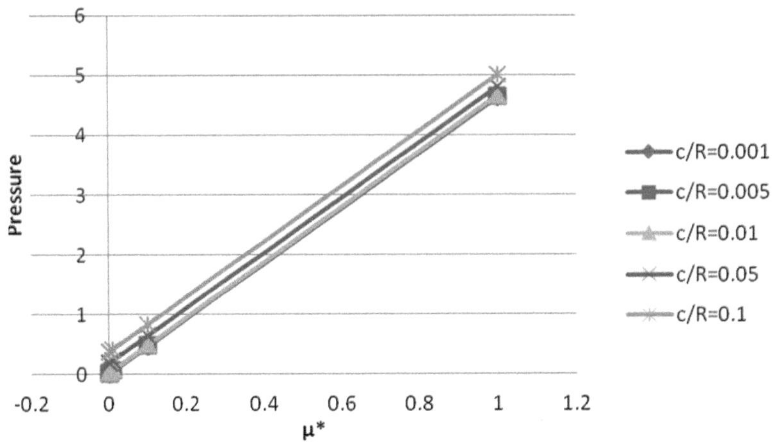

Fig. 6 Non-dimensional load versus μ^* for different values of ε

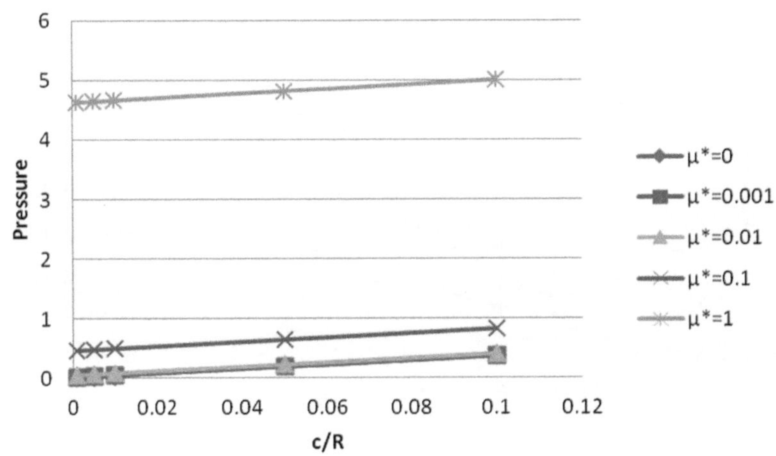

Fig. 7 Non-dimensional pressure distribution P versus μ^* for different values ε

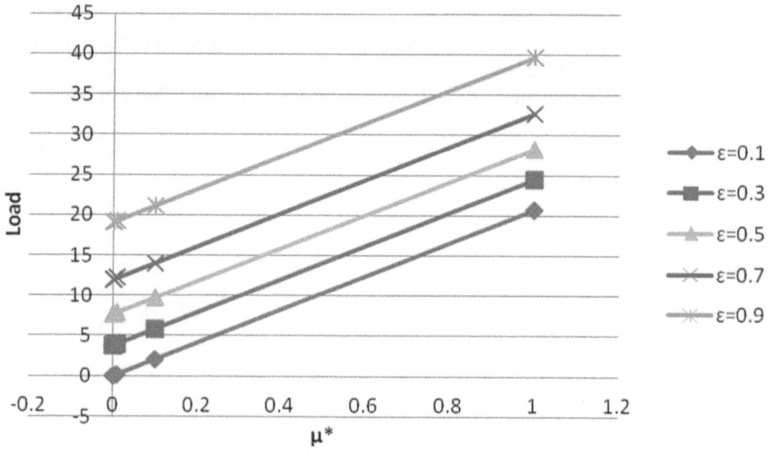

Fig. 8 Non-dimensional pressure distribution P versus μ^* for different values of c/R

2 Conclusion

This type of bearing system becomes more favorable in the industry as the magnetization induces an enhanced performance for almost all geometrical configurations. Even in the absence of flow this type of bearing system sustains certain amount of load which is far from being true in the case of traditional lubricants.

References

1. Cameron A (1966) Principles of lubrication. Longmans, London
2. Majumdar BC (2010) Introduction to tribology of bearings. S Chand & Company Ltd
3. Hemrock BJ (1994) Fundamental of fluid film lubrication. McGraw-Hill, New York
4. Prakash J, Gururajan K (1999) Effect of velocity slip in an infinitely long rough porous journal bearing. Tribol Trans 42(3):661–667
5. Wang JK, Khonsari MM (2008) Effects of oil inlet pressure and inlet position of axially grooved infinitely long journal bearings Part I: analytical solutions and static performance. Tribol Int 41:119–132
6. Shah RC, Bhat MV (2004) Ferrofluid squeeze film in a long journal bearing. Tribol Int 37:441–446
7. Patel NS, Vakharia DP, Deheri GM (2012) A study on the performance of a magnetic-fluid-based hydrodynamic short journal bearing. ISRN Mech Eng 7, doi:10.5402/2012/603460
8. Patel NS, Vakharia DP, Deheri GM (2012) A Study on the performance of a magnetic fluid based hydrodynamic short porous journal bearing. J Serb Soc Comput Mech 6(2):28–44
9. Aggrawal N, Chawla D, Kumar V, Tayal A, Mittal S, Kumar P(2013) Analysis of hydrodynamic lubrication in infinitely long journal bearing with oscillating velocity. IJIEASR, 2(2):45–48
10. Agrawal VK (1986) Magnetic fluid based porous inclined slider bearing. WEAR 107:133–139

11. Neuringer JL, Rosensweig RE (1964) Phys Fluids 7:19–27
12. Bhat MV, Deheri GM (1991) Squeeze film behavior in porous annular discs lubricated with magnetic fluid. Wear 151:123–128
13. Bhat MV (1978) Hydrodynamic lubrication of a porous composite slider bearing. Jpn J Appl Phys 17:479–481

Thermo Hydrodynamic Lubrication Characteristics of Power Law Fluids in Rolling/Sliding Line Contact

Dhaneshwar Prasad and S. V. Subrahmanyam

Abstract Hydrodynamic lubrication of rollers having the same dimension and moving with different velocities is studied assuming the consistency of the non-Newtonian incompressible power law lubricants to vary with pressure and the mean film temperature. The equations of motion, continuity, and momentum energy are solved first analytically and then numerically by Runge–Kutta Fehlberg method. Some important bearing characteristics are analyzed and displayed in the form of some graphs to study their behaviors.

Nomenclature

H	Film thickness at $x = -x_1$
h	Lubricant film thickness
h_o	Minimum film thickness
$\bar{h}$	h/h_o etc.
m	Lubricant consistency
m_o	Initial consistency temperature
n	Consistency index of the power law lubricant
p	Hydrodynamic pressure
R	Radius of the equivalent cylinder
T	Lubricant temperature
T_{11}	Film temperature for $y \geq \delta$ in region-I etc.
T_m	Mean film temperature
T_0	Ambient temperature
$\overline{T}_{Fh+}$	Traction force ($= -(2\alpha T_{Fh}/h_o)$) etc.
U_1, U_2	Cylinders velocities at $y = -h$ and $y = h$ respectively
u	Velocity of the lubricant in x-direction

D. Prasad (✉)
Dr. S.R.K. Government Arts College, Yanam 533464, Pondicherry, India
e-mail: rpdhaneshwar@gmail.com

S. V. Subrahmanyam
K. L. University, Vaddeswaram, Guntur District, Andhra Pradesh, India

H. C. Patel et al. (eds.), *Proceedings of International Conference on Advances in Tribology and Engineering Systems*, Lecture Notes in Mechanical Engineering,
DOI: 10.1007/978-81-322-1656-8_11, © Springer India 2014

u_m	The mean velocity of the lubricant
v	Velocity of the lubricant in y-direction
W	Load in y-direction
$\overline{W}$	Dimensionless load (= $\alpha W/(Rh_o)^{1/2}$)
$\bar{x}$	$x/(2Rh_o)^{1/2}$) etc.
x_1	Point of maximum pressure
x_2	Cavitation point
φ	$\frac{\rho c u_m}{k}\left(\frac{dT_m}{dx}\right)$
α, β	Pressure and temperature coefficients

1 Introduction

Hydrodynamic bearings play vital roles in lubrication mechanism as successful performance of mechanical components such as gears, roller bearings, came etc. When these bearings run at heavy loads and high speeds, the lubricant material properties vary with pressure [1]. Further, high pressure generation causes viscous heating, so the temperature of the lubricant increases significantly along with pressure [2]. Hence the effect of temperature on the lubricant cannot be neglected [3].

Furthermore, lubrication of the line contact of two cylindrical rollers is assumed to be symmetric. More research work has not been done by ignoring the symmetric condition except a few where closed form or semi analytical solutions were found. Savage [4] gave a new direction to anti symmetric line contact problem and obtained a semi analytical solution for isoviscous fluid. Later, Prasad et al. [5] analyzed this problem while adding thermal effects and obtained a semi analytical solution. Tsong-Rong Lin [6] studied this problem in detail with thermal compressibility including EHD effects and obtained fully numerical solution of the problem.

As such the classical hydrodynamic lubrication of heavily loaded cylindrical line contact problems were studied by various researchers. One of the most famous contributors among them is Dowson and his associates. Dowson and Higginson [7] studied this problem where the viscosity and density of the lubricant were taken to be pressure and temperature dependent. Cheng [8] studied the problem with thermal effects and developed a refined solution for the problem. These results, particularly frictional forces, were in good agreement with the experimental results (Fig. 1).

In most of the problems, lubricant is assumed to be Newtonian. Since, the lubricant is subject to high pressure and heavy load, the Newtonian fluid characteristics ceases to exist [9] i.e. the lubricant becomes non-Newtonian. The non-Newtonian characteristics have been invariably served in various lubrication problems and this is due to the shear rate and the high pressure gradient, or due to

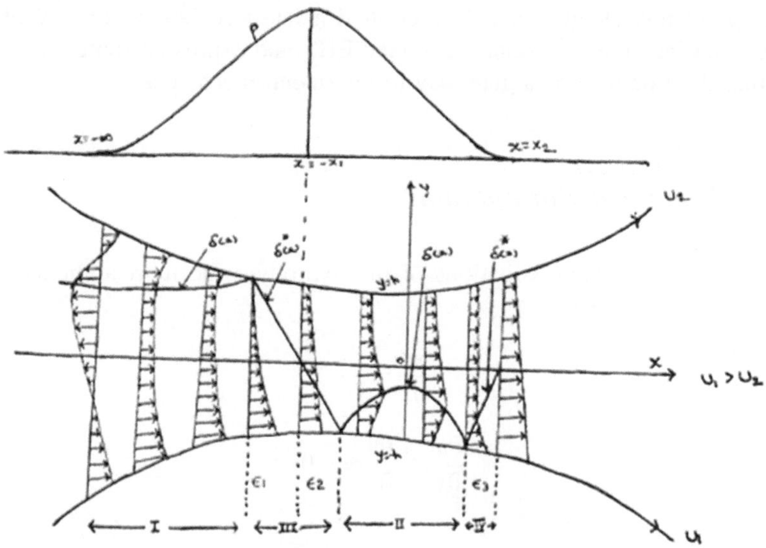

Fig. 1 Lubrication of asymmetric cylinders

the additives [10]. Hence the effect of non-Newtonian characteristic of the lubricant is also to be incorporated along with the effects of pressure and temperature. Hsiao and Hamrock [11] developed a numerical computational algorithm for non-Newtonian fluids and derived the temperature profile for this model lubricant throughout the region for pure rolling, rolling/sliding and pure sliding conditions. Punit and Khonsari [12] applied Carreau viscosity model for high viscosity oil to study the combined effects of shear thinning and temperature rise on the EHL behavior of rolling/sliding line contact problem. Later, Punit and Khonsari [13] exhibited their views on the role of the lubricant rheology and Piezo-viscous properties in EHL line/point contact where a comprehensive study of number of non-Newtonian fluid models have been proposed and applied for analyzing shear thinning and limiting shear stress behaviors.

The Power law non-Newtonian fluid model has got attention in the recent years because of their simplicity and potential to describe many lubricants like polymer solutions, silicon fluids etc. In fact, it characterizes two different types of non-Newtonian fluids i.e. viscoelastic and dilatants plus Newtonian as well when index n of the power law model is unity [14]. Recently, Punit et al. [15] dealt with a similar problem on isothermal EHL behavior of line contact using Newtonian fluid and power law fluid as base oil and additive oil respectively for the effect of polymeric fluid additives. Very recently, Hasim Khan et al. [16] described a computational algorithm for the solution of thermal EHL problem of infinite line contact and observed that the EHL of rough surfaces is significantly influenced by the consideration of thermal effects, and this scheme can also be extended to the analysis of two-dimensional thermal EHL problems such as finite line/point contact problems. Dong Zhu [17] presented a latest review on history of EHL research, outlined the importance of mixed EHL and expressed some views about its further development as a gate way to interfacial mechanics.

1.1 Mathematical Formulation

The fluid flow governing equations of the hydrodynamic lubrication with some usual assumptions are [5]:

$$\frac{dp}{dx} = \frac{\partial \tau}{\partial y} \tag{1}$$

$$\frac{\partial u}{\partial x} + \frac{\partial v}{\partial y} = 0 \tag{2}$$

where

$$m = m_0 e^{\alpha p - \beta (T_m - T_0)} \tag{3}$$

$$T_m = \frac{1}{2h} \int_{-h}^{h} T \, dy \tag{4}$$

$$h = h_0 + x^2/2R. \tag{5}$$

$$\tau = m \left| \frac{\partial u}{\partial y} \right|^{n-1} \frac{\partial u}{\partial y}, \tag{6}$$

The boundary conditions for the problem are:

$$u = U_1 \text{ at } y = -h \tag{7}$$

$$u = U_2 \quad \text{at} \quad y = h \tag{8}$$

$$p = 0 \quad \text{at} \quad x = -\infty \tag{9}$$

$$p = 0 \; ; \; \frac{dp}{dx} = 0 \text{ at } x = x_2 \tag{10}$$

The velocity gradient conditions for the geometry under consideration are:

$$\frac{\partial u_1}{\partial y} \geq 0, \quad \delta \leq y \leq h; \tag{11}$$

$$\frac{\partial u_2}{\partial y} \leq 0\,, \quad -h \leq y \leq \delta; \; I : -\infty < x < -x_1\,; \tag{12}$$

$$\frac{\partial u_3}{\partial y} \geq 0\,,\ -h \leq y \leq \delta \tag{13}$$

$$\frac{\partial u_4}{\partial y} \leq 0\,,\ \delta \leq y \leq h;\ \text{II}: -x_1 < x < x_2 \tag{14}$$

Now the volume flux Q for the region:

$$-\infty < x < -x_1$$

$$Q = \int_{-h}^{h} u\,dy \tag{15}$$

$$= U_1(h+\delta) + U_2(h-\delta) - \left(\frac{n}{2n+1}\right)\left(\frac{1}{m_1}\frac{dp_1}{dx}\right)^{1/n}\left[(h+\delta)^{\frac{2n+1}{n}} + (h-\delta)^{\frac{2n+1}{n}}\right] \tag{16}$$

The flux Q through the point $x = -x_1$ can be evaluated directly from Eq. (1) yielding:

$$Q(-x_1) = (U_1 + U_2)H \tag{17}$$

where H is film thickness at $x = -x_1$.

Reynolds equation:

Equating the flux Eq. (16) and (17), and simplifying them, it can be obtained as

$$\frac{dp_1}{dx} = m_1\left(\frac{2n+1}{n}\right)^n\left[\frac{(U_1+U_2)(h-H)+(U_1-U_2)\delta}{(h+\delta)^{\frac{2n+1}{n}}+(h-\delta)^{\frac{2n+1}{n}}}\right]^n, \quad -\infty < x \leq -x_1 \tag{18}$$

$$\frac{dp_2}{dx} = -m_2\left(\frac{2n+1}{n}\right)^n\left[\frac{(U_1+U_2)(H-h)+(U_1-U_2)\delta}{(h+\delta)^{\frac{2n+1}{n}}+(h-\delta)^{\frac{2n+1}{n}}}\right]^n, \quad -x_1 \leq x \leq x_2 \tag{19}$$

Using $u_1 = u_2$ and $u_3 = u_4$ at $y = \delta$ one can get

$$(U_1 - U_2) + \left(\frac{n}{n+1}\right)\left(\frac{1}{m_1}\frac{dp_1}{dx}\right)^{1/n}\left[(h-\delta)^{\frac{n+1}{n}} - (h+\delta)^{\frac{n+1}{n}}\right] = 0, \quad -\infty < x \leq -x_1 - \epsilon_1 \tag{20}$$

$$(U_1 - U_2) + \left(\frac{n}{n+1}\right)\left(\frac{1}{m_2}\frac{dp_2}{dx}\right)^{1/n}\left[(h-\delta)^{\frac{n+1}{n}} - (h+\delta)^{\frac{n+1}{n}}\right] = 0, \quad -x_1 + \leq \epsilon_2\ x \leq -x_2 - \epsilon_3 \tag{21}$$

Eliminating $\frac{dp_1}{dx}$ and $\frac{\mathrm{dp}_2}{\mathrm{dx}}$ from Eq. (20) and (21) and using Reynolds Eq. (18) and (19), one can obtain a single relationship as:

$$(U_1 - U_2) + \left(\frac{2n+1}{n+1}\right)\left[\frac{(U_1 + U_2)(h - H) + (U_1 - U_2)\delta}{(h + \delta)^{\frac{2n+1}{n}} + (h - \delta)^{\frac{2n+1}{n}}}\right]\left[(h-\delta)^{\frac{n+1}{n}} - (h+\delta)^{\frac{n+1}{n}}\right] = 0 \tag{22}$$

1.2 Heat Energy Equation

The heat energy equation with usual assumptions is considered to be [12, 18]:

$$\frac{\rho c}{k} + \left(u_m \frac{dT_m}{dx}\right) = \frac{\partial^2 T}{\partial y^2}\left(\frac{m}{k}\right)\left|\frac{\partial u}{\partial y}\right|^{n-1}\left(\frac{\partial u}{\partial y}\right)^2 \tag{23}$$

The boundary conditions for temperature:

$$T = T_{-h} \text{ at } y = -h \text{ and } T = T_h \; at \; y = h \tag{24}$$

Now the lubricant film temperatures T_{11} *and* T_{12}, separated by δ, are calculated for the region $-\infty < x \leq -\, x_1$ as:

$$T_{11} = \varphi(x)y^2/2 + S_1\,(y-\delta)^{\frac{3n+1}{n}} + c_1 y + d_1\,, \quad \delta \leq y \leq h \tag{25}$$

$$T_{12} = \varphi(x)y^2/2 + S_1\,(\delta - y)^{\frac{3n+1}{n}} + c_2 y + d_2\,, \quad -h \leq y \leq \delta \tag{26}$$

Using the temperature matching condition Eliminating $\frac{dp_1}{dx}$ and $\frac{\mathrm{dp}_2}{\mathrm{dx}}$ from Eq. (20) and (21) and using Reynolds Eq. (18) and (19), one can obtain a single relationship as:

$T_{11} = T_{12}$ at $y = \delta$ and the matching heat flux condition $k\frac{\partial T_{11}}{\partial y} = k\frac{\partial T_{12}}{\partial y}$ at y $= \delta$, one may get

$c_1 = c_2 = c\,(say)$ and , $d_1 = d_2 = d\,(say)$. The boundary conditions (24) in (25) and (26) gives

$$c = \frac{T_h - T_{-h}}{2h} - \left(\frac{S_1}{2h}\right)\left[(h-\delta)^{\frac{3n+1}{n}} - (h+\delta)^{\frac{3n+1}{n}}\right] \tag{27}$$

$$d = \frac{T_h + T_{-h}}{2} - \left(\frac{S_1}{2}\right)\left[(h-\delta)^{\frac{3n+1}{n}} + (h+\delta)^{\frac{3n+1}{n}}\right] - \frac{h^2}{2}\varphi(x) \tag{28}$$

Similarly $-x_1 \le x \le x_2$ in the region one can get

$$T_{21} = \varphi(x)y^2/2 + S_2\,(\delta - y)^{\frac{3n+1}{n}} + a\,y + b \tag{29}$$

$$T_{22} = \varphi(x)y^2/2 + S_2\,(y - \delta)^{\frac{3n+1}{n}} + a\,y + b \tag{30}$$

where

$$a = \frac{T_h - T_{-h}}{2h} - \left(\frac{S_2}{2h}\right)\left[(h-\delta)^{\frac{3n+1}{n}} - (h+\delta)^{\frac{3n+1}{n}}\right] \tag{31}$$

$$b = \frac{T_h + T_{-h}}{2} - \left(\frac{S_2}{2}\right)\left[(h-\delta)^{\frac{3n+1}{n}} + (h+\delta)^{\frac{3n+1}{n}}\right] - \frac{h^2}{2}\varphi(x) \tag{32}$$

Thus, T_{11}, T_{12}, T_{21}, and T_{22} are explicitly known functions of x and y analytically. Finally, the mean temperature T_{m1}, and T_{m2} as defined in Eq. (4), can be written as :

$$T_m = \frac{1}{2h}\int_{-h}^{h} T\,dy$$

or

$$T_{m1} = \varphi(x)\frac{h^2}{6} + \frac{nS_1}{2h(4n+1)}\left[(h-\delta)^{\frac{4n+1}{n}} + (h+\delta)^{\frac{4n+1}{n}}\right] + d\,, \quad -\infty < x \le -x_1 \tag{33}$$

$$T_{m2} = \varphi(x)\frac{h^2}{6} + \frac{nS_2}{2h(4n+1)}\left[(h-\delta)^{\frac{4n+1}{n}} + (h+\delta)^{\frac{4n+1}{n}}\right] + b\,, \quad -x_1 \le x \le x_2 \tag{34}$$

where $S_1 = \left(\frac{m_1}{k}\right)\left(\frac{1}{m_1}\frac{dp_1}{dx}\right)^{\frac{n+1}{n}}\left(\frac{-n^2}{(2n+1)(3n+1)}\right)$

$$S_2 = \left(\frac{m_2}{k}\right)\left(-\frac{1}{m_2}\frac{dp_2}{dx}\right)^{\frac{n+1}{n}}\left(\frac{-n^2}{(2n+1)(3n+1)}\right)$$

Now, using the following dimensionless scheme:

$$\bar{x} = \frac{x}{\sqrt{2Rh_0}},\ \bar{p} = \alpha p,\ \bar{m} = 2mc_n\alpha,$$
$$c_n = \frac{1}{2}\left(\frac{2n+1}{n}\right)^n\left(\frac{U}{h_0}\right)^n\sqrt{\frac{2R}{h_0}},\ \bar{h} = 1 + \bar{x}^2 \tag{35}$$
$$\bar{T}_m = \beta T_m\,,\ \bar{\delta} = \delta/h_0\,,\ \bar{U} = U_1/U_2\,,\ \bar{h} = h/h_0$$

Eq. (18), (19), (22), (25), (26), (29), (30), (33) and (34) can be rewritten as

$$\frac{d\overline{p}_1}{d\overline{x}} = \overline{m}_1\,\overline{(f_x)}^n \tag{36}$$

$$\frac{d\overline{p}_2}{d\overline{x}} = -\overline{m}_2\,\overline{(-f_x)}^n \tag{37}$$

$$(\overline{U}-1)+\left(\frac{2n+1}{n+1}\right)\left[\overline{(h-\delta)}^{\frac{n+1}{n}}-\overline{(h+\delta)}^{\frac{n+1}{n}}\right]\overline{f_x} = 0 \tag{38}$$

$$\overline{T}_{11}=\overline{\lambda}+\overline{\psi}_{1x}-\frac{\overline{\gamma m_1 n}}{(3n+1)}\,\overline{(f_x)}^{n+1}\,\overline{g}_{1x} \tag{39}$$

$$\overline{T}_{12}=\overline{\lambda}+\overline{\psi}_{1x}-\frac{\overline{\gamma m_1 n}}{(3n+1)}\,\overline{(f_x)}^{n+1}\,\overline{g}_{2x} \tag{40}$$

$$\overline{T}_{21}=\overline{\lambda}+\overline{\psi}_{2x}-\frac{\overline{\gamma m_2 n}}{(3n+1)}\,\left(-\overline{f_x}\right)^{n+1}\overline{g}_{2x} \tag{41}$$

$$\overline{T}_{22}=\overline{\lambda}+\overline{\psi}_{2x}-\frac{\overline{\gamma m_2 n}}{(3n+1)}\,\left(-\overline{f_x}\right)^{n+1}\overline{g}_{1x} \tag{42}$$

$$\begin{aligned}\frac{p_e\overline{h}^2}{3}\left(\overline{u_m}\frac{dT_{m1}}{d\overline{x}}\right)+\overline{T}_{m1}&=\frac{1}{2}\left(\overline{T_h}+\overline{T_{-h}}\right)\\&-\frac{\overline{\gamma}\,\overline{m}_1 n}{3n+1}\sqrt{\overline{(f_x)}}^{\,n+1}\;\overline{g_x}\end{aligned} \tag{43}$$

$$\frac{p_e\overline{h}^2}{3}\left(\overline{u}_m\,\frac{d\overline{T}_{m2}}{d\overline{x}}\right)+\overline{T}_{m2}=\frac{1}{2}\left(\overline{T_h}+\overline{T_{-h}}\right)-\frac{\overline{\gamma}\,\overline{m}_2 n}{3n+1}\left(-\overline{f_x}\right)^{n+1}\overline{g}_x \tag{44}$$

$$\overline{m}=\overline{m}_0\,e^{\left(\overline{p}_1-\overline{T}_{m1}+\overline{T_0}\right)}\quad\text{etc.} \tag{45}$$

$$\overline{\gamma}=\left(\frac{U_2h_0\beta}{2k\alpha}\right)\sqrt{\frac{h_0}{2R}} \tag{46}$$

$$\overline{f_x}=\frac{(\overline{U}+1)(\overline{h}-\overline{H})+(\overline{U}-1)\overline{\delta}}{\overline{(h+\delta)}^{\frac{2n+1}{n}}+\overline{(h-\delta)}^{\frac{2n+1}{n}}} \tag{47}$$

$$\overline{\lambda}=\left(\frac{\overline{T}_h+\overline{T}_{-h}}{2}\right)+\frac{\overline{T}_h-\overline{T}_{-h}}{2}\left(\frac{\overline{y}}{\overline{\overline{h}}}\right) \tag{48}$$

$$\overline{\psi}_{1x}=\frac{\overline{p_e}}{2}\,\overline{u}_m\,\frac{d\overline{T}_{m1}}{d\overline{x}}\left(\overline{y}^2-\overline{h}^2\right) \tag{49}$$

$$\overline{\psi}_{2x}=\frac{\overline{p_e}}{2}\,\overline{u}_m\,\frac{d\overline{T}_{m2}}{d\overline{x}}\left(\overline{y}^2-\overline{h}^2\right) \tag{50}$$

where $\overline{p_e} = \frac{\rho c U_2 h_0}{k}\sqrt{\frac{h_0}{2R}}$

$$\overline{g}_x = \frac{n}{(4n+1)\overline{h}}\left[(\overline{h}+\overline{\delta})^{\frac{4n+1}{n}} + (\overline{h}-\overline{\delta})^{\frac{4n+1}{n}}\right] - \left[(\overline{h}-\overline{\delta})^{\frac{3n+1}{n}} + (\overline{h}+\overline{\delta})^{\frac{3n+1}{n}}\right] \tag{51}$$

$$\overline{g}_{1x} = 2(\overline{y}-\overline{\delta})^{\frac{3n+1}{n}} - \left[(\overline{h}+\overline{\delta})^{\frac{3n+1}{n}} + (\overline{h}-\overline{\delta})^{\frac{3n+1}{n}}\right] + (\overline{y}/\overline{h})\left[(\overline{h}+\overline{\delta})^{\frac{3n+1}{n}} - (\overline{h}-\overline{\delta})^{\frac{3n+1}{n}}\right] \tag{52}$$

$$\overline{g}_{2x} = 2(\overline{\delta}-\overline{y})^{\frac{3n+1}{n}} - \left[(\overline{h}+\overline{\delta})^{\frac{3n+1}{n}} + (\overline{h}-\overline{\delta})^{\frac{3n+1}{n}}\right] + (\overline{y}/\overline{h})\left[(\overline{h}+\overline{\delta})^{\frac{3n+1}{n}} - (\overline{h}-\overline{\delta})^{\frac{3n+1}{n}}\right] \tag{53}$$

Load and Traction: The dimensionless load

$$\overline{W} = \frac{w\alpha}{\sqrt{2Rh_0}} \tag{54}$$

$$\overline{W} = \int_{-\infty}^{\overline{x}_2} \overline{p}\, d\overline{x} = -\int_{-\infty}^{\overline{x}_2} \overline{x}\, \frac{d\overline{p}}{d\overline{x}}\, d\overline{x} \tag{55}$$

Dimensionless tractions is

$$\overline{T}_{Fh-}\left(= \alpha\frac{T_{F_0}}{h_0}\right) = -\int_{-\infty}^{\overline{x}_2} \overline{\tau}_{\overline{y}=-\overline{h}}\, d\overline{x} \tag{56}$$

$$\overline{T}_{Fh+} = -\int_{-\infty}^{\overline{x}_2} \overline{\tau}_{\overline{y}=-\overline{h}}\, d\overline{x} \tag{57}$$

1.3 Result and Discussion:

A semi analytical solution of the Reynolds equations (36, 37) and the energy Eqs. (39-42, 43, 44) are obtained for the following representative values: $U_2 = 400$ cm/s, $h_0 = 4 \times 10^{-4}$ cm, $\alpha = 1.6 \times 10^{-9}$ dyne^{-1} cm^2, R = 3 cm, $\overline{\gamma} = 1000.5$, $\overline{T}_h + \overline{T}_{-h} = 3.2$.

The Reynolds and the energy Eq. (36, 37, 43 and 44) coupled through $\overline{m}$ contain two unknowns $\overline{\delta}$ and $\overline{x_1}$. These unknowns are also present in Eq. (38). In order to solve Eq. (38), first of all, an initial value of $\overline{x}$ is replaced by a large but a finite negative value i.e. $\overline{x} = -5$. For solution details, see Ref. [5]. Runge–Kutta Fehlberg method is used to solve the Eq. (38). As noted earlier [5], $\overline{\delta}$ does not exist in the neighborhood of $\overline{x} = -\overline{x_1}$ and lying in the interval: $-\overline{h} \leq \overline{\delta} \leq \overline{h}$. Hence in the neighborhood of $\overline{x} = -\overline{x_1}\ \overline{\delta}\left(=\overline{\delta*}\right)$ say has to be determined solely on the basis of physical considerations. The $\in -$ neighborhood $(-\overline{x_1}- \in_1 \leq \overline{x} \leq -\overline{x_1}+ \in_2)$ of $\overline{x} = -\overline{x_1}$ is to be determined as the region where there does not exist any $\overline{\delta}$ lying in the interval$-\overline{h} \leq \overline{\delta} \leq \overline{h}$ and satisfying Eq. (38). To ease the mathematical complexity, a linear profile for $\overline{\delta}^*$ chosen as below:

$$\overline{\delta}^* = -\overline{h}_{b1}\left(\overline{x} + \overline{x_1}\right)/ \in_1, \quad -\overline{x_1}- \in_1 \leq \overline{x} \leq -\overline{x_1} \tag{58}$$

$$\overline{\delta}^* = -\overline{h}_{h2}\overline{(x + x_1)}/ \in_2, \quad -\overline{x_1} \leq \overline{x} \leq -\overline{x_1}+ \in_2 \tag{59}$$

$\overline{h}_{b2} = 1 + (-\overline{x_1} - \in_1)^2\ \overline{h}_{b2} = 1 + (-\overline{x_1} - \in_1)^2$ is assumed. Having determined $\overline{\delta} = \overline{\delta}^*$ using Eq. (58) and (59) in the neighborhood of $\overline{x} = -\overline{x_1}$, the same procedure is followed in [5] with

$$\overline{\delta}^* = \overline{h}_{b3}\left(\overline{x} - \overline{x_2}\right)/ \in_3 \tag{60}$$

where $\overline{h}_{b3} = 1 + \overline{(x_2} - \in_3)^2\ \overline{\delta}^*$, in the region $\overline{x_2} - \in_3 \leq \overline{x} \leq \overline{x_2}$, is calculated using Eq.(60) together with $\overline{p}_2$ and $\overline{T}_{m2}$.

The lubricant pressure $\overline{p}$ is presented in Figs. (2) and (3) i.e. $\overline{p}$ increases with $\overline{x}$ and then decreases till last. A similar type of profile was also obtained by Wang et al. [19]. The qualitative behavior of $\overline{p}$ for different values of $\overline{U}$ (for fixed n) is identical (see Fig. 2). Further, $\overline{p}$ increases as $\overline{U}$ decreases. This is somewhat contrary to the results of Prasad et al. [5], Jang and Khonsari [20], and Rong-Tsong and Hamrock [21], For fixed $\overline{U}$, pressure increases as n increases [22, 23].

The mean temperature $\overline{T}_m$ *is* shown in Fig. (4) and Fig. (5), as a function of n, and $\overline{U}$ respectively. It may be observed that the qualitative behavior of $\overline{T}_m$ versus $\overline{x}$ is very similar to the temperature profile obtained in Prasad et al. [5], Saini et al. [24] and Liu et al. [25]. Further, it may be noted from Fig. (4) that for fixed values of $\overline{U}$and $\overline{P_e}$, $\overline{T}_m$ increases with n showing that the temperature for dilatants fluid is higher than that of Newtonian and pseudo plastic fluids both. For fixed values of n and $\overline{P_e}$, $\overline{T}_m$ increases with $\overline{U}$, indicates that the sliding temperature is higher than that of pure rolling [26].

The calculated values of the normal load carrying capacity $\overline{W}$, the traction force $\overline{T}_{Fh+}$ are calculated and presented in Figs. 6 and 7 respectively. Results are in agreement with the previous findings [6]. All these have same characteristics, i.e., both increase with n and decreases as $\overline{U}$ increases. The trend of the traction forces with n is quite similar to Punit and Khonsari [13],

Fig. 2 Pressure $\overline{p}$ verses $\overline{X}$ var rining $\overline{U}$

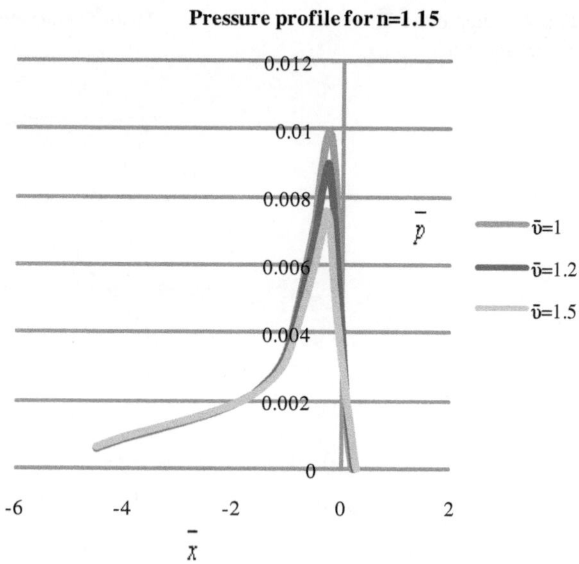

Fig. 3 Pressure $\overline{p}$ verses $\overline{X}$ var rining n

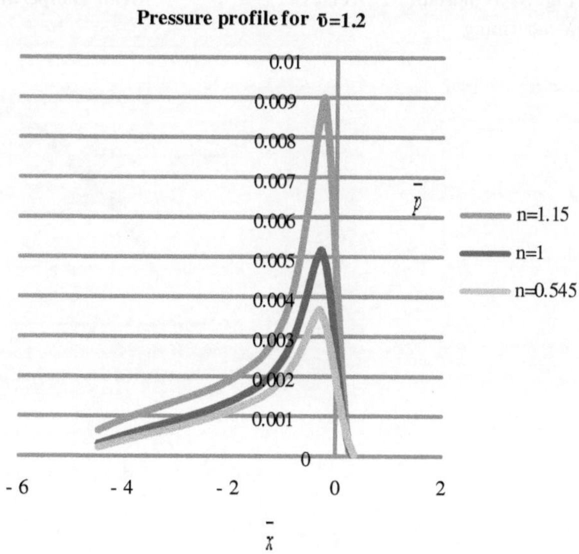

Fig. 4 Tempreture $\overline{T_m}$ verses $\overline{X}$ var rining n

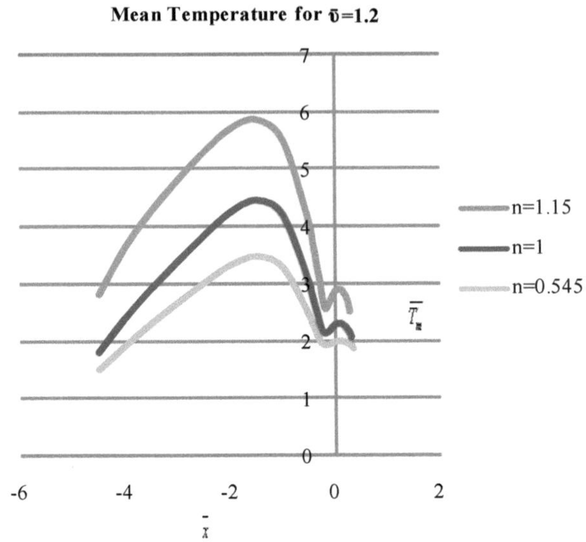

Fig. 5 Tempreture $\overline{T_m}$ verses $\overline{X}$ var rining

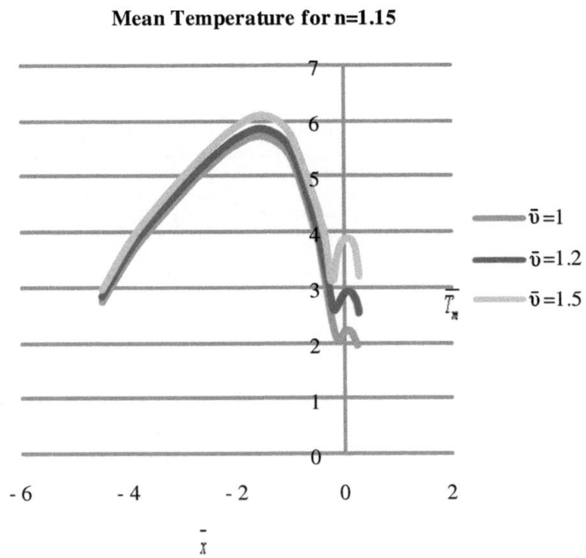

Fig. 6 Load $\overline{W}$ verses $\overline{U}$ var rining n

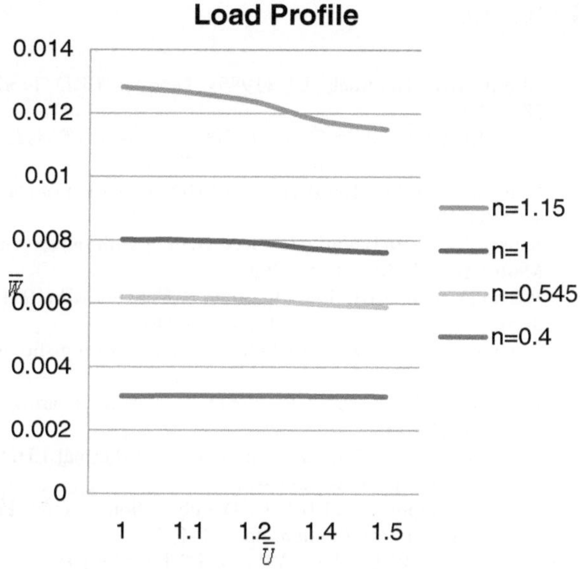

Fig. 7 Traction $\overline{T_{F+}}$ verses$\overline{U}$ var rining n

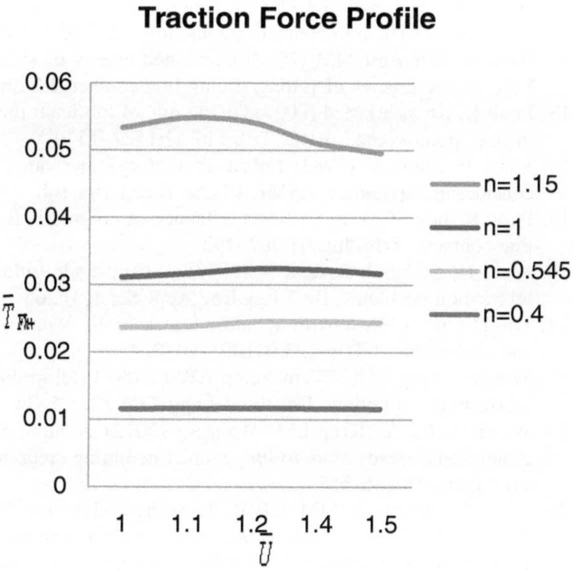

2 Conclusion

(i) There is a significant increase in pressure, hence the load and the traction, with the flow index n for a fixed value of $\overline{U}$. (ii) There is also a significant increase in the mean film temperature with n and $\overline{U}$ both. Hence it is justifiable to treat $\overline{m}$ as a function of $\overline{p}$ and temperature.

Reference:

1. Ghosh MK, Hamrock BJ (1985) Thermal EHD lubrication of line contacts. ASLE 28:159–171
2. Yung-Kuang Y, Ming-Chang J (2004) Analysis of viscosity interaction on the misaligned conical-cylindrical bearing. Trib Intl 37:51–60
3. Sadeghi F, Sui PC (1990) Thermal EHD lubrication of rolling/sliding contacts. ASME J Trib 112:189–195
4. Savage MD (1983) Variable speed coating with purely viscous non-newtonian fluid. J Appl Math Phy (ZAMP) 34:358–369
5. Prasad D, Shukla JB, Singh P, Sinha P, Chhabra RP (1991) Thermal effects in lubrication of asymmetrical rollers. Trib Intl 24:239–246
6. Lin T-R (1992) Thermal EHD lubrication of rolling/sliding contacts with a power-law fluid. WEAR 54:77–93
7. Dowson D, Higginson GR (1961) New roller bearing lubrication formula. Engineering (London) 192:158
8. Cheng HS (1965) A refined solution to the thermal EHD lubrication of rolling and sliding cylinders. ASLE Trans 8:397–410
9. Hirst W, Moore, AJ (1978) EHD Lubrication of High Pressure. In: Proceedings of Royal Society A, vol 360. London, pp 403–425
10. Chu H-M, Li W-L, Chang YP (2006) Thin film EHL – a power law fluid model". Trib Intl 39:1474–1481
11. Hsiao HS, Hamrock BJ (1994) Temperature distribution and thermal degradation of the lubricant in EHL line contact conjunctions. ASME J Trib 116:794–798
12. Punit K, Khonsari MM (2008) Combined effects of shear thinning and viscous heating on EHL characteristics of rolling/sliding line contacts.J Trib vol 130. 041505-1–041505-13
13. Punit K, Khonsari MM (2009) On the role of lubricant rheology and piezo viscous properties in line- point contact EHL. Trib Intl 42:1522–1530
14. Sinha P, Singh C (1982) Lubrication of cylinder on a plane with a non-newtonian fluid considering cavitation. ASME J Lubr Techn 104:168
15. Punit K, Jain SC, Ray S (2008) Influence of polymeric fluid additives in EHL rolling/sliding line contacts. Trib. Intl 41:482–492
16. Khan H, Sinha P, Saxena A (2009) A simple algorithm for thermo-elasto-hydrodynamic lubrication problems. Int J Res Rev Appl Sci 1(3):265
17. Zhu D (2011) Elastohydrodynamic Lubrication: A gateway to interfacial mechanics-review and prospectus. J Trib 133:041001–041014
18. Morales-Espejel GE, Wemekamp AW (2008) Ertel-grubin methods in elasto hydrodynamic lubrication – a review. IMechE J Engg Trib 222:15–34
19. Wang Z, Jin X, Keer LM, Wang Q (2012) A numerical approach for analyzing three-dimensional steady-state rolling contact including creep using a fast semi-analytical method. Trib Trans 55:446–457
20. Jang JY, Khonsari MM (2010) Elastohydrodynamic line contact of compressible shear thinning fluids with consideration of the surface roughness. ASME J Trib 132,034501-1
21. Rong-Tsong L, Hamrock BJ (1989) "Squeezing and entraining motion in non-conformal line contacts: part-I-hydrodynamic lubrication. ASME J Trib 111:1–7
22. Wang SH, Hua DY, Zang HH (1988) A full numerical EHL solution for line contacts under pure rolling condition with a non- newtonian rheological model. ASME J Trib 110:583–586
23. Bruyere V, Fillot N, Morales-Espejel GE, Vergne P (2012) Computational fluid dynamics and fully elasticity model for sliding line thermal EHD contact. Tribo Inter 46:3–13
24. Saini PK, Kumar P, Tandon P (2007) Thermal elastohydrodynamic lubrication characteristics of couple stress fluids in rolling/sliding line contacts. IMechE, J.Engg. Trib vol 221. pp.141–153

25. Liu X, Cul J, Yang P (2012) Size effect on the behavior of thermal EHL of roller pairs. J. Trib vol 134. pp 011502-1–011502-10
26. Sadeghi F, Dow TA (1987) Thermal effect in rolling/sliding contacts–II: analysis of thermal effects in fluid films. ASME J Trib 109:512–518

Transient Analysis of Plain Circular Bearing with Micropolar Fluid

Rajiv Verma and Puneet Mathur

Abstract The paper presents the static and dynamic performance characteristics in conjunction with linear and nonlinear trajectories for a plain circular journal bearing operating with newtonian and micropolar fluid. A modified form of Reynolds equation is used which is solved by the finite difference technique. The static and dynamic performance characteristics includes evaluation of bearing load, stiffness coefficients, damping coefficients, critical mass, threshold speed and whirl frequency ratio for different operating conditions. Linear and nonlinear motion trajectories are obtained by solving the equations of motion using the Runge–Kutta fourth order method. It is concluded that the nonlinear trajectory becomes unstable even when its mass is equal to the critical mass as obtained from the linear study for both newtonian and micropolar fluid.

Keywords Circular bearing · Micropolar fluid · Static and dynamic characteristics · Linear and nonlinear trajectories

1 Introduction

The classical theory of hydrodynamic lubrication assumes that the lubricant behaves as a newtonian viscous fluid. However, characteristics of lubricants are often modified using additives in order to meet the specific requirements of many engineering applications. Polymer thickened oils behave as pseudo-plastic or dilatants fluids. The viscosities of these lubricants follow nonlinear relationships between shear stresses and shear strain rates. With respect to micropolar lubrication theory, Eringen [1] observes that the motion of colloidal fluids is not

R. Verma · P. Mathur (✉)
Department of Mechanical Engineering, NIT-Kurukshetra,
Kurukshetra 136119, Haryana, India
e-mail: puneet1802@gmail.com

H. C. Patel et al. (eds.), *Proceedings of International Conference on Advances in Tribology and Engineering Systems*, Lecture Notes in Mechanical Engineering,
DOI: 10.1007/978-81-322-1656-8_12,

governed by basic equations of newtonian fluid. Therefore, the equations of momentum and energy conservation were modified for such fluids. Allen and Kline [2] discussed the theory of lubrication with a micropolar fluid and presented solution for an inclined slider bearing. The analysis shows that for a micropolar fluid, load capacity and frictional force were increased. Zaheruddin and Isa [3] provided the expressions for different hydrodynamic characteristics such as the load carrying capacity, the volume flow flux and the frictional force for one dimensional journal bearings, both infinitely long and infinitely short with micropolar lubrication. An enhanced performance of bearing was observed for one-dimensional squeeze film journal bearings for cyclic load [4]. Similar enhancement in terms of load capacity, frictional moment and frictional coefficient were observed in micropolar lubrication for long [5] and short bearing [6]. Huang and Weng [7] obtained the dynamic characteristics of finite width journal bearing using linear stability theory. The effectiveness of micropolar fluids in a dynamically loaded journal bearing was studied by Wang and Zhu [8]. The analysis showed that the effects depend on the size of material characteristic length and the coupling number. In addition to this, Guha et al. [9] applied the first order perturbation of the film thickness and obtained steady state film pressure, the performance characteristics, critical mass parameter and whirl ratio for varying eccentricity and slenderness ratios. Jain et al. [10] determined the journal motion trajectories using nonlinear equations of motion for flexible bearing shell. Sinhasan and Goyal [11] observed that the nonlinear trajectories for two-lobe journal bearing system becomes unstable even at values of non-dimensional journal mass less than the critical journal mass computed using the linearized equations of motion and Routh-Hurwitz stability criteria.

The linear and nonlinear stability analysis has been carried out in the present paper for a plain circular bearing using micropolar fluid. Linear stability analysis is used to evaluate the critical mass parameter and is utilize to predict the critical mass for nonlinear transient analysis.

2 Analysis

General form of governing equation for a micropolar fluid is obtained as:

Conservation of mass for an incompressible fluid is shown below:

$$\nabla.\mathrm{V} = 0 \tag{1}$$

Conservation of linear momentum for an incompressible fluid

$$(\lambda + 2\mu)\nabla(\nabla.\mathrm{V}) - \frac{(2\mu + \mathcal{X})}{2}\nabla * (\nabla * \mathrm{V}) + \mathcal{X}\nabla * \upsilon - \nabla.\pi + \mathrm{F_B} = \rho\frac{\mathrm{DV}}{\mathrm{Dt}} \tag{2}$$

Conservation of angular momentum for an incompressible fluid

$$(\alpha+\beta+\gamma)\nabla(\nabla.\upsilon)-\gamma\nabla*(\nabla*\upsilon)+\mathcal{X}\nabla*V-2\mathcal{X}\upsilon+C_B=\rho j\frac{D\upsilon}{Dt} \quad (3)$$

where, ρ is the mass density, V is the velocity vector, υ is the micro-rotational velocity vector. π is the thermodynamic pressure and is to be replaced by the hydrodynamic film pressure, ρ, since, $\pi=-\left[\frac{\partial E}{\partial \rho^{-1}}\right]=P$. E is the internal energy and P is to be determined by the boundary conditions. λ and μ are the familiar viscosity coefficients of the classical fluid mechanics, while α, β and γ are the new viscosity coefficients derived as the combinational effects of the gyroviscosities for the micropolar fluid as defined by Eringen [1]. $\mathcal{X}$ is also a new viscosity coefficient for micropolar fluid, termed as spin viscosity, which establishes the link between velocity vector and the microrotational velocity vector. F_B is the body force per unit mass, C_B is the body couple per unit mass and j is the microinertia constant. D/Dt represents the material derivative.

Now consider the flow of thin lubricant layer in journal bearing and assume the components of velocity vector and microrotation vector have the following form:

$$V=\left(V_x,V_y,V_z\right)\text{and } \upsilon=(\upsilon_x,0,\upsilon_z)$$

Using the usual postulations for lubricant theory [13] and ignoring infinitesimal quantities with high order we obtain the following governing differential equations for lubricant flow:

$$\frac{(2\mu+\mathcal{X})}{2}\frac{\delta^2 V_x}{\delta y^2}+\mathcal{X}\frac{\delta \upsilon_3}{\delta y}-\frac{\delta p}{\delta x}=0 \quad (4a)$$

$$\frac{(2\mu+\mathcal{X})}{2}\frac{\delta^2 V_z}{\delta y^2}-\mathcal{X}\frac{\delta \upsilon_1}{\delta z}-\frac{\delta p}{\delta z}=0 \quad (4b)$$

$$\gamma\frac{\delta^2 \upsilon_1}{\delta y^2}+\mathcal{X}\frac{\delta V_z}{\delta y}-2\mathcal{X}\upsilon_1=0 \quad (4c)$$

$$\gamma\frac{\delta^2 \upsilon_3}{\delta y^2}+\mathcal{X}\frac{\delta V_x}{\delta y}-2\mathcal{X}\upsilon_3=0 \quad (4d)$$

For the lubricant layer the boundary conditions are as follows:

$$\left.\begin{array}{r}\text{At } y=0: V_x=U_1, V_y=V_z=0, \upsilon_1=\upsilon_3=0\\ \text{At } y=h: V_x=U_2, V_y=\frac{\delta h}{\delta t}+U_2\frac{\delta h}{\delta x}, V_z=0, \upsilon_1=\upsilon_3=0\end{array}\right\} \quad (5)$$

where h represents the thickness of oil film, U_1, the tangent velocity of surface one, U_2 and V_2, the tangent and normal velocity of surface two respectively.

Velocity components can be obtained by solving Eqs. (4a–d). Substituting all these velocity components into Eq. (1) and integrating across the film thickness. Modified Reynolds equation for a micropolar lubricant is obtained as:

$$\frac{\delta}{\delta x}\left[\Phi(N,l_m,h)\frac{\delta p}{\delta x}\right]+\frac{\delta}{\delta z}\left[\Phi(N,l_m,h)\frac{\delta p}{\delta z}\right]=6U\mu\frac{\delta h}{\delta x}+12\mu\frac{\delta h}{\delta t} \tag{6}$$

where,

$$\Phi(N,l_m,h)=h^3+12l_m^2h-6Nl_mh^2coth\left(\frac{Nh}{2l_m}\right)$$

and $N=\sqrt{\frac{\mathcal{X}}{2\mu+\mathcal{X}}}$, $l_m=\sqrt{\frac{\gamma}{4\mu}}$

N^2 and l_m are two important parameters distinguishing a micropolar lubricant from a newtonian lubricant. N^2 is a dimensionless parameter called the coupling number, which couples the linear and angular momentum equations arising from the micro-rotational effects of the suspended particles in the lubricant. l_m represents the interaction between the micropolar lubricant and the film gap and is termed as the characteristic length of the micropolar lubricant. Non-Dimensional form of modified Reynolds equation is given as:

$$\frac{\delta}{\delta\theta}\left(\Phi(N,\overline{l_m},\bar{h})\frac{\delta\bar{p}}{\delta\theta}\right)+\frac{\delta}{\delta z}\left(\Phi(N,\overline{l_m},\bar{h})\frac{\delta\bar{p}}{\delta\bar{z}}\right)=6\frac{\delta\bar{h}}{\delta\theta}+12\frac{\delta\bar{h}}{\delta\bar{t}} \tag{7}$$

where, $\theta=\frac{x}{R};\bar{h}=\frac{h}{c};\bar{z}=\frac{z}{R};\bar{p}=\frac{pc^2}{\mu UR};\ \overline{l_m}=\frac{c}{l_m};U=\omega R;\ \bar{t}=\omega t$

$$\Phi(N,\overline{l_m},\bar{h})=\bar{h}^3+\frac{12\bar{h}}{\overline{l_m}^2}-\frac{6N\bar{h}^2}{\overline{l_m}}coth\left(\frac{\bar{h}N\overline{l_m}}{2}\right)$$

3 Boundary Conditions

The following boundary conditions, pertinent to the flow field, are defined in terms of pressure and pressure gradient:

$$\bar{p}=0 \quad \text{at} \quad \theta=0,\theta_2$$

$$\frac{\delta\overline{p}}{\delta\theta}=0 \quad \text{at} \quad \theta=\theta_2$$

$$\bar{p}=0 \quad \text{at} \quad \bar{z}=0,2$$

where θ_2 is the location of the trailing edge.

4 Film Thickness and its Gradient

For a plain circular bearing, the film thickness and its gradient is given by

$$\bar{h} = 1 - \overline{X_j}\cos\theta - \overline{Y_j}\sin\theta;$$

$$\bar{\dot{h}} = -\overline{\dot{X}_j}\cos\theta - \overline{\dot{Y}_j}\sin\theta;$$

where $\overline{X_j}$ and $\overline{Y_j}$ indicates the centre position of the journal. $\overline{\dot{X}_j}$ and $\overline{\dot{Y}_j}$ indicates the rate of change of its position.

5 Static Characteristics

5.1 Load Carrying Capacity

Bearing forces are calculated by using:

$$\left.\begin{aligned} \overline{F_x} &= \int_0^2\int_0^{\theta_2} \bar{p}\cos\theta \, d\theta \, d\bar{z} \\ \overline{F_y} &= \int_0^2\int_0^{\theta_2} \bar{p}\sin\theta \, d\theta \, d\bar{z} \end{aligned}\right\} \tag{8}$$

Resultant load is given as:

$$\bar{F} = \sqrt{\overline{F_x}^2 + \overline{F_Y}^2}$$

where $\overline{F_x}$ and $\overline{F_Y}$ are radial and tangential components of dimensionless hydrodynamic force.

5.2 Attitude Angle

The attitude angle is defined as the angle between the line of action of external force and line of centres.

6 Dynamic Characteristics

6.1 Stiffness Characteristics

The non-dimensional coefficients of stiffness are given as:

$$\begin{bmatrix} \overline{K_{xx}} & \overline{K_{xy}} \\ \overline{K_{yx}} & \overline{K_{yy}} \end{bmatrix} = \begin{bmatrix} -\frac{\delta\overline{F_x}}{\delta\overline{X}} & -\frac{\delta\overline{F_x}}{\delta\overline{y}} \\ -\frac{\delta\overline{F_y}}{\delta\overline{X}} & -\frac{\delta\overline{F_y}}{\delta\overline{y}} \end{bmatrix} \tag{9}$$

The first subscript denotes the direction of force and the second, the direction of displacement.

6.2 Damping Characteristics

The non-dimensional coefficients of damping are given as:

$$\begin{bmatrix} \overline{C_{xx}} & \overline{C_{xy}} \\ \overline{C_{yx}} & \overline{C_{yy}} \end{bmatrix} = \begin{bmatrix} -\frac{\delta\overline{F_x}}{\delta\dot{\overline{x}}} & -\frac{\delta\overline{F_x}}{\delta\dot{\overline{y}}} \\ -\frac{\delta\overline{F_y}}{\delta\dot{\overline{x}}} & -\frac{\delta\overline{F_y}}{\delta\dot{\overline{y}}} \end{bmatrix} \tag{10}$$

The first subscript denotes the direction of force and the second, the direction of velocity.

7 Stability Analysis

Journal bearing experiences forces when disturbed from its equilibrium position resulting in the journal rotating around its static equilibrium position. The motion trajectories can be obtained by the linear and nonlinear equations of motion of the journal. The linearized equations of motion of a journal centre are:

$$\begin{bmatrix} M_j & 0 \\ 0 & M_j \end{bmatrix} \begin{bmatrix} \ddot{\bar{x}} \\ \ddot{\bar{y}} \end{bmatrix} + \begin{bmatrix} \overline{C_{xx}} & \overline{C_{xy}} \\ \overline{C_{yx}} & \overline{C_{yy}} \end{bmatrix} \begin{bmatrix} \dot{\bar{x}} \\ \dot{\bar{y}} \end{bmatrix} + \begin{bmatrix} \overline{K_{xx}} & \overline{K_{xy}} \\ \overline{K_{yx}} & \overline{K_{yy}} \end{bmatrix} \begin{bmatrix} \bar{x} \\ \bar{y} \end{bmatrix} = \begin{bmatrix} 0 \\ 0 \end{bmatrix} \tag{11}$$

The nonlinear equations of motion of a journal centre are:

$$\begin{bmatrix} M_j & 0 \\ 0 & M_j \end{bmatrix} \begin{bmatrix} \ddot{\bar{x}} \\ \ddot{\bar{y}} \end{bmatrix} = \begin{bmatrix} \Delta\overline{F}_x \\ \Delta\overline{F}_y \end{bmatrix} \tag{12}$$

The assumption of complementary solution to Eq. (11) of the form $\overline{x} = \overline{X}e^{\lambda t}$ and $\overline{y} = \overline{Y}e^{\lambda t}$ leads to following polynomial

$$A_0\lambda^4 + A_1\lambda^3 + A_2\lambda^2 + A_3\lambda + A_4 = 0$$

where $A_0 = \overline{M}_j^2$

$$A_1 = \overline{M}_j\left(\overline{C_{xx}} + \overline{C_{yy}}\right)$$

$$A_2 = \overline{M}_j * \overline{K_{xx}} + \overline{M}_j * \overline{K_{yy}} + \overline{C_{xx}} * \overline{C_{yy}} - \overline{C_{yx}} * \overline{C_{xy}}$$

$$A_3 = \overline{K_{xx}} * \overline{C_{yy}} + \overline{C_{xx}} * \overline{K_{yy}} - \overline{K_{yx}} * \overline{C_{xy}} - \overline{K_{xy}} * \overline{C_{yx}}$$

$$A_4 = \overline{K_{xx}} * \overline{K_{yy}} - \overline{K_{yx}} * \overline{K_{xy}}$$

Making the use of the polynomial equation into Routh-Hurwitz stability criteria, the stability periphery of the journal bearing system is obtained.

7.1 Critical Mass

The critical mass is evaluated by using the stiffness and damping coefficients incorporated with Routh-Hurwitz stability criteria in following equation:

$$\overline{M}_c = \frac{a_0}{b_0 - c_0} \tag{13}$$

where, $a_0 = \overline{C_{xx}} * \overline{C_{yy}} - \overline{C_{yx}} * \overline{C_{xy}}$

$$b_0 = \frac{\left(\overline{C_{xx}} + \overline{C_{yy}}\right) * \left(\overline{K_{xx}} * \overline{K_{yy}} - \overline{K_{yx}} * \overline{K_{xy}}\right)}{\overline{K_{xx}} * \overline{C_{yy}} + \overline{C_{XX}} * \overline{K_{yy}} - \overline{K_{yx}} * \overline{C_{xy}} - \overline{K_{xy}} * \overline{C_{yx}}}$$

$$c_0 = \frac{\overline{K_{xx}} * \overline{C_{xx}} + \overline{C_{yy}} * \overline{K_{yy}} + \overline{K_{yx}} * \overline{C_{xy}} + \overline{K_{xy}} * \overline{C_{yx}}}{\overline{C_{xx}} + \overline{C_{yy}}}$$

A journal bearing system is stable when its mass ($\overline{M}_j$) is less than the critical mass ($\overline{M}_c$). For any negative value of $\overline{M}_c$, the journal bearing system will always be stable.

7.2 Whirl Frequency Ratio

It is the ratio of the rotor whirl or precessional frequency to the rotor onset speed at instability.

$$\overline{\varpi} = \sqrt{\frac{\left(\overline{K_{xx}} - \overline{K_{eq}}\right) * \left(\overline{K_{yy}} - \overline{K_{eq}}\right) - \overline{K_{yx}} * \overline{K_{xy}}}{\overline{C_{xx}} * \overline{C_{yy}} - \overline{C_{yx}} * \overline{C_{xy}}}} \tag{14}$$

where,

$$\overline{K_{eq}} = \frac{\overline{K_{xx}} * \overline{C_{yy}} + \overline{C_{xx}} * \overline{K_{yy}} - \overline{K_{yx}} * \overline{C_{xy}} - \overline{K_{xy}} * \overline{C_{yx}}}{\overline{C_{xx}} + \overline{C_{yy}}} \quad (15)$$

$\overline{K_{eq}} \leq 0$ implies absence of whirl.

7.3 Threshold speed

Threshold speed is the utmost value of speed at which the bearing remains stable. It is obtained using the relation given below

$$\overline{\Omega_s} = \sqrt{\frac{\overline{F} * K_{eq}}{M_c}} \quad (16)$$

8 Solution Procedure

8.1 Linear Analysis

A computer programme is developed to solve the modified Reynolds equation for an assumed eccentricity ratio and characteristic length ($\overline{l_m}$) and coupling number (N^2). The modified Reynolds Eq. (7) is discretized by using FDM and the solution obtained for the boundary conditions mentioned in Sect. 3. Static forces along radial and tangential directions are calculated in Eqs. (8–10) are used to obtain the stiffness and damping coefficients and are used to get the critical mass, threshold speed and whirl frequency ratio Eqs. (13–16). Linear motion trajectories of the journal centre are obtained by solving Eq. (11) using the fourth-order Runge–Kutta method. Initial values of displacements and velocities of the journal centre are $\overline{x} = 0.001, \overline{y} = 0, \dot{\overline{x}} = 0, \dot{\overline{y}} = 0$ and time step = 0.001 is employed to obtain the journal centre motion trajectories.

8.2 Nonlinear Analysis

Nonlinear motion trajectories of the journal centre are obtained by solving Eq. (12) using fourth order Runge-Kutta method. After each time step the fluid film thickness is modified using the computed values of horizontal and vertical components of journal centre. The current velocities are used for calculating the components of the fluid film forces. The procedure is repeated for each time step (Fig. 1).

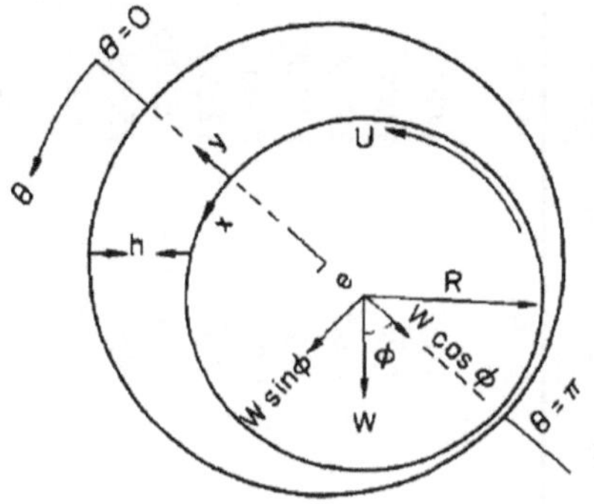

Fig. 1 Schematic diagram of circular hydrodynamic journal bearing

9 Results and Discussion

The results from the computer program have been compared with [12] for Newtonian fluid and found to be accurate. Table 1 shows the comparison of performance characteristics for a newtonian and micropolar fluid. The results are for a journal bearing with L/R = 2 with eccentricity = 0.5, $\overline{l_m} = 10$, 40 and $N^2 = 0.1$, 0.3, 0.5. It is observed that as the coupling number (N^2) tends to zero or the characteristic length ($\overline{l_m}$) increases, the properties of the micropolar fluid approach those of a newtonian fluid. On the basis of results obtained we scrutinize that load carrying capacity of plain circular bearing increases for a micropolar fluid as compared to the newtonian lubricant. Bearing load is higher at smaller value of the characteristic length ($\overline{l_m}$) for any value of coupling number (N^2). It is also observed that critical mass ($\overline{M}_c$) and threshold speed ($\overline{\Omega_s}$) increases with increase in coupling number while the whirl frequency ratio decreases with increase in coupling number. A micropolar fluid thus exhibits better stability than a newtonian fluid.

Transient responses of a plain circular journal bearing are presented in Table 2. For a newtonian fluid at $M_j = M_c$, motion of a journal for linear analysis exhibit limit cycle. However for the same mass, nonlinear analysis shows instability of journal. When $M_j = 0.92M_c$, journal motion achieve limit cycle for nonlinear analysis while journal is stable for linear analysis. It has observed that if the mass of journal is taken below to this limit then the journal shows stability for both the analysis. There are several cases presented in Table 2 for micropolar fluid. On the basis of results obtained in Table 2 and to support the above discussion, trajectories are plotted, publicized in Figs. 2, 3, and 4.

Table 1 Comparison of performance characteristics of plain circular journal bearing for eccentricity = 0.5 and L/R = 2

Coupling number (N^2)	Newtonian fluid	Micropolar fluid					
		($\overline{l_m} = 10$)			($\overline{l_m} = 40$)		
	0	0.1	0.3	0.5	0.1	0.3	0.5
Attitude angle(ϕ)	56.875	56.641	55.781	54.688	56.563	56.094	55.859
Load ($\overline{F}$)	3.5581	3.9068	4.6178	5.4207	3.7363	3.9227	4.0633
K_{xx}	6.7467	7.4255	8.9418	10.817	7.1416	7.5829	7.9422
K_{xy}	3.6037	3.9419	4.5646	5.1728	3.7524	3.8873	3.9838
K_{yx}	−11.418	−12.553	−15.015	−18.056	−12.067	−12.813	−13.393
K_{yy}	6.993	7.7472	9.5538	11.943	7.4718	8.0561	8.496
C_{xx}	11.256	12.325	14.527	17.085	11.82	12.411	12.893
$C_{xy} \approx C_{yx}$	−7.3447	−8.0935	−9.8384	−12.068	−7.7995	−8.3062	−8.7215
C_{yy}	21.871	24.133	29.157	35.309	23.153	24.595	25.669
Critical mass ($\overline{M}_c$)	21.915	24.432	29.751	36.389	23.475	25.029	26.23
Threshold speed ($\overline{\Omega}_s$)	4.7022	4.9429	5.4545	6.0323	4.8452	5.0029	5.1215
Whirl frequency ($\overline{\varpi}$)	0.4809	0.4804	0.4785	0.476	0.48027	0.47954	0.479

Table 2 Critical mass and Transient response for a plain circular journal bearing at eccentricity = 0.5 and L/R = 2

N^2	$\overline{l_m}$	$\overline{M_j}$	Transient response	
			Linear analysis	Nonlinear analysis
Newtonian		21.9155	Limit cycle	Unstable
		20.1622	Stable	Limit cycle
0.1	10	24.4319	Limit cycle	Unstable
		22.5506	Stable	Limit cycle
	40	23.475	Limit cycle	Unstable
		21.597	Stable	Limit cycle
0.5	10	36.3888	Limit cycle	Unstable
		32.7499	Stable	Limit cycle
	40	26.230	Limit cycle	Unstable
		23.607	Stable	Limit cycle

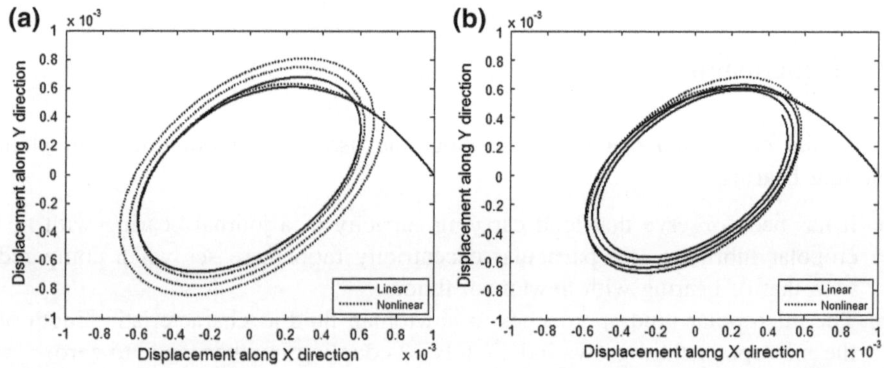

Fig. 2 Linear and nonlinear trajectories for a plain circular bearing with newtonian fluid for eccentricity = 0.5 and L/R = 2. **a** $\overline{M_j} = \overline{M_c} = 21.9155$. **b** $\overline{M_j} = 0.9\ \overline{M_c} = 20.16226$

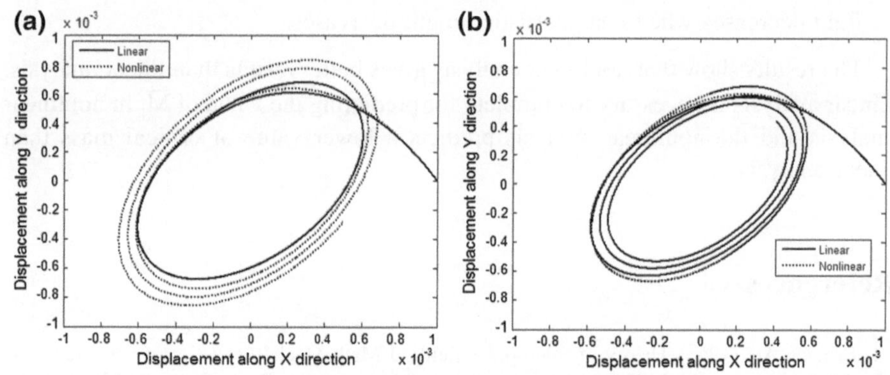

Fig. 3 Linear and nonlinear trajectories for a plain circular bearing with micropolar fluid for eccentricity = 0.5, L/R = 2, $\overline{l_m} = 10$ and $N^2 = 0.1$. **a** $\overline{M_j} = \overline{M_c} = 24.4319$. **b** $\overline{M_j} = 0.9\overline{M_c} = 22.5506$

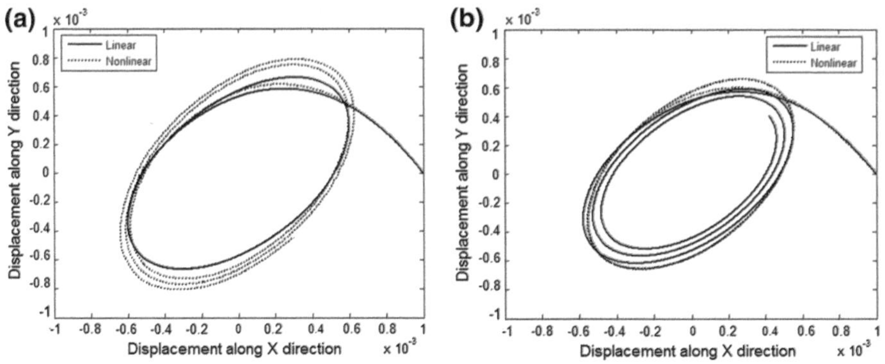

Fig. 4 Linear and nonlinear trajectories for a plain circular bearing with micropolar fluid for eccentricity = 0.5, L/R = 2, $\overline{l_m} = 10$ and $N^2 = 0.5$. **a** $\overline{M_j} = \overline{M_c} = 36.3888$. **b** $\overline{M_j} = 0.9\overline{M_c} = 32.74992$

10 Conclusion

Important conclusion can be made about the use of micropolar fluid in plain circular bearing

a. It has been observe that load carrying capacity of a journal bearing with micropolar lubricant at a particular eccentricity ratio increases when compared with that of bearing with newtonian fluid.
b. The micropolar fluid approaches to newtonian fluid as characteristic length of the micropolar fluid grows indefinitely or coupling number tends to zero.
c. The critical mass and threshold speed for a circular bearing under micropolar fluid is increases for high coupling number. Hence, stability of a bearing increases for high coupling number.
d. The critical mass and threshold speed for a circular bearing under micropolar fluid decreases when characteristic length decreases.

The results show that nonlinear analysis gives better results than linear analysis. Linear analysis is necessary to complete for predicting the value of $\overline{M_j}$ in nonlinear analysis and the nonlinear analysis predicts a lower value of critical mass than linear analysis.

References

1. Eringen AC (1966) Theory of micropolar fluids. J Math Mech 16(1):1
2. Allen SJ, Kline KA (1971) Lubrication theory for micropolar fluids. Trans ASME 38:646
3. Zaheeruddin Kh, Isa M (1978) Micropolar fluid lubrication of one-dimensional journal bearings. Wear 50:211

4. Isa M, Zaheeruddin Kh (1980–1981) Characteristics of squeeze film porous journal bearings with micropolar lubricant. Trans ASME 6(3):146
5. Tipei N (1979) Lubrication with micropolar liquids and its application to short bearings. J Lubr Technol Trans ASME 101:356
6. Khonsari MM, Brewe DE (1989) On the performance of finite journal bearings lubricated with micropolar fluids, STLE. Tribology Trans 32(2):155
7. Huang TW, Weng CI (1990) Dynamic characteristics of finite width journal bearings with micropolar fluids. Wear 140:24
8. Wang XL, Zhu KQ (2004) A study of the lubricating effectiveness of micropolar fluids in a dynamically loaded journal bearing (T1516). Tribol Int 37:481
9. Das S, Guha SK, Chattopadhyay AK (2005) Linear stability analysis of hydrodynamic journal bearings under micropolar lubrication. Tribol Int 38:500
10. Jain SC, Sinhasan R, Pilli SC (1989) A study on the dynamic response of compliant shell journal bearings. Tribol Trans 32(3):297
11. Sinhasan R, Goyal KC (1995) Transient response of a two-lobe journal bearing lubricated with non-newtonian lubricant. Tribol Int 28(4):233
12. Sinhasan R, Goyal KC (1992) Transient response of a circular journal bearing lubricated with non-newtonian lubricant. Wear 156:385
13. Gan QZ, Ji LZ (1987) Lubrication theory for micropolar fluids and its application to a journal bearing with finite length. Appl Math Mech 8(7):655

Part II
Friction and Wear

Effect of Cryogenic Treatment on Hardness, Microstructure and Wear Behavior of Hot Die Steel Grade AISI-H13

Sanjeev Katoch, Rakesh Sehgal and Vishal Singh

Abstract In this experimental work effect of deep cryogenic treatment (C1, done at −155 °C) on the micro hardness, morphology of microstructure and dry sliding wear behavior of hot die steel grade AISI H13 studied in comparison to conventionally heat treated (A3T) samples of the same grade. Microstructure, Hardness and wear rate of CI and A3T group samples performed SEM analysis, Micro Vicker Hardness and Block-on-Ring dry sliding wear test against the hardened cold work die steel, grade AISI D3, roller (counter-face). Improved morphology of microstructure, micro hardness and higher resistance to wear evidenced in deep cryogenic treated samples.

Keywords Cryogenics treatment · Die steel · Retained austenite · ε-carbide · η-carbide · Lath martensite

1 Introduction

High speed steel (HSS), Hot die steel (HDS), Cold work steel (CWS) and Tungsten carbides (WC) mostly used materials for tools and dies manufacturing. Tools and dies must have the characteristics like hot red hardness, toughness, wear resistance, chemical stability and inertness, thermal stability and good thermal conductivity. To attain these properties in tools and dies materials conventional heat treatment is well-known [1]. AISI H13 is one of grade of Chromium hot die steel

S. Katoch (✉) · V. Singh
Center for Material Science and Engineering, National Institute of Technology, Hamirpur 177005, Himachal Pradesh, India
e-mail: katochsanjeev@gmail.com

R. Sehgal
Department of Mechanical Engineering, National Institute of Technology, Hamirpur 177005, Himachal Pradesh, India

H. C. Patel et al. (eds.), *Proceedings of International Conference on Advances in Tribology and Engineering Systems*, Lecture Notes in Mechanical Engineering, DOI: 10.1007/978-81-322-1656-8_13, © Springer India 2014

family which has these properties and is used for die casting of aluminum, zinc and magnesium casting, forging dies and inserts, punches, piercers, mandrels and shear blade for hot works [2]. To get the better tribological properties of tools and dies various surface treatment techniques like PVD, CVD, Nitriding, Plasma-Nitriding utilized since long. But these Surface treatments only improve the surface mechanical properties of treated materials such as hardness, wear resistance, fatigue and friction [3–5]. Nowadays, one of the latest techniques which are used in the industry to enhance the materials physical and mechanical properties is cryogenic treatment. In contrast to surface treatment in cryogenic treatment bulk properties of the materials as well as surface properties affected hence its effects are permanent [6]. Last two decades several researchers cited the usefulness of cryogenic treatment on the modification of the morphology of microstructure and enhancement of wear resistance of tools and die materials [7–16]. The morphology of microstructure and resistance to wear of tools and dies are crucial stricture in production and quality control of components [17–19]. In the present work attempt has been made to study the effect of cryogenic treatment on the hardness, microstructure and dry sliding wear behavior of hot die steel, grade AISI-H13 against hardened cold work steel, grade AISI-D3 roller (counter face). The wear performance of the cryogenic treated and conventional treated AISI-H13 has been compared under identical test conditions.

2 Material and Methodology

Experiments were conducted on commercially available hot die steel, grade AISI-H13. Material was received in the shape of rolled, spheroidized annealed bar of 16 mm × 20 numbers, manufactured with a conventional metallurgical route (EAF-LRF-VD). Chemical composition of the received material conformed to the optical spark emission Spectrometer (Model: DV6, Make: Baird, USA), and is presented in Table 1.

2.1 Research Methodology

The received bar was machined to block of 6.35 × 6.35 × 9 mm for the purpose of dry sliding wear and metallographic study. The samples were divided into two groups namely A3T (Conventional heat treated) and C1 (Cryogenic treated

Table 1 Result of chemical analysis of AISI H13 raw material in weight %

Elements	C	Si	Mn	P	S	Ni	Cr	Mo	V
Observed	0.40	0.86	0.36	0.018	0.007	0.04	5.05	1.30	0.98

Table 2 Detail of heat treatment sequence followed for hot die steel H13

S. No.	Nomenclature	Depiction of treatment
1.	A 3T	VFA = 1040 °C, ST = 30 min, nitrogen gas quench, quench pressure = 5 bar, Three T = 550, 580, 620 °C respectively for 2 h .
2.	A T C1(6)T	VFA = 1040 °C, ST = 30 min, nitrogen gas quench, quench pressure = 5 bar, T = 550 °C for 2 h, C1, ST = 6 h, T = 620 °C for 2 h.

Abbreviation
VFA vacuum furnace austenization, *ST* soak time, *T* tempering, *C1* cryogenic treatment at −154 °C, number in parentheses shows the soaking times in hours at cryogenic temperature

at −155 °C for 6 h). Both groups contained 18 samples. All the samples were vacuum heat treated in a front loading, horizontal electrical heated vacuum furnace, (Make HHV, Bangalore, India, Capacity: 600 × 600 × 900 mm) at a austenization temperature 1040 °C, Vacuum maintained during the heating cycle was 10^{-2} mbar, soaked for 30 min, nitrogen gas quench to 27 °C at a gas pressure of 5 bar. Samples were taken out from the furnace for the subsequent treatment. A3T group specimen were tempered at 550, 580 and 620 °C respectively for 2 h, C1 group specimen were deep cryogenically treated at −155 °C, for soak time of 6 h with pre and post tempering at 550 and 620 °C respectively for 2 h. Detailed sample treatment conditions and nomenclature are given in Table 2.

The cryogenic treatment was performed in a computer controlled cryogenic processor with tempering facility up to 200 °C (Make: Primero Enserve, Chennai, India). To keep the stresses at minimum level and to avoid thermal soaking of the material due the abrupt temperature gradient of case and core of material following steps were taken.

Step 1: The slow cooling rate (1 °C/Min) from ambient to deep cryogenic treatment soak temperature and a slow heating rate (1 °C/Min) from cryo treatment soak temperature to ambient temperature was selected.

Step 2: Deep cryogenic treatment was done in stage cooling; at every stage 30 min soak time was given to maintain the case and core temperature rate.

3 Experimental Procedure

3.1 Microstructural Analysis

Samples for micro structural analysis were prepared as per ASTM standards designation E3-01 (Reapproved 2007) [20]. Specimen molds were prepared with cold setting resins, grinding of molds was done with silicon carbide emery paper of grit size 120, 200, 600, 800, 1000, 1600, 2000, 3000 and fine polished using diamond slurry, particle size 6 μm, with white kerosene oil as a suspension media on rotating velvet cloth. Polished specimen were then etched with 3 % Nital (97 % Nitric Acid + 3 % Ethanol) and dried in hot air. Micro structural features of A3T and C1 groups specimens were studied under the scanning electron microscope model: Quanta FEG450 make: FEI, Holland.

3.2 Micro Indentation Hardness Test

This test was performed on Micro Vicker hardness tester, model: MVK-H2, make: Akashi, Japan, as per ASTM standards designation E384-08a [21]. The hardness test was performed with a 1000 gf (9.8 N) load with dwell time of 15 s, on polished un-etched specimen. To evaluate any change in hardness value, five hardness readings were taken at different points to estimate the average value of hardness for each of the samples. The Micro Vicker hardness number obtained for the A3T and C1 groups samples as a function of treatment condition plotted as a Box-Plot and mean value comparison is given in Fig. 3.

3.3 Wear Test

Block-on-ring dry sliding wear tests were performed to assess the wear rate as per the ASTM standards designation G77-05 (Reapproved 2010) [22] using a Multi Tribo Tester, model: TR-25-M4, make: DUCOM Instrument Pvt. Ltd. Bangalore, India. To attain the surface roughness in the range of 0.152–0.305 μm, faces of the blocks were grounded manually using silicon carbide emery paper of grit size 100, 220, 800, 1000, 1200, 1400 and fine polished using diamond slurry of grit size 6 μm with white kerosene oil as a suspension media on rotating velvet cloth. Specimens were cleaned in hexane prior to wear test. These rectangular blocks

Fig. 1 SEM image of conventional heat treated (A3T) sample of AISI-H13 sample

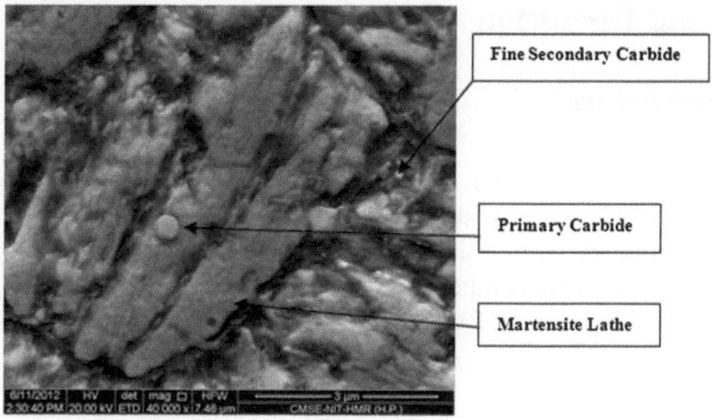

Fig. 2 SEM image of cryogenic treated (C1) sample of AISI-H13 sample

were used as static block against the hardened cold work die steel, grade AISI D3 roller (Ø60 × 20 mm) having hardness 52 HRc and Ra < 0.2 μm as rotating counter face. The gap between static block sample and counter face of roller was maintained 2 mm for each experiment. After each test roller was polished using silicon carbide emery paper of grit size 100, 400 and 1000 and then cleaned with hexane solution. Before and after each experiment weight of specimens were taken using a precision electronic analytical balance (Model: HM-200, Make: A and D) with an accuracy of 0.00001 gm to calculate weight loss. The wear tests were carried out using normal loads of 60, 80, 100, 120, 140 N at sliding velocity of 0.628 m/s, 0.942 m/s, 1.257 m/s, 1.571 m/s and 1.885 m/s respectively. The wear rate of each block was calculated from the weight loss and each test was repeated three times for each condition to get the average value of weight loss. A comparison of cryogenic treatment with respect to conventional heat treatment has been made to identify its effects on dry sliding wear behavior.

Fig. 3 Box-plot and mean value comparison of hardness data of conventional heat treated and cryogenic treated groups

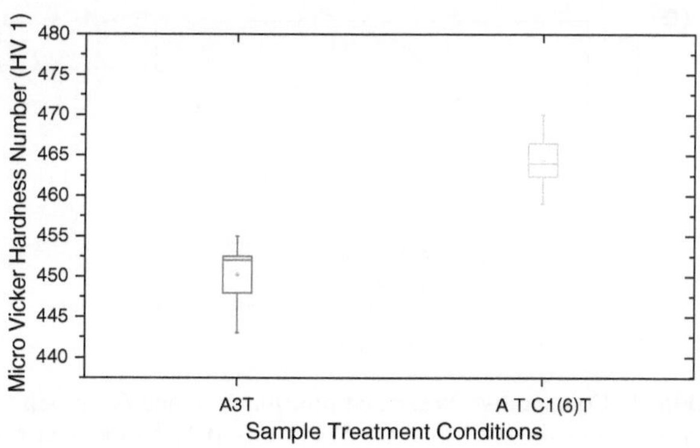

4 Result and Discussion

4.1 Microstructure

SEM Micrograph of A3T and C1 groups samples are shown in Figs. 1 and 2 respectively. Figure 2 clearly shows the more density and even distribution of precipitated secondary carbide in comparison of (A3T) sample shown in Fig. 1. This Result suggests that number of secondary carbide increases after the cryogenic treatment. The results of this study is in concurrence with previous studies

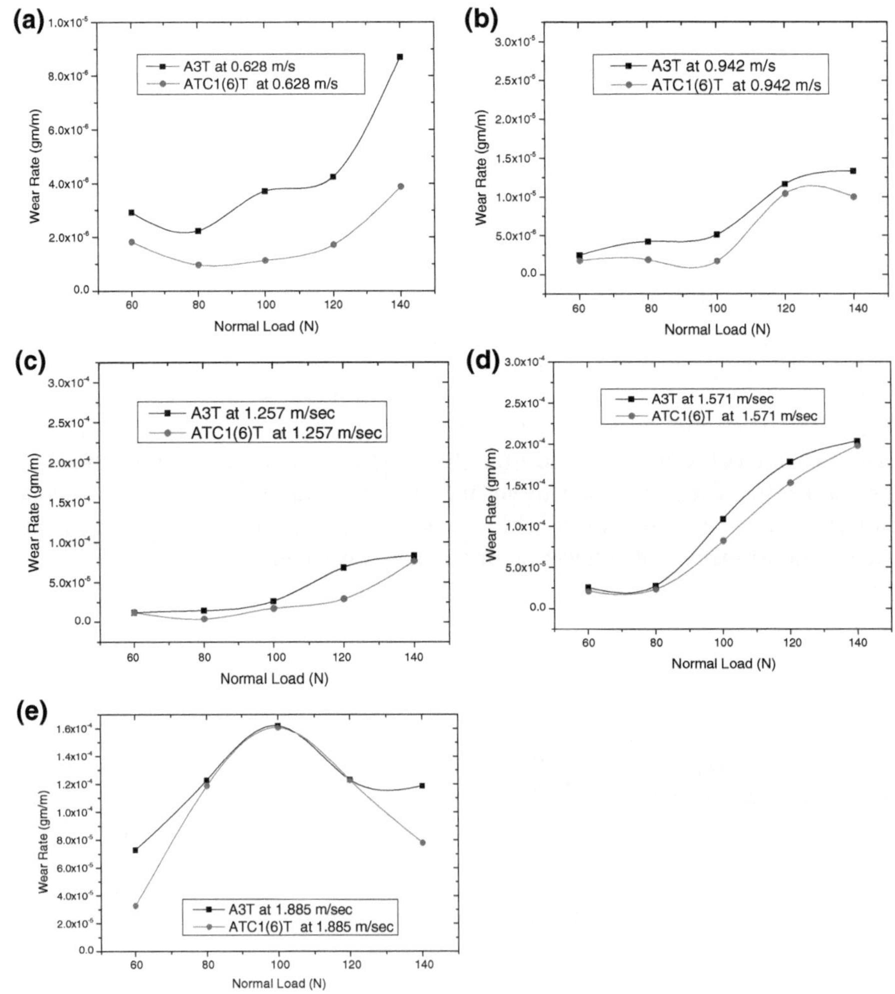

Fig. 4 Comparative Wear rate plot for A3T and C1 group samples: effect of sliding velocity, **a** 0.628 m/s, **b** 0.942 m/s, **c** 1.57141 m/s, **d** 1.257 m/s and **e** 1.885 m/s

that reported the enhancement of number of secondary carbides, which is responsible for the strengthening of matrix, load bearing capacity and wear resistance of tools and dies [6–8, 23].

4.2 Micro Indentation Hardness

The Micro Vicker hardness numbers obtained for the A3T and C1 groups samples are shown in Box-Plot and mean value comparison graph—Fig. 3. It indicates that there is marginal increase in the mean hardness value (i.e. 14.2 HV_1) of the C1 group samples. Obtained results infer that application of cryogenic treatment enhanced the hardness value of specimen. Resistance of material to wear and load bearing capacity increases with the hardness number [8].

4.3 Wear Test

The result obtained for the dry sliding wear experiments for both treatment conditions (A3T and C1 groups) are shown in Fig. 4a–e as wear rate versus normal load. Figure 4a–e reveals that wear resistance of the cryogenic treated specimen is significantly higher in comparison to the conventional heat treated specimen at all levels of normal loads and sliding velocities.

5 Conclusion

The study concluded that the deep cryogenic treatment marginal improve the micro hardness number and better dry sliding wear behavior of the hot die steel grade AISI H13 in comparison to the conventional heat treatment normally used for tools and dies for this grade in the industries. The augmentation of micro Vicker hardness number and wear resistance are ascribed to the increase in number of secondary carbide after the cryogenic treatment.

References

1. Totten GE (2006) Steel heat treatment handbook (metallurgy and techniques). Taylor and Francis, New York
2. Roberts G (1998) Tools steels, 5th edn. ASM International, Materials Park
3. Holmberg K, Matthews A (1998) Coating tribology: properties, techniques and applications in surface engineering. Elsevier Science BV, Amsterdam

4. Leite MV, Figueroa CA, Gallo SC, Rovani AC, Basso RLO, Mei PR, Baumvol IJR, Sinatora A (2010) Wear mechanisms and microstructure of pulsed plasma nitrided AISI H13 tool steel. Wear 269:466–472
5. Jawing XJ, Peng Q, Fan H, Wang Y, Li G, Shen B (2009) Effects of DC plasma nitriding parameter on microstructure and properties of 304L stainless steel. Mater Char 60:197–203
6. Lal DM, Renganarayanan S, Kalanidhi A (2001) Cryogenic treatment to augment wears resistance of tool and die steel. Cryogenics 41:149–155
7. Collins DN (1998) The cryogenic treatment of tool steels. Adv Mater Process 12:23–29
8. Molinari A, Pellizzari M, Gialanella S, Straffelini G, Stiasny KH (2001) Effect of deep cryogenic treatment on the mechanical properties of tool steels. J Mater Process Technol 118:350–355
9. Das D, Dutta AK, Ray KK (2010) Sub-Zero treatments of AISI D2 steel: Part I. Microstructure and hardness. Mater Sci Eng A 527:2182–2193
10. Straffenlini G, Bizzotto G, Zanon V (2010) Improving the wear resistance of tools for stamping. Wear 269:693–697
11. Koneshlou M, Meshinchi K, Khomamizadeh F (2011) Effect of cryogenic treatment on microstructure, mechanical and wear behaviors of AISI H13 hot work tool steel. Cryogenics 51:55–61
12. Amini K, Nategh S, Shafyei A (2010) Influence of different cryo-treatments on tribological behavior of 80CrMo12 5 cold work tool steel. Mater Des 31:4666–4675
13. Dhokey NB, Nirbhavne S (2009) Dry sliding wear of cryo treated multiple tempered D-3 tool steel. J Mater Process Technol 209:1484–1490
14. Das D, Dutta AK, Ray KK (2009) Influence of varied cryo-treatment on the wear behavior of AISI D2 steel. Wear 266:297–309
15. Mehtedi ME, Ricci P, Drudi L, Mohtadi SE, Cabibbo M, Spigarelli S (2012) Analysis of the effect of deep cryogenic treatment on the hardness and microstructure of X30 CrMoN 15 1 steel. Mater Des 33:136–144
16. Gill SS, Singh R, Singh H, Singh J (2009) Wear behavior of cryogenically treated tungsten carbide inserts under dry and wet turning conditions. Int J Mach Tools Manuf 49:256–260
17. Ho LJ, Hwan JJ, Don JB, Myung SY, Hoon MY (2009) Laser surface hardening of AISI H13 tool steel. Trans Nonferrous Met Soc China 19:917–920
18. Chen ZW, Fraser DT, Jahedi MZ (1999) Structure and intermetallic phases formed during immersion of H13 tool steel in Al-Si-3Cu die casting alloy melt. Mater Sci Eng A 260:188–196
19. Khan TI, Rizvi SA, Matsuura K (2000) The effect of wear behavior of H13 tool steel surface modified using a tungsten arc heat source. Wear 244:154–164
20. ASTM E3-01 (Reapproved 2007) (2009) Standard guide for preparation of metallographic specimens. ASTM annual book of standards, vol 3.01. ASTM International, West Conshohocken
21. ASTM E384-08a (2009) Standard test method for micro indentation hardness of materials. ASTM annual book of standard's, vol 3.01. ASTM International, West Conshohocken
22. ASTM G77-05 (Reapproved 2010) (2010) Standard test method for ranking resistance of materials to sliding wear using block-on-ring wear test. ASTM annual book of standards, vol 3.01. ASTM International, West Conshohocken
23. Fanju M, Kohsuke T, Ryo A, Hidea K (1994) Role of η-carbide precipitation in the wear resistance improvement of Fe-12Cr-Mo-V-1.4C tool steel by the cryogenic treatment. ISIJ Int 34(2):205–210

Experimental Study to Measure Piston Ring Assembly Friction of Multicylinder I.C. Engine (S.I.) on Motorized Engine Test Rig: A Case Study

Atul S. Shah and D. V. Bhatt

Abstract Tribological parameter is more important in reducing friction of an I.C. engine. The frictional losses in Piston Ring Assembly system is approximately 40–50 % of total mechanical losses as reported in the literature. A motorized multicylinder engine test rig is used for experimental work at laboratory scale to measure the PRA friction by the measuring the "Power consumption" under the different operating parameter (speed, lubricant). The experimental results and observations are carried with standard make five different lubricant and the tests were experimented at 600–2,400 rpm range The results shows that lubricant properties and grades of lubricant have a considerable effect on engine performance. The ranking of lubricant is done considering all the parameters and most suitable oil for this particular test rig is predicted.

Keywords Tribology · PRA · Lubricant · Power

1 Introduction

The phenomena considered in tribology are among most fundamental and most common of those encountered by humans in interaction with their largely solid environment. Although the name tribology is new, the constituent parts of tribology encompassing friction and wear are as old as history. The economic aspects of tribology are significant. Investigations by a number of countries arrived at figures of savings of 1.0–1.4 % of the GNPs, obtainable by the application of

A. S. Shah (✉)
Department of Mechanical Engineering, Government Polytechnic Waghai, Waghai 394730, Dist.-Dang, India
e-mail: asshah97@rediffmail.comsonigra_bhaskar@yahoo.com

D. V. Bhatt
Mechanical Engineering Department, S.V. National Institute of Technology, 395007 SURAT, India

H. C. Patel et al. (eds.), *Proceedings of International Conference on Advances in Tribology and Engineering Systems*, Lecture Notes in Mechanical Engineering, DOI: 10.1007/978-81-322-1656-8_14, © Springer India 2014

tribological principles. The reciprocating internal combustion engine accounts for the vast majority of power units used in petrol and diesel engines however it is remarkably reliable and versatile. Global economic development has boost up the automobile market in India for Two/Four wheelers vehicles in last decay with a growth rate of 15–20 %. It is need of present era to have engines with more fuel efficiency and less impact on environment.

The knowledge of triblogical factors are important in design to reduced friction and wear. Technocrats have focused to upgrade technology to improve the efficiency and performance of an automotive I.C. engine. As per the referred literature the mechanical losses in the piston ring assembly (PRA) vary from 18 to 25 % [1] and the major percentage in these mechanical losses is of piston ring assembly (PRA) friction are about 45–51 % [2].There for the tribological factors are important to reduce friction losses, emission level and to improve the fuel economy in an I.C. engine. So in order to improve efficiency of engine, PRA friction is one of the area which has great potential to help in improvement of fuel efficiency of engine. Castleman [3] was first to do calculation of piston ring and cylinder lubrication in 1936. These early studies are based on simple theory of hydrodynamic lubrication. Mufti and Priest [4] have carried out experiments on single cylinder Ricardo Hydra gasoline engine to measure piston assembly friction in both motored and fired conditions. Lubricant inlet temperatures are maintained 24, 40, 60, and 80 °C. Mukesh Gupta and et al. [5] have used of multigrade oil in engine lubrication is being advocated to consumers to minimize viscosity-temperature effect. Bhatt and Mistry [6] have experimented on motorized piston cylinder system with application of different lubricants such as SAE-15, SAE-20, SAE-30 & 2T along with different piston ring geometry at a RPM range from 500 to 1,850.

To reduce frictional losses, various parametric approaches are made particularly at design stage and at experimental level. There are many types of four stroke multi cylinder automotive petrol engines are available in market with fuel efficiency of 10 to 24 km/lt. More than 50 % of market share is cover by 800 to 1,000-cc vehicle and there for it is preferable to select the same engine system for the study. The research study is mainly focused on experimental work. Authors have put efforts to find out PRA friction with different standard make lubricants under different operating parameters on a developed multi cylinder I.C. engine motorized test rig [8]. A set of experiments were carried out to measure the PRA friction in terms of power consumption of multi cylinder I.C. engine system [Table 1, 2, 3, 4, -various specifications of test rig elements]. The experimental observations have taken under different operating variables (i.e. Speed, Lubricants, Table 5).

Table 1 Engine specification

Type	4 Stroke cycle, water cooled
No. cylinder	3
Lubricating system	Wet pump High pressure and splash system
Cylinder bore size	68.505–68.520
Stroke length	72 mm
Piston displacement	796 C.C
Compression ratio	8.7 : 1

Table 2 Motor specification

• 3 ph 440 V induction motor
• 7.5 k W. 2800 rpm

Table 3 VFD drive specifications

Model number	DELTA Electronics Inc.
Applicable motor output (kW)	7.5
Rated output capacity (kVA)	13.4
Rated output current (A)	13.5
Output frequency (Hz)	0.1–400 Hz
Rated input current (A)	13.8
Rated input voltage (V)	3-Phase 500–600 V

Table 4 Multifunctional Watt meter specification

Company name	Phasetrack
Current/transformer ratio	30/1
Category	Digital

Table 5 Lubricants used for experiments

Sr. No	Name of lubricants	Viscosity (cp)	Code
1	Maruti Genuine oil (Servo)	167.2	A
2	Pennzoil	220.5	B
3	Gulf	235	C
4	Valvoline	240	D
5	Castrol GTX	220	E

2 Fabrication of Test Rig

The fabricated test rig, Engine system of 800-cc multi cylinder internal combustion engine with crank mechanism, piston cylinder head, and engine lubrication system, with engine cooling system, without gear box is used to fabricate the test rig. Crank shaft is coupled with induction motor to drive the engine. For varying the speed of test rig the variable frequency drive is used. The multi functional watt meter is used to measure the performance in terms of power consumption [7] (Refer Fig. 1).

3 Experiments Method

The experimental work is carried out on developed multi cylinder I.C. engine motorized test rig under different variables i.e. speed, different lubricants with standard cooling conditions. The test sequences to conduct experiments are as under.

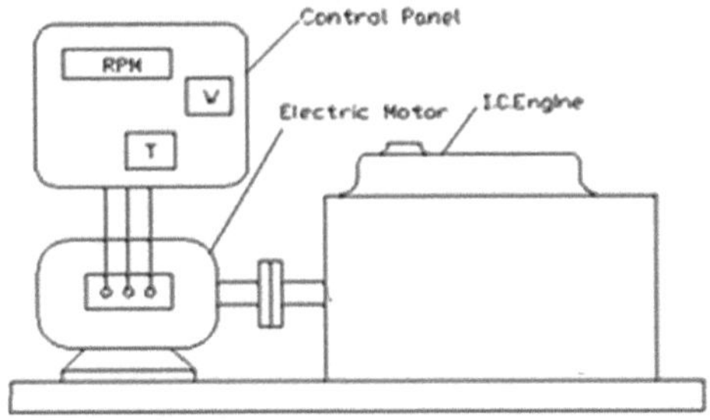

Fig. 1 Block diagram of motorized test rig

1. First of all select coolant and lubricating oil and prepare the engine
2. Check all electrical connection of test rig including VFD & watt meter etc.
3. Switch on the power supply and set the frequency on VFD to required rpm
4. Initially the system is to be run for at least 5–10 min, so that the system get stabilize and the lubricating oil can reach properly up to the surface of piston rings and cylinder liner
5. After getting the stable condition of the system, record the rpm of the system and also the power consumption an engine
6. Now for the next measurement change the frequency on VFD to change the rpm of the system and allow time to stabilize
7. Repeat the cycle in descending order.

4 Results and Observations

The experiments were carried out on developed multi cylinder engine test rig. Experimentally observed readings for power consumption, for different lubricant at different engine speed are recorded. Following key observations are made from the plotted graph

From Fig. 2 it is observed that with oil "A" initially the power consumption is higher, but after 900 rpm it increases with increase in speed of the engine.

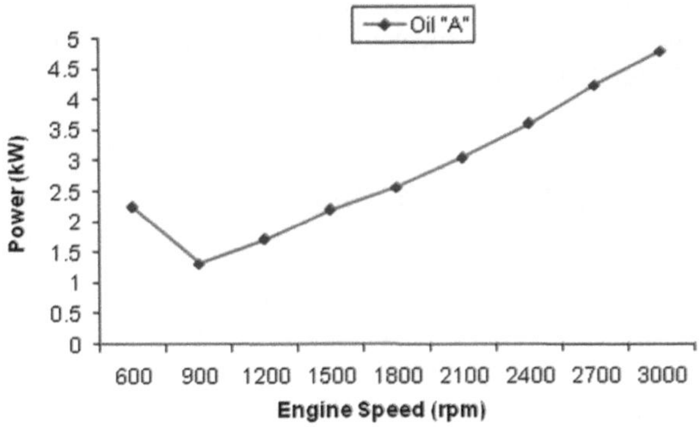

Fig. 2 Engine speed versus power consumption

Fig. 3 Engine speed versus power consumption

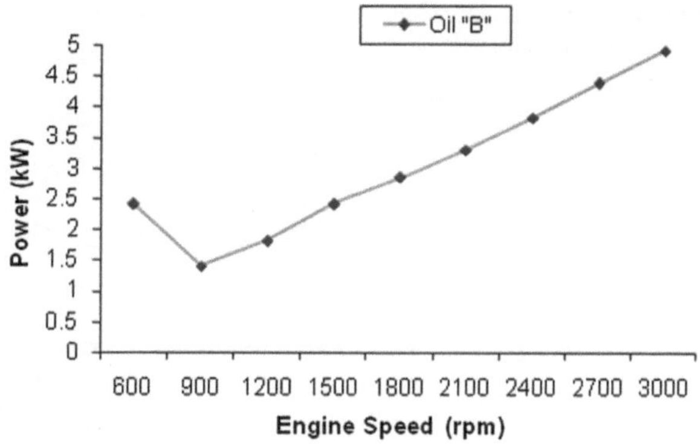

Fig. 4 Engine speed versus power consumption

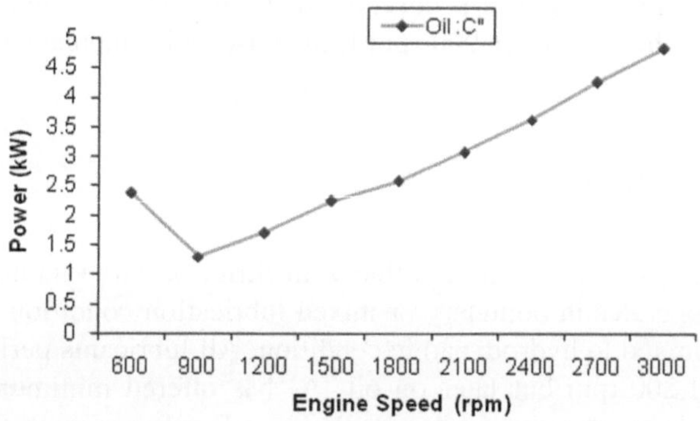

Fig. 5 Engine speed versus power consumption

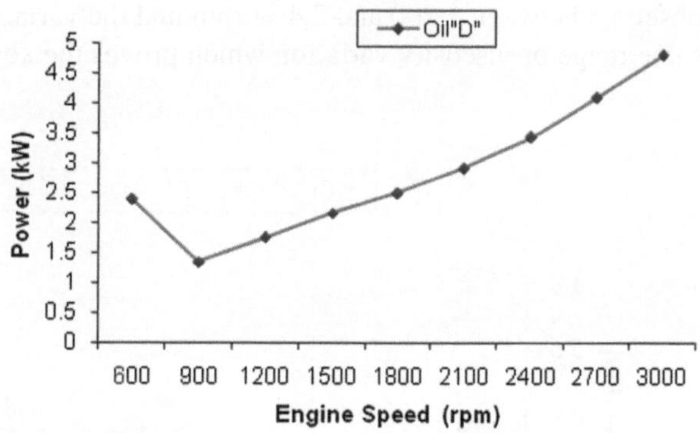

From Fig. 3 it is observed that with oil "B" initially the power consumption is higher, but after 900 rpm it increases with increase in speed of the engine.

From Fig. 4 it is observed that with oil "C" initially the power consumption is higher, but after 900 rpm it increases with increase in speed of the engine.

From Fig. 5 it is observed that with oil "D" initially the power consumption is higher, but after 900 rpm it increases with increase in speed of the engine.

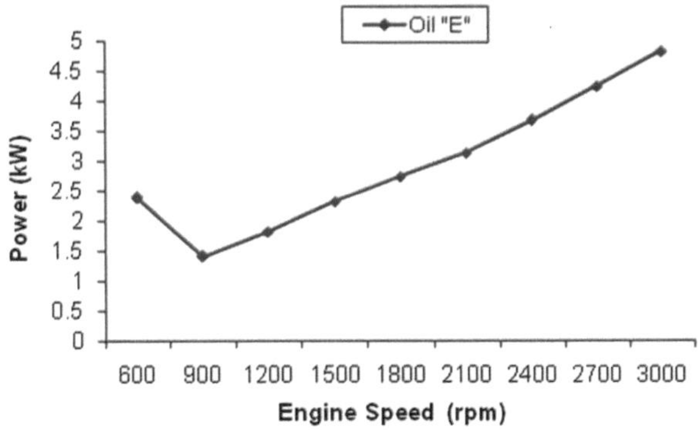

Fig. 6 Engine speed versus power consumption

From Fig. 6 it is observed that with oil "E" initially the power consumption is higher, but after 900 rpm it increases with increase in speed of the engine.

5 Discussion

Nature of the curve inline with Stribeck curve nature means initially the system operates in boundary or mixed lubrication condition and later on (after 900 rpm) mixed to hydrodynamic condition. All lubricants perform more or less equally till 1,200 rpm but later on oil 'A' has offered minimum power consumption at all observed speed while oil 'B' has offered maximum (see Fig. 7). This may due to higher viscosity value at all the observed speed. The effect of different lubricant is observed between 1,200 and 2,400 rpm and the variation is also more or less in the same range of viscosity variation which proves the key role of lubricants viscosity.

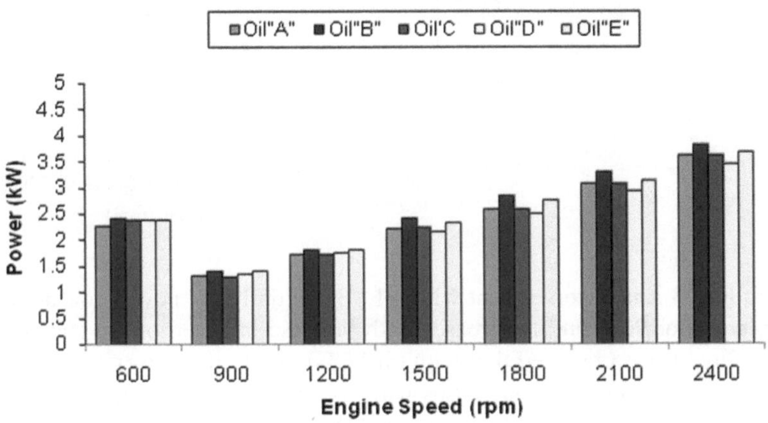

Fig. 7 Engine speed versus power consumption

6 Conclusions

1. The behavior of the different lubricant observed more or less seemlier with respect to engine speed v/s power consumption.
2. Initially the power consumption is reduces till 900 rpm but then it increases with increase in speed of the engine.
3. Same specified lubricants under different brand offers variation in PRA friction.
4. Power consumption variation is observed at all different speed, means that there is a variation in viscosity of lubricants though the technical specification is same for all lubricants. It means that lubricants properties varies with different manufacture.
5. Performance range of lubricants is observed in order of "A-C-D-E-B" at all range of speed experimented.

References

1. Haywood John B (1998) A text book of Internal Combustion Engine Fundamentals (Automotive Technology Series). McGraw-Hill, New York
2. Taylor CM (1998) Automobile engine tribology—design considerations for efficiency and durability. Tribol Int Wear 221:1–8
3. Castleman RA (1936) A hydrodynamic theory of piston ring lubrication. Physics 6:364–367
4. Mufti R, Priest M (2005) experimental evaluation of piston-assembly friction under motored and fired conditions in a gasoline engine. Trans ASME 127:826–836
5. Gupta M, Singhal AS, Biswas AS (1994) Analytical investigation on the effect of multigrade oil in piston ring lubrication. Tribol Trans 37:719–726
6. Bhatt DV (2002) Performance study of tribological parameters of singlecylinder 2 stroke petrol engine. PhD thesis, S. G. University
7. Shah AS (2008) Development of PRA friction measurement test rig for multi cylinder engine system and experimental study of tribological parameters. M.Tech thesis, SVNI
8. Sharma RP (2003) Practical consideration inengine tribology. In: Proceedings of workshop in current trends in I. C. Enginedevelopment, Hydrabad, April 2003

Influence of Surfactant in Hexaferrites as Wear Resistance

N. R. Panchal and R. B. Jotania

Abstract $SrFe_{12}O_{19}$ hexaferrite particles were synthesized by a chemical co-precipitation technique with a precipitator $NH_3{\cdot}H_2O$. To improve the wear resistance property of a material, another set of $SrFe_{12}O_{19}$ hexaferrites was synthesized containing non-ionic surfactant polyoxyethelene (20) sorbitan monolate (Tween-80). The prepared Sr-M hexaferrite precipitates were heat treated at 950 °C for 4 h in a muffle furnace. Both the sets of obtained Sr-M powders were characterized by using various instrumental techniques, like FTIR, XRD, and SEM. It is observed from SEM that surfactant plays significant role in controlling the grain size and size distribution of $SrFe_{12}O_{19}$ hexaferrite.

Keywords Surfactant · Hexaferrite · Wear resistance · Tween-80 · Nano particles

1 Introduction

Nanotechnology and nanomechanics studies are valuable in the fundamental understanding of interfacial phenomena in macrostructures and provide a bridge between science and engineering.

In order to select a material for a particular component, one must consider the materials properties like strength, hardness, toughness etc. Among these, one of the important properties is wear resistance. In the selection process of material, properties of material are determined according to its application where it is to be used.

N. R. Panchal (✉)
Department of Physics, M. G. Science Institute, Ahmedabad 380009, Gujarat, India
e-mail: nital_panchal@yahoo.co.in

R. B. Jotania
Department of Physics, University School of Sciences, Gujarat University, Ahmedabad 380009, Gujarat, India

H. C. Patel et al. (eds.), *Proceedings of International Conference on Advances in Tribology and Engineering Systems*, Lecture Notes in Mechanical Engineering, DOI: 10.1007/978-81-322-1656-8_15,

'Wear' means material loss, degradation of surface and decreasing functionality. Tribology is the science of finding solution related to friction and wear problems.

Exceptional properties of ferrite in the areas of mechanical strength and extreme hardness make these a versatile class of materials. In combination with other properties such as dimensional stability and temperature and corrosion-resistance, both traditional and advanced ferrites can be used in a wide range of wear resistance applications.

For iron oxides, used in tribological applications, adhesion behavior is significantly affected by both structure and surface chemistry. Basic factor affecting wear deformation is influenced by micro structural characteristics such as pores and grains. Amorphous materials with wide particle size distribution undergo deformation that can increase surface roughness and surface friction. Wear increases with increasing grain size. Matters become even more complicated if grain size varies. Material properties can be tailored by manipulation or alteration.

The hexaferrites can be synthesized by different routes including co-precipitation [1], hydrothermal [2], mechanical alloying [3] and sol-gel auto-combustion methods [4]. In present work, hexaferrites are synthesized by co-precipitation method in two sets. First set is prepared via conventional co-precipitation route, which gives non-uniform particles which may produce the deformation as stated above. To resolve this problem, we have introduced a non-ionic surfactant Tween-80 during the reaction in second set. Results show that the grain size and distribution of particles are controlled by addition of surfactant without the phase change. So we can say that hexaferrite provides excellent wear resistance, especially when synthesized with surfactant.

2 Experimental

2.1 Materials

All starting materials are of analytic grade and used without further purification. Strontium Nitrate; $Sr(NO_3)_3$ (99 % pure, Loba Chemie), Ferric Nitrate; $FeN_3O_9{\cdot}9H_2O$ (99 %, Sigma Aldrich), Polyoxyethelene (20) sorbitan monoolate (Tween-80) (95 % pure, S.D. Fine Chem), Ammonia solution (30 %, Merck) as precipitating agent, deionised water to make the solution and distilled water was used throughout the sample preparation and washing the samples.

2.2 Sample Preparation

A stoichiometric amount of $Sr(NO_3)_3$ and $FeN_3O_9{\cdot}9H_2O$ were dissolved in deionised water to prepare the aqueous solution. Precipitation was carried out with NH_4OH having pH of reaction above 10. Solution is stirred throughout the process.

Fig. 1 Method of preparation

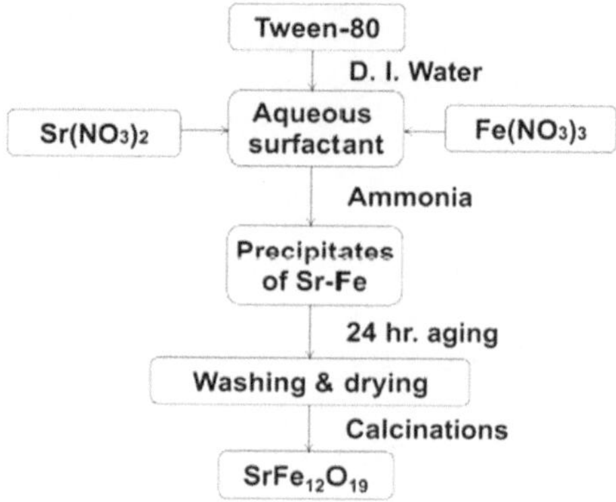

After precipitation, solution is kept for aging for 24 h and then washed several times with methanol: acetone (1:1). It is further washed with distilled water until the washings are free from nitrate ions. The product was dried in furnace at 100 °C overnight to remove water contents and then calcined at 950 °C for 4 h then slowly cooled to room temperature to obtain Sr-M hexaferrite first set.

For second set, 0.5 M Tween-80, a non-ionic surfactant was dissolved in deionised water and stirred rigorously for 1 h to obtain an aqueous surfactant. Stoichiometric proportion of Strontium Nitrate and Ferric Nitrate taken for first set was now dissolved in this aqueous surfactant solution. After this, repeating the procedure as for set-1, we obtained the Sr-M hexaferrites with surfactant. Figure 1 shows the flow chart of preparation of $SrFe_{12}O_{19}$ hexaferrite with use of surfactant Tween-80.

2.3 Characterization

Ferrite formation of the samples was determined from the Fourier Transform Infrared (FTIR) spectroscopic studies. FTIR spectra of both the samples, with and without were carried out between wave number ranges 4000–400 cm^{-1}.

Phase identification and structure analysis is performed on X-ray powder diffractometer (XRD PW 1830) with nickel filtered CuKα radiation ($\lambda = 1.54056$ Å) at room temperature. Particle size, d is calculated from the X-ray line broadening, using the Debye–Scherrer equation [5].

$$d = \frac{k\lambda}{\beta \cos\theta} \tag{1}$$

where, λ is the wavelength of X-ray,

k is the shape factor $\sim$0.89,

β is the full width at half-maximum expressed in units of 2θ, and

θ is the Bragg angle (°).

The particle size and morphology was also confirmed by SEM images.

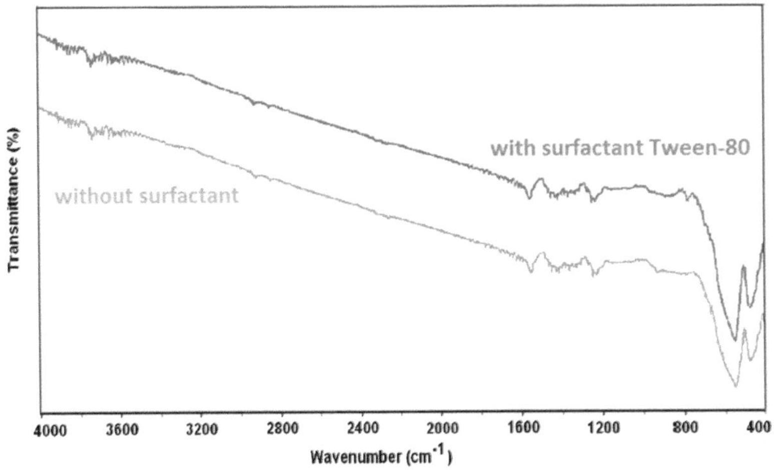

Fig. 2 FTIR spectra of Sr-M calcined at 950 °C

3 Results and Discussions

3.1 FTIR

Figure 2 shows the FTIR spectra of both the samples. The absorption bands around 440–580 cm^{-1} is due to the M–O vibrations [6] which gives the evidence of ferrite formation. FTIR for sample prepared with surfactant has more intensity of bands, which can be relate with formation of pure phase.

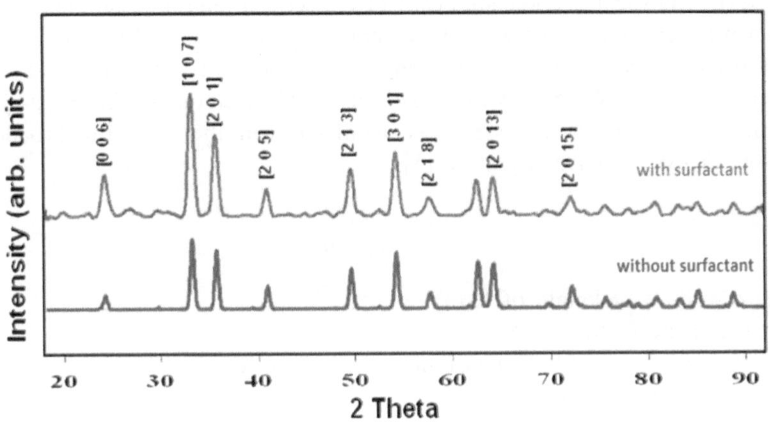

Fig. 3 XRD pattern

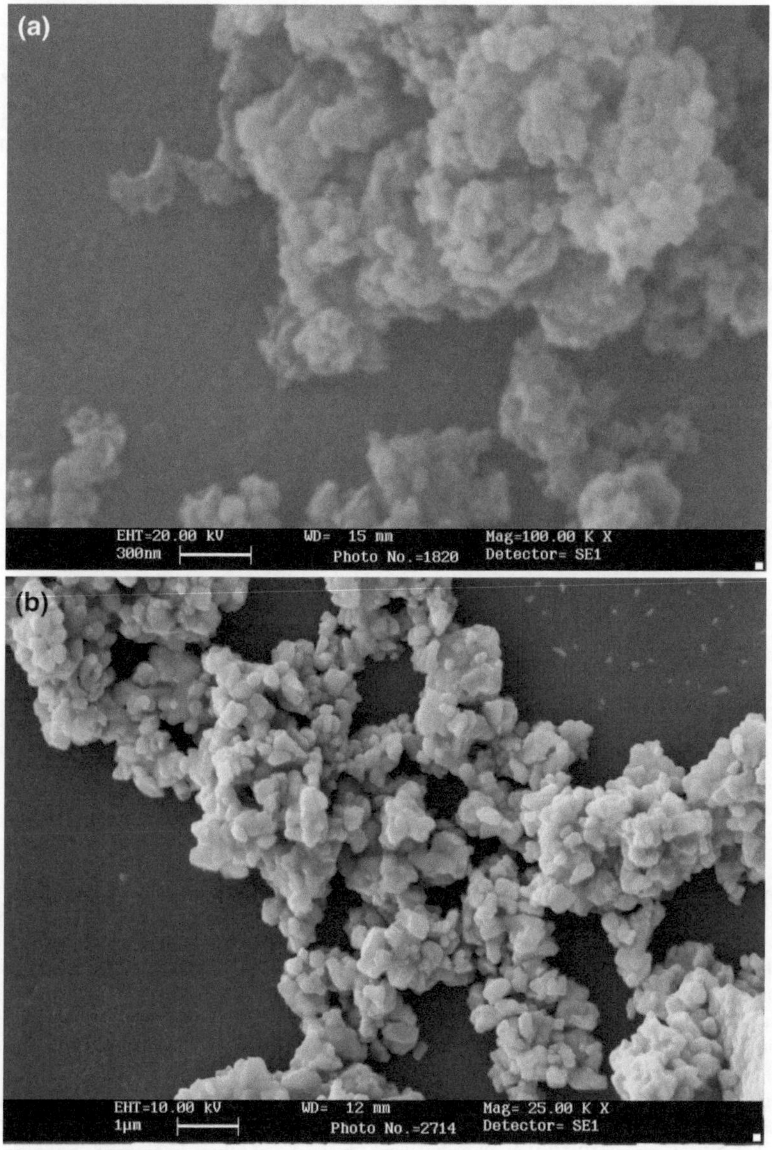

Fig. 4 SEM images of $SrFe_{12}O_{19}$ hexaferrite prepared **a** without surfactant and **b** with surfactant Tween-80

3.2 XRD

X-ray Diffraction patterns for both the samples are recorded at room temperature. Figure 3 represents XRD pattern of Strontium hexaferrite. All peaks are identified and matched with standard JCPDS data. Particle size is calculated for major peak [1 0 7] with Debye–Scherrer formulae. The peaks broadening for sample prepared with surfactant indicate the reduction in grain size. Average particle size for sample without surfactant is around 78 nm whereas sample with surfactant is 34 nm. Also no any impurity is added in the sample with surfactant. That means phase remains single and only we achieve the smaller particle size.

3.3 SEM

Figure 4a, b represent Scanning Electron Micrographs which gives information of particle size and particle size distribution.

From Fig. 4a, it is clear that the particle size varies and not distributed uniformly. In Fig. 4b, particles are well distributed. They are uniform in size. And particles are small compared to the hexaferrite prepared without surfactant. This result is in well agreement with the XRD results.

4 Conclusion

It is observed that the particles prepared by co-precipitation route using surfactant are uniform, nano sized and separated. This property is very useful in the application of ferrites as wear resistance.

References

1. Pankov VV, Pernet M, Germi P, Mollard P (1993) Fine hexaferrite particles for perpendicular recording prepared by the coprecipitation method in the presence of an inert component. J Magn Magn Mater 120:69–72
2. Ataie A, Harris I, Ponton C (1995) Magnetic properties of hydrothermally synthesized strontium hexaferrite as a function of synthesis condition. J Mater Sci 30:1429–1433
3. Ding J, Miao W, McCormick P, Street R (1998) High-coercivity ferrite magnets prepared by mechanical alloying. J Alloy Compd 281:32–36
4. Panchal NR, Jotania RB (2012) Physical properties of strontium hexaferrite nano magnetic particles synthesized by a sol-gel auto-combustion process in presence of non ionic surfactant. Nanosci Nanotechnol Lett 4:623–627
5. Constantino V, Pinnavaia T (1995) Basic properties of $Mg^{2+}_{1-x}Al^{3+}_{x}$ layered double hydroxides intercalated by carbonate, hydroxide, chloride, and sulfate anions. Inorg Chem 34:883–892
6. Wang S, Zhai Y, Li X, Li Y, Wang K (2006) Coprecipitation synthesis of MgO-doped ZrO_2 nano powder. J Am Ceram Soc 89:3577–3581

Investigations of Friction and Wear in Pultruded Glass Fibre Epoxy Composites

Piyush P. Gohil, Hiral H. Parikh and Vimal B. Patel

Abstract Polymer composites are well known for offering high strength-to-weight ratios and flexibility in material design. Physical properties of a composite can be satisfied various functional requirements of a target application, including stiffness and strength, thermal and electrical transport, and wear resistance. Composites are designed to fulfill several functions simultaneously. Friction and wear experiments of pultruded glass fibre epoxy composite (GFEC) were carried out at ambient conditions using Pin on disc (POD) machine arrangement. Tribological characterization of material determining wear and friction coefficient at different operating parameters. This paper outlines a methodology based on Taguchi's experimental design approach to make a parametric analysis of sliding wear and friction behaviour. The systematic experimentation leads to determination of significant process parameters that predominantly influence on the wear and friction co-efficient.

Keywords Composite · Wear · Friction · Putrusion · Pin on disk (POD)

1 Introduction

Tribology is a language of engineers covering friction wear and lubrication science. Today it extends over the scientific fields of physics, chemistry, solid mechanics, fluid mechanics, heat transfer, materials science and lubricant [1]. In material selection and designing process tribology is very important element to be considered with physical and mechanical properties [2, 3]. Wear is the progressive

P. P. Gohil (✉) · V. B. Patel
Department of Mechanical Engineering, CSPIT-Charusat, Changa 388421, Gujarat, India
e-mail: piyushgohil.me@charusat.ac.in

H. H. Parikh
Department of Mechanical Engineering, DJMIT, Mogar 388340, Gujarat, India

H. C. Patel et al. (eds.), *Proceedings of International Conference on Advances in Tribology and Engineering Systems*, Lecture Notes in Mechanical Engineering,
DOI: 10.1007/978-81-322-1656-8_16,

material loss from solid surfaces in contact which occurs as a result of friction. Most of failure occurred in designed component is due to wear progress during operation [4]. Friction is the resistance to motion whenever one solid body moves over another. It is one of the oldest problems in physics and is of great practical importance in many of the industrial operations.Therefore, attention has been paid by many engineering designers in investigation the friction and wear behaviour of materials under different operating parameters like applied load, sliding distance and sliding speed, etc. [5, 6].

Fiber reinforced polymer (FRP) composites which have been established as the one of the most promising modern materials to replace conventional metals and alloys in numerous structural and tribological applications. FRP materials developed using thermoplastic and thermosets as matrices, natural and synthetic fibres as reinforcements and organic and inorganic materials as fillers. They have tremendous potential owing to their high strength-to-weight ratio, tailoring potential and resistance to wear, corrosion and impact. Synthetic fibers such as glass, carbon, aramid, etc. [7], have been used largely with both thermoset and thermoplastic to develop FRP materials.

Polymer matrix composite (PMC) reinforced with discontinuous, continuous and woven fabric of fibers have always been considered as good structural materials [8]. PMC are promising as tribological materials because of their inherent properties such as self-lubrication, low cost, lightweight, quiet operation, better friction properties, ease of fabrication and resistance to wear, corrosion and organic solvents. They are used as seals, bearings, gears (low fabrication, low wear), conveyer belts (low wear), turbine or pump blades (low wear), brakes, tyres (low wear and moderate friction), dental applications (low wear) and hip replacements in which the substitute material should have low wear and low friction coefficient. Based on type (short fiber, unidirectional long fiber, woven fabric), content and orientation (parallel, anti- parallel and perpendicular to sliding direction) of fibrous reinforcement, the wear resistance of composites materials is either enhanced or lowered. Appropriate filler along with fiber reinforcement generally reduces the wear rate, increases the thermal conductivity and creep resistance and modifies the wear mechanism including interfacial wear phenomena. There are number of investigations exploring the influence of test conditions, contact geometry and environmental condition on the friction and wear behavior of polymers and composites [9, 10].

Many of the researchers have carried out no. of experiments using hand layup and resin transfer moulding (RTM) techniques which has restriction for fibre volume fraction. While in pultrusion we can achieve higher fibre volume fraction. Pultrusion is a continuous process for manufacturing of composite materials with constant cross-section. Reinforced fibers are pulled through a resin, possibly followed by a separate pre forming system, and into a heated die where the resin undergoes polymerization. Many resin types may be used in Pultrusion including polyester, polyurethane, vinyl ester and epoxy. Pultrusion is a continuous, automated closed-moulding process that is cost effective for high volume production of constant cross section parts. Due to uniformity of cross-section, resin dispersion, fibre distribution, excellent composite structural materials can be fabricated by Pultrusion [11]. The basic process usually involves pulling of continuous fibres through a bath of resin, blended with a section is partially pre-shaped and excess resin is removed. It is then passed through a heated die, which determines the sectional geometry and finishing of the final product.

2 Research Review

Friction and wear behaviour of GFEC were reviewed and results shows that in general, friction and wear are strongly influenced by all the test parameters such as applied load, sliding speed, sliding distance and fiber orientations [12]. Pihtili and Tosun [13] shows that applied load and sliding speed play significant role on the wear behavior of polymer and composites. They also show that applied load has more effect on the wear than the speed for composites. Several authors [14] observed that the friction coefficient of polymers and its composites rubbing against metals decreases with the increase in load though some other researchers have different views. Stuart [15] and other researchers show that value of friction coefficient increases with the increase in load. Friction coefficient and specific wear rate values for different combinations of polymer and its composite were obtained and compared by Mimaroglu [16].

3 Materials and Methods

Pultruded glass fibre epoxy composite (GFEC) rod 12 mm in diameter was procured. It has 75 % glass fibre and unidirectional fibre orientation. Test pins were cut from pultruded rod where each has 32 mm length. Both top and bottom surface were produce by grinding and polishing as shown in Fig. 1. The Tribometer uses a POD system to measure wear (Fig. 2). The unit consists of a gimballed arm to which the pin is attached, a fixture which accommodates disks up to 165 mm in diameter and 8 mm thickness, an electronic force sensor for measuring the friction force, and a computer software (on Lab view platform) for displaying the parameters, printing, or storing data for analysis. Users have to specify the turntable speed, the load, and any other desired test variables such as time limit and number of rotations.

Test Parameters

Load (L)	Values of the force in Newton's at the wearing contact
Speed (N)	the relative sliding speed between the contacting surfaces in metres per second
Sliding Distance (SD)	the accumulated sliding distance in meters
Temperature	temperatures of one or both specimens at locations close to the wearing contact

Fig. 1 Pin of pultruded GFEC (75 % glass fibre)

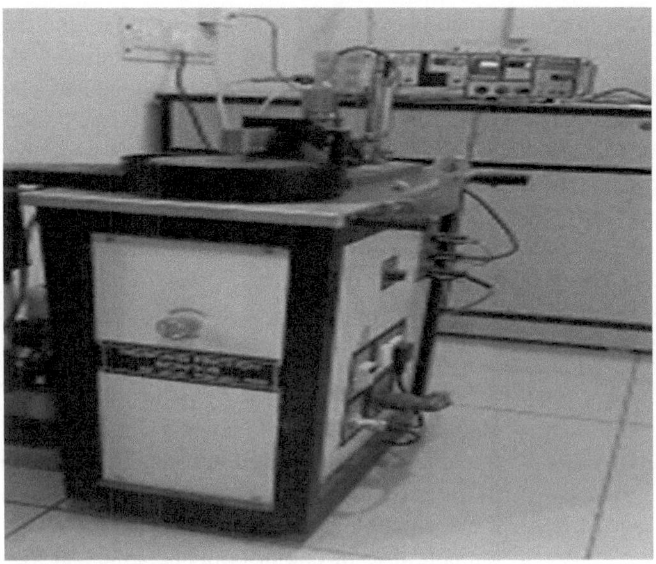

Fig. 2 Pin on disk (POD) test machine

Table 1 Process parameters with their values at four levels

Parameter	Level 1	Level 2	Level 3	Level 4
Sliding dist. (m)	1,000	1,500	2,000	2,500
Load (kg)	1	2	3	4
Speed (rpm)	500	750	1,000	1,250

3.1 Experiment Design

The most important stage in the design of experiment lies in the selection of the control factors. In this work by using Minitab 16 software, three factors and four levels, (shown in Table 1) are selected to build a proper orthogonal array for Taguchi experiments. Since, conventional full factorial experiment design would have required $4^3 = 64$ runs to study three factors each at four levels. As per Taguchi methodology it reduces to 16 runs, and from the L16 orthogonal array 16 experiments are required (shown in Tables 2, 3). In this Taguchi design and no interaction is incorporated [17]. The experimental observations are transformed into signal-to-noise (S/N) ratios. There are several S/N ratios available depending on the type of characteristics. The S/N ratio for minimum wear rate is (smaller-the-better) characteristic, which can be calculated as logarithmic transformation of the loss function as shown below *Smaller-the-better characteristic*:

$$\frac{s}{n} = -10\log\frac{1}{n}\sum_{i=2}^{n} y^2 \tag{1}$$

> Equations

$$\text{Sliding distance(SD)} = \frac{\pi DNt}{60,000}\ (\text{m}) \tag{2}$$

$$\text{co} - \text{efficient of friction}(\mu) = \frac{f_f}{f_n} \tag{3}$$

$$\text{wear rate } (W_r) = \frac{\text{wear} \times \pi \times (D^2)}{4 \times \text{ load} \times \text{SD} \times \text{time}}\ \left(\text{mm}^3/\text{Nm}\right) \tag{4}$$

4 Results and Discussion

The experiments were conducted with an aim of relating the influence of sliding speed (N), applied load (L) and sliding distance (SD) on wear and friction of GFEC. To evaluate the friction and sliding wear performance, wear tests were carried out on POD type friction and wear monitoring test ring (DUCOM) as per ASTM G 99. The counter disc body is made of hardened ground steel (EN-32,

Table 2 Experimental conditions

Material of pin	Glass fibre with epoxy resin
Fibre content	75 %
Fibre orientation	Unidirectional
Pin length and diameter	Length 32 mm, dia 12 mm
Density	2.6 (gm/cm^3)
Environment condition	Dry
Experiment time	600 s

hardness 72 HRC, surface roughness 0.7 μ Ra). The specimen is held stationary and the disc is rotated while a normal force is applied through a lever mechanism.

During the test, friction force was measured by transducer mounted on the loading arm. The friction force readings are taken as the average of 60 readings for every 60 s. For this purpose a microprocessor controlled data acquisition system is used. On conducting the experiments as per the orthogonal array, the dry sliding wear results for various combinations of parameters were obtained and shown in Table 4.

Table 5 shows calculated wear rate and co-efficient Of friction after the test conducted. here wear rate in terms of height loss which is calculated by the equation [4].

Table 3 L_{16} orthogonal array based on taguchi design

SD (m)	L (kg)	N (rpm)
1,000	1	500
1,000	2	750
1,000	3	1,000
1,000	4	1,250
1,500	1	750
1,500	2	500
1,500	3	1,250
1,500	4	1,000
2,000	1	1,000
2,000	2	1,250
2,000	3	500
2,000	4	750
2,500	1	1,250
2,500	2	1,000
2,500	3	750
2,500	4	500

Table 4 Observed data from POD

SD (m)	L (kg)	N rpm	W (microns)	f_f (N)
1,000	1	500	27	0.4
1,000	2	750	59	4.4
1,000	3	1,000	110	16.6
1,000	4	1,250	52	20
1,500	1	750	29	1.1
1,500	2	500	39	2.4
1,500	3	1,250	121	14
1,500	4	1,000	70	18.1
2,000	1	1000	17	3.8
2,000	2	1,250	87	13.8
2,000	3	500	58	7.6
2,000	4	750	98	20
2,500	1	1,250	41	3.2
2,500	2	1,000	80	6.7
2,500	3	750	46	9.8
2,500	4	500	125	10

Table 5 Calculated result after test

SD (m)	L (kg)	N (rpm)	$W_r(10^{-5})$ (mm^3/Nm)	μ
1,000	1	500	0.46	0.04
1,000	2	750	7.28	0.22
1,000	3	1,000	5.01	0.56
1,000	4	1,250	1.174	0.51
1,500	1	750	0.12	0.11
1,500	2	500	0.16	0.12
1,500	3	1,250	5.46	0.48
1,500	4	1,000	3.44	0.46
2,000	1	1,000	4.503	0.39
2,000	2	1,250	7.26	0.7
2,000	3	500	0.2	0.26
2,000	4	750	0.11	0.51
2,500	1	1,250	8.68	0.33
2,500	2	1,000	0.13	0.34
2,500	3	750	9.2	0.33
2,500	4	500	0.42	0.25

4.1 Effect of Operating Parameters on Friction

Co efficient of friction for glass epoxy composites were tested under normal loads of 1–4 (kg), sliding distance of 1,000, 1,500, 2,000, and 2,500 (m) and speed 500, 750, 1,000 and 1250 (rpm) are shown in Figs. 3 and 4

Results show that the coefficient of friction increases with an increase in applied normal load under dry sliding condition. The maximum coefficient of friction was

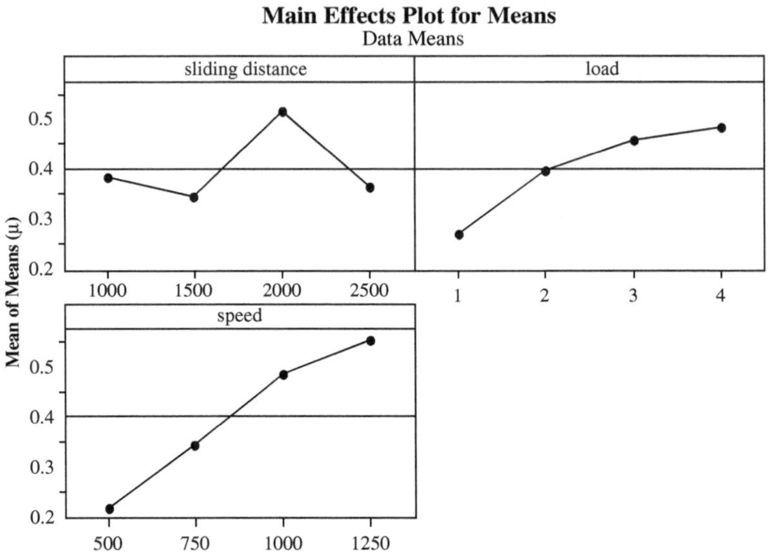

Fig. 3 Plots for parameter versus friction co-efficient (μ)

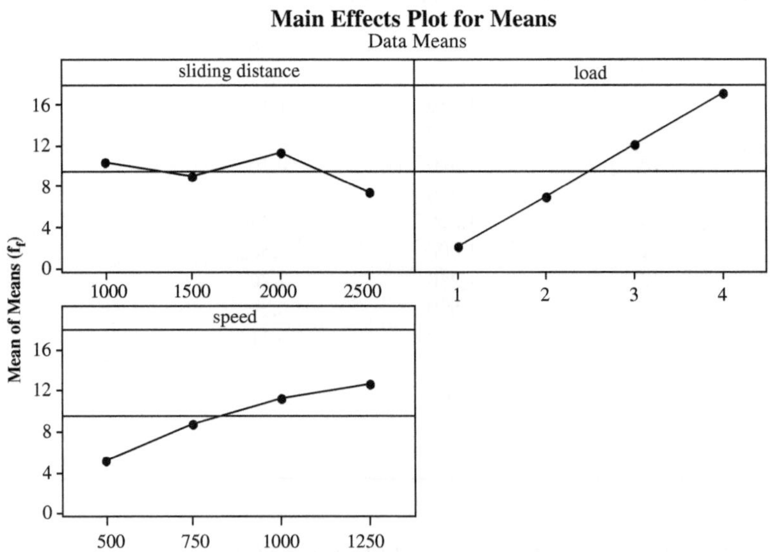

Fig. 4 Plots for parameter versus friction force (f_f)

found for epoxy composites at sliding speed of 1,250 rpm and applied normal load of 20 N and minimum at sliding speed of 500 rpm and applied normal load of 10 N. It is observed that coefficient of friction increases with increase in sliding speed under both dry conditions. At initial stage, friction force is low due to contact between superficial layer of pin and disc and then, friction coefficient increases due to ploughing effect which causes roughening of the test disc surface. The increase in friction coefficient with the increase in normal load is due to the detachment and removal of worn materials and more contact with reinforced fibers. Higher the normal load time to reach constant friction is less due to the fact that the surface roughness and other parameters attain a steady level at shorter period with the increase in normal load. When the applied normal load increases to the limit load values of the polymer the friction will increase due to the critical surface energy. These findings are in agreement with the findings of Nuruzzaman D.M., Chowdhury M.A [18].

Further it can be explained as the frictional power increases the temperature of the steel surface which leads to relaxation of polymer moleculе chains and bond at fibre-matrix gets weakened. As a result, fibres are broken into fragments and form debris with matrix particles. At starting of the rubbing, the value of friction coefficient is low which increases for few minutes to a certain value and then decreases almost linearly over some duration of rubbing and after that it remains constant for the rest of the experimental time. In these cases, transfer film formed on the stainless steel couterface and the transfer film has important effects on the tribological behaviour of a material.

Friction behaviour of polymer sliding against a metal is strongly influenced by its ability to form a transfer film on the counterface.The transfer film formed on a non-polymer counterface is controlled by the counterface material, roughness, and sliding conditions.

Comparison of the variation of friction coefficient with sliding speed for different materials. Results show that friction coefficient increase almost linearly with sliding speed. With the increase in sliding speed, the frictional heat may decrease the strength of the materials and high temperature results in stronger or increased adhesion with pin. The increase of friction coefficient with sliding speed can be explained by the more adhesion of counterface pin material on disc.

4.2 *Effect of Operating Parameters on Wear*

Wear rate for GFEC tested under normal loads of 1–4 (kg), sliding distance of 1,000, 1,500, 2,000 and 2,500 (m) and speed 500, 750, 1,000 and 1,250 (rpm) are shown in Fig. 5. Figure 6 shows wear in microns measured from POD.

Wear rate with normal load for GFEC are shown in Fig. 5. This figure indicates that wear rate increases with the increase in normal load up to 30 N. The shear force and frictional thrust are increased with the increase in applied load and these increments accelerate the wear rate at initial stage then wear rate decrease. At higher load it shows less specific wear rate. Hence these composites have good wear properties at higher loads. Here in Fig. 5 wear in microns are increasing with increasing load and speed.

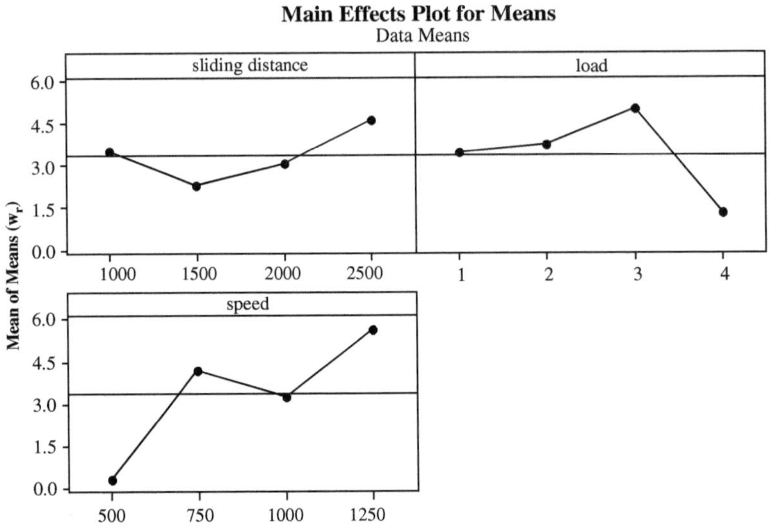

Fig. 5 Plots for parameter versus wear rate (w_r)

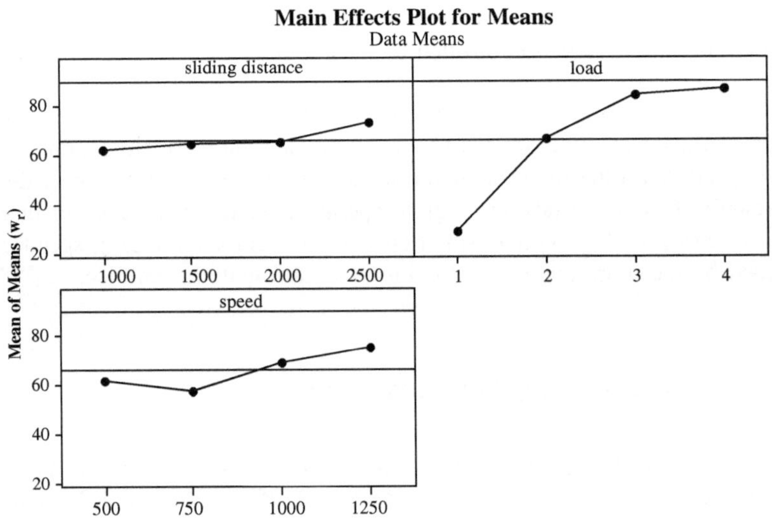

Fig. 6 plots for parameter versus wear (w_r)

At low load, frictional heat generation is low resulting in less extent of back transfer polymer film into the composite pin. In Fig. 5 shows that at higher load larger frictional heat generation resulted in large extent of back transfer patches of polymer film, which were intermittently spread over the surface and shield the composite surface from further damage. This may explain the improvement in

Table 6 Response table for signal to noise ratios

Level	SD	L	N
1	−6.42	−1.66	11.04
2	2.21	−0.20	0.26
3	0.71	−8.50	−5.09
4	−3.19	−3.64	−13.01
Delta	8.68	12.15	24.07
Rank	3	2	1

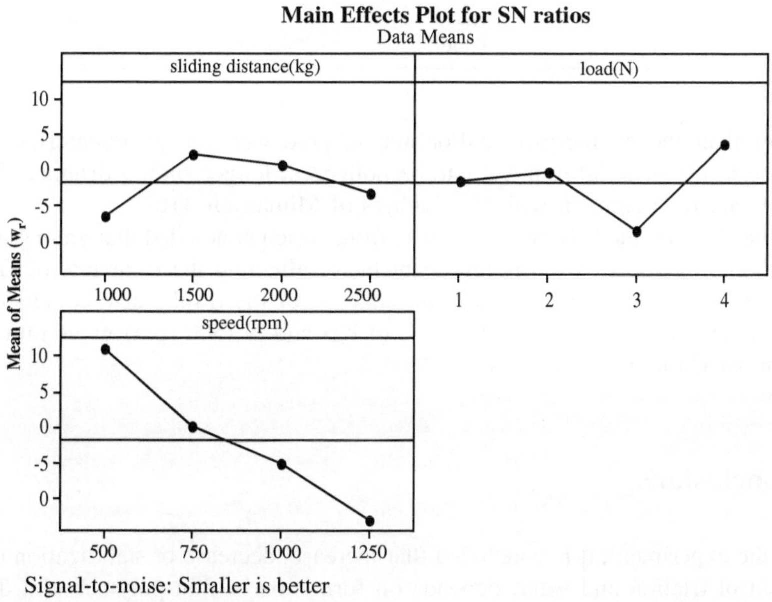

Fig. 7 Plots for parameter versus wear rate (W_r)

wear resistance at higher load. Secondly, at higher load the relatively soft non-abrasive GF were exposed to a larger extent of transfer film of polymer on the metal counterface. Thus, the increasing wear resistance with increasing the load could be possibly due to formation of a fibre-rich surface.

It is also noticed that under dry sliding condition the specific wear rate increase with increase in sliding speed shown in Fig. 5. At high speed fatigue effects and frictional heating are intensified causing a strong surface damage in the form of fibre pull-outs and matrix fracture, resulting in very high wear rate. High rotation speeds also intensify the debris removal by centrifugal forces avoiding the formation of protective layers in open tribological systems. At lower sliding speeds there are minimum value in specific wear rates. It is highly dependent on the film generated on the counterface. If the film transfer is smooth and adhered well on the counterface, the steady state can be achieved. Meanwhile, if the film transfer is

Nomenclature

Parameter	Particulars	Units
W	Wear	μm
w_r	Wear rate	mm^3/Nm
f_f	Friction force	N
μ	Co-efficienf of friction	–
SD	Sliding distance	m
L	Load	kg
D	Track diameter	mm
N	Speed	rpm
T	Time	sec

rougher than the counterface and/or not adhered well on the counterface then increase in the material removal can be noticed at longer sliding distance. These findings are in agreement with the findings of Mimaroglu [16].

Table 6 shows the S/N ratio response, from which concluded that among all the factors, sliding speed is most significant factor affecting the wear rate of GFEC. Figure 7 shows s/n cure for wear rate, analysis of the results also conclude that factor combination of SD (1,500 m), L (4 kg) and N (500 rpm) gives minimum specific wear rate.

5 Conclusions

From the experiment, it is concluded that increase, decrease or stabilization of co-efficient of friction and wear, depends on formation of thin polymer film during process.

Friction coefficient increases with increasing load and speed (rpm). For sliding distance it increase up to 30 N, then friction co-efficient decrease. This may be due to formation thin film.

Wear rate also increasing with increasing speed and sliding distance. Also observed that at higher load GFEC provides good wear resistance.

The current trends of these experimental and analytical results can be used in future to design different tribological and mechanical components. The researchers can use these results to innovate some design strategies for improving different concerned mechanical processes. It is expected that the research findings of tribological behaviour of polymer and composites discussed in this paper will also be used for future research and development.

References

1. Bhushan B, Gupta BK (1991) Handbook of tribology: materials, coatings and surface treatments. McGraw-Hill, New York Reprinted with corrections, Krieger Publishing Co., Malabar
2. Yousif BF, Nirmal U, Wong KJ (2010) Three-body abrasion on wear and frictionalperformance of treated betelnut fibre reinforced epoxy (T-BFRE) composite. Mater Des 31(9):4514–4521
3. Difallah Ben (2012) Mechanical and tribological response of ABS polymer matrix filled with graphite powder. Mater Des 34:782–787
4. El-Tayeb NSM, Yousif BF, Yap TC (2008) An investigation on worn surfaces of chopped glass fibre reinforced polyester through SEM observations. Tribol Int 41(5):331–340
5. Bansal DG, Streator JL (2009) Corrigendum to a method for obtaining the temperature distribution at the interface of sliding bodies. Wear 266:721–732. http://dx.doi.org10.1016/j.wear.2008.08.019
6. Narish S, Yousif BF, Rilling D (2011) Adhesive wear of thermoplastic composite based on kenaf fibres. Proc Inst Mech Eng [J]: J Engl Tribol 225(2):101–109
7. Xue Q, wang Q (1997) Wear mechanisms of polyetheretherketone composites filled with various kinds of SiC. Wear 213:54–58
8. Zhang SW (1998) State-of-the-art of polymer tribology. Tribol Int 31(1–3):49–60
9. Tanaka K (1982) Transfer of semicrystalline polymers sliding against smooth steel surface. Wear 75:183–199
10. Bahadur S, Tabor D (1985) Role of fillers in the friction and wear behavior of highdensitypolyethylene. In: Lee LH (ed) Polymer wear and its control, vol 287. ACS Symposium Series, Washington DC, pp 253–268
11. Kendall KN, Rudd CD, Owen MJ, Middleton V (1992) Characterisation of the resin transfer moulding process. Compos Manufact 3(2):235–249
12. El-Tayeb NSM, Yousif BF, Yap TC (2006) Tribological studies of polyester reinforced with CSM 450-R-glass fiber sliding against smooth stainless steel counterface. Wear 261:443–452
13. Pihtili H, Tosun N (2002) Investigation of the wear behavior of a glass fiber-reinforced composite and plain polyester resin. Compos Sci Technol 62:367–370
14. Anderson JC (1986) The wear and friction of commercial polymers and composites. In: Friedrich K (ed) Friction and wear and polymer composites. Composite materials series, vol 1. Elsevier, Amsterdam, pp 329–362
15. Stuart BH (1998) Tribological studies of poly(ether ether ketone) blends. Tribol Int 31(11):647–651
16. Mimaroglu A, Unal H, Arda T (2007) Friction and wear performance of pure and glass fiber reinforced Poly-Ether-Imide on polymer and steel counterface materials. Wear 262:1407–1413
17. Glen SP (1993) Taguchi methods: a hand on approach. Addison-Wesley, NY
18. Nuruzzaman DM, Chowdhury MA, Rahaman ML (2011) Effect of duration of rubbing and normal load on friction coefficient for polymer and composite materials. Ind Lubr Tribol 63:320–326

Morphology and Friction Characterization of CVD Grown Graphene on Polycrystalline Nickel

M. Tripathi, G. Paolicelli and S. Valeri

Abstract Nanoscale friction properties of graphene produced by Mechanical Exfoliation (ME) of graphite crystal and grown by CVD have been studied by Atomic Force Microscopy (AFM). In particular we have analyzed ME graphene deposited on SiO_2 (300 nm oxide) with respect to CVD graphene grown on polycrystalline Ni. We use the AFM in Friction Force Mode (FFM) with force resolution in the nano-newton range and lateral resolution in the nanometer scale. The detailed morphology of the samples, that may strongly influence the friction response at these length scales, has been analyzed by Raman spectroscopy and SEM imaging. We confirm that on ME graphene on SiO_2 friction force decreases film thickness (i.e. increasing the effective number of layers that compose the film). Moreover we verify that the same behavior appears for graphene grown by CVD on polycrystalline Ni substrate indicating that this characteristic is a specific mechanical properties of a few layer film.

Keywords Friction force microscopy · Graphene · Morphology · Mechanical exfoliation · CVD

1 Introduction

Graphene is a unique material composed of sp^2-bonded carbon atoms forming a stable and self sustaining sheet only one atom thick. Graphene has exceptional properties which include large electron mobility, high thermal conductivity,

M. Tripathi (✉) · G. Paolicelli · S. Valeri
Dipartimento di Scienze Fisiche Informatiche e Matematiche (FIM),
Università di Modena e Reggio Emilia, Via Campi 213, Modena, Italy
e-mail: manoj.triapthi@unimore.it; manojtriapthi83@gmail.com

M. Tripathi · S. Valeri
Centro S3, Istituto Nanoscienze CNR, Via Campi 213, Modena, Italy

H. C. Patel et al. (eds.), *Proceedings of International Conference on Advances in Tribology and Engineering Systems*, Lecture Notes in Mechanical Engineering,
DOI: 10.1007/978-81-322-1656-8_17,

optical transparency and elevated mechanical strength [1–3]. In addition, according to recent microscale tribology measurements, graphene is a promising candidate as a thin solid lubricant [4] and as anti-wear coating agent due to its low friction coefficient, elastic characteristic and ability for high pressure resistance at ambient conditions [5]. The friction coefficient obtained in these experiments is similar for both Mechanical Exfoliated (ME) graphene deposited on SiO_2 (300 nm oxide) and CVD graphene grown on polycrystalline Ni. Nevertheless, it is well known that nanoscale morphology and crystalline structure of the two films are different because of the different fabrication procedure.

The details of structural film characteristics may influence tribological properties of graphene sheet. For example it has been demonstrated for ME graphene film deposited on SiO_2 that nanoscale friction forces measured by FFM decreases with increasing "graphene thickness." We define as "graphene thickness" the effective number of layers that compose a graphene film on a given region. In other word the friction of bi-layer region, is almost half respect to that measured on a single layer region and further decreases with 3–4 layers [6, 7].

Mechanical exfoliation of graphite crystal [8] results in large single domain flakes (1–100 μm), with very numbers of low defects but random variable thickness. With this method the first single layer graphene has been produced and still ME is widely utilized for scientific application but definitively it cannot be implemented for industrial requirements.

A number of methods have been developed and optimized to grow graphene sheets for industrial purpose which includes: thin film heteroepitaxial grown by graphitization of Silicon carbide surface [9], dissociation of ethylene gas on metal surface (Ni (111)) in UHV [10] and chemical vapor deposition (CVD) in ambient pressure over metal surfaces [11].

For wide scale applications, the CVD growth of graphene on polycrystalline metal is considered one of the more interesting methods. The films obtained with this procedure usually consist of a large number of graphene domains of few layer thickness separated by small disordered regions [12]. Single or bi-layer regions can be upto 10 μm in lateral size but the films are continuous over much larger areas (few cm^2).

In our present investigation, we have localized specific regions on graphene produced by ME deposited on SiO_2 and graphene directly grown by CVD on polycrystalline Ni. These regions have been characterized by Micro-Raman spectroscopy, a technique that gives information about the local thickness of the graphene film. Their morphology has been studied by AFM (Atomic Force Microscopy) and SEM (Scanning Electron Microscopy). Finally the friction characteristics have been measured with FFM technique in ambient conditions.

One of the major concerns of this analysis is to verify whether graphene obtained from different procedures possess different or similar tribological properties. This analysis is useful to investigate friction properties of few layer graphene for future technological applications as thin solid lubricant but also it could be a benchmark to better understand tribological studies of graphene based on its effective thickness, stacking behavior, homogeneity, and grain boundaries presence.

2 Analysis

2.1 Optical and Raman Characterization

CVD grown graphene on polycrystalline Nickel (Gr/poly-Ni) is a commercial sample obtained from Graphene Laboratories Inc., Calverton, USA. Mechanical Exfoliated graphene (Gr/SiO_2) has been prepared in our lab starting from HOPG (Highly Oriented Pyrolytic Graphite) and using silicon wafer with oxide layer (300 ± 30 nm) as a substrate. Figure 1a shows optical image of Gr/poly-Ni and Fig. 1b is the optical image of Gr/SiO_2.

In Gr/poly-Ni, regions that appear dark correspond to thick carbon over layer and generally are disordered areas covering the grain boundaries of the polycrystalline Ni substrate. Bright regions indicate thin carbon over layer and usually correspond to graphene sheets grown on the crystalline grains of Ni substrate [11].

Gr/SiO_2 (Fig. 1b) shows better contrast due to presence of ~300 nm oxide layer. On bottom of Fig. 1b a large graphite flake (grey) is clearly visible while on the right top corner a composition of graphene sheets is present. In both optical images the darker region represents thicker carbon over layer respect to brighter regions but the exact correspondence with the number of layer require a Micro-Raman analysis.

Micro-Raman analysis has been carried out by Horiba spectrometer (model: Labram, Jobin–Yvon) at 632.8 nm wavelength under ambient condition. The diameter of laser spot is nearly 1 μm with maximum power of 20 mW. The region marked 3–5 on Fig. 1a and those marked by black and red arrows in Fig. 1b are the locations where Raman analysis has been performed.

Figure 1c shows Raman spectra of CVD grown graphene. Line colors and numbers to the right of the spectra correspond to the regions marked on Fig. 1a while the black and red spectra (labeled 1 and 2) correspond to regions outside Fig. 1a. The black is a bare Ni regions and the red is an intense bright region similar to region 3. The inset shows Raman spectra from Gr/SiO_2. MEBL indicates a Mechanical Exfoliated Bi-Layer region and MEML is Mechanical Exfoliated Mono-Layer region.

For Gr/SiO_2 layers, the principles peaks are observed at ~1,595 cm^{-1} (G) and ~2,665 cm^{-1} (2D). Only in case of bi-layer we observe a very low intensity peak at ~1,330 cm^{-1} (D) probably because we were close to the edge of the flake. The confirmation of the film thickness, which is a mono-layer for the black spectra and a bi-layer in the case of red spectra was obtained by fitting 2D peak with one and four lorentzian component respectively with peak width of 28 cm^{-1} [13]. We also show a reference spectra corresponding to the large graphite flakes.

For Gr/poly-Ni, the Raman spectrum (Fig. 1c) at location no. 1 (bare Ni) clearly shows no carbon signal. Spectra on location 2 (not in image) and 3 only show negligible D and G signals and no 2D peak. In agreement with recent results [14] we believe that these regions correspond to very thin graphene film. It has been found that the Raman analysis in this condition (few layers on top of Nickel

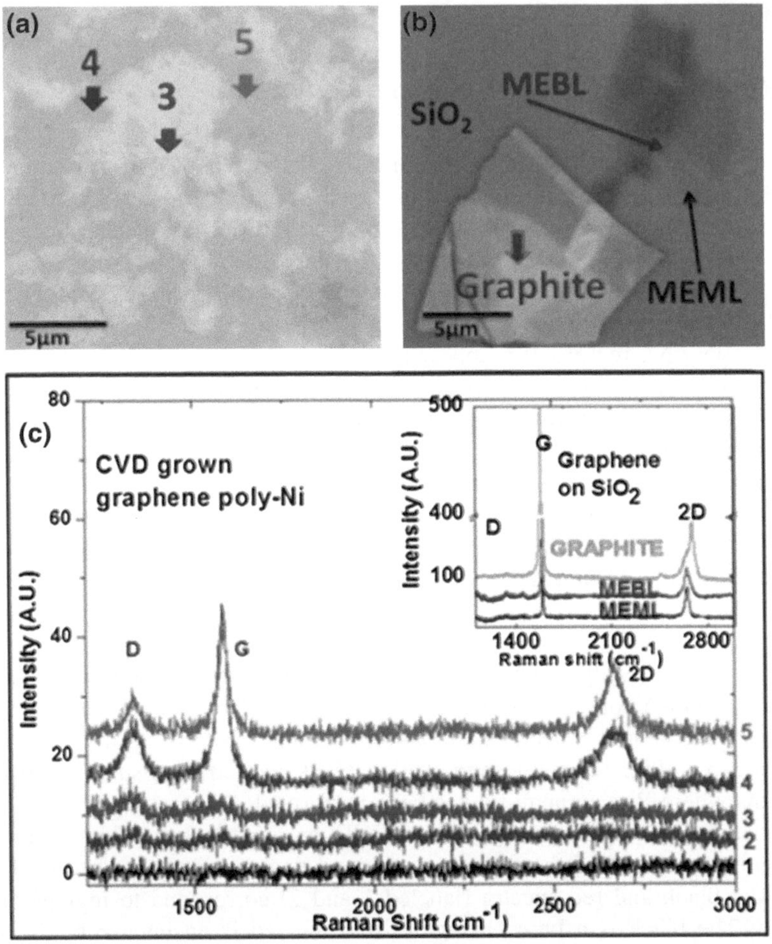

Fig. 1 **a** Optical image Gr/poly-Ni showing *bright* and *dark* regions. Marked locations *3–5* are regions where Raman analysis has been carried out. **b** Optical image of Gr/SiO_2 showing Mono-Layer (*MEML*), Bi-Layer (*MEBL*) regions and bulk graphite. **c** Raman spectra of Gr/poly-Ni measured on five different locations. Location no. *3–5* are marked in the optical image (**a**) while location no. *1* and *2* are out of the image frame. Location no. *1* is bare Ni metal zone used as a reference and location no. *2* is a large *bright* region. Inset of panel **c** shows Raman spectra of *MEML* (*black*), *MEBL* (*red*) and bulk graphite flake (*cyan*)

substrate) do not reveals peculiar signature of graphene presence. One of the possible explanation is that the first graphene layer on Ni surface is strongly bonded to the Ni surface atoms and phonons dispersion differs from that of intrinsic graphene [14]. Spectra on location no. 4 and 5 (Fig. 1c) shows 2D peaks at 2662.77 cm^{-1} (FWHM 119 cm^{-1}) and 2665.63 cm^{-1} (FWHM: 80 cm^{-1} with single lorentzian fitting) respectively and correspond to the presence of thick graphene sheets.

Moreover, for Gr/poly-Ni the thickness and shape of 2D peak does not provide measurements of the thickness in terms of number of layers. Raman spectra of CVD grown graphene on polycrystalline Ni are typically less intense and the peaks are broader compared to Gr/SiO_2 which is an indication that sample is inhomogeneous with domains having comparable or smaller size respect to laser beam diameter (1 μm). Actually, a visible disorder-induced D band ($\sim$1,330 cm^{-1}) is observed left to that of G-band ($\sim$1,583 cm^{-1}).

2.2 Topographical Analysis

The thickness evaluation by Raman analysis of ME graphene on SiO_2 has been confirmed by AFM topography images. Figure 2a is a large (10 × 10 μm^2) height image obtained in amplitude modulation mode in air. Figure 2b is (1.8 × 1.8 μm^2) obtained in contact mode in dry condition (vacuum 10^{-5} torr followed by N_2 gas). The region indicated by the red arrow in optical image (Fig. 1b) is now presented in details (Fig. 2a). This AFM image is rotate nearly 45° clockwise respect to the optical one and the monolayer sheet (MEML) is clearly visible in the left down quadrant. Figure 2b is a zoom of the black marked region in panel (a) showing a clear view of graphene layers and their edge boundaries.

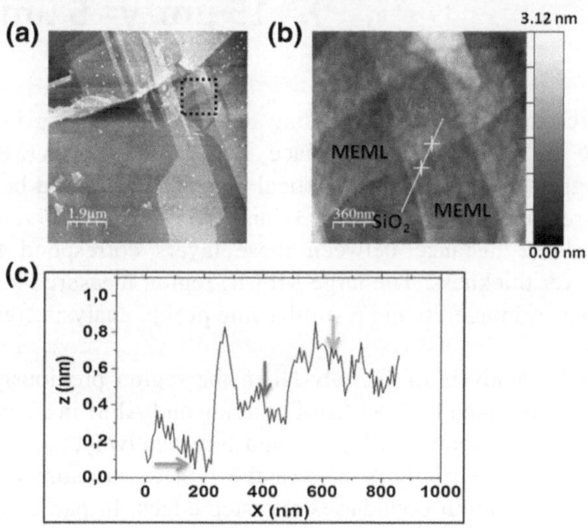

Fig. 2 **a** AFM large (10 × 10 μm^2) height image obtained in amplitude modulation mode in air. The image is rotate nearly 45° clockwise respect to the optical image on Fig. 1**b**, the monolayer sheet (*MEML*) is clearly visible in the *left down* quadrant. **b** Is a zoom (1.8 × 1.8 μm^2) of the marked region in **a** obtained in contact mode in dry condition. It shows the monolayer, bi-layer and Silica regions. **c** Line profile along the *blue line* drawn in **b**. The profile contains the Silica surface, then a monolayer (*MEML*) and bilayer (*MEBL*) graphene areas

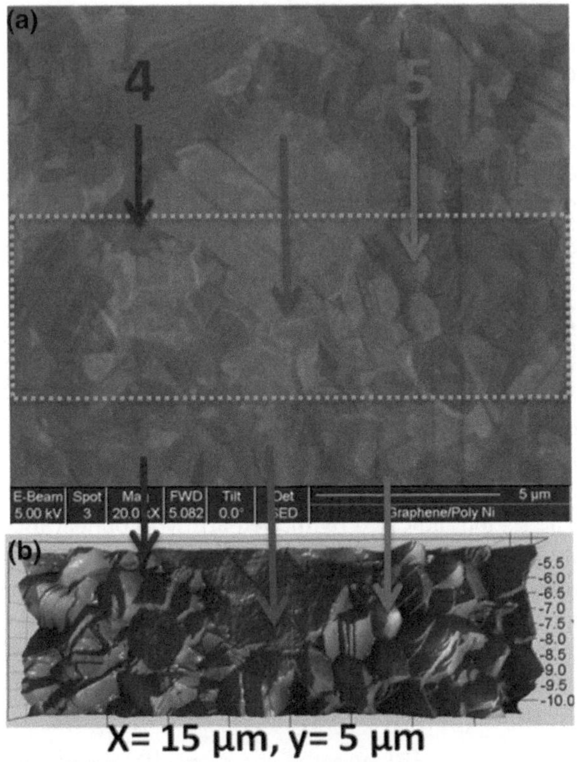

Fig. 3 **a** SEM image of graphene on polycrystalline Ni substrate centered on the region previously analyzed by Raman spectroscopy and optical imaging (region 3–5) terraces, grain boundaries and step edges are clearly visible. In particular we recognize the location no. *3* which belongs to a large terrace. **b** 3-D topographic image ($15 \times 5\ \mu m^2$) by AFM that correspond to the *marked rectangle* in **a**

Figure 2c present the line profile along the blue line drawn in Fig. 2b. This profile contains the silicon dioxide surface, then a mono-layer (MEML) and bi-layer (MEBL) graphene region. The vertical distances measured between marked arrows (green–red and red–blue) are 0.37 and 0.35 nm respectively and confirm that the inter-planar distance between these layers correspond to the typical graphene one layer thickness. The large MEML region measured by Raman starts on the left corner of image 2b and a similar line profile analysis confirm its single layer thickness.

The topography analysis of Gr/poly-Ni in the region previously explored by Raman has been performed by both SEM imaging and AFM in air conditions. The resulting images are presented in Fig. 3a and b respectively.

The SEM image of Gr/poly-Ni shows the typical structure of this material composed of terraces, grain boundaries and step edges. In particular the location no. 3 appears to belong to a large terrace ploughed by two diagonal step in the middle. This analysis is confirmed by the AFM 3D image which clearly shows a flat zone corresponding to location 3 while location 4 and 5 are rough at the micrometer scale.

Table 1 Roughness analysis on bare SiO_2, graphene on SiO_2 and graphene on polycrystalline Ni (regions 3–5). Sampled area is 1 × 1 μm^2

Gr/poly-Ni					
	Sio_2	Gr/Sio_2	"3"	"4"	"5"
RMS (nm)	0.292	0.267	7.79	17.9	18.2
Ra (nm)	0.232	0.217	6.55	14.2	21.6

A line profile analysis is impossible in this case because is difficult to define a reference plane like the bare SiO_2 surface for the ME deposited graphene, anyhow the results from Micro Raman analysis are qualitatively confirmed by the topography images.

To further characterize the location 3–5 we have calculated roughness parameters on these areas and on the ME graphene (see Table 1). The roughness measurements are performed on a 1 × 1 μm^2 square area. MEML graphene on SiO_2 conforms closely to underline substrate morphology and it reveals a surface roughness slightly smaller than the substrate. In the case of Gr/poly-Ni locations 3 confirm to be a flat region having, even at sub-micrometer scale, a roughness two time smaller respect to the locations 4 and 5.

2.3 Friction Analysis

Friction analysis has been carried out in air condition using the AFM in Friction Force Mode (FFM).

Figure 4a is a friction map on Gr/SiO_2 sample generated by applying TMR analysis (Trace Minus Retrace). This map is measured in parallel with the topography presented on Fig. 2b. The very bright zone in the bottom correspond to friction on SiO_2 substrate while the dark part is related to the region composed of mono-layer, bi-layer and thick layer (TL) graphene. The friction dependence with respect to thickness is well represented by the line profile (blue line) drawn across the three different regions and presented in panel (b). In agreement with recent results obtained on different lamellar solids [6], friction on few layer graphene decreases with increasing graphene thickness. This behavior was attributed to the dependence of out-of-plane deformation on the number of layers onto a rigid substrate (SiO_2) which induces a variation of the real contact area between AFM tip apex and graphene film.

Figure 5a represent the friction map on Gr/poly-Ni sample generated by TMR analysis measured in the area described in Fig. 3. The map reveals that, even in this case, the friction response of the graphene film shows a contrast as a function of graphene thickness. The bright region located in the central part of the image is the large terrace that we image with SEM and AFM. The terrace contains the

Fig. 4 **a** Friction map 1.8 × 1.8 μm^2 of Gr/SiO_2. The friction force varies with number of graphene layers. The *bright zone* corresponds to SiO_2 region and shows the highest friction force. **b** Line profile drawn in **a** from monolayer to thick layer shows decreasing of friction force and only at the edges higher forces are recorded

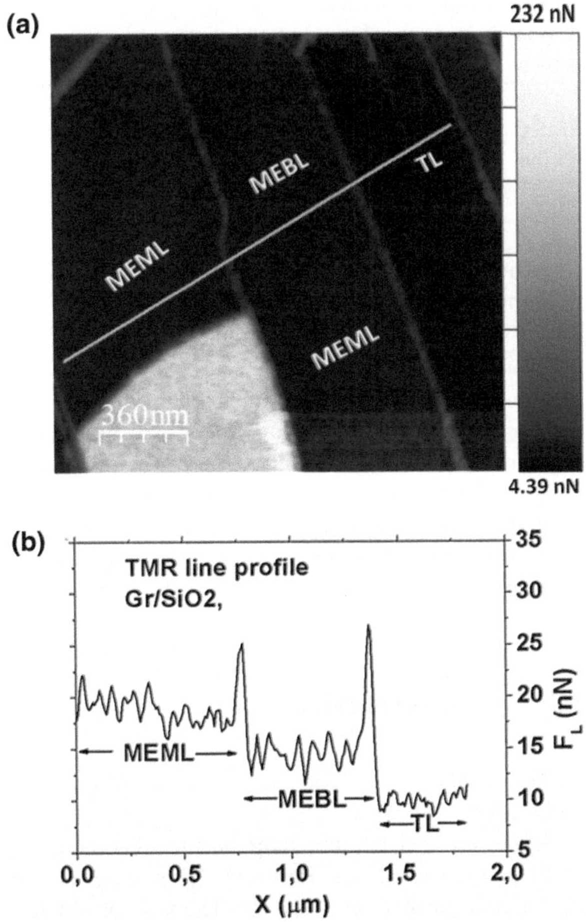

location 3 that we prove by Raman analysis to be covered by a thin graphene film respect to the surrounding location 4 and 5.

A line profile from location 4–5 (blue line, Fig. 5a) that crosses the thin film region is presented in Fig. 5b. The profile appears relatively noisy because of the intrinsic higher disorder characteristic of the CVD grown graphene. Nevertheless the marked increase of friction force on the central thin film region is clearly visible whereas the locations 4 and 5, containing thicker carbon layer, shows lower friction force.

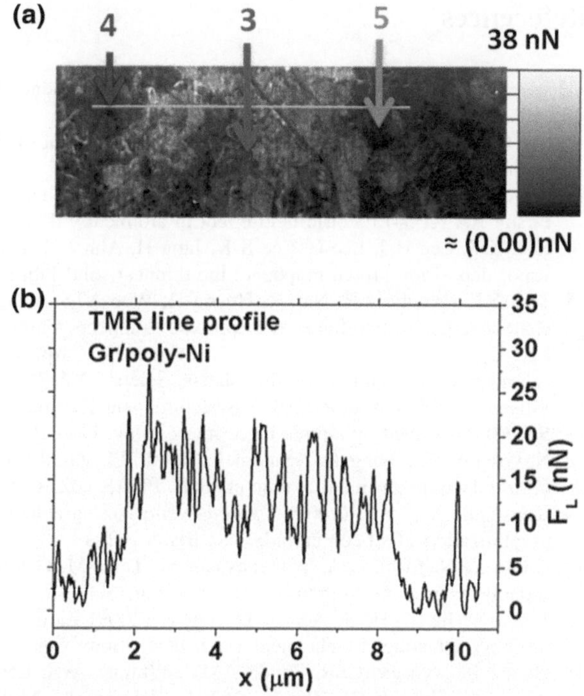

Fig. 5 **a** Friction map 15 × 5 μm^2 of Gr/poly-Ni surface. The color contrast shows higher (*bright*) and lower (*dark*) friction force corresponding to thinner and thicker graphene regions. The *marked arrows* correspond to locations no. "–5" where Raman analysis has been carried out. **b** Line profile carried out close to the *marked regions* reveals quantitative values of friction force

3 Conclusion

Graphene film produced by Mechanical Exfoliation (ME) of graphite crystal and grown by CVD on polycrystalline Ni substrate has been studied to address their local tribological properties.

Measurement on ME graphene on SiO_2 confirm recent results obtained on this system and other different lamellar solids lubricant. Friction measured at the nanoscale with a FFM set-up decreases with increasing film thickness (i.e. increasing the effective number of layers that compose the film).

In addition, we prove that the same mechanism works for graphene grown by CVD on polycrystalline Ni substrate. In this case Raman analysis allows only a qualitative evaluation of film thickness. Nevertheless the increase of friction force on thin film region respect to thicker one is clearly visible in our friction maps.

Our results suggest that this particular behavior depends by the peculiar mechanical properties of few layer graphene arrangement and it is less influenced by the morphological characteristics induced by deposition technique.

References

1. Wang X, Zhi L, Müllen KT (2008) Conductive graphene electrodes for dye-sensitized solar cells. Nano Lett 8:323
2. Lee C, Wei X, Kysar JW, Hone J (2008) Measurement of the elastic properties and intrinsic strength of monolayer graphene. Science 321:385
3. Novoselov KS, Geim AK, Morozov SV, Jiang D, Zhang Y, Dubonos SV, Grigorieva IV, Firsov AA (2004) Electric field effect in atomically thin carbon films. Science 306:666
4. Kim K-S, Lee H-J, Lee C, Lee S-K, Jang H, Ahn J-H, Kim J-H, Lee H-J (2011) Chemical vapor deposition-grown graphene: the thinnest solid lubricant. ACS Nano 5:5107
5. Shin YJ, Stromberg R, Nay R, Huang H, Wee ATS, Yang H, Bhatia CS (2011) Frictional characteristics of exfoliated and epitaxial graphene. Carbon 49:4059
6. Lee C, Li Q, Kalb W, Liu X-Z, Berger H, Carpick RW, Hone J (2010) Frictional characteristics of atomically thin sheets. Science 328:76
7. Filleter T, Bennewitz R (2010) Structural and frictional properties of graphene films on SiC(0001) studied by atomic force microscopy. Phys Rev B 81:155412
8. Novoselov KS, Jiang D, Schedin F, Booth TJ, Khotkevich VV, Morozov SV, Geim AK (2005) Two-dimensional atomic crystals. PNAS 102:10451
9. Konstantin VE et al (2009) Towards wafer-size graphene layers by atmospheric pressure graphitization of silicon carbide. Nat Mater 8:203
10. Gamo Y, Nagoshima A, Wakabayashi M, Terai M, Oshima C (1997) Atomic structure of monolayer graphite formed on Ni (111). Surf Sci 41:61
11. Reina A, Jia X, Ho J, Nezich D, Son H (2009) Large area: few-layer graphene films on arbitrary substrates by chemical vapor deposition. Nano Lett 9:30
12. Huang PY, Vargas CSR, Zande AMV, Whitney WS, Levendorf MP, Kevek JW, Garg S, Alden JS, Hustedt CJ, Zhu Y, Park J, McEuen PL, Muller DA (2011) Grains and grain boundaries in single-layer graphene atomic patchwork quilts. Nat Lett 469:389
13. Malard LM, Pimenta MA, Dresselhaus G, Dresselhaus MS (2009) Raman spectroscopy in graphene. Phys Rep 473:51
14. Takahashi K, Yamada K, Kato H, Hibino H, Homma Y (2012) In situ scanning electron microscopy of graphene growth on polycrystalline Ni substrate. Surf Sci 606:728

Optimization of Reciprocating Friction and Wear Test Rig Operating Parameters for Segmented Piston Ring: Liner Assembly

B. M. Sutaria and D. V. Bhatt

Abstract The hydrodynamic mechanism of piston ring-cylinder liner is complex phenomena. It is not easy to predict performance at the design stage. The friction coefficient plays important role for the friction losses in piston ring assembly (PRA) under different operating parameters. The small cross section area of piston ring and its profile decides the friction losses under variation of loads, speeds and operating temperature of different lubricants. In the present work, effects of significant factors on friction coefficient and friction force are analyzed on reciprocating friction and wear test rig (RFWTR) by using Taguchi approach. The Taguchi method (TM) is employed for evaluate the experimental work, analyzed the effect of working parameters for the PRA and decides the influence of optimum parameters. It is found that these parameters have a significant role on PRA performance. The analysis of variance (ANOVA) is performed to identify the statically significant operating parameters. The optimum level of operating parameters is obtained from the analysis. All the parameters have equal contributions in the performance of engine efficiency, friction coefficient etc. Experimental results are provided to validate the suitability of the proposed approach.

Keywords Friction coefficient · Piston ring cylinder assembly · Taguchi method · ANOVA · DOE

1 Introduction

Development of I. C. Engines with low frictional losses are requires knowledge of the tribological interaction between fundamental design element such as component geometry and material, roughness of surfaces and selection lubricant.

B. M. Sutaria (✉) · D. V. Bhatt
Department of Mechanical Engineering, S.V. National Institute of Technology, Surat 305007, Gujarat, India
e-mail: bms@med.svnit.ac.in

H. C. Patel et al. (eds.), *Proceedings of International Conference on Advances in Tribology and Engineering Systems*, Lecture Notes in Mechanical Engineering, DOI: 10.1007/978-81-322-1656-8_18, © Springer India 2014

Increases in fuel efficiency can be achieved in designs that minimize hydrodynamic shear losses in high speed engine components through low viscosity of oils. However, in low speed operating modes at elevated temperatures and loads, the used of low viscosity oils can result in more frequent boundary contact between components [1, 2]. Under theses situation the performance of engine can be improved by the proper selection of lubricant and other operating parameters.

Approximately 30–50 % of the total mechanical friction in a I. C. Engines has been reported to be at the ring-cylinder wall interface [3, 4]. Therefore, it is essential to investigate experimental designs to minimize this power loss at the ring-cylinder wall interface. To minimize the friction, a sufficient amount of oil is required to formed a proper oil film, but excess oil left on the cylinder wall causes not only loss of the oil but also environmental pollutions. The surface topography of the ring is constructed in such a way that the hydrodynamic pressure is generated at both ends during bi-directional motion. However, in actuality the hydrodynamic pressure is generated only at the converging clearance while diverging clearance dropped the pressure which does not contribute the support of external force.

An early example of modeling and analysis of the frictional losses at the piston ring cylinder wall contact is attributed to Rohde et al. [5]. This mixed friction model was based on the average flow Reynolds equation with half-Sommerfeld boundary condition obtained by Patir and Cheng [6]. Asperity interactions were included through the Greenwood-Tripp [7] model. Dowson et al. [3] used hydrodynamic lubrication model to study film thickness, lubricant transport and viscous friction at the piston ring cylinder wall contact. They used the Reynolds boundary condition in their analysis.

Ting [8, 9] developed a reciprocating test rig to measure the friction coefficient between rings and cylinder liners taken from actual engines. Friction results were obtained using piezoelectric type load cell. The results were presented in a Stribeck-type relationship. Dearlove and Cheng [10] developed a reciprocating test rig based on actual single cylinder engine. A floating liner segment was instrumented to provide frictional data. The lubricant film thickness at mid stroke was measured using the laser fluorescence technique. Arcoumanis et al. [11] measured film thickness throughout the stroke using a purpose-made capacitance transducer. Their reciprocating test rig had maximum stroke length of 50 mm. Frictional data were obtained using a deflection-based measurement. Aklinn and Newaz [12, 13] employed a deflection based force approach to measure friction using actual engine rings and cylinder liner. Tung and Mcmillan [14] were studied using a modified Cameron-plint high frequency friction machine, the cut cylinder bore section was stationary and segmented piston ring reciprocates on inside the cylinder block. Among all engine oil used for bench tests, the engine oil containing MoDTC friction modifier shows the lowest friction compared with the engine oils with organic friction modifier. Furthermore pointed that mineral base engine oils of the same viscosity grade and formulation had slightly lower friction than synthetic-based engine oil. Peggy Lin et al. [15] measured friction coefficient of segmented piston ring-cylinder liner on a reciprocating lubricant tester with variation of phosphorus percentage in the base oil. The oils were evaluated under two different sets of condition i.e., frequency of 20 Hz and a load of 60 N and a frequency of 40 Hz and a load of 120 N. The friction coefficients were continuously recorded, and the wear depths on the cylinder bore segments were measured at the conclusions of each experiment. Mistry et al. [16] conducted a series of tests using a reciprocating tribometer to evaluate the tribological performance of a plasma electrolytic oxidation treated eutectic aluminum-silicon alloy and was compared with a more conventional nickel-ceramic coating ring.

In the present work, a RFWTR is used for the experimental analysis of friction force and friction coefficient for the 74.6 CC single cylinder gasoline engine components (TVS pep model). Recently several mathematical models have been developed to correlate tribological performance with operating parameters. Furthermore, numerous experiments are required to be performed to validate these models. Some optimization techniques are also being used for proper selection of tribo tests operating parameter with constraints. A systematic and efficient technique such as Taguchi method is used to determine the optimal parameters for minimum friction coefficient and friction force. Therefore a comprehensive study of the effects of PRA friction parameters (lubricant, speed, load and elevated temperature of lubricant) on the tribo characteristic such friction and wear on RFWTR. Although study of these parameters has been performed by many researchers, most of the studies do not much consider both engineering philosophy (DOE) and mathematical formulation (AVOVA) [17, 18]. The experimental details when using the Taguchi method are described.

2 Design of Experiment

In PRA friction measurement, there are many operating parameters such as selection of lubricant, applied load, speed and lubricant temperature, which affects friction force and friction coefficient. If one uses traditional experimental methodology. It is impossible to study all the operating parameters and determine main influencing factors in single experiment. In this study four factors are selected for analysis. Taguchi Method (TM) is used to optimize the operating parameters on the PRA friction for considered reciprocation systems.

2.1 Variables and Constant Factors

As per the requirement for determine the friction coefficient and friction force of PRA for considered reciprocating system, lubricant, operating speed, applied load and elevated lubricant temperature are selected as variable factors. The ring width, ring tension, and stroke length are kept constant. Table 1 shows the variable factors and their levels.

Table 1 Variable factors and their levels

Factor	Column	Level 1	Level 2	Level 3
Lub. Oil (L)	A	10W30	20W40	SAE40
Speed (N)	B	300	900	1500
Load (W)	C	60	80	100
Temp (T)	D	40	70	100

2.2 Taguchi Method

In recent year, TM has proven to be a practical method of experimental design for increasing efficiency of the systems. It is uses orthogonal arrays to reduce number of experiments. Orthogonal array of L_9 (3^4) is selected and experiments designed accordingly. Table 2 shows standard format of orthogonal array and Table 3 shows orthogonal array for PRA friction experimentation. The results of experiments are transformed into signal-to-noise (*S/N*) ratio, which is used to analyze the friction coefficient and friction force.

Table 2 L_9 orthogonal array

Experiment	Column			
	A	B	C	D
1	1	1	1	1
2	1	2	2	2
3	1	3	3	3
4	2	1	2	3
5	2	2	3	1
6	2	3	1	2
7	3	1	3	2
8	3	2	1	3
9	3	3	2	1

Table 3 L_9 orthogonal array for PRA friction

Exp. No.	Lub. Oil (L)	Speed (N)	Load (W)	Temp. (T)
1	10W30	300	60	40
2	10W30	900	80	70
3	10W30	1,500	100	100
4	20W40	300	80	100
5	20W40	900	100	40
6	20W40	1,500	60	70
7	SAE40	300	100	70
8	SAE40	900	60	100
9	SAE40	1,500	80	40

Table 4 Specification of 74.6 CC single cylinder engine

Bore	47 mm
Stroke	43 mm
Compression ratio	10.1:1
Max power	4.1 BHP @6500 rpm
Max torque	5 Nm @4000 rpm
Idle speed	1400 ± 100 rpm
Cylinder pressure	10–13 kgf/cm^2
Piston displacement	74.6 cm^3
Compression ring	2 no.
Oil control ring	1 no
Piston ring diameter (O.D)	48 mm
Piston ring diameter (I.D)	47 mm
Ring tension	70 N

2.3 Engine Specifications

In this experimental work, 74 CC single cylinder engine of make TVS pep is used. It is observed in the rural and city transportation that this engine widely used with good caring capacity. The specifications of this engine are as shown in Table 4.

3 Results and Discussion

3.1 Sample Calculation of S/N Ratio

The results obtained for the first DOE and corresponding *S/N* ratio are shown in Table 6, respectively. The results are analyzed by calculating Signal/Noise (*S/N*) ratio. There are nine different combinations of experiments. Three replicates for all the factor level settings as per the Taguchi Orthogonal Array (OA) were performed. Y_1, Y_2 and Y_3 refer to the response (friction coefficient) in the first second and third replications. The friction coefficient should be minimum thus TM refers to "lower is the better" quality characteristics; the *S/N* ratio is calculated using the formula (1),

$$\frac{S}{N} = -10 \log \left[\frac{1}{r}\sum_{i=1}^{r} y_i^2\right] \quad (1)$$

where, r is the number of replicates and Y_i is the response at ($i = r$). *S/N* Ratio for experiment No. 1–9 are computed and shown in Table 5.

Table 5 S/N Ratio for analysis of friction coefficient

No.	Expt. set Y_1	Expt. set Y_2	Expt. set Y_3	S/N ratio (Y_i)
1	0.170	0.153	0.162	15.820
2	0.165	0.164	0.161	15.738
3	0.107	0.0967	0.112	19.546
4	0.169	0.196	0.194	14.575
5	0.121	0.117	0.109	18.728
6	0.037	0.051	0.043	27.124
7	0.188	0.186	0.186	14.579
8	0.168	0.156	0.159	15.860
9	0.0325	0.0303	0.022	30.865

3.2 Response Table and Response Diagram for Friction Coefficient

The response diagram is shown in Fig. 1 and Table 6, which contains the sums of the *S/N* ratio for each level and for each factor.

From Table 6 and Fig. 1 revels that all level totals are compared and combination yielding the lowest S/N ratio is obtained as A_1–B_1–C_3–D_3. This is the optimal levels combination of factors for the improvement in the friction coefficient during PRA friction measurement on RFWTR. The actual parameters are shown in Table 7.

To know the effects of the four operating parameters the overall mean values of S/N ratio is calculated by Eq. (2)

$$M = \frac{1}{9}\sum n_j \tag{2}$$

Now all three levels of every factor are equally represented in the nine experiments. Thus '*M*' is a balanced overall mean over the experimental region. The 'effect of factor level' is defined as the deviation it causes from the overall mean. The average *S/N* ratio for these experiments which is denoted by M_{A1} as given by Eq. (3),

Fig. 1 Response diagram for friction coefficient

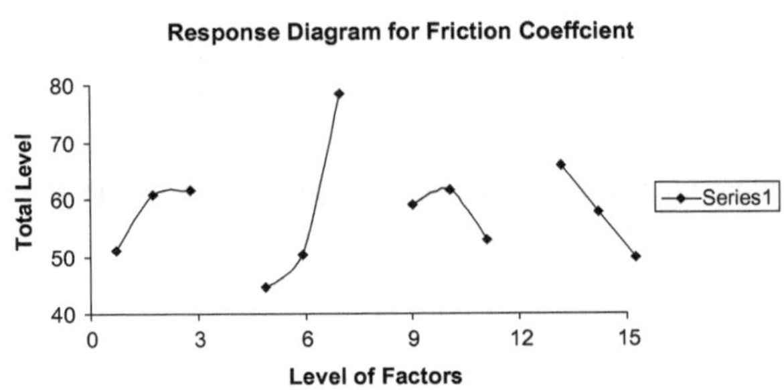

Table 6 Level totals of *S/N* ratio for each factor

Factors	A	B	C	D
Level 1	51.104	44.974	58.804	65.413
Level 2	60.427	50.326	61.178	57.441
Level 3	61.304	77.535	52.853	49.981
Difference	10.20	32.561	8.325	15.432
Total	172.835	172.835	172.835	172.835

Table 7 Optimal value for friction coefficient

Lub. Oil A_1	Speed B_1	Load C_3	Elevated temperature D_3
SAE10W30	600 r/min	100 N	100 °C

Table 8 Average *n* by factor levels

Factor	Level 1	Level 2	Level 3
A	17.034	20.1423	20.434
B	14.991	16.775	25.845
C	19.601	20.392	17.617
D	21.804	19.147	16.660

$$\begin{aligned} M_{A1} &= 1/3(n_1 + n_2 + n_3) \\ &= 1/3(15.820 + 15.738 + 19.546) = 17.034 \end{aligned} \tag{3}$$

The numerical values for n_j cal be substituted in the above expression to obtain values of the average n for each level of the four factors. These averages are shown in Table 8.

3.3 Analysis of Variance (ANOVA) for Friction Coefficient

The purpose of the ANOVA is to investigate operating parameters, which significantly affect the quality characteristic. This is accomplished by separating total availability of the multi response Signal-To-Noise ratios, which is measured by the sum of squared deviations from the total mean of the multi response Signal-To-Noise ratio, into contributions by each of the operating parameters and the error. First the total sum of the squared deviation (SS_T) from the total mean of the multi response Signal-To-Noise ratio can be calculated by using Eq. (4)

$$SS_T = \sum (n_j - n_m) \tag{4}$$

The total sum of the squared deviations SS_T is decomposed into three sources, the sum of squared deviations SS_d due to each operating parameters and the sum of the squared error SS_e. The percentage contribution by each of the operating parameters in the total sum of the squared deviations SS_T can be used to evaluate importance of the operating parameter change on quality characteristic. Then mean squared is calculated by dividing the degrees of the freedom of a factor to sum squares due to each factor. The number of independent parameter associated with an entry such as matrix experiment or factor or sum of squares is called as degrees of freedom.

Sum of squares due to a factor is calculated as follows:

Sum of squares due to factor

$$\begin{aligned} \text{A } (SS_d) &= 3(M_{A1} - M)^2 + 3(M_{A2} - M)^2 + 3(M_{A3} - M)^2 \\ &= 3(17.034 - 19.2038)^2 + 3(20.1423 - 19.2038)^2 \\ &\quad + 3(20.434 - 19.2038)^2 \end{aligned}$$

Mean squares due to factor A (SS_d): 21.3064
Mean squares due to factor A (MS): Sum of square due to factor/degree of freedom
Mean squares due to factor A (MS): 21.3064/2 = 10.6532
Similarly for other factors sum of squares and mean square values are calculated.

3.4 Estimation of Error Variance

$$\text{Error variance} = \frac{\text{Sum of square due to error}}{\text{Degrees of freedom for the error}} \tag{5}$$

In this present example only one factor (*C*) is used to estimate the error variance or errors mean square value than other factors.

$$\text{Error variance} = \sigma_e^2 = (12.8793/2) = 6.4396$$

3.5 F-Test

F-Test is used to determine which process parameters have a significant effect on the quality characteristic. The variance ratio denoted by *F* is given by Eq. (6),

$$\text{F} = \frac{\text{Mean Square due to a factor}}{\text{Error mean variance}} \tag{6}$$

From the ANOVA Table 9, it is found that factor '*B*' (operating speed) has the largest degree of contribution and is the most significant factor for PRA friction.

Table 9 ANVOA for friction coefficient

Factor	S.S	D.F	M.S	F	% P
A	21.3064	2	10.6532	1.6543	7.673
B	203.7629	2	101.8814	15.8210	73.387
C	12.8793	2	6.4396	1	4.6386
D	39.7045	2	19.8527	3.0829	14.3
Error	0	0	0	–	–
Total	277.6531	8	138.3255	–	100
Pooled error	12.8793	2	6.4396	–	–

Factor '*D*' (Elevated temperature) is the second most significant factor and factor '*A*' (SAE10W30 oil) is third. For the DOE, only the significant factors—lubricants used, operating speed, applied load and elevated lubricant temperature, were taken for analysis. The ANOVA for the DOE is shown in Table 9. The results show that the significance of the factors prevails in the following order of importance: (1) Operating speed, (2) Elevated lubricant temperature, (3) Type of lubricant used, (4) Applied normal loads. The response data of S/N ratio for main factors of DOE are plotted in Fig. 1. Considering minimum value of S/N ratio, the optimum levels of the significant parameters are as follows: (1) Operating speed 600 rpm, (2) SAE 10W30 oil, (3) Elevated temperatures 100 °C, (4) Applied load 100 N.

4 Summary

In the experimental performance, Taguchi method is employed to determine the main influence factors affecting the main characteristic of PRA friction. Following conclusions are derived.

An attempt has been made to investigate the significance parameters and their interactions over the PRA friction measurement on RFWTR with using 74 CC TVS pep models segment of piston ring-liner pair is made up of 73.38 % of operating speed and to determine the optimum levels by using Taguchi's Design of Experiment Technique.

The second significant factor is the elevated temperatures and it contributes 14.3 % for the minimization of PRA friction.

The third significant factor is the selection of proper lubricant in PRA friction measurements and it contributes in PRA friction is 7.673 %.

The applied load shows only little significance and its contribution on the PRA friction is 4.638 % only.

References

1. Sorab J, Korcek S (1998) Engine oil and surfaces effect in journal bearing lubrication. SAE 981408, May 1998
2. Johnson MD, Jensen RK, Korcek S (1997) Base oil effects on friction reduction capabilities of molybdenum dialkyldithiocarbamate containing engine oils. SAE 972860
3. Dowson D, Economou PN, Ruddy BL, Strachan PJ, Baker AJS (1979) Piston ring lubrication Part-II: Theoretical analysis of a single ring and complete ring pack. In: Proceedings of energy conservation through fluid film lubrication technology: frontier in research and design. ASME, New York, pp 23–52
4. Hwu CJ, Weng CI (1991) Elastohydrodynamic lubrication of piston rings. Wear 151:203–215
5. Rohde SM, Whitaker KW, McAllister GT (1990) A mixed friction model for dynamically loaded contacts with application to piston ring lubrication in surface roughness effects in hydrodynamic and mixed lubrications. In: Proceeding of the ASME winter annual meeting, pp 19–50
6. Patir N, Cheng HS (1978) An average flow model for determining effects of three dimensional roughnesses on partial hydrodynamic lubrication. Trans ASME J Lubric Technol 100:12–17
7. Greenwood JA, Tripp JH (1971) The contact of two nominal flat rough surfaces. Proc Inst Mech Eng 185:625–633
8. Ting LL (1993) Development of a reciprocating test rig for tribological studies of piston engine moving components. Part-I: Rig design and piston ring friction coefficients measuring method. SAE Paper 930685
9. Ting LL (1993) Development of a reciprocating test rig for tribological studies of piston engine moving components. Part-II: Rig design and piston ring friction coefficients measuring method. SAE Paper 930686
10. Dearlove J, Cheng WK (1995) Simultaneous piston ring friction and oil film thickness measurements in reciprocating test rig. SAE Paper 952470
11. Arcoumanis C, Duszynski M, Flora H, Ostovar P (1995) Development of piston ring lubrication test rig and investigation of boundary conditions for modeling lubrication film properties. SAE Paper 952468
12. Akalin O, Newaz GM (2001) Piston ring cylinder bore friction modeling in mixed lubrication regime. Part I: Analytical results. Trans ASME J Tribol 123:211–218
13. Akalin O, Newaz GM (2001) Piston ring cylinder bore friction modeling in mixed lubrication regime. Part II: Correlation with bench test data. Trans ASME J Tribol 123:219–223
14. Tung SC, Mcmillan ML, Goa H, Baradasz E (2005) Engine oil effects on friction and Wear using 2.2 L direct injection diesel engine components foe bench testing Part: II—Tribology bench test results and surface analyses. SAE International, 2004-01-2005
15. Peggy L, Barber G, Zou Q, Anderson JR, Tung SC, Angloe Q (2008) Friction and Wear of low-phosphorus engine oils with additional molybdenum and boron compound, measured on a reciprocating lubricant tester. Tribol Trans 51:659–672
16. Mistry KN, Priest M, Shrestha S (2010) The potential of plasma electrolytic oxidized eutectic aluminium-silicon alloy as a cylinder wall surface for lightweight engine blocks. Proc IMechE J J Eng Tribol 224:221–229
17. Roy RK (1996) A premier on the Taguchi method. Van Nostrand Reinhold, New York
18. Roy RK (2001) Design of experiment using the Taguchi approach. Wiely, New York

Strengthening Fiber-Matrix Adhesion to Improve Tribo-Performance of Composites

Nidhi Dureja, Ajay Kumar Kadiyala and Jayashree Bijwe

Abstract Polyetherimide (PEI) composites reinforced with graphite faric (GrF) were developed. For improved adhesion between the matrix and fiber, the fibers were treated with 0.3 % lanthanum fluoride (LaF_3) and nano sized particles of ytterbium fluoride (YbF_3) based suspension in ethanol. The specific wear rate (K_0) showed improvement of $\approx$50 % and the coefficient of friction (μ) reduced by $\approx$13 % due to treatment. YbF_3 treated fabric composite showed higher wear resistance compared to LaF_3 μ of LaF_3 treated composites on the other hand was lower than the composite treated with YbF_3.

Keywords Rare earth elements · Graphite fabric · Anti-friction · Nano-YbF_3 · Surface treatment · Wear

1 Introduction

Fibers/fabrics of carbon, graphite, glass, aramid etc. are widely used as reinforcements for development of composites [1]. Amongst these, graphite has higher thermal and thermo-oxidative stability, specific strength, thermal conductivity and self-lubricity leading to an ideal choice for developing tribo-composites in spite of its high cost [2]. In order to utilize the superior properties of GrF fully proper load transfer from the matrix to the fiber has to occur. For the selected matrix material surface of reinforcement plays a key role for determining final strength of a composite. GrFs are known to be chemically inert leading to poor fiber matrix adhesion. Inclusion of functional groups on the surface, surface topography and

N. Dureja · A. K. Kadiyala · J. Bijwe (✉)
Industrial Tribology Machine Dynamics and Maintence Engineering Centre (ITMMEC), Indian Institute of Technology Delhi, Hauz Khas, New Delhi 110016, India
e-mail: jbijwe@gmail.com

H. C. Patel et al. (eds.), *Proceedings of International Conference on Advances in Tribology and Engineering Systems*, Lecture Notes in Mechanical Engineering, DOI: 10.1007/978-81-322-1656-8_19,

increase in roughness of surface of a fiber can improve the fiber-matrix interaction [1, 3–5].

The fiber matrix adhesion can be enhanced by applying various surface treatments to the fibers surface. These include electrochemical, chemical, thermal, discharge plasma, rare earth salt (RES) treatment etc. The large effective nuclear charge of RES and trivalent nature of lanthanide series elements increase their ability to attract electrons of other atoms around it. Rare earth (RE) metals are highly reactive, which readily react with oxygen to form oxides. RE metal also chemically react with functional groups of polymers and carbon fiber to form Ionic and coordination bonds [6].

Most of the work based on RES treatment of fiber has been with $LaCl_3$ or praseodymium nitrate {Pr $(NO_3)_3$} based alcoholic suspension. The dip time and concentration were the selected variables. It was found the RES treatment improved the tribo-properties significantly compared to other traditional methods [4–11]. Tiwari et al. [4, 5] used nano YbF_3 in varying concentrations and found that 0.3 wt% YbF_3 showed the optimum tribo and mechanical properties. Tiwari [12] found that among plasma oxidation, acid oxidation, RES and gamma treatment, RES was the most effective in improving both friction and wear behavior of composite. At higher load the performance of these composites was substantially better than other the composites.

In the present work, different RES viz. LaF_3 and nano-YbF_3 were selected for treatment of graphite fabric.

2 Experimental Details

Graphite fabric (twill weave 2 × 2) was procured from SGL Technologies GmbH, Germany. Polyetherimide PEI ULTEM 1000 was selected as a matrix for developing composites and was procured from GE Plastics, USA. LaF_3 was procured from HPLC Pvt. Ltd., India. Rare earth salt Ytterbium fluoride (YbF_3) was supplied by Nano-Amor, USA in a 40–80 nm size.

2.1 Treatment of Fabric

0.3 wt% of nano/micro suspension of RES were prepared by probe sonication for 10 min duration with ethanol as a medium. GrF rolls of 3.5 × 8 inch were de-sized by refluxing in acetone at 55 °C for 3 h. After drying the fabrics were refluxed in nano/micro suspensions at 78 °C for ½ h duration, followed by cooling, and washing in water to remove any loosely attached particles and then drying in vacuum oven for 1 h at 100 °C to remove the ethanol and water.

2.2 Development of Composites

Composites were developed by impregnation method. The loose sides of the fabric were sealed with PTFE tape to avoid misalignment during impregnation process. Viscous solution of PEI using dichloromethane as solvent was prepared. The GrF were impregnated for 12 h in a sealed condition to avoid solvent evaporation. The plies were then removed from the container in stretched conditions and dried to remove the any solvent at 70 °C for 24 h. Fourteen piles were then stacked together in 3.5 × 8 inch mold and compression molded at 340 °C. These were kept at this temperature under 4 MPa pressure for 20 min followed by cooling at ambient temperature and release of pressure.

2.3 Characterization of Composites

2.3.1 Physical Characterization

The designations and physical properties of the developed composites are shown in Table 1.

2.3.2 Tribo-Characterization

Adhesive wear test was carried out using pin on disc configuration on Tribometer-3MT supplied by CETR, USA.

The pins of dimensions 10 × 10 × 3.5 mm were cut using ISOMat 1000 precision diamond cutter. Prior to test, bedding was carried out on C-1000 emery paper at low speed till 80 % of conformal contact was achieved. Pin was then cleaned in ultrasonic bath using petroleum ether followed by drying. Initial and final weight were measured (on a weighing balance with an accuracy of 0.0001 g) for weight loss calculation. Test was carried out against mild steel disc with Ra 0.15–0.2 μm as counter surface.

The input parameters during test were

- Load: Loads (100 N and 200 N)
- Speed: 2 m/s (600 rpm)
- Sliding time: 1 h

Table 1 Physical properties of composite

Composites	Density (g/cc)	Fiber content
G_O (Containing untreated graphite fabric)	1.538	52
G_L (Containing LaF_3 treated graphite fabric)	1.547	50
G_Y (Containing YbF_3 treated graphite fabric)	1.472	48

- Sliding distance: 7.68 km

The sp. wear rate K_0, of composite was calculated by following equation

$$K_0 = \Delta m/\rho L d \ m^3/N - m \quad (1)$$

where, K_0 is specific wear rate, Δm, weight loss of pin; ρ, density; L, Load; d, sliding distance.

3 Results and Discussion

Figures 1 and 2 show the K_0 and μ under two loads, slid under identical conditions.

Following are the observations based on the experiment:

- Wear rate of the composites were low and in the range of 1–2 × 10^{-15} m^3/NM.
- μ of the composites was little higher in the range of (0.2–0.32).

Fig. 1 Variation in K_0 with load

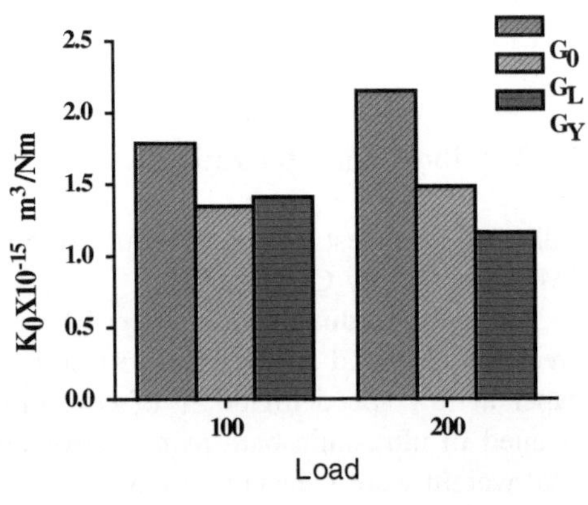

Fig. 2 Variation in μ with load

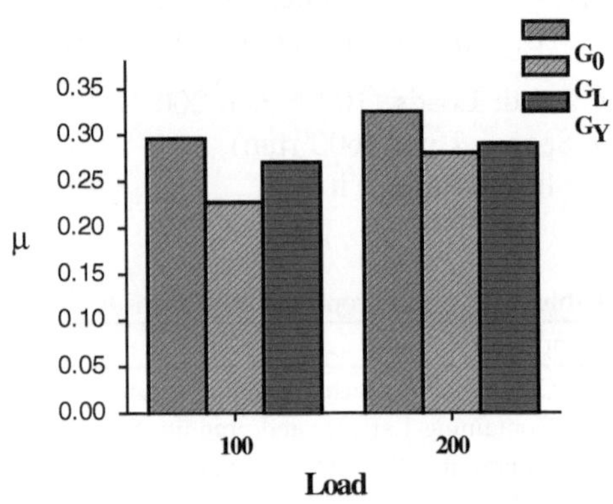

- With increase in load the K_0 of treated composites decreased, while for untreated composite it increased.
- μ increased with increase in load for all composites.
- Treated GrF composites showed improvement in Sp. wear rate.
- Higher load G_Y showed the best result. With $\approx$ improvement in sp. wear rate. This might be due to enhance fiber matrix bonding and hence matrix supported the fibers very firmly. This led to decrease in wear rate.
- The treated fabric showed reduction of μ compared to that of untreated fabric. With increase in load the μ decreased. LaF_3 showed the best performance in terms of μ.
- As compared to LaF_3, YbF_3 showed more promise.

4 Conclusions

From the tribo-analysis of the composites following conclusions were drawn.

- Treatment proved effective in decreasing the sp. wear rate of the composites compared to untreated. Nano-sized YbF_3 treatment showed $\approx$50 % improvement.
- Treated composites showed slight improvement in μ compared to untreated.

Acknowledgments The author is grateful to Department of Science and Technology India, for providing funding to carry out the present work.

References

1. Bijwe J, Rekha R (2007) Influence of weave of carbon fabric in polyetherimide composites in various wear situations. Wear 263(7–12):984–991
2. Kadiyala AK, Bijwe J, Surface lubrication of graphite fabric reinforced epoxy composites with nano- and micro-sized hexagonal boron nitride. Wear. doi: 10.1016/j.wear.2012.11.069
3. Jr. Pittman CU, Jiang W, Yue ZR, Gardner S, Wang L, Toghiani H (1999) Surface properties of electrochemically oxidized carbon fibers. Carbon 37:1797–807
4. Tiwari S, Bijwe J, Panier S (2013) Strengthening of a fibre-matrix interface: a novel method using nanoparticles. Nanomater Nanotechnol 3(3):1–8
5. Tiwari S, Bijwe J, Panier S (2011) Enhancing the adhesive wear performance of polyetherimide composites through nano-particle treatment of the carbon fabric. J Mater Sci 47:2891–2898
6. Li J, Cheng XH (2006) Effect of rare earth solution on mechanical and tribological properties of carbon fiber reinforced thermoplastic polyimide composite. Tribol Lett 25:207–214
7. Zhiwei X, Yudong H, Yuanjun S, Chunhua Z (2007) Surface characteristics of rare earth treated carbon fibers and interfacial properties of composites. J Rare Earths 25:462–468
8. Shang-guan QQ, Cheng XH (2008) Effect of rare earth surface treatment on tribological properties of carbon fibers reinforced PTFE composite under oil-lubricated condition. J Rare Earths 26:584–589

9. Bao D, Cheng X (2006) Evaluation of tribological performance of PTFE composite filled with rare earths treated carbon fibers under water-lubricated condition. J Rare Earths 24:564–568
10. Cheng XH, Shang-guan QQ (2006) Effect of rare earths on mechanical and tribological properties of carbon fibers reinforced PTFE composite. Tribol Lett 21:153–160
11. Shang-guan QQ, Cheng XH (2007) Friction and wear of rare earths modified carbon fibers filled PTFE composite under dry sliding condition. Appl Surf Sci 253:9000–9006
12. Tiwari S (2011) Role of fiber-matrix interface on mechanical and tribological properties of carbon fabric-polyetherimide composites. Ph.D. Thesis, Indian Institute of Technology Delhi

Part III
Lubricant

Ferrofluid Lubrication of Squeeze Film in Curved Circular Plates with Assorted Porous Structures

N. D. Patel, G. M. Deheri and S. S. Mehta

Abstract This paper aims to analyze the squeeze film performance of the magnetic fluid lubrication in curved porous circular plates with different porous structures. The globular sphere model due to Kozeny-Carman and capillary fissures model of Irmay for porous structure have been subjected to investigations. Here the film thickness is determined by the hyperbolic expression. The external magnetic field is considered to be oblique to the lower plate. It is noticed that the load carrying capacity increases with increasing values of magnetization. It is observed that there is a significant increase in the load carrying capacity in the case of concave plates and in the case of globular sphere model. Further, it is established that the effect of porosity in the case of capillary fissures model is nominal.

Keywords Circular plates · Ferrofluid · Porosity · Squeeze film · Load carrying capacity

1 Introduction

By now, the use of magnetic fluid as a lubricant modifying the performance of a bearing system has been well recognized. In fact, the research interest in suspension of magnetic nano particles in appropriate carrier liquid has been generated because of the fact that such fluids combine normal liquid behavior with the possibilities to control their flow and properties with moderate magnetic fields. This is the peculiarity of ferrofluids. Another distinguishable property of the magnetic fluid is that the magnetic fluids can be retained at a desired location by an external magnetic field. The bio medical application and the investigations of the

N. D. Patel (✉) · G. M. Deheri · S. S. Mehta
Department of Mathematics, S.P. University, V.V. Nagar, Gujarat 388120, India
e-mail: ndpatel2002@gmail.com

H. C. Patel et al. (eds.), *Proceedings of International Conference on Advances in Tribology and Engineering Systems*, Lecture Notes in Mechanical Engineering, DOI: 10.1007/978-81-322-1656-8_20, © Springer India 2014

rheological properties of ferrofluid considerable importance during the last few years only. A recent development in the field of ferrolfuid found in Odenbach [1] emphasized that the formation of structures of magnetic nano particles has considerable influence on the magneto viscous behavior of ferrofluids. Bhat and Deheri [2] considered the squeeze film behavior in porous annular disks under the presence of a magnetic fluid lubricant. Here the load carrying capacity and response time were found to be increased with increasing magnetization. Bhat and Deheri [3] studied the behavior of magnetic fluid based squeeze film in curved porous circular disks. It was shown that the effects due to magnetization were independent of the curvature of the upper disks. Shah and Bhat [4] analyzed the effect of rotational inertia on the magnetic fluid based squeeze film in curved porous circular plates. The rotational inertia significantly affected the performance of the bearing system. However, the response time depended on rotational inertia and the speed of the rotations of the plates. Shah and Bhat [5] discussed the ferrofluid squeeze film behavior between curved annular plates taking the rotation of magnetic particles into consideration. Load carrying capacity and response time increased when the volume concentration of the solid phase, Langevin's parameter or the curvature of the upper plate were increased. Deheri et al. [6] studied the effect of transverse roughness on the magnetic fluid based squeeze film behavior in curved circular plates. It was proved that the transverse roughness affected the bearing system adversely, the situation was found to be relatively better in the case of negatively skewed roughness. For the better performance the curvature parameters played seminal role. Patel et al. [7] dealt with the effect of surface roughness on the performance of a magnetic fluid based squeeze film in circular plates with porous matrix of variables thickness. This investigation suggested that by choosing properly the thickness ratio parameter and the strength of the magnetic field the adverse effect of roughness on the bearing system could be minimized in the case of negatively skewed roughness. Lin [8] derived the lubrication equation in cylindrical coordinates for ferrofluid squeeze film including the effects of convective inertia forces and the presence of transverse magnetic field for engineering applications to parallel circular disks. The ferrofluid circular squeeze film considering fluid inertia effect provided a higher load capacity and a longer elapse time as compared to the non inertia non magnetic case. Naduvinamani et al. [9] observed that the combined effect of MHD and surface roughness on couple stress squeeze film lubrication between porous circular stepped plates was significant. It was observed that the effect of applied magnetic field on the squeeze film lubrication between circular stepped plates with couple stress fluid was to increase the load carrying capacity significantly and to delay the time of approach. Recently, Abhangi and Deheri [10] discussed the numerical modeling of squeeze film performance between rotating transversely rough curved circular plates under the presence of a magnetic fluid lubricant. This investigation showed that although the bearing suffered on account of transverse surface roughness, there were sufficient scopes to get a relatively better performance in the case of negatively skewed roughness by suitably choosing curvature parameters and rotation ratio. Lin et al. [11] investigated the parallel circular squeeze films with a non

Newtonian ferrofluid in the presence of transverse magnetic field. The modified Reynolds equation for obtaining the pressure distribution was obtained by applying the Stokes's micro continuum theory. The non Newtonian ferrofluid lubricated squeeze film gave a higher load carrying capacity.

In this paper efforts have been made to study and discuss the performance of a ferrofluid based squeeze film in curved porous circular plates with different porous structures adopting Kozeny-Carman and Irmay models.

2 Analysis

The geometrical configuration of the bearing system presented in Fig. 1 has two circular plates each of radius a.

The thickness of the porous facing of the upper plate is l1. A solid wall backs the porous housing. It approaches normally towards an impermeable and flat lower plate with uniform velocity $\dot{h}_0 = \frac{dh_0}{dt}$, h_0 being the central film thickness. The film thickness h admits the form Bhat [13].

$$h = h_0 \frac{1}{(1 + cr)} \tag{1}$$

where r (meter) is the radial coordinate.

Assuming axially symmetric flow of the ferrofluid between the plates under an oblique magnetic field $H = (H(r) \cos\varphi, 0, H(r) \sin\varphi)$ whose magnitude H vanishes at $r = 0$, a the modified Reynolds's type equation governing the film pressure P (Shah and Bhat [12]) turns out to be

$$\frac{1}{r}\frac{d}{dr}\left\{\left(h^3 + 12\psi l_1\right) r \frac{d}{dr}\left(p - \mu_0 \bar{\mu} H^2\right)\right\} = 12\eta \dot{h}_0 \tag{2}$$

where ψ is the permeability of the porous region, μ_0 is the free space permeability, $\bar{\mu}$ is the magnetic susceptibility and η is the absolute fluid viscosity.

Case I (a globular sphere model as given in Fig. 2). Here globular particles (of mean particle size D_c) fill a porous material

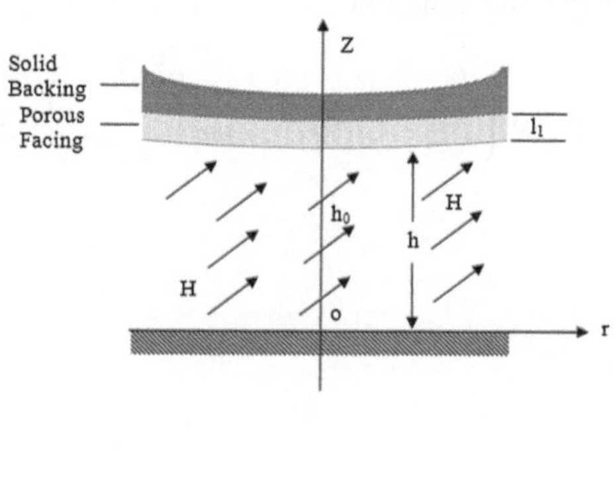

Fig. 1 Configuration of the bearing system

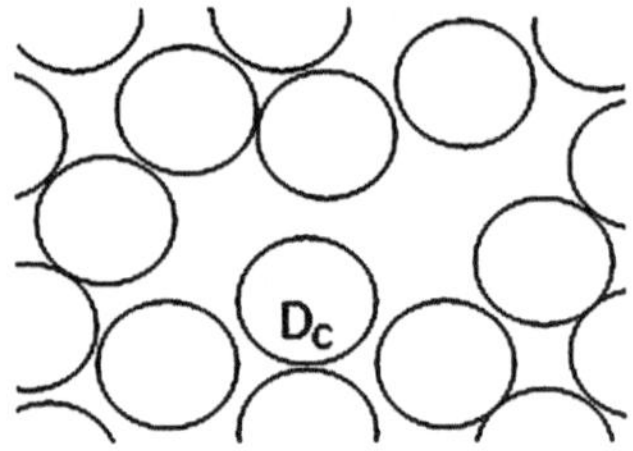

Fig. 2 Structure model of porous sheets given by Kozeny-Carman

Taking recourse to Kozeny-Carman model [14], one gets

$$\psi = \frac{D_c^2 \varepsilon^2}{180\,(1 - \varepsilon)^2} \tag{3}$$

where ε is the porosity.

The dimensionless quantities;

$$P = \frac{-ph_0^3}{\eta a^2 \dot{h}_0}\, R = \frac{r}{a}\ \bar{\psi} = \frac{D_c^2 l_1;}{h_0^3}, C = ca,$$
$$\mu^* = \frac{-\mu_0 \bar{\mu} h_0^3 k}{\eta \dot{h}_0} \text{ are considered.} \tag{4}$$

Following Bhat [13] the magnitude of the magnetic field is

$$H^2 = \frac{kr^2(a - r)}{a} \tag{5}$$

where $k = 10^{14} - 10^{16}\, A^2/m^4$ is chosen so as to have a magnetic field of strength between the order of 105 and 106.

Since the magnetic field arises out of a potential, the angle $\varphi = \varphi(r, z)$ can be obtained from

$$\cot\varphi \frac{\partial\varphi}{\partial r} + \frac{\partial\varphi}{\partial z} = \frac{3r - 2a}{2r(a - r)^3} \tag{6}$$

with the help of elliptic integrals depending on the value of a.

Using Eqs. (1)–(5), one concludes that the dimensionless film pressure P satisfies the equation

$$\frac{1}{R}\frac{d}{dR}\left\{\left(\frac{R}{(1 + CR)^3} + \frac{R\bar{\psi}\varepsilon^2}{15(1 - \varepsilon)^2}\right)\frac{d}{dR}\left(P - 0.5\mu^* R^2(1 - R)\right)\right\} = -12 \tag{7}$$

Solving Eq. (7) under the boundary conditions

$$P = 0 \text{ when } R = 1 \text{ and } \frac{dP}{dR} = 0 \text{ when } R = 0 \tag{8}$$

One arrives at

$$P = 0.5\,\mu^* R^2(1 - R) + \frac{15(1 - \varepsilon)^2}{\bar{\psi} C\varepsilon^3} \ln\left\{\frac{15(1 - \varepsilon)^2 + \frac{\bar{\psi}\varepsilon^3}{(1 + C)^3}}{15(1 - \varepsilon)^2 + \frac{\bar{\psi}\varepsilon^3}{(1 + CR)^3}}\right\} \tag{9}$$

The load capacity W of the bearing can be expressed in dimensionless form as

$$W = \frac{\mu^*}{40} + \int_0^1 \frac{15R(1 - \varepsilon)^2}{\bar{\psi} C\varepsilon^3} \ln\left\{\frac{15(1 - \varepsilon)^2 + \bar{\psi}\varepsilon^3/(1 + C)^3}{15(1 - \varepsilon)^2 + \bar{\psi}\varepsilon^3/(1 + CR)^3}\right\} dR \tag{10}$$

Case II (a capillary fissures model as shown Fig. (3)). The model consists of three sets of mutually orthogonal fissures (a mean solid size D_s). Assuming no loss of hydraulic gradient at the junctions, Irmay [15] derived the permeability

$$\psi = \frac{\left(1 - m^{2/3}\right) D_s^2}{12m} \tag{11}$$

where $m = 1 - \varepsilon$, ε being the porosity

Introducing the dimensionless quantities

$$P = \frac{-ph_0^3}{2\dot{h}_0}\; R = \frac{r}{a}\; \bar{\psi} = \frac{D_s^2 l_1}{h_0^3},\; C = ca \text{ and}$$
$$\mu^* = \frac{-\mu_0 \bar{\mu} h_0^3 k}{\eta \dot{h}_0} \tag{12}$$

Fig. 3 Structure model of porous sheets given by Irmay

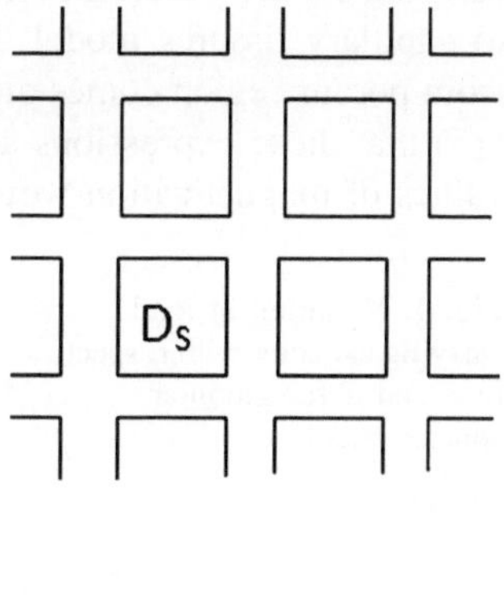

and making use of Eqs. (1), (2), (5), (11) and (12), the dimensionless squeeze film pressure P satisfies the modified Reynolds's equation

$$\frac{1}{R}\frac{d}{dR}\left\{\left(\frac{R}{(1+CR)^3}+\frac{R\psi(1-m^{2/3})}{m}\right)\frac{d}{dR}(P-0.5\mu^*R^2(1-R))\right\} = -12 \tag{13}$$

Solution of Eq. (13) under the boundary conditions (8), one finds the expression for non-dimensionless pressure distribution

$$P = 0.5\mu^*R^2(1-R)+\frac{m}{\bar{\psi}C(1-m^{2/3})}\ln\left\{\frac{m+\frac{(1-m^{2/3})\psi}{(1+C)^3}}{m+\frac{(1-m^{2/3})\psi}{(1+CR)^3}}\right\} \tag{14}$$

The load capacity W of the bearing in dimensionless form is determined from

$$W = \frac{\mu^*}{40}+\int_0^1\frac{mR}{\bar{\psi}C(1-m^{2/3})}\ln\left\{\frac{m+\frac{(1-m^{2/3})\psi}{(1+C)^3}}{m+\frac{(1-m^{2/3})\psi}{(1+CR)^3}}\right\}dR \tag{15}$$

3 Results and Discussion

It is observed that Eq. (10) presents the expression of load carrying capacity in dimensionless form for globular sphere model given by Kozeny-Carman while the load carrying capacity for capillary fissures model by Irmay is provided by Eq. (15). The corresponding non magnetic cases can be obtained by taking the magnetization parameterμ^* to be zero from Fig. 4 and 5. It is seen that the load carrying capacity increases more in the case of globular sphere model as compared to capillary fissures model. Probably this may be due to fact that the ferrofluid from porous region comes out relatively slowly. It can be seen from Eqs. (10) and (15) that these expressions are linear with respect to μ^*. Accordingly, increasing values of magnetization would result in increased load carrying capacity.

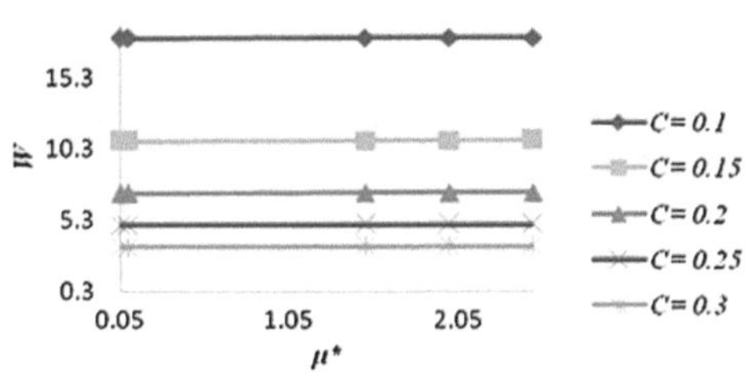

Fig. 4 Variation of load carrying capacity with respect to $\mu^*$and C for globular sphere model

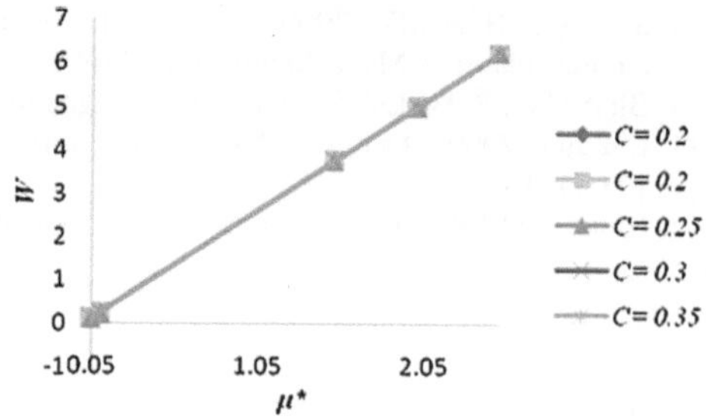

Fig. 5 Variation of load carrying capacity with respect μ^* and C for capillary fissures model

4 Conclusions

This investigation establishes that the Globular Sphere model of Kozeny-Carman provides a better alternative as compared to the capillary fissures model of Irmay, for a large range of magnetic strength. This article may provide an additional degree of freedom from design point of view. Even when there is no flow, this type of bearing system tends to support a load unlike what one finds in the case of traditional lubricants.

References

1. Odenbach S (2004) Recent progress in magnetic fluid research. J Phys Condens Matter 16:1135–1150
2. Bhat MV, Deheri GM (1991) Squeeze film behavior in porous annular discs lubricated with magnetic fluid. Wear 151:123–128
3. Bhat MV, Deher GM (1993) Magnetic fluid based squeeze film in curved porous circular discs. J Magn Magn Mater 127:159–162
4. Shah RC, Bhat MV (2000) Squeeze film based on magnetic fluid in curved porous rotating circular plates. J Magn Magn Mater 208:115–119
5. Shah RC, Bhat MV (2005) Ferrofluid squeeze film between curved annular plates including rotation of magnetic particles. J Eng Math 51:317–324
6. Deheri GM, Patel RM, Abhangi ND (2010) Magnetic fluid based squeeze film behavior between transversely rough curved plates. Adv Trib 54–55
7. Patel RM, Deheri GM, Patel HC (2011) Effect of surface roughness on the behavior of a magnetic fluid based squeeze film between circular plates with porous matrix of variable thickness. Acta Polytechnica Hungarica 8:171–190
8. Lin JR (2012) Derivation of ferrofluid lubrication equation squeeze films with convective fluid inertia forces and application to circular disks. Tribol Int 49:110–115
9. Naduvinamani NB, Hanumagowda BN, Syeda Tasneem Fathima (2012) Combined effects of MHD and surface roughness on couple stress film lubrication between porous circular stepped plates. Tribol Int 56:19–29
10. Abhangi ND, Deheri GM (2012) Numerical modeling of squeeze film performance between rotating transversely rough curved circular plates under the presence of a magnetic fluid lubricant. ISRN Mech Eng, p 9
11. Lin JR, Lu RF, Lin MC, Wang PY (2013) Squeeze film characteristics of parallel circular disks lubricated by ferrofluids with non-Newtonian couple stresses. Tribol Int 61:56–61

12. Shah RC, Bhat MV (2000) Squeeze film based on magnetic fluid in curved porous rotating circular plates. J Magn Magn Mater 208:115–119
13. Bhat MV (2003) Lubrication with a magnetic fluid. Team Spirit (India) Pvt. Ltd, Hyderabad
14. Liu Jun (2009) Analysis of a porous elastic sheet damper with a magnetic fluid. J Tribol 131:11–15
15. Irmay S (1955) Flow of liquid through cracked media. Bull Res Counc Isr 1(5A), p 84

Lubrication for Microsystems

Sujeet K. Sinha

Abstract Tribology plays an important role in the performance and life of any mechanical machine. This is even more so for micro-scale devices such as micro-electro-mechanical systems (MEMS) because of the large surface area to volume ratios of their components. Large surface area leads to high surface adhesion and friction. Devices fail by stiction or wear as a result. In order to keep up the progress in micro- and nano-technology, it is vital that nanotribological research is advanced further, and appropriate lubrication solutions are found which are specific to microsystems. In this talk, we will present recent research results on several solutions for nanolubrication. A number of chemical and surface textural modifications of the surface of materials such as Si/SiO_2 and SU-8, which are the most used structural materials for MEMS, have been employed with the aims of reducing friction and increasing wear life.

Keywords Nanolubrication · MEMS

1 Introduction

Reducing adhesion, friction and wear are important issues in the running of any machine including micro-machines. These tribological issues lot depend on the materials that are used in making the components of the machine and the type of lubrication solutions employed. Materials used for making MEMS components are generally Si, SU-8, PDMS, PMMA etc. These materials are tribologically poor materials as they show high adhesion, friction and wear against themselves or other materials. Therefore, it is important that the materials themselves are

S. K. Sinha (✉)
Department of Mechanical Engineering, Indian Institute of Technology,
Kanpur 208016, India
e-mail: sujeet@iitk.ac.in

H. C. Patel et al. (eds.), *Proceedings of International Conference on Advances in Tribology and Engineering Systems*, Lecture Notes in Mechanical Engineering,
DOI: 10.1007/978-81-322-1656-8_21,

properly treated either on the surface or in the bulk so that adhesion, friction and wear are drastically reduced.

1.1 Analysis

The surface force, F, due to van der Waals interactions is given for two flat surfaces as [1],

$$F = -\frac{A}{6\pi D^3} \tag{1}$$

where A is a constant known as Hamaker constant (typically in the range of 0.4–4 × 10^{-19} J for solids) and D is the intervening distance between the two surfaces. As the distance between two bodies closes, the force will increase to very high values. For example, at D = 0.2 nm the attractive adhesive pressure is in the range of 7 × 10^3 atm or 709 MPa [1]. However, as the distance increases, for example because of the surface roughness, the value of the surface force will drastically decrease. Therefore, for smooth surfaces and small/light components, the surface forces control the functioning of the machine by controlling adhesion, friction and wear which are interrelated.

The attractive force, F_o, (also known as pull-off force) between two surfaces can also be obtained from the surface energy calculations using the Bradley's equation (for a ball in contact with a flat surface) as,

$$F_0 = 2\pi R(\gamma_1 + \gamma_2 - \gamma_{12}) \tag{2}$$

where R is the diameter of the spherical surface. γ_1, γ_2 and γ_{12} are the surface energies of solid 1, solid 2 and the interfacial surface energy of 1 and 2, respectively. Clearly F_0 will be high if the surface energies of the solid surfaces are high. An empirical relation has been established between the pull-off force and the initial coefficient of friction between two solids. The relation is given as [2],

$$\mu_i = \left[\frac{C\exp(nF_o)}{L^{0.25}}\right] + \alpha \tag{3}$$

where C is a constant to be found out experimentally. n is an exponent to be found out from the relation between the initial shear stress and the pull-off force. α is a coefficient for the pressure dependence term of the contact shear stress as defined by the classical equation, $\tau = \tau_0 + \alpha P$ [3, 4]. Equation (3) clearly shows that for low loads, the initial coefficient of friction is an exponential function of the pull-off force. This means that the coefficient of friction will exponentially increase if the adhesion forces are increased. This conclusion has direct bearing on the tribology of microsystems such as MEMS where tradition liquid lubrication is often impractical and the choice of structural material is very limited.

Table 1 Mechanical and tribological properties of MEMS materials

Materials	Elastic modulus, GPa	Hardness, MPa	Coefficient of friction against silicon nitride
Si/SiO_2	~130–173	~11 x 10^3	~0.4–0.6
SU-8	~3.8–4.29	~0.271	~0.8
h-PDMS	~10 MPa	–	~0.54

Source various technical papers and reports available on the Internet

1.2 Material Issues

There is a limitation in the selection of material for making MEMS because of two main reasons. The first is that the material used must be compatible with the micro-fabrication process for cost-effective and fast manufacturability requirements of the industry. Secondly, the material must satisfy mechanical property requirements such as strength and stiffness. Table 1 compares mechanical and tribological properties of common materials used in MEMS.

The above two requirements limit the choice of materials to Si/SiO_2, SU-8, PDMS and PMMA etc. which have been used for making MEMS. Therefore, with the limitations in the choice of the structural material, it is vitally important to improve the tribological performances of the given material for MEMS fabrication. The tribological solution must be compatible with the process of manufacturing of MEMS, cost-effective and highly durable against wear.

Because of the above requirements and constraints, and the fact that tradition tribological solutions such as liquid film lubrication are not practical, developing suitable strategies for the lubrication of microsystems acquires extra importance.

We have proposed and tested several types of surface modification and coating techniques with excellent tribological performances, which have prospects of further development into industrial practice. Some of the solutions are presented here.

2 Lubrication Solutions

2.1 Application of Self-Assembled Monolayers

A number of studies have been published to show the effectiveness of self-assembled monolayers (SAMs), which are hydrophobic and thus of low surface energy. SAMs are effective in reducing stiction and initial friction. However, the wear lives of SAMs have limitations when applied on a surface such as Si/SiO_2. The SAM must have hydrophobic end group to be effective and there must be a possibility of chemical or physical bonding between the substrate (Si) and the SAM molecule. Our works have shown that despite low initial friction, SAMs do not show long wear life. This is because SAMs are solid molecules and hence the contact shear stresses in sliding can be high leading to the detachment of the molecules and the exposure of the substrate. This problem can be alleviated if we overcoat a SAM surface with a mobile nanolubricant. An ultra-thin layer of perfluoropolyether (PFPE) is very effective in increasing the wear life by several orders. Figure 1 presents wear lives of two types of SAMs overcoated with an ultra-thin layer of PFPE. It was found that a hydrophilic SAM (APTMS) performs better than a hydrophobic SAM (OTS) when lubricated with a nanolubricant.

2.2 Ultra-Thin Polymer Film

Linear thermoplastics such as polyethylene (PE), polytetrafluoroethylene (PTFE) and ultra-high molecular weight polyethylene (UHMWPE) are low friction polymers (against many surfaces), and hence a thin layer of such a polymer could provide durable coating on Si or other materials. Application of functionalized PE film with the help of an intermediate layer of benzophenone has helped to increase the wear-life considerably when the film is overcoated with a nanolubricant (PFPE) [6]. Figure 2 presents results for the composite layer of benzophenone/fPE/PFPE on Si substrate for comparison. The role of the intermediate layer is to provide strong bonding of the PE molecules through functionalize groups and chemical reaction. A slightly thicker layer of UHMWPE is also applicable in this context.

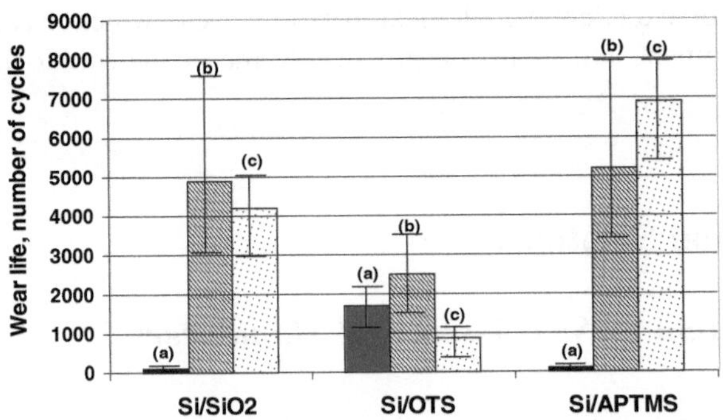

Fig. 1 Wear life (in terms of the number of cycles at which the interfacial friction reached a value of 0.3), of bare Si, OTS SAM and APTMS SAM with and without PFPE overcoat and with thermal treatment after PFPE overcoating. *a* No PFPE overcoat, *b* PFPE overcoat and *c* PFPE overcoat followed by thermal treatment. Data scatter bars are also presented for each case. Counterface was 4 mm diameter silicon nitride ball (*OTS* octadecyltrichlorosilane; *APTMS* 3-aminopropyltrimethoxysilane, figure taken from Ref. [5])

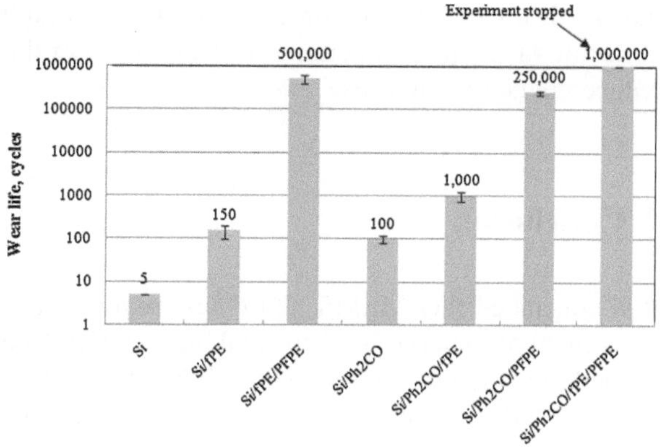

Fig. 2 The wear durability of Si against 4 mm diameter silicon nitride ball with different modifications under the application of 40 mN applied normal load and 500 rpm rotational speed. *fPE* = functionalized PE and Ph_2CO = benzophenone, figure taken from Ref. [6])

2.3 Nano-Texturing and Nano-Lubrication of Polymers

In an attempt to improve the tribological performances of polymer materials used for MEMS, we have applied oxygen plasma treatment coupled with nanolubrication by PFPE on the surface of PMMA and SU-8 [7, 8]. In both cases, the reduction in friction was many folds and the wear-life improved by several orders. Figure 3 provides wear durability data for PMMA surface before and after O_2 plasma treatment and PFPE overcoat.

It was observed that O_2 plasma treatment performs several functions on the surface of a polymer. It cleans the surface of any contaminant (oxides, hydrocarbons, dirt etc.), generates functional groups such as hydroxyl and carboxyl [9], and creates nanoscale texturing on the surface of the polymer. The former two helps in the bonding of the nanolubricant molecules to the polymer surfaces whereas nano-texturing has effect on reducing the contact area and adhesion which

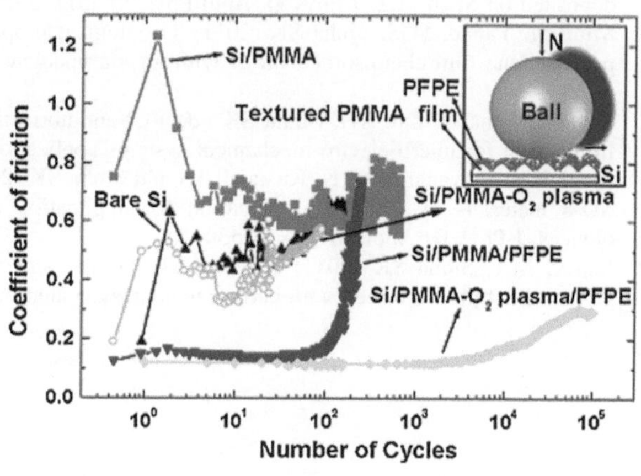

Fig. 3 Coefficient of friction versus number of cycles for a typical run for all the surfaces studied viz. PMMA film on Si/SiO_2, PMMA/PFPE and PMMA-O_2 plasma/PFPE. The normal load and sliding velocities used are 30 g and 0.042 ms^{-1}, respectively. Inset shows a graphical model of the contact between the ball and PFPE coated O_2-PMMA islands (the drawing is not to scale), figure is taken from Ref. [7])

directly leads to low friction and low wear. In addition to the bonded lubricant molecule layer, it is imperative to have an ultra-thin layer of mobile fraction to reduce contact shear stress.

3 Conclusion

It is concluded that MEMS lubrication requires a suitable strategy depending on the structural material used for making the MEMS and the scheme of integration of the lubrication solution into the MEMS fabrication process. Si/SiO_2 MEMS will require SAM or a polymer coating before lubrication with PFPE. Polymers such as SU-8 and PMMA can be lubricated and wear life increased by a simple treatment of O_2 plasma treatment and nanolubrication by PFPE.

Acknowledgments [Note: The author would like to acknowledge the help of Dr. Myo Minn, Dr. N. Satyanarayana, Dr. M. A. Samad and Dr. Arvind R. Singh who conducted the researches mentioned in this extended abstract. The readers can read the published papers mentioned in the reference for further information on the data provided in this extended abstract].

References

1. Israelachvili JN (2011) Intermolecular and surface forces, 3rd edn. Elsevier, Amsterdam, p 255
2. Minn M, Sinha SK (2010) The frictional behavior of UHMWPE films with different surface energies at low normal loads. Wear 268:1030–1036
3. Bowden FP, Tabor D (1958) The friction and lubrication of solids. Oxford University Press, Oxford
4. Briscoe BJ, Scruton B, Willis RF (1973) The shear strength of thin lubricant films. Proc R Soc Lond A 333:99–114
5. Satyanarayana N, Sinha SK (2005) Tribology of PFPE overcoated self-assembled monolayers deposited on Si surface. J Phys D: Appl Phys 38:3512–3522
6. Minn M, Yanadi SGS, Sinha SK (2011) Tribological properties of ultra-thin functionalized polyethylene film chemisorbed on Si with an intermediate benzophenone layer. Tribol Lett 41:217–226
7. Satyanarayana N, Lau KH, Sinha SK (2008) Nanolubrication of poly(methyl methacrylate) films on Si for micro-electro-mechanical systems applications. Appl Phys Lett 93:261906
8. Singh RA., Satyanarayana N, Kustandi TS and Sinha SK (2011) Tribo-functionalizing Si and SU-8 materials by surface modification for application in MEMS/NEMS actuator-based devices. J Phys D: Appl Phys 44:015301
9. Samad MA, Sinha SK (2011) Mechanical, thermal and tribological characterization of a UHMWPE film reinforced with carbon nanotubes coated on steel. Tribol Int 44:1932–1941

Study on the Effect of Blend Oil with Based Oil Analysis

B. M. Sutaria, M. K. Bhatt, B. K. Sonigra and D. V. Bhatt

Abstract Modern lubricants are careful balance of mineral or synthetic base stock, coupled with oil additives to enhance an existing property of the based oil such as oxidation resistance. In used oil analysis, nonconforming test data related to lubricant physical and chemical properties; contamination and wear debris is frequently encountered. The paper presents shear stability and extreme pressure test in order to characterize base stock oil and effect of blend low viscosity oil in the based oil. In present analysis, the oil used in sewing machine for the purpose of lubrication is used as blend oil in the Castrol GTX oil. The ASTM performance tests were conducted on the four ball test rig for comparative study which establishes the way that the mechanical solicitations during service life of the oils affect the viscosity modification and thus the quality of the oil performance.

Keywords Sewing machine oil · GTX oil · Shear stability · Four ball test rig

1 Introduction

For conventional machine elements various types of lubricants are available and user has to select any one of them. In case lubricant does not work effectively, the failures may increase. It is essential to check the physic-chemical properties of the oils while selecting.

The viscosity is the basic property of the lubricant for a given grade and class, such as industrial, automotive, metal working etc. User must ask for performance evaluation test in all cases.

B. M. Sutaria · M. K. Bhatt · B. K. Sonigra (✉) · D. V. Bhatt
Department of Mechanical Engineering, S.V. National Institute of Technology, Surat, Gujarat 395007, India
e-mail: sonigra_bhaskar@yahoo.com

H. C. Patel et al. (eds.), *Proceedings of International Conference on Advances in Tribology and Engineering Systems*, Lecture Notes in Mechanical Engineering, DOI: 10.1007/978-81-322-1656-8_22, © Springer India 2014

It is observed that viscosity of oil is decreasing rate during service life. The viscosity of oil is influenced mainly by temperature variations. The most important goal is to obtain oils with as constant viscosity index overall functioning under thermal domain. In this regard some polymeric materials are used as oil additives [1]. The main problem encountered in polymer added oils is the loss of viscosity due to long polymeric chains fracture during the oil life cycle [2]. The establishing of the viscosity loss in such cases is compulsive for a long oils' service life. The tests could be made on the four ball test rig. The four ball machine was proposed for the first time by Boerlage [3] in order to determine some lubricant properties. Because of its design simplicity and simple geometry, it is widely used in research and industrial laboratories [4].

2 Four Ball Test Rig

The four ball apparatus is widely used for determination of the tribological characteristics of lubricant such as extreme pressure and anti-wear properties. From the apparatus, it can be tested for load carrying capacity, coefficient of friction and anti-wear properties of lubricating oils under standard operating conditions.

The schematic view of four ball tester is shown in Fig. 1. In the four balls test rig top steel ball rotated through variable frequency drive electric motor and three steel balls are fixed in ball pot assembly. The load is given to the bottom of these

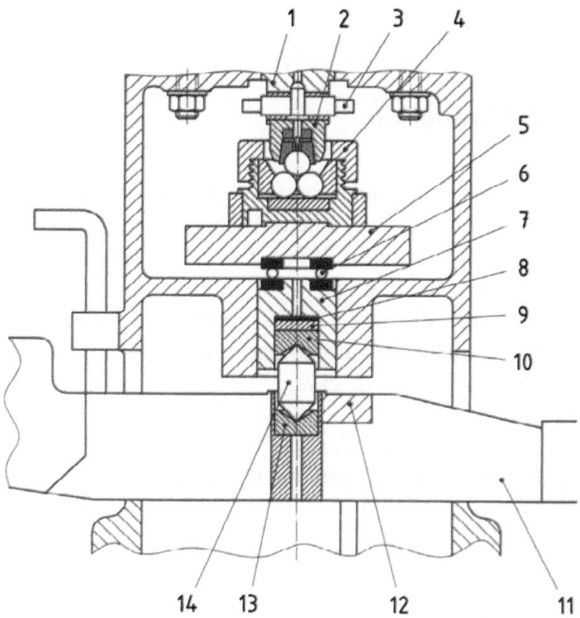

Fig. 1 Four ball tester—mechanical design of the testing area [4]. *1*—ball chuck holder, *2*—ball chuck, *3*—cam for removing ball chuck, *4*—ball pot assembly, *5*—ball pot mounting disc, *6*—trust bearing, *7*—cross head, *8*—brass shims, *9*—rubber disc, *10*—step bearing, *11*—counter-weighted lever arm, *12*—fulcrum, *13*—step bearing, *14*—pressure pin

fixed balls pot assembly through Pneumatic systems. The oil to be tested is filled in the ball cup, so that all balls properly deep. For application of loading, there is enough contact with all four balls and parameters that could be measured like the friction torque which acts on the fixed balls. The rotational speed and operating conditions are selected as per the ASTM standards. The load could be constant or progressively applied, in a discontinuous or continuous way [5, 6]. The sliding motion between balls conducts to a shearing onto lubricant. The four ball tester is used for evaluating the anti-wear and extreme pressure properties of lubricants. The procedures for the determination of tribo physical properties are described by standards i.e. ASTM D2783, ASTM D4172, ISO 20623:2003, EN ISO 20623:2003, DIN 51350, etc.

3 Operating Test Parameters and Specimen

In the present performance test work, ASTM D-4172 standard has been used for determine the tribo properties of the lubricant oil. This test method covers procedure for making a preliminary evaluation of the anti-wear properties of lubricants in sliding contact by means of the four-ball Tester Machine. The operating conditions were maintained as; rotational speed 1200 rpm, Temperature 75 ± 2 °C, load 147 ± 2 N, and test run for 60 min. Test Balls are made of chrome alloy steel from AISI standard steel E-52100, having diameter of 12.7 mm and Grade 25 EP (Extra Polish). Such balls are described in ANSI B3.12. The extra-polish finish is not described in that specification. The hardness of steel is in the range of 64–66 RC. Good cleaning fluids used to clean balls and ball pot assembly for the test, which should be approved as nontoxic, capable of removing antirust coatings from the balls, eliminating test-oil carryover from one test to the next, and not contribute to wear or anti-wear of the test lubricant. There are four different blending ratio oil considered for the study of performance test. The low viscosity sewing machine oil is mixed in the proportion of 10–40 % with based stock Castrol GTX oil [7].

4 Experiment Results and Discussions Viscosity

Viscosity of oil is being measured with the help of Redwood viscometer. The kinematic viscosity (ν) of liquid and time (t) required to pass 50 ml of liquid are correlated by following equation [8]

$$\nu = 0.26t - 172/t \tag{1}$$

where,

ν = kinematic viscosity in centi-stokes
t = time in second to collect 50 ml

Fig. 2 Variation of viscosity of blended oils with temperature

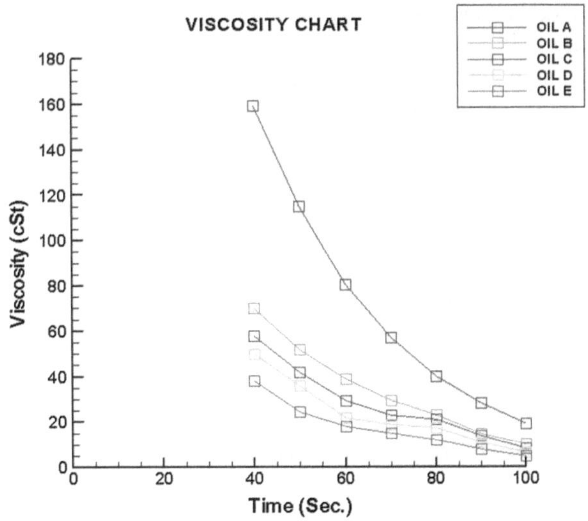

All different blending ratio oils are used to determine the viscosity in the span of temperature from 40 to 100 °C. Figure 2 shows the variation of different blending oil viscosity with the temperature. It is seen from the Fig. 2 that as the percentage of blend ratio increases the viscosity of oil decreases with variation of temperature from 40 to 100 °C. It is also observed that fresh blend oil shows comparatively low viscosity and the trend of evaluated viscosity of oil is found decreases exponentially.

4.1 Analysis of Blending Oil at 15 kg Load

Figure 3 represents the comparison of different ration of blending oils friction torque versus time. It is observed that initial period of time sudden peak of frictional torque is found in all the oils. These could be reason of an application of loading all balls have proper stick each other and then sliding occur. It is also seen from Fig. 3 that at the lower load application, frictional torque found more uniform and the value is less. While in case of higher load application the friction torque behavior is non uniform. Frictional torque is obtained higher in case of lower ratio of blending oil compared to higher ratio of blending with based stock Castrol GTX oil. This is due to the lower co-efficient of friction of higher blend oil.

4.2 Coefficient of Friction

As per the Coulomb's law, which came in 1785, the kinetic friction is nearly independent of speed of sliding. But this law has smaller range of applicability.

Fig. 3 Frictional torque versus time for various blend oils at 15 kg

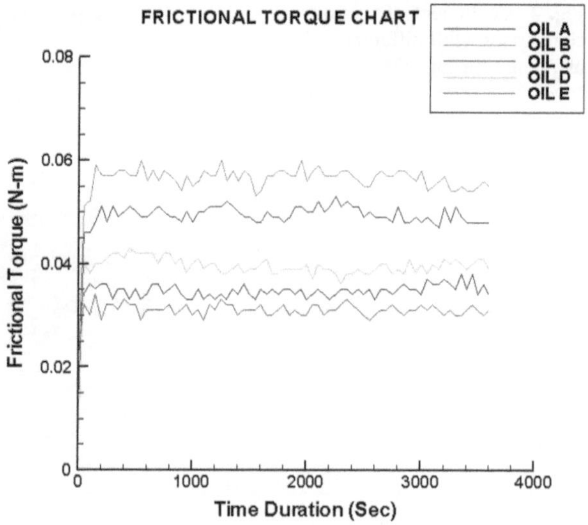

During rolling of four surfaces relative to each other, any relative motion can be regarded as combination of rolling, sliding and spin [9]. Consider two nonconforming bodies which touch at single points. The coefficient of friction (μ) between two solid surfaces is defined as the ratio of the tangential force (F) which required sliding or rolling, and the normal force applied between the surfaces (N). It can be presented mathematically as, $F = \mu N$.

Blending oil is pure viscous liquid which contain no additives with it. Because of this as we increases the quantity of blend oil, the resultant oil will have a less additives in it. Due to this, the resistance towards friction will be less and hence the co-efficient of will be decreasing, as we increases the percentage of blend oil. In a frictional torque chart, we can see that the torque diagram of higher percentage blend oil is more uniform than lower percentage blend oil (i.e. 40 % blend oil will have a smoother torque curve than 30 %).

In the four ball test rig, the coefficient of friction was recorded with the help of latest lab view software. The data were sorted in the data acquisition system and average coefficient of friction value with different blending ratio oil is shown in the Fig. 4. It is seen from the Fig. 4 that lower ratio of blending with based oil indicates higher coefficient of friction. Further as blending ratio increases the coefficient of friction decreases.

4.3 Wear Scar Diameter

Figure 5 shows comparison of different ratio of blending oil of wear scar diameter.

It is seen from Fig. 5 that wears scar diameter increases as the viscosity of oil decreases. The blending ratio with based oil varies from 10 to 40 % during the

Fig. 4 Coefficient of friction comparison with different blending ratio of oils

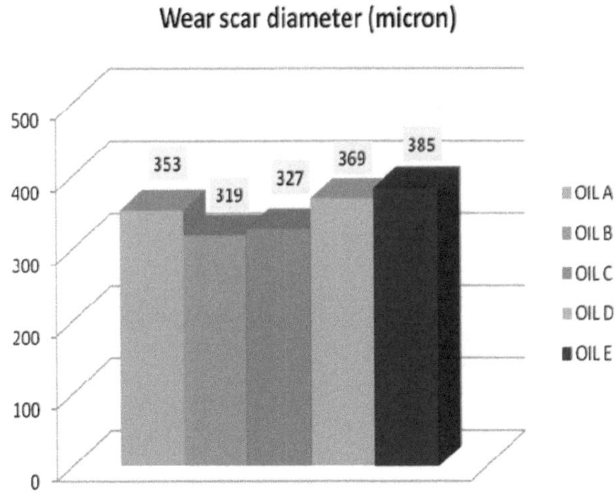

experimentations. From the wear scar diameters, it was noted that as blending ration of based oil increases the wear scar diameter increases.

Co-efficient of friction is very less for higher blend oil. Due to this frictional resistance provided by oil film between contact surfaces of ball will be very less. This will result into a higher wear and ultimately higher scar diameter.

4.4 Analysis of Blending Oil at 40 kg Load

The analysis shows that stability of 10 % blend oil at higher load (i.e., 40 kg) is exists and it is satisfies the ASTM standard. Fresh oil is more viscous than blend

Fig. 5 Wear scar diameter of blend oil

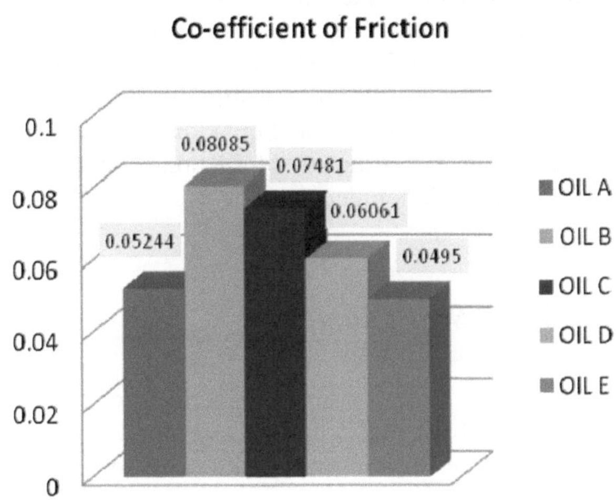

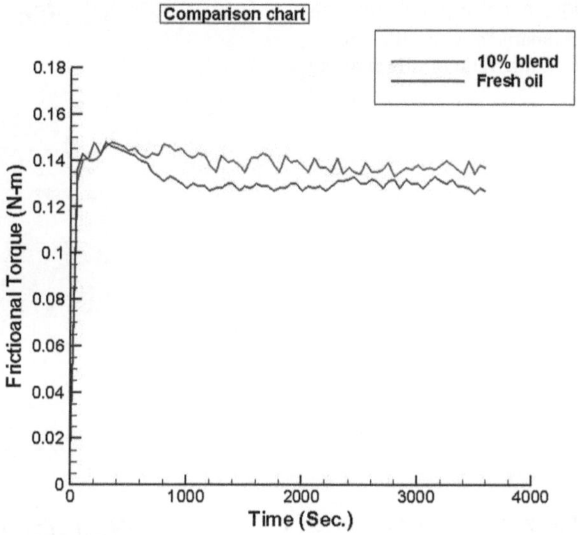

Fig. 6 Frictional torque versus time for blend oils at 40 kg

oil so frictional torque is higher during the beginning stage. We can see from the diagram that as the analysis approaches towards the end, frictional torque curve tends to get closer. This shows us that at higher operating time, the frictional torque value is almost same for either oil or that are having a very little difference in it. And more smoothness is found in blend oil curve (Fig. 6).

4.5 Co-Efficient of Friction

Co-efficient of friction is found to be slightly higher in fresh oil but value of both is with very narrow margin (Fig. 7).

4.6 Wear Scar Diameter

Figure 8 shows the comparison chart of wear scar diameter of blend oil and Castrol oil. It is clear seen from that wear scar diameter is slightly higher in case of blend oil due to lower viscosity of blend oil (Fig. 8).

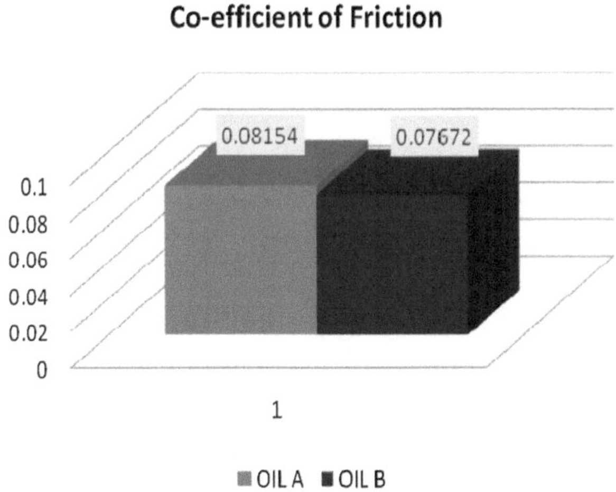

Fig. 7 Coefficient of friction comparison with different blending ratio of oils at 40 kg

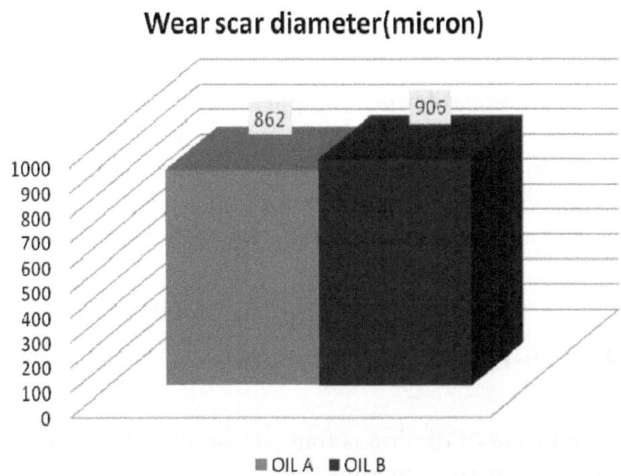

Fig. 8 Wear scar diameter of blend oil at 40 kg

5 Conclusions

In this work, frictional torque, wear scar diameter and coefficient of friction test for steel ball bearing with two different loads and five different blending ratios of oils were tested, using four ball test apparatus, under fully lubricated conditions. The coefficient of friction, wear scar diameter and frictional torque were plotted and from that the following conclusions were drawn.

1. As blend ratio of oil increases the viscosity of oil decreases in case of increasing the temperature.
2. In case of higher percentage of blend oil application friction torque found uniform in nature while at lower percentage of blend oil application the uniformity is not observed.

3. Wear scar diameter found increases as the blending ratio of oil with based oil increases.
4. Coefficient of friction is found to be decreasing as the blending oil ratio increases.
5. From the analysis of oil at both loads (i.e., at 15 and 40 kg), it is found that 10 % blend oil shows the characteristic at higher load, while other oils of different percentage of blending oils are not shown good characteristics at higher load. It can be concluded that 10 % blend oil at 40 kg load gives good tribological performance.

References

1. Abdel-Azim A et al (2001) Polymeric Additives for improving the flow properties and viscosity index of lubricating oils. J Polym Res 8:111–118
2. Yorulmaz Y (1983) Alkylated polymers as lube oil additives. J Mater Sci 18:1638–1640
3. Boerlage GD (1933) Four-ball testing apparatus for extreme pressure lubricants. Engineering, 1 CXXXVI(3522):46
4. Ripa M, Spanu C et al, (2008) Characterization of hydraulic oils by shear stability and extreme pressure tests tribology in industry, 30(3&4)
5. Troyer DD (1999) The sequential four ball test. Practicing Oil Analysis Magazine, November 1999
6. Perez J et al (1999) Sequential four-ball study of some lubricating oils. Lubrication Engineering, Sept 1999, pp 28–30
7. ASTM D4172-94 (2010) Standard test method for wear preventive characteristics of lubricating fluid (four-ball method) Developed by subcommittee: D02.L0.11 Book 07 stemdends volume: 05.02
8. Majumdar BC (1999) Introduction to tribology of bearings. Wheeler Publishing, New Delhi
9. Bhushan B (1999) Principles and applications of tribology. Wiley, New York

Tribological Testing of Electrolyzed 17-4 PH Steel Against SS 440C Under Water Lubricated Condition

Neelima Khare, P. K. Limaye, N. L. Soni and R. J. Patel

Abstract Tribological performances of electrolyzed 17-4 PH steel (using HEEF-25 Salt) were evaluated under water lubricated environment. Tests were carried out on Plint TE 70SLIM reciprocating sliding wear and friction machine. Test conditions were 3–11 N load and 5–15 Hz sliding frequency and counterface material was a ball of SS 440C. Results indicated that, above 7 N loads, coating removal from the substrate occurred during prolonged sliding. Rate of removal increased with increased load above 7 N.

Keywords HEEF-25 · 17-4 PH SS · SS 440C · Friction · Wear

1 Introduction

HEEF® 25 is proprietary high-speed chromium plating process that deposits bright, shiny and smooth functional (hard) chromium up to four times faster than conventional chromium processes and up to three times faster than other high-speed baths. This speed comes from an increase in current efficiency from the bath chemistry and a significantly higher limiting current density before burning occurs. Unlike other high-speed chromium plating processes, such as those based on mixed catalyst systems, the HEEF-25 process does not cause cathodic low current density etching of parts. Deposits have micro-hardness values in the 1,000–1,150 KHN 100 range and are micro-cracked at greater than 1,000 cracks per inch.

HEEF-25 was selected as an alternative to conventional electrolyzing process due to un-availability of the conventional salt and due to stricter environmental norms. Hexavalent chromium salts are classified as hazardous substances (toxic,

N. Khare (✉) · P. K. Limaye · N. L. Soni · R. J. Patel
Refueling Technology Division, Bhabha Atomic Research Centre, Mumbai 400085, India
e-mail: neelima.khare@gmail.com

H. C. Patel et al. (eds.), *Proceedings of International Conference on Advances in Tribology and Engineering Systems*, Lecture Notes in Mechanical Engineering, DOI: 10.1007/978-81-322-1656-8_23, © Springer India 2014

sensitizing and carcinogenic); they are environmentally and toxicologically hazardous. It would therefore be desirable to replace these products with a suitable "environmentally friendly" commercially acceptable alternative.

2 Experimental

2.1 Materials

17-4 PH stainless steel samples were prepared using HEEF-25, with coating thickness ranging from 17–20 μm, and were supplied for carrying out sliding wear and friction tests under water lubricated conditions.

2.2 Tribological Testing

The wear and friction tests were carried out on Plint TE 70SLIM reciprocating sliding wear and friction machine under water lubricated conditions. Mating specimen was a ball 1/2″ dia (grade 25) of SS 440C having hardness of 60 HRC.

The tests were carried out at 3, 5, 7, 9, 11 N load, at sliding frequency of 5, 10, 15 Hz and reciprocating distance of 2 mm for durations of 1–2 h. Total no of 30 tests were carried out. Each test was repeated on different samples to check the repeatability of the results. Wear scars were inspected through microscope and their dimensions were measured. Wear scar depth was measured using 3D-profilometer, which was used to calculate wear volume. Wear rate is calculated by normalizing wear volume by applied load and the sliding distance, hence the unit of wear rate is mm^3/NM Frictional force data was recorded online continuously. Average coefficient of friction is reported here for comparison.

3 Results and Discussion

Friction force data for representative tests is plotted in Fig. 1 to depict the progress of frictional regime. In all the tests COF starts with a low valued of typically 0.25–0.30 and saturates to the constant value after certain amount of sliding to remain constant throughout. Saturation value is different for different loads and speeds. This saturation value was averaged over the time to arrive at dynamic COF.

This average COF is compared for different loads at constant speed and for different speeds at constant load and represented in Figs. 2 and 3. It is evident from the graphs that average COF reduces with increase in load and increases with increase in sliding speed. Maximum COF of 0.70 was measured at 3 N load and

Fig. 1 Typical graphs of coefficient of friction versus time in sec (HEEF coated 17-4 Ph plate V/S SS 440C ball)

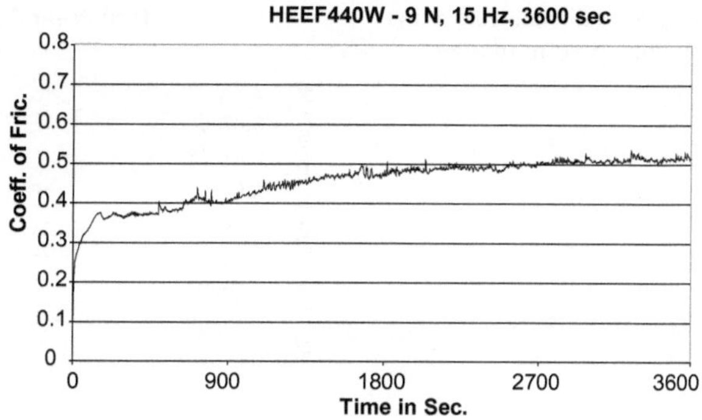

Fig. 2 Average coefficient of friction versus load in N

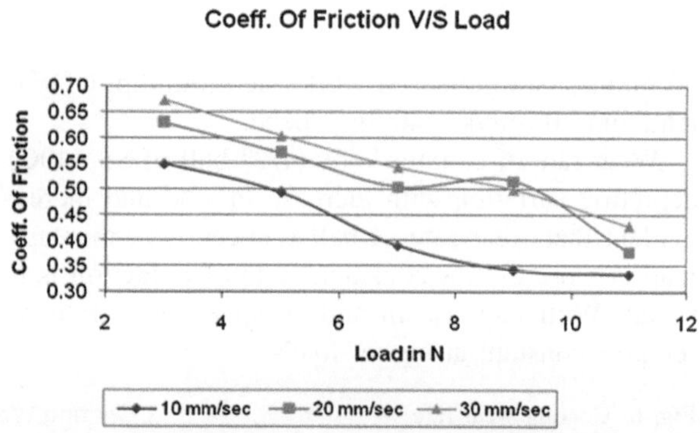

Fig. 3 Average coefficient of friction versus sliding speed in mm/sec

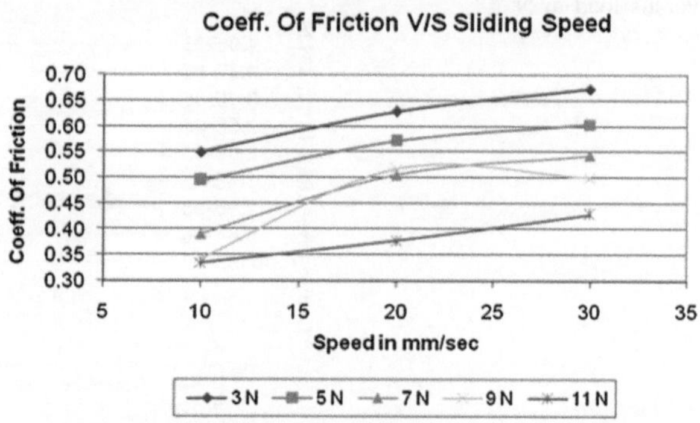

Fig. 4 Ball wear rate versus load in N

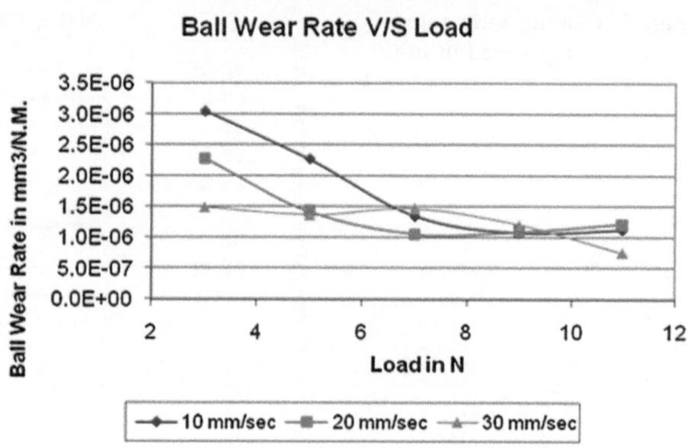

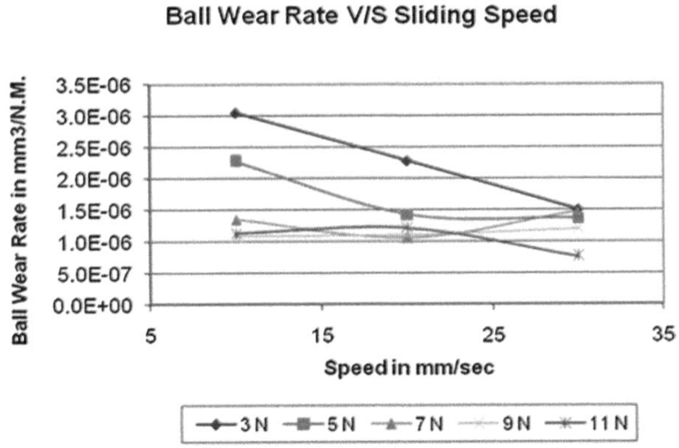

Fig. 5 Ball wear rate versus sliding speed in mm/sec

30 mm/sec sliding speed while the minimum COF of 0.33 was measured at 11 N load and 10 mm/sec sliding speed.

Wear rate of counter body (1/2″ ball of SS 440C) is presented in Figs. 4 and 5 depicting variation with increase in load and increase in speed respectively. It is evident that wear rate of ball reduces with increase in load. Wear rate at higher load is more or less constant. Rate of reduction is higher with increase in sliding speed. With increase in sliding speed wear rate reduce at lower loads while it remains constant at higher loads.

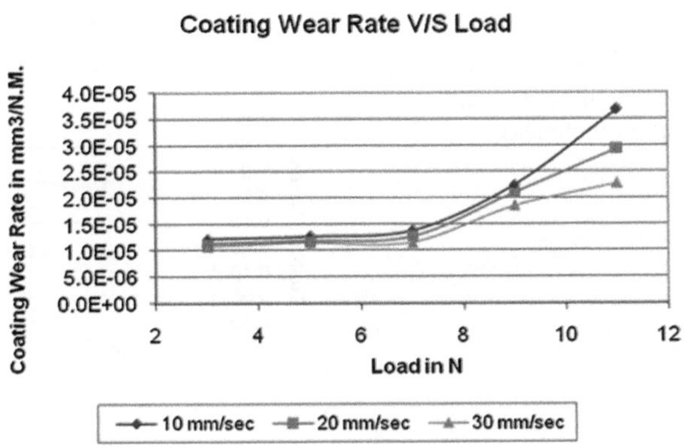

Fig. 6 Coating wear rate versus load in N

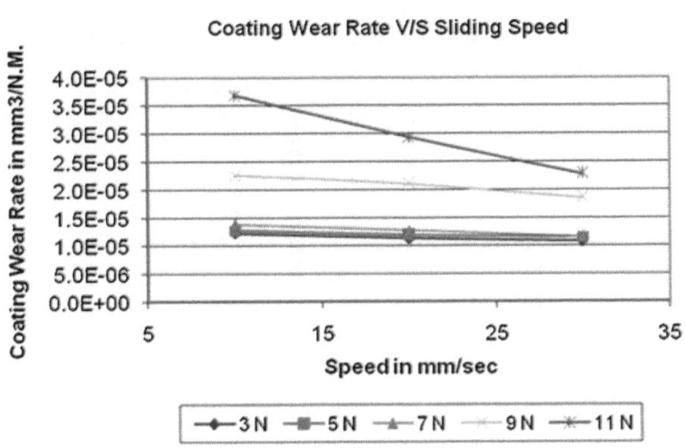

Fig. 7 Coating wear rate versus sliding speed in mm/sec

Fig. 8 Typical photographs indicating ball wear and plate wear (HEEF coated 17-4 Ph plate vs. SS 440C ball). **a** HEEF440CC (11 N, 15 Hz, 3,600 s, water)-ball-5X. **b** HEEF440A (3 N, 5 Hz, 3,600 s, water)-ball-5X. **c** HEEF440D (7 N, 10 Hz, 3,600 s, water)–Plate-5X. **d** HEEF440AA (11 N, 10 Hz, 3600 s,)

Wear rate of coating is presented in Figs. 6 and 7. Wear rate of coating is more or less constant with increase in load up to 7 N and then suddenly increase with increase in load at all speeds. Microscopic studies reveal that at loads higher then 7 N failure of coating takes place resulting in generation of wear particles which causes 3-Body abrasion resulting in removal the coating from the surface. This is also evident from the micrographs (Fig. 8) which show scratch marks and substrate at sliding at higher loads. Hence 7 N may be considered as critical load value for the coating. There is an insignificant reduction in wear rate with increase in sliding speed below critical load.

4 Conclusion

Wear and friction studies of Electrolyzed (HEEF-25) coated 17-4 PH were carried out in reciprocating sliding wear and friction machine under water lubricated environment against SS 440C.

- Wear of coating was minimal up to 7 N of load where only few scratch marks are visible in the micrographs.
- At loads above 7 N coating removal from the substrate has taken place during prolonged sliding. Rate of removal increased with increase in load beyond this critical load.
- Wear of counter body (SS 440C) is very low and further reduced with increase in load as well as speed.
- Coefficient of friction (COF) reduces with increase in load and increase with increase in speed.
- Below 7 N load COF remained as high as 0.5–0.7 indicating actual dynamic COF between the coating and SS 440C neglecting effect of substrate.
- Further evaluation needed to establish critical load for failure, which includes adhesion test, SEM and other hard counter body or configuration.
- Critical load of 7 N shall be considered as mean contact pressure of 0.45 GPa neglecting effect of substrate hardness. (Young's modulus of Chromium plating 157 GPa and that of SS 440C 200 GPa, Poisson's Ration 0.29—Calculation by Hertzian point contact).

Part IV
Machine Components and Bearing Design

Design and Development of Rotary Fixture for CNC with Tribology Considerations of Mechanics Analysis and Dynamic Balancing as Pre-mortem Tool

Nirav P. Maniar and D. P. Vakharia

Abstract Tribology is the science and engineering of interacting surfaces in relative motion. It includes the study and application of the principles of friction, lubrication and wear. Rotary fixture is also a similar but novel concept which poses problems of friction and especially dynamic balancing in case of asymmetrical rotary components. Various areas related to tribology and fixture are already been described by renowned authors, still there is need to apply these research works to an industrial application. This paper presents design of rotary fixture with tribology considerations of mechanics analysis and computer aided mass balancing for real industrial component-Flow TEE body of petroleum refinery. The operations to be performed are front facing, outside diameter turning, grooving, boring and back facing. Tribology plays an important role in manufacturing especially in case of dynamic balancing of rotary components such as rotary fixture. In such cases, friction and dynamic imbalance problems if not resolved, increase tool wear and the power required to work a piece. This results in increased costs due to more frequent tool replacement, loss of tolerance as tool dimensions shift, and greater forces required to shape a piece. Methodology for mass balance of rotary components mostly act as post-mortem tool; calculating unbalanced mass after fixture is manufactured. In the present work, a pre-mortem tool is developed to predict unbalanced mass before manufacturing with three alternate methods for mass balancing.

Keywords Tribology · CNC · Rotary fixture · Mechanics analysis · Dynamic Balancing · Design

N. P. Maniar (✉)
Department of Mechanical Engineering, Dharmsinh Desai Universtiy, Nadiad, Gujarat 387001, India
e-mail: niravpmaniar@rediffmail.com

D. P. Vakharia
Department of Mechanical Engineering, Sardar Vallabhbhai National Institute of Technology, Surat, Gujarat 395007, India

H. C. Patel et al. (eds.), *Proceedings of International Conference on Advances in Tribology and Engineering Systems*, Lecture Notes in Mechanical Engineering, DOI: 10.1007/978-81-322-1656-8_24,

1 Introduction

Machines and their constituent elements like aircraft and ships to micro pumps and gears are parts of modern society everywhere. Wherever machine elements exist, tribological problems present themselves. Designers' understanding of tribology is sufficient to provide solution occasionally, but this is not true in every case. Most of the times, the designer is unaware of valuable tribological research, or else the research base is inadequate [1]. Areas of interest which are high-lighted in tribological research include: technology transfer; friction, lubrication and wear research; tribo-testing; computers in tribology; failure diagnosis; plant condition monitoring; tribology in hostile environments (nuclear, off-shore, space); surface treatments and surface coatings; plastics; ceramics; and bio-engineering [2]. Much limited work is reported on applying tribology principles in dealing with rotary components. And this has excited this research work on solving tribological problem of mass balancing of rotary components such as rotary fixture using mechanics analysis and computer aided mass balancing method.

Methodology developed by most of the researchers mostly act as post-mortem tool, calculating and determining unbalanced mass after fixture is manufactured followed by unbalanced mass removal or counterweight addition and both these means add cost to work a piece. Alternately, there is a need to develop a tool that could predict unbalanced mass during fixture design stage but such a tool is not yet developed. The present volume of this paper fills this gap and proposes unique method of use of Creo Elements/Pro 5.0, which would enable prediction of unbalanced mass during design stage well before manufacturing. This approach would be highly useful in the shop floor, saving material cost, increasing the productivity and decreasing the human labor. In this work, fixture is balanced by adding counterweight equal in magnitude and opposite in direction as that of resultant unbalanced mass. The object of the work presented here is to develop the study and to satisfy basic aim of tribology in design, manufacturing, static analysis with force and moment balancing of fixture. The present research proposes alternate methods of 4 Quadrant, 8 Quadrant and 8 Diamond Quadrant Computer Aided Mass Balancing Method (CAMBM) for mass balancing of rotary fixture.

The fixture designing and manufacturing is considered as complex process that demands the knowledge of different areas, such as geometry, tolerances, dimensions, procedures, tribology and manufacturing processes. While designing this work, a good number of literature and titles written on the subject by renowned authors are referred. All findings and conclusions obtained from the literature review and the interaction with fixture designers are used as guide to develop the present research work. As stated by Koji Teramoto, Masahiko Anasoto and Kazuaki Iwata [4], Fixturing Plan (FP) and Machining Plan (MP) are mutually dependent. Implicit to this conclusion, paper coordinates MP and FP by coupling a fixture design with manufacturing and tribology considerations. For this research, a relevant issue when considering requirements, taking this as a general concept, is to make explicit the meaning of two main terms: Functional Requirement (FR) and

Constraint (C) [5]. Functional Requirement (FR), as it stated by different authors, 'represents what the product has to or must do independently of any possible solution'. Constraint (C) can be defined as 'a restriction that in general affects some kind of requirement, and it limits the range of possible solutions while satisfying the requirements'.

Though some contributions have been made in several areas related to design of fixture like knowledge model for fixture design process, workpiece location, tribology in design, computer aided fixture design, fixture analysis under dynamic machining etc. [3–12], but there is a great deal of urgency and importance to couple all these research works to an industrial application. This paper reviews all these research works and transforms the theoretical knowledge of tribological fixture design to practical application.

2 Design and Development of Fixture

2.1 *Statement of Problem*

"Design and development of rotary fixture for machining flow TEE body on CNC turning centre. The operations to be performed are front facing, outside diameter turning, grooving, boring and back facing. The fixture being rotary in nature has to be dynamically balanced with mechanics analysis with respect to tribological considerations."

2.2 *Component Details*

The component is Flow TEE body, made up mild steel, weighing 46.5 kg and is one of the components of petroleum refinery [Figs. 1, 2]. The component is used as a joint or coupler for pipes through which petroleum liquid products flows and mixes. The component in raw material form is forged, proof machined with 3 mm machining allowance on conventional lathe with 24 inch swing over diameter. The operations to be performed on component, using designed fixture set up, are front facing, outside diameter turning, grooving, boring and back facing. Three set up of fixture are to be designed to perform required operations.

2.3 *Set Up-1 Fixture Design*

The complete location and clamping of Set Up-1 fixture is accomplished by using 3 V-blocks and latch clamp. The important parts of fixture used here are V-block, latch clamp, base plate, vertical plate, adapter plate, locator and rib [Fig. 3]. The

Fig. 1 3D view of raw component

Fig. 2 3D view of finished part

fixture uses three V-blocks to locate and a latch clamp to hold the component. The latch clamp consists of two M 6 bolts to directly clamp the workpiece. The chuck of CNC turning centre will be replaced with complete fixture set up using an adapter plate. The adapter plate holds the same dimensions of chuck plate. The locator locates the vertical plate in correct position with adapter plate. The base plate serves to hold the complete fixture assembly. The ribs are clamped to base plate and provide the holding arrangement for latch clamp. The complete fixture assembly rotates with 550 rpm while performing operations on CNC turning centre. The specification of spindle nose of CNC turning centre used in this work is A2-8, which can carry weight of 450 kg. The fixture is directly mounted on spindle nose.

2.4 Tribology Consideration: Computer Aided Mass Balancing of Set Up-1 Fixture

As the fixture is asymmetrical, it has to be mass balanced. The fixture rotates around one axis; hence it has to be balanced about other two perpendicular axis.

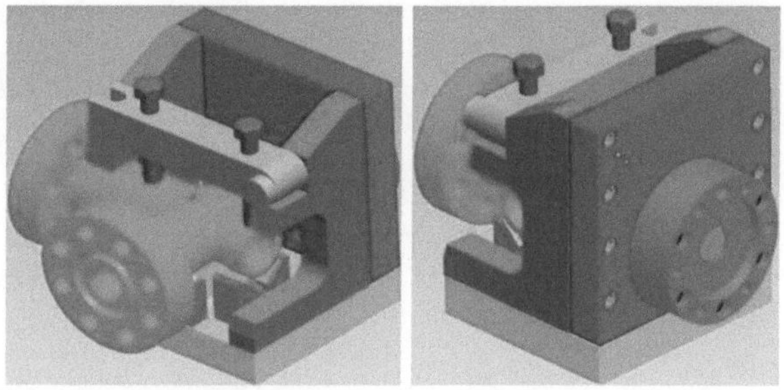

Fig. 3 3D views of set up-1 fixture

Here x-axis is the axis of rotation. Three innovative methods are proposed to mass balance the fixture—4 Quadrant, 8 Quadrant and 8 Diamond Quadrant Computer Aided Mass Balancing Method. Comparison of results obtained from three different methods used is also reported in relevant section. The results and outputs from Creo Elements/Pro 5.0 with solution of balancing are shown below.

2.5 8 Quadrant Computer Aided Mass Balancing of Set Up-1 Fixture

Step I: C. G., weight and offset distance of C. G. of Set Up-1 fixture from axis of rotation are determined [Fig. 4]. The important results from the above output are as follows: weight of fixture with component, without balancing mass = 233.12 kg. C.G. is offset from axis of rotation in x-axis by -130.56 mm, in y-axis by -1.11 mm and in z-axis by 2.38 mm.
Step II: Now Set Up-1 fixture is cut in 4 quadrants about 2 axis, perpendicular to each other and perpendicular to axis of rotation below [Fig. 5].
Step III: The weight and C. G. of Set Up-1 fixture in each quadrant are determined. [Figs. 6, 7, 8, 9].
Step IV: The outputs of weight and C. G. of Set Up-1 fixture in each quadrant are summarized [Table 1].
Step V: According to principles of mechanics, $\Sigma F = 0$ and $\Sigma M = 0$ for mass balancing. The sum of unbalanced mass in horizontal direction ΣF_H and in vertical direction ΣF_V are calculated [Table 2].

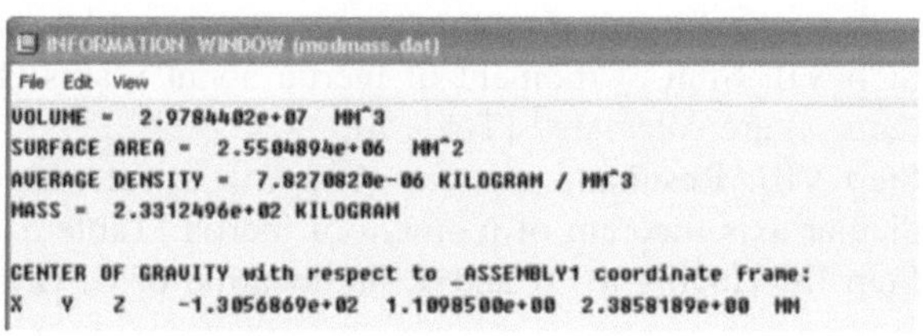

Fig. 4 Mechanics analysis of set up-1 fixture

Fig. 5 3D view of set up-1 fixture in 4 quadrants

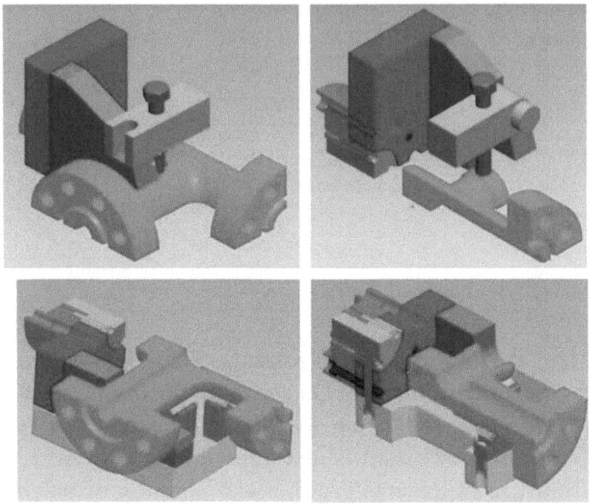

Fig. 6 Weight and C. G. of set up-1 fixture in I quadrant

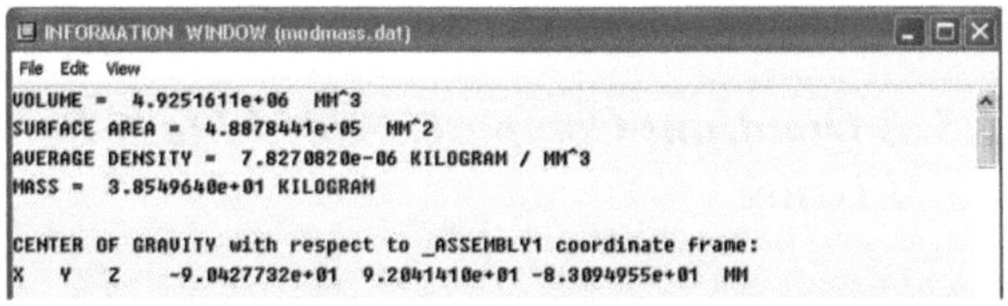

Fig. 7 Weight and C. G. of set up-I fixture in II quadrant

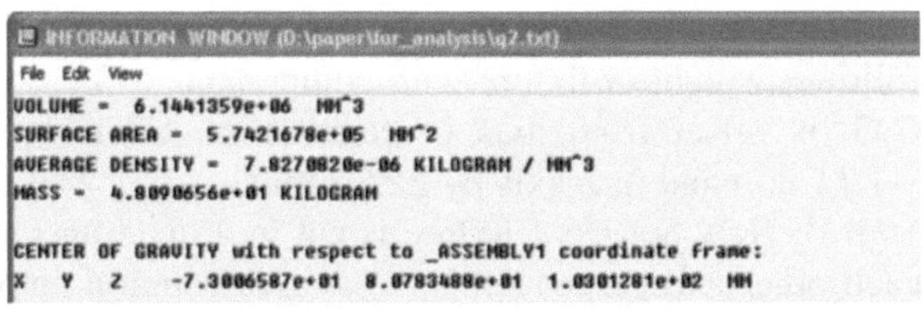

Fig. 8 Weight and C. G. of set up-I fixture in III quadrant

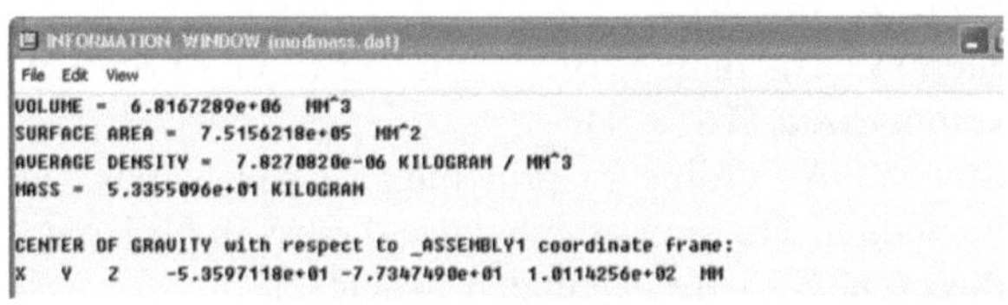

Step VI: Resultant unbalanced mass (R) and its line of action in terms of angle (α) with x-axis are calculated using parallelogram law of forces [Table 3].
Step VII: Sum of moment of inertia about x-axis ($\Sigma m_i x_i^2$) and that about y-axis ($\Sigma m_i y_i^2$) are calculated [Table 4].
Step VIII: Resultant moment of inertia is calculated using principle of perpendicular axis theorem of moment of inertia [Table 5].
Step IX: Having M, R and α, the location of C. G. (r_{cm}) of R is determined.

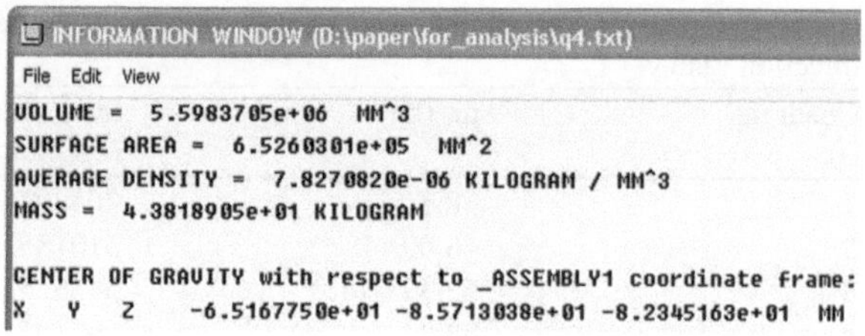

Fig. 9 Weight and C. G. of set up-IV fixture in IV quadrant

Table 1 Summary of C. G. of set up-1 fixture in all quadrants

Quadrant (i)	Co-ordinates of C. G. (mm)		$\tan \theta_i$	θ_i (Degree)
	x_i	y_i		
1	83.09	92.04	1.10	47.92
2	−103	80.78	−0.78	38.11
3	−101.14	−77.35	0.76	37.41
4	82.35	−85.71	−1.04	46.14

Table 2 Calculation of resultant mass in horizontal direction (ΣF_H) and in vertical direction (ΣF_V)

Quadrant (i)	m_i (kg)	$F_H = x_i = m_i \cos\theta_i$ (kg)	$F_V = y_i = m_i \sin\theta_i$ (kg)
1	38.55	25.79868396	28.5775769
2	48.09	−37.8405504	29.6772782
3	53.36	−42.38538755	−32.415559
4	43.82	30.35986498	−31.598591
	Σ	−24.06738901	−5.7592963

Table 3 Calculation of resultant unbalanced mass, R

ΣF_H^2	579.2392137 kg^2
ΣF_V^2	33.16949445 kg^2
$\Sigma F_H^2 + \Sigma F_V^2$	612.4087082 kg^2
Resultant, $R = \sqrt{(\Sigma F_H^2 + \Sigma F_V^2)}$	24.7468929 kg
$\tan \alpha$	0.23929876
α	13.45773737°

$$M = R\ rcm^2$$
$$r_{cm}^2 = M/R$$
$$r_{cm} = 342.33\ \text{mm}$$

Thus the unbalanced mass is found to be 24.75 kg and its C. G. is situated at an angle of 13.45° with x-axis at a distance of 342.33 mm in III quadrant. Hence the fixture can be balanced by placing the counterweight equal in magnitude and opposite in direction as that of unbalanced mass.

Table 4 Calculation of sum of moment of Inertia about X-direction ($\Sigma m_i x_i^2$) and that of about Y-direction ($\Sigma m_i y_i^2$)

Quadrant (i)	m_i (kg)	$m_i x_i^2$ (kg mm^2)	$m_i y_i^2$ (kg mm^2)
1	38.5	265802.0019	326147.421
2	48.09	510186.81	313806.89
3	53.36	545835.4267	319254.080
4	43.82	297166.316	321910.663
	Σ	1618990.554	1281119.05

Table 5 Calculation of resultant moment of inertia, M

$I_{xx} = \Sigma\ m_i x_i^2$	1618990.554 kg mm^2
$I_{yy} = \Sigma\ m_i y_i^2$	1281119.056 kg mm^2
$I_{zz} = I_{xx} + I_{yy}$	2900109.61 kg mm^2
$\therefore\ M = \Sigma\ m_i x_i^2 + \Sigma\ m_i y_i^2$	

3 8 Quadrant Computer Aided Mass Balancing of Set Up-1 Fixture

In this method, Set Up-1 fixture is cut in 8 quadrants about 4 axis at angle of 45° to each other and perpendicular to axis of rotation. Following same steps as 4 Quadrant Computer Aided Mass Balancing Method with difference of 8 Quadrants instead of 4 Quadrants Mechanics Analysis, the unbalanced mass is found to be 59.78 kg and its C. G. is situated at an angle of 72.22° with x-axis at a distance of 159.11 mm in II quadrant. Hence the fixture can be balanced by placing the counterweight equal in magnitude and opposite in direction as that of unbalanced mass.

4 Diamond Quadrant Computer Aided Mass Balancing of Set Up-1 Fixture

In this method, Set Up-1 fixture is cut in 8 quadrants in diamond cutting method and perpendicular to axis of rotation [Fig. 10].

Again following same steps for mass balancing, the unbalanced mass is found to be 59.51 kg and its C. G. is situated at an angle of 77.82° with x-axis at a distance of 164.79 mm in II quadrant. Hence the fixture can be balanced by placing the counterweight equal in magnitude and opposite in direction as that of unbalanced mass. Next section reports relative comparison of results obtained from three methods used for Computer Aided Mass Balancing of Set Up-1 fixture.

Comparison of Results obtained from three methods used for Computer Aided Mass Balancing of Set Up-1 Fixture

Parameters	Mass balancing method		
	Quadrant	Quadrant	8 Diamond Quadrant
Unbalanced Mass (kg)	24.75	59.78	59.51
Angle at which C.G. of unbalanced mass is situated (degree)	13.45	72.22	77.82
Distance at which C.G. of unbalanced mass is situated (mm)	342.33	159.11	164.79
Quadrant in which C.G. of unbalanced mass is situated	III	II	II
Mass of component (kg)	46.5	46.5	46.5
Mass of fixture including mass of component, excluding unbalanced mass (kg)	183.46	183.46	183.46
Total mass of fixture including mass of component and unbalanced mass (kg)	208.21	243.24	242.97
Actual mass of fixture including mass of component and unbalanced mass (kg)	233.12	233.12	233.12
Absolute error	24.91	10.12	9.85
Relative error	0.10685	0.04341	0.04225
Percentage error	10.685	4.341	4.225

Three alternate methods of Computer Aided Mass Balancing are presented and 8 Diamond Quadrant Computer Aided Mass Balancing Method is found more accurate with the result of decrease in percentage error by almost 6 %.

4.1 Set Up-2 Fixture Design

Set Up-2 Fixture has to locate and hold the component for performing same machining operations on another circular face of the component. Almost same methodology of locating and clamping is used in this set up with minor change in location method. The locating is accomplished by using rest pad and two cylindrical pins. Very basic principle of fixture design suggests that machined surface should be used as locating surface. The face of the component machined using Set Up-1 fixture is used as locating surface in Set Up-2 fixture. The bigger diameter hole and one of the smaller diameter holes are used for location using cylindrical pin locator and rest pad. The clamping is accomplished using four C washers and latch clamp [Fig. 11]. The latch clamp consists of one M 6 bolt to directly clamp the workpiece. The function of other components of fixture remain the same as that of set up 1 fixture.

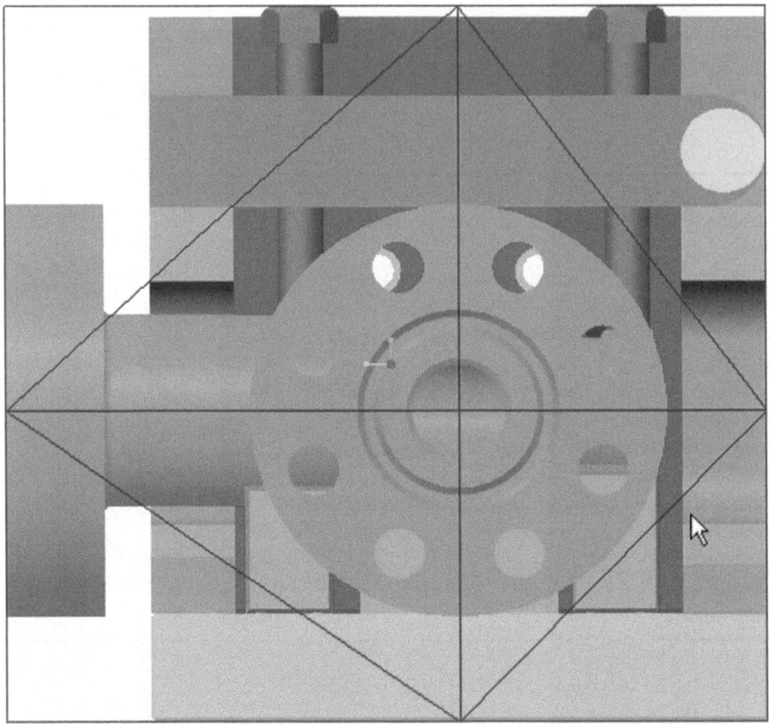

Fig. 10 3D view of set up-1 fixture in 8 diamond quadrants

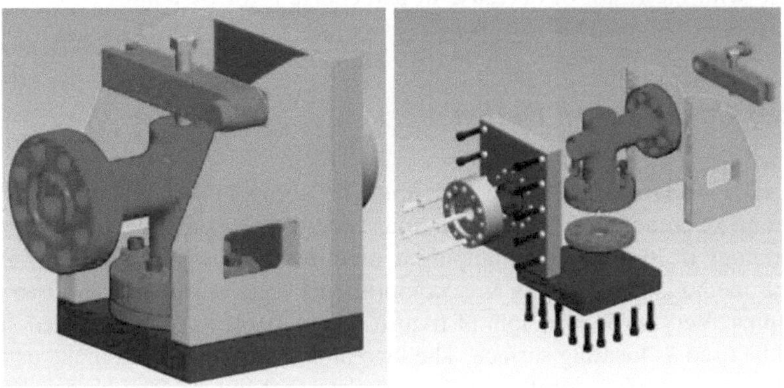

Fig. 11 3D views of set up-2 fixture

4.2 Set Up-3 Fixture Design

Face 2 machined in set up 2 fixture is used as locating surface for machining face 3 in set up-3 Fixture [Fig. 12].

Same three methods of 4 Quadrant, 8 Quadrant and 8 Diamond Quadrant are used for Computer Aided Mass Balancing of Set Up-2 and Set Up-3 Fixture. Comparison of results obtained from three methods used is reported in next sections.

Comparison of Results obtained from three methods used for Computer Aided Mass Balancing of Set Up-2 Fixture

Parameters	Mass balancing method		
	4 Quadrant	8 Quadrant	8 Diamond Quadrant
Unbalanced mass (kg)	94.71	32.06	26.69
Angle at which C.G. of unbalanced mass is situated (degree)	89.76	89.23	89.20
Distance at which C.G. of unbalanced mass is situated (mm)	268.61	156.76	176.07
Quadrant in which C.G. of unbalanced mass is situated	IV	I	I
Mass of component (kg)	46.5	46.5	46.5
Mass of fixture including mass of component, excluding unbalanced mass (kg)	221.27	221.27	221.27
Total mass of fixture including mass of component and unbalanced mass (kg)	315.98	253.33	247.96
Actual mass of fixture including mass of component and unbalanced mass (kg)	239.78	239.78	239.78
Absolute error	76.2	13.55	8.18
Relative error	0.2411545	0.0534875	0.03298919
Percentage error	24.115450	5.3487545	3.29891918

The above comparison shows that 8 Diamond Quadrant Computer Aided Mass Balancing Method gives more accurate results compared with experimental results. Percentage error in this method is reduced by almost 20 % in comparison to 4 Quadrant Computer Aided Mass Balancing Method and by almost 2 % in comparison to 8 Quadrant Computer Aided Mass Balancing Method.

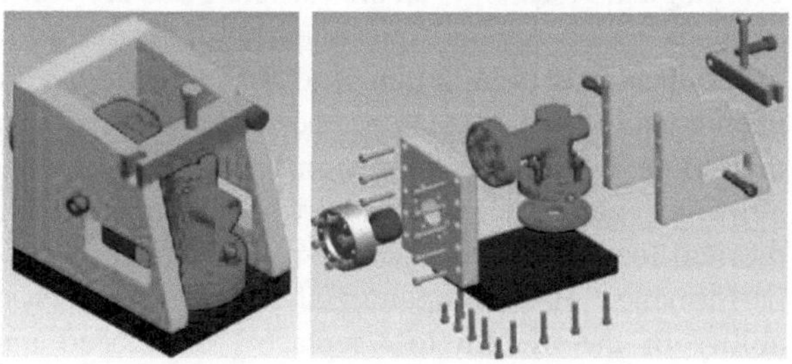

Fig. 12 3D views of set up-3 fixture

Comparison of Results obtained from three methods used for Computer Aided Mass Balancing of Set Up-3 Fixture

Parameters	Mass Balancing Method		
	4 Quadrant	8 Quadrant	8 Diamond Quadrant
Unbalanced Mass (kg)	126.07	39.39	30.64
Angle at which C.G. of unbalanced mass is situated (degree)	89.8	89.31	89.03
Distance at which C.G. of unbalanced mass is situated (mm)	275.59	169.61	197.47
Quadrant in which C.G. of unbalanced mass is situated	IV	I	I
Mass of component (kg)	46.5	46.5	46.5
Mass of fixture including mass of component, excluding unbalanced mass (kg)	274.4	274.4	274.4
Total mass of fixture including mass of component and unbalanced mass (kg)	400.47	313.79	305.04
Actual mass of fixture including mass of component and unbalanced mass (kg)	294.31	294.31	294.31
Absolute Error	106.16	19.48	10.73
Relative Error	0.265088	0.062079	0.0351757
Percentage Error	26.50885	6.207973	3.5175714

The above comparison shows that 8 Diamond Quadrant Computer Aided Mass Balancing Method gives more accurate results compared with experimental results. Percentage error in this method is reduced by almost 23 % in comparison to 4 Quadrant Computer Aided Mass Balancing Method and by almost 3 % in comparison to 8 Quadrant Computer Aided Mass Balancing Method.

5 Conclusion

This paper presents contribution of tribology to machine design. It proves that Tribological research is an area where there are considerable opportunities for new methods to be applied, particularly in the rational analysis of extensive field experience data, which can lead to major advances in the understanding of the phenomena involved such as the rotary fixture in this case. It is also recognized that computerized fixture design with tribology considerations can result in high efficiency, stable accuracy, short set-up time, and low cost due to proposed pre-mortem tool of computer aided mass balancing and mechanics analysis. The real performance of a computerized fixture design system is rooted in a powerful ability of the system to ''replace'' or exceed mass balancing which is done manually by a fixture designer. Also, an increasing research interest is using various methods to obtain optimal fixture layout solutions for mass balancing, which often requires much precise calculations in geometry and mechanics. The present research work proposes Computer Aided Mass Balancing Method (CAMBM) which ease fixture designer from tedious and time consuming work of finding offset distance and C.G. of irregular shape parts and also solving mass balancing problem. Three alternate methods of Computer Aided Mass Balancing are presented and 8 Diamond Quadrant Computer Aided Mass Balancing Method is found more accurate in an analysis of all three set up fixture design with the result of significant decrease in percentage error. Hence, it proves that 8 Quadrant Computer Aided Mass Balancing Method has a better performance on this kind of work. The findings of unbalanced mass and its location of C. G. is remarkably same as with experimental results. This approach is of crucial importance in real manufacturing environment as it does not apply only to symmetrical parts but also to asymmetrical parts as the case in this work. However, this approach of solving the balancing problem and stress analysis is expected to have more flexibility in its application.

As such tribology is truly multidisciplinary and this extraordinary breadth of scientific interest is reflected.

References

1. Hamrock BJ, Jacobson B, Schmid SR (2000) The role of tribology in the design of machine elements. Tribol Ser 38:217–227
2. Roberts WH (1986) Some current trends in tribology in the UK and Europe. Tribol Int 19(6):295–311
3. Neale M (2003) Tribology in design. Tribol Ser 41:3–11
4. Teramoto K, Anasoto M, Iwata K (1998) Coordinative generation of machining and fixturing plans by a modularized problem solver. CIRP Annu, Manuf Technol 47:437–440
5. Hunter R, Rios J, Perez JM, Vizan A (2006) A functional approach for the formalization of the fixture design process. Int J Mach Tools Manuf 46:683–697
6. Hunter R, Vizan A, Perez J, Rios J, (2005) Knowledge model as an integral way to reuse the knowledge for fixture design process. J mater process technol 164–165:1510–1518
7. Archibald FR (1956) Load capacity and time relations for squeeze films", Trans ASME 78:231–245
8. Li B, Melkote SN (1999) Improved workpiece location accuracy through fixture layout optimization. Int J Mach Tools Manuf 39:871–883
9. Hargrove SK, Kusiak A (1994) Computer-aided fixture design: a review. Int J Prod Res 32:733–753
10. Meyer RT, Liou FW (1997) Fixture analysis under dynamic machining. Int J Prod Res 35:1471–1489
11. Arakelian V, Dahan M (2000) Dynamic balancing of mechanisms. Mech Res Commun 27:1–6
12. Deiab IM, Elbestawi MA (2005) Experimental determination of the friction coefficient on the workpiece-fixture contact surface in workholding applications. Int J Mach Tools Manuf 45:705–712

Design and Kinematic Analysis of an Automatic Tool Changing Mechanism Used in VMC

M. B. Vaghela, V. J. Savsani and S. B. Jadeja

Abstract Design analysis of high speed automatic tool changing mechanism for high speed machines like VMC, and minimize tool changing cycle time as compare with current used ATC. To design tool changing mechanism of ATC, do the design of critical components like radial and globoidal cam and take it exciting design of some components from the company and make a final assembly design in Creo Parametric 1.0 software. To get the result of displacement and velocity at every connecting point for tool changing cycle time 2 s by using mechanism module in Creo Parametric 1.0 software. Main goal of this paper is find the position and velocity at every connecting point by the kinematic analysis of tool changing mechanism using Creo parametric 1.0 software. If kinematic analysis is run successfully than can be do dynamic analysis with considering gravity and find acceleration and relative force for every connecting point of tool changing mechanism.

Keywords ATC-automatic tool changer · Creo parametric 1.0 · Dynacam 10 · Kinematic analysis · Motion analysis · Tool changing mechanism · VMC–vertical machining centre

---

M. B. Vaghela (✉) · S. B. Jadeja
Department of Mechanical Engineering, B. H. Gardi College of Engineering and Technology, Rajkot, Gujarat 360005, India
e-mail: manoj0085@gmail.com

V. J. Savsani
Department of Mechanical Engineering, Pandit Deendayal Petroleum University, Gandhinagar, Gujarat 380015, India

H. C. Patel et al. (eds.), *Proceedings of International Conference on Advances in Tribology and Engineering Systems*, Lecture Notes in Mechanical Engineering, DOI: 10.1007/978-81-322-1656-8_25,

1 Introduction

Tool changing mechanism of ATC for VMC type of machine having simultaneously two types of motion which is occurred at shaft, downward and upward motion and rotary motion of shaft. For downward and upward motion of shaft used radial cam and for rotary motion of shaft used globoidal cam. To design and analysis of radial cam used graph [1] of arm downward and upward movement versus cam rotation angle and for globoidal cam used graph [1] of arm rotating chart versus cam shaft rotation angle with the help of Dynacam 10 software [2]. To design and develop high speed tool changing mechanism, do the design and analysis of some critical components like radial cam, globoidal cam [3, 4] and take existing design of shaft and linkage mechanism from the company [5] and also design the other components which is required for make a complete assembly of tool changing mechanism of ATC [3, 4].

Design and prepare final assembly of tool changing mechanism of ATC and prepare a final kinematic model [6, 7] for kinematic analysis. Obtain the results of displacement by position analysis, velocity by kinematic analysis and acceleration by dynamic analysis for each defined connecting points with using mechanism module of Creo parametric 1.0 software. Compare result of displacement for shaft with actual required movement and by kinematic analysis interferences for joints and for complete assembly at motion condition.

2 Design and Analysis of Cams

In tool changing mechanism of ATC two motions which is occurred at shaft, downward and upward motion of shaft and rotary motion of shaft. For downward and upward motion of shaft used radial cam and for rotary motion of shaft used globoidal cam.

2.1 Design and Analysis of Radial Cam

To design and analysis of radial cam used Dynacam 10 software [2] and graph of arm downward and upward movement versus cam rotation angle [1]. Radial cam design for maximum displacement 115 mm. From Dynacam 10 software getting results of displacement, velocity, acceleration and jerk for each degree 0° to 360° and also get cam profile of radial cam. Based on this cam profile prepare solid model of radial cam [8] using Creo Parametric 1.0 software as shown in Fig. 1.

Fig. 1 Radial cam

2.2 Design and Analysis of Globoidal Cam

To design and analysis of globoidal cam used Dynacam 10 software [2] and graph of arm rotation angle versus cam rotation angle [1]. From Dynacam 10 software getting results of displacement, velocity, acceleration and jerk for each degree 0° to 360°. Prepare solid model of globoidal cam using Creo Parametric 1.0 software as shown in Fig. 2.

Fig. 2 Globoidal cam

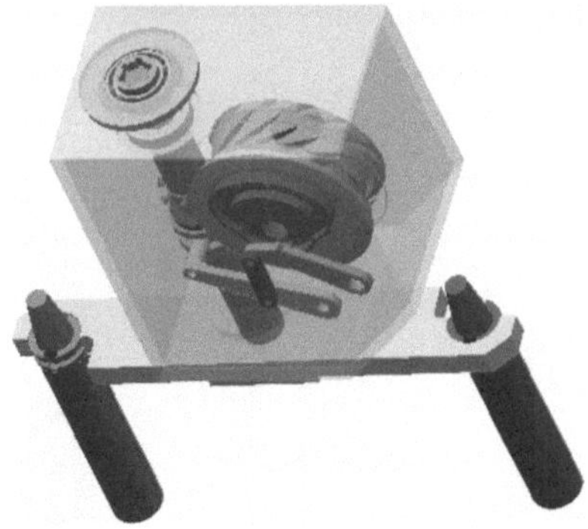

Fig. 3 Tool changing mechanism of ATC

3 Design of Tool Changing Mechanism of ATC

To design final tool changing mechanism assembly [9] of ATC make a design of radial and globoidal cam [10] and take existing design of shaft, linkage from company [5] and also design other components required to make an assembly using Creo Parametric 1.0 software. Prepare design of final tool changing mechanism assembly of ATC in Creo Parametric 1.0 software as shown in Fig. 3.

4 Motion Analysis

In Creo Parametric 1.0 software powerful mechanism module is given for motion analysis. There can be prepare mechanism assembly [6, 7, 9] as per connecting joint and motion; also do the analysis for position, kinematic, dynamic, static and force balance and measure definition for position, velocity, acceleration, connection reaction, net load and may more in mechanism module. Do the motion analysis [11] of tool changing mechanism of ATC for position, kinematic and dynamic analysis for tool changing cycle time 2 s.

4.1 Position Analysis

For position analysis using Creo Parametric 1.0 software, after completion of final assembly of tool changing mechanism select analysis definition type "position" in mechanism module and run the analysis for tool changing cycle time 2 s. To measure the position of each connecting point, select measure results graph type measure versus time and measure definition type position in mechanism module and get the results of displacement for each connecting point as shown in Fig. 4.

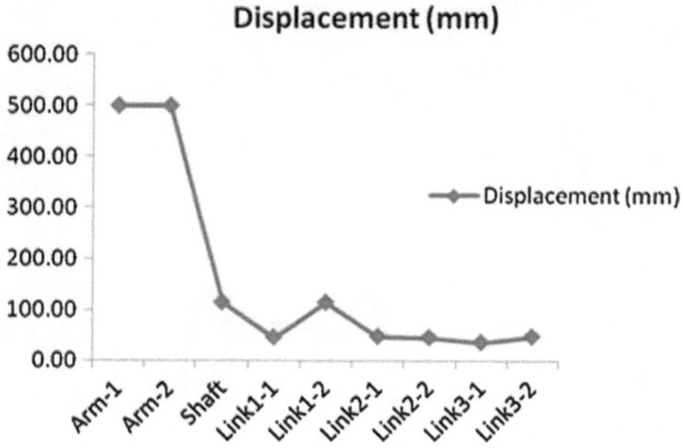

Fig. 4 Displacement chart of tool changing mechanism

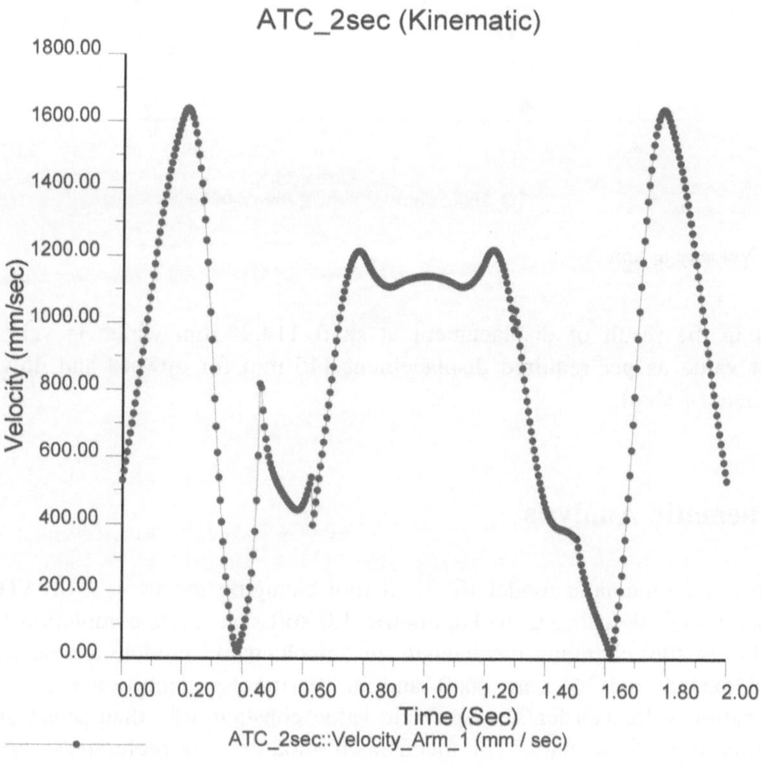

Fig. 5 Velocity at arm-1

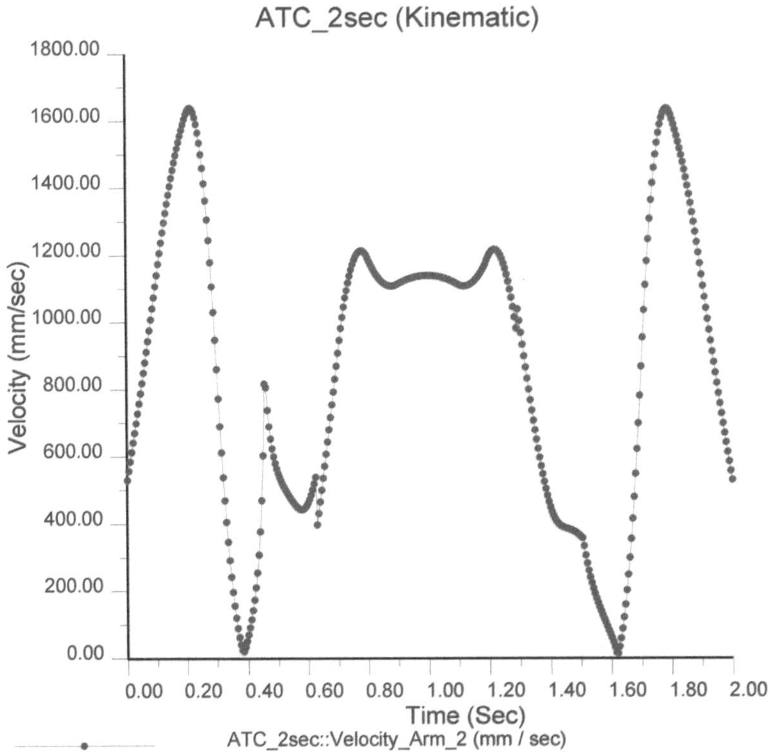

Fig. 6 Velocity at arm-2

Obtain the result of displacement at shaft 114.95 mm which is very much nearest value as per required displacement 115 mm for upward and downward movement of shaft.

5 Kinematic Analysis

First prepare kinematic model [6, 7] of tool changing mechanism of ATC. For kinematic analysis using Creo Parametric 1.0 software, after completion of final assembly of tool changing mechanism, in "mechanism" module add servomotor motion axis at axis of cam shaft and in servomotor definition select profile specification velocity (deg/s) magnitude value constant 180, than select analysis definition type "kinematic" in mechanism module and preferences graphical display start time 0 and end time 2 for tool changing cycle time 2 s, and run the kinematic analysis.

After complete the kinematic analysis, to measure the velocity at each connecting point, select "measure results", graph type "measure versus time" and

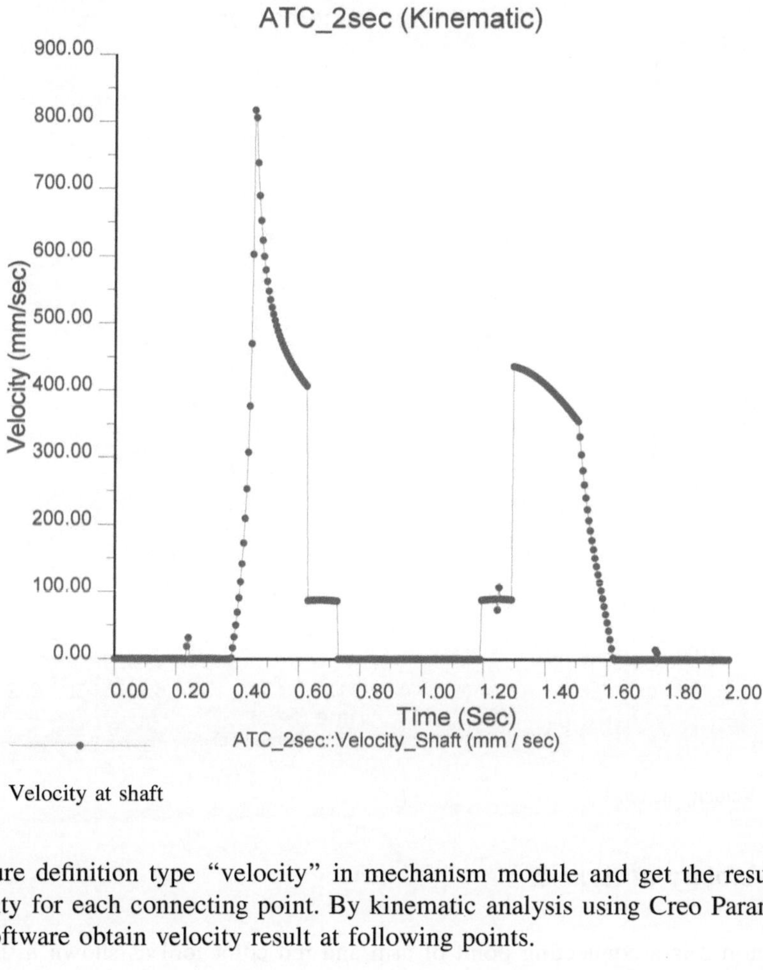

Fig. 7 Velocity at shaft

measure definition type "velocity" in mechanism module and get the results of velocity for each connecting point. By kinematic analysis using Creo Parametric 1.0 software obtain velocity result at following points.

5.1 Velocity at Arm-1

Point arm-1 is a connecting point of yellow color arm and green color tool as shown in Fig. 3. From kinematic analysis of tool changing mechanism for tool changing cycle time 2 s, result of velocity at arm-1 obtain as shown in Fig. 5.

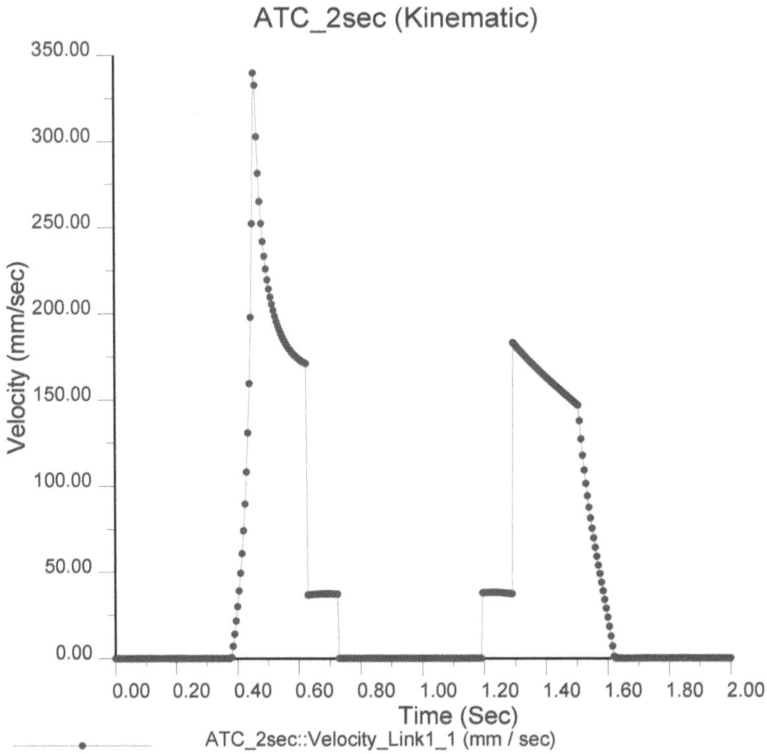

Fig. 8 Velocity at link1-1

5.2 Velocity at Arm-2

Point arm-2 is a connecting point of arm and red color tool as shown in Fig. 3. From kinematic analysis of tool changing mechanism for tool changing cycle time 2 s, result of velocity at arm-1 obtain as shown in Fig. 6.

5.3 Velocity at Shaft

Shaft point is a connecting point of arm and shaft as shown in Fig. 3. From kinematic analysis of tool changing mechanism for tool changing cycle time 2 s, result of velocity at shaft obtain as shown in Fig. 7.

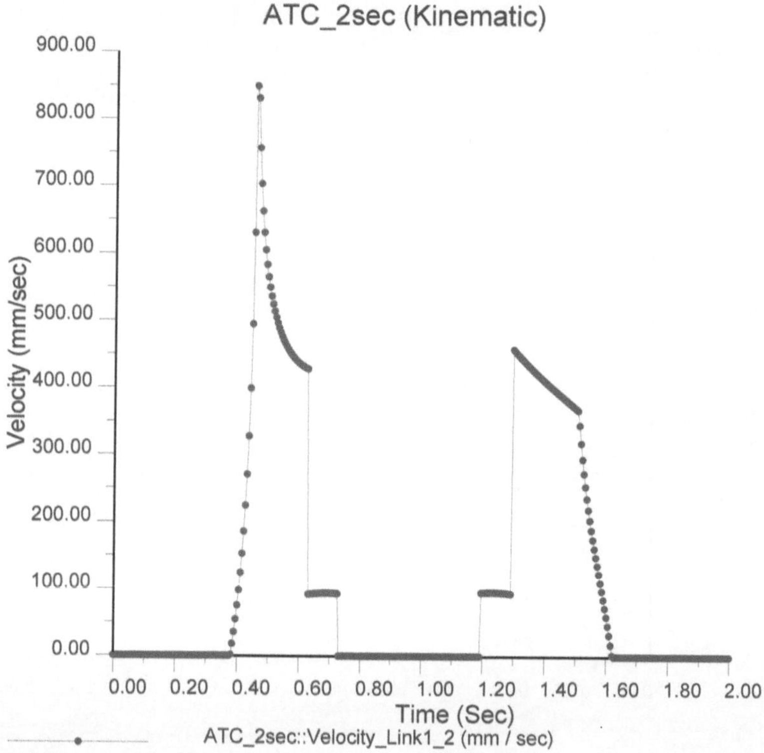

Fig. 9 Velocity at link1-2

5.4 Velocity at Link1-1

Link1-1 point is a connecting point of purple color link2 and yellow color link1 as shown in Fig. 3. From kinematic analysis of tool changing mechanism for tool changing cycle time 2 s, result of velocity at link1-1 obtain as shown in Fig. 8.

5.5 Velocity at Link1-2

Link1-2 point is a connecting point of shaft and yellow color link1 as shown in Fig. 3. From kinematic analysis of tool changing mechanism for tool changing cycle time 2 s, result of velocity at link1-2 obtain as shown in Fig. 9.

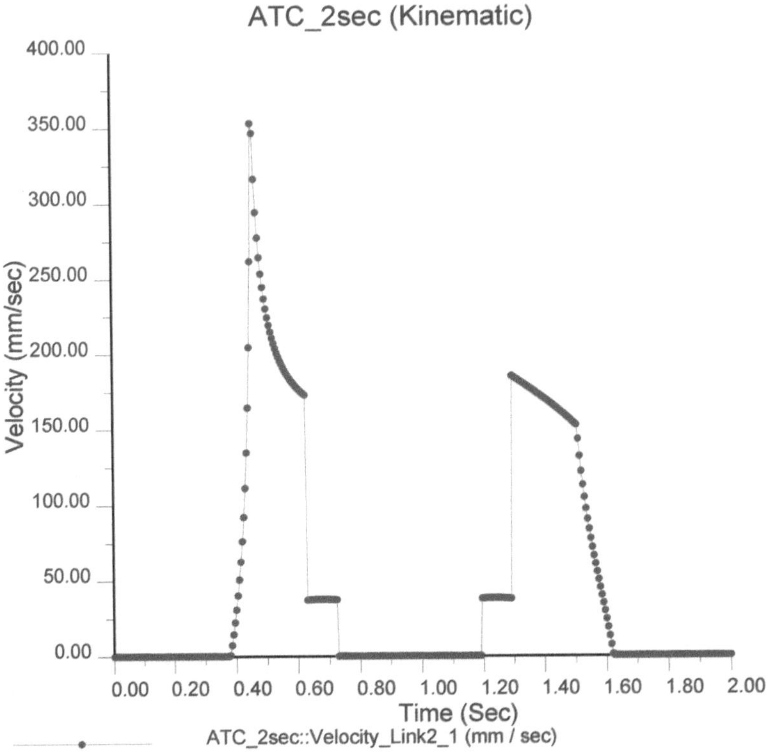

Fig. 10 Velocity at link2-1

5.6 Velocity at Link2-1

Link2-1 point is a connecting point of purple color link2 and green color link3 as shown in Fig. 3. From kinematic analysis of tool changing mechanism for tool changing cycle time 2 s, result of velocity at link2-1 obtain as shown in Fig. 10.

5.7 Velocity at Link2-2

Link2-2 point is a connecting point of purple color link2 and yellow color link1 as shown in Fig. 3. From kinematic analysis of tool changing mechanism for tool changing cycle time 2 s, result of velocity at link2-2 obtain as shown in Fig. 11.

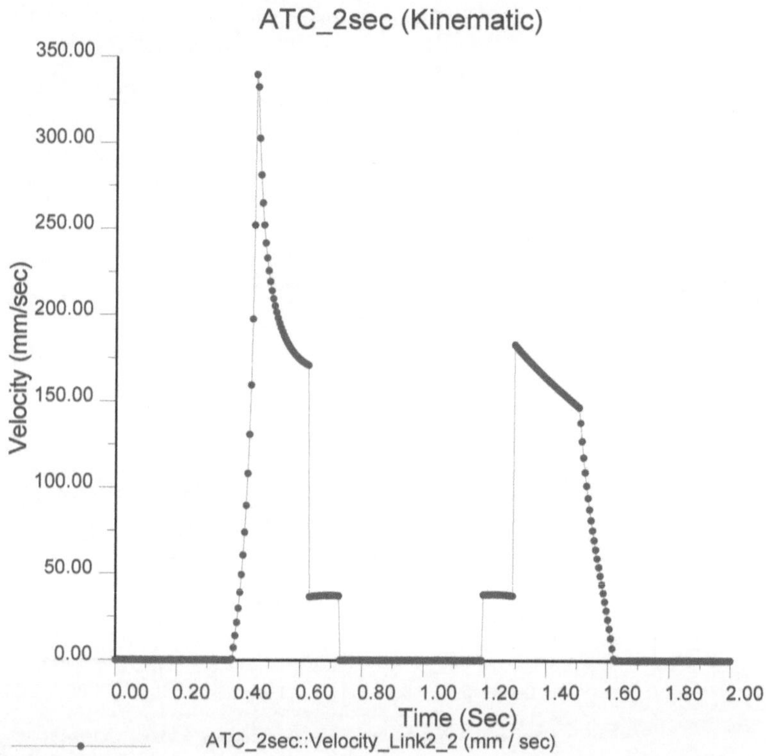

Fig. 11 Velocity at link2-2

5.8 Velocity at Link3-1

Link3-1 point is a connecting point of radial cam and green color link3 as shown in Fig. 3. From kinematic analysis of tool changing mechanism for tool changing cycle time 2 s, result of velocity at link3-1 obtain as shown in Fig. 12.

5.9 Velocity at Link3-2

Link3-2 point is a connecting point of purple color link2 and green color link3 as shown in Fig. 3. From kinematic analysis of tool changing mechanism for tool changing cycle time 2 s, result of velocity at link3-2 obtain as shown in Fig. 13.

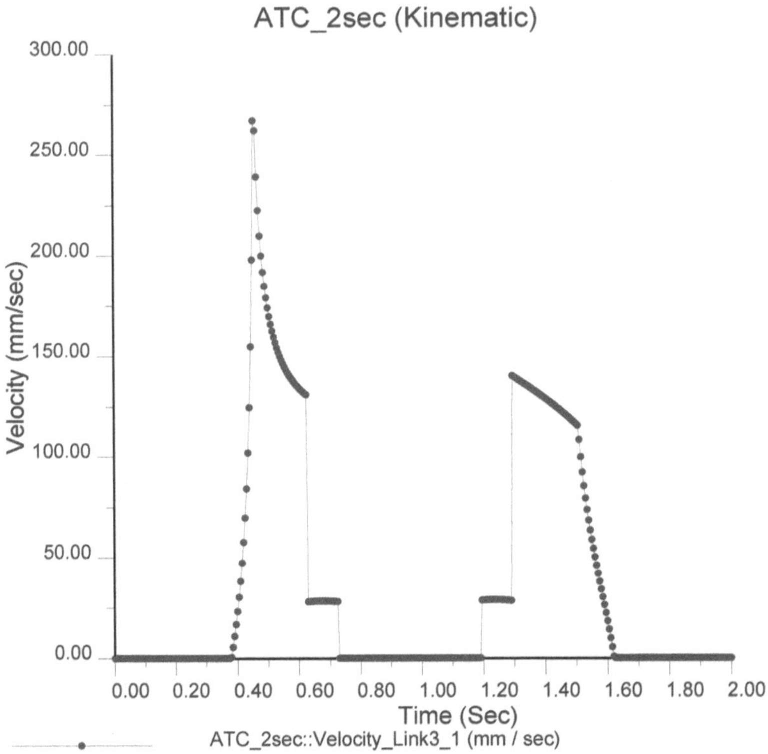

Fig. 12 Velocity at link3-1

6 Result Discussion

From kinematic analysis using Creo Parametric 1.0 software gets the result of velocity at each connecting point of tool changing mechanism of ATC for tool changing cycle time 2 s, as shown in Fig. 14.

From the kinematic analysis of tool changing mechanism of ATC for tool changing cycle time 2 s, maximum velocity obtain 1638.76 mm/s at arm-1 and arm-2.

For tool changing cycle time 1 s by kinematic analysis get the result as shown in Fig. 15.

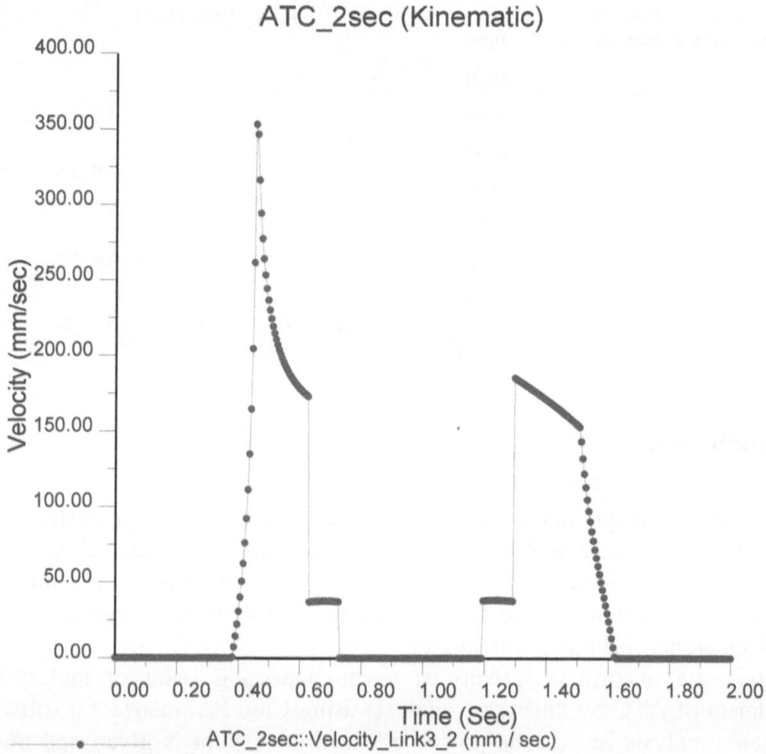

Fig. 13 Velocity at link3-2

Fig. 14 Velocity chart of tool changing mechanism

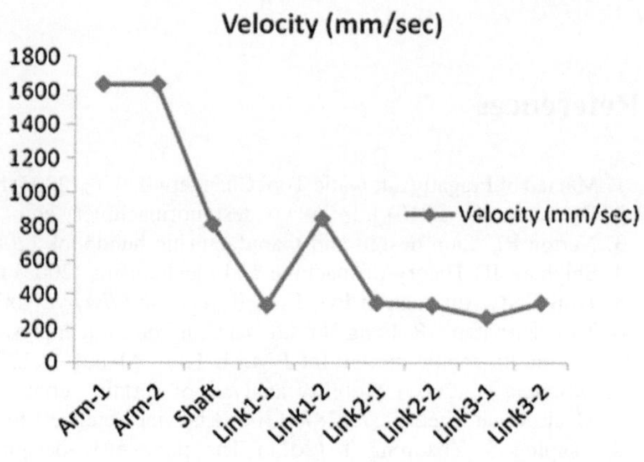

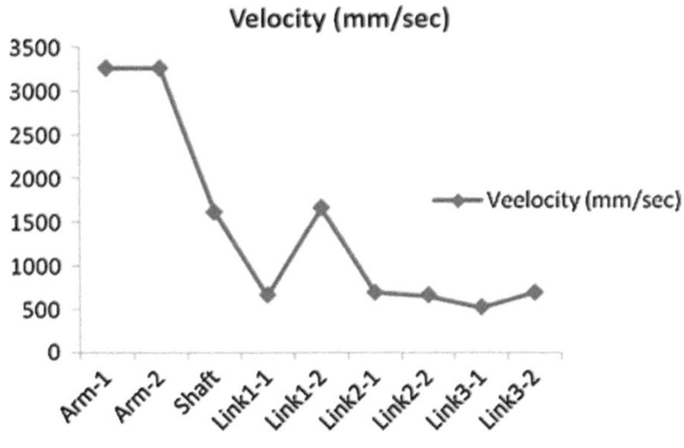

Fig. 15 Velocity chart of tool changing mechanism

7 Conclusion

Due to lengthy analytical calculation used Creo Parametric 1.0 powerful software for position, kinematic analysis. By position analysis obtain very close displacement at shaft of tool changing mechanism, actual displacement is 115 mm and by position analysis obtain 114.95 mm displacement at shaft. So design and assembly of tool changing mechanism is proper.

Obtain the maximum velocity at each connecting point of tool changing mechanism of ATC by kinematic analysis using Creo Parametric 1.0 software. If kinematic analysis run means proper connection or joint is given and no interferences is there in tool changing mechanism assembly. So now add gravity and applied force, by dynamic analysis can get the acceleration and relative force at each connecting point, so can be optimize the design of component and tool changing mechanism.

References

1. Manual of Pragati Automatic Tool Changer 4020 V. (2007) http://www.pragati-automation.com
2. Dynacam 10. (2010) http://www.designofmachinery.com
3. Norton RL Cam design and manufacturing handbook (2009) Pearson Education, Delhi
4. Shighley JE Theory of machine and mechanisms (2003) Oxford University Press, Oxford
5. Jyoti CNC Automation Pvt. Ltd., Rajkot. http://www.jyoti.co.in
6. Tutunea-Fatan OR, Feng HY (2004) Configuration analysis of five axis machine tools using a generic kinematic model. Int J Mach Tools Manuf 44:1235–1243 (Elsevier Science Ltd.)
7. Dasgupta A (2002) Mobility analysis of certain geometries of a RPSPR kinematic chain. Mech Mach Theory 37:1287–1306 (Elsevier Science Ltd.)
8. Jianping S, Zhaoping T (2011) The parametric design and motion analysis about line translating tip follower cam mechanism based on model datum graph. Procedia Eng 23:439–444 (Elsevier Science Ltd.)
9. Chang WT, Wu LL (2013) Tolerance analysis and synthesis of cam-modulated linkages. Math Comput Model 57:641–660 (Elsevier Science Ltd.)

10. Chang Z, Xu C, Pan T, Wang L, Zhang X (2009) A general framework for geometry design of indexing cam mechanism. Mech Mach Theory 44:2079–2084 (Elsevier Science Ltd.)
11. Butler O, Paranamana B, Powers W, Srinivasan A (2011) Project report on design of a servo driven, adjustable pick and place mechanism. Worcester Polytechnic Institute, Worcester

Effect of Elliptical Shaft Geometry on Non-dimensional Pressure and Load in Hydrodynamic Journal Bearings

P. C. Chhotani and D. P. Vakharia

Abstract Hydrodynamic journal bearings are critical power transmission components which carry high loads. Therefore, knowing the true operating conditions of hydrodynamic journal bearings is essential to machine design. Oil film pressure is one of the key operating parameters describing the operating conditions in hydrodynamic journal bearings. In the analysis of fluid-film bearings the basic equation employed is known as the Reynolds equation. Using Reynolds equation, one-dimensional analysis is done for plane tilted pad and journal bearing conventionally. Finite Difference approach is used to solve the Reynolds equation for plane tilted pad and journal bearing (one dimensional and two dimensional). Various approaches like Gauss elimination, Gauss–Seidel, Liebmann with Over-relaxation are applied to solve the equations produced on implementing Finite Difference method. Following same methodology, a special case of deformed shaft (elliptical geometry) has been studied.

Keywords Elliptical geometry · Pressure · Load · Journal bearings · Fluid film

Nomenclature

H	Oil film thickness
C	Radial clearance
E	Eccentricity
ε	Eccentricity ratio $= \frac{e}{c}$
p^*	Non-dimensional pressure
W^*	Non-dimensional load
C_r	Clearance ratio $= \frac{c}{R}$

P. C. Chhotani (✉)
Department of Mechanical Engineering, C. G. Patel Institute of Technology, Bardoli, Gujarat 394350, India
e-mail: paresh.c.chhotani@gmail.com

D. P. Vakharia
Department of Mechanical Engineering, S. V. N. I. T, Surat, Gujarat 395007, India

H. C. Patel et al. (eds.), *Proceedings of International Conference on Advances in Tribology and Engineering Systems*, Lecture Notes in Mechanical Engineering,
DOI: 10.1007/978-81-322-1656-8_26,

θ Angular co-ordinate measured from maximum film thickness

h_val $= (1 + \varepsilon \cos\theta) + \frac{1}{c_r}\left\{1 - \frac{e}{\sqrt{\{e^2\cos^2\theta + \sin^2\theta\}}}\right\}$

h_der $= -\varepsilon \sin\theta + \frac{e\sin\theta\cos\theta(1-e^2)}{c_r(e^2\cos^2\theta+\sin^2\theta)^{\frac{3}{2}}}$

1 Introduction

Theyse [1] emphasized the need of making better use of available knowledge about full film lubricated bearings in design-engineering. Kvitnitsky et al. [2] discussed the solution of Reynolds equation for hydrodynamic pressures of the load-carrying lubricating layer in journal bearings. Jakeman [3] presented a numerical method of hydrodynamic bearing analysis which is simple in concept, yet capable of development to handle complex situations such as dynamic misalignment. Chandrawat and Sinhasan [4] presented algorithms for two solution techniques, namely the Gauss-Siedel iterative scheme and the linear complementarity problem (LCP) approach, for finding the positive pressure region in a hydrodynamic journal bearing lubricant flow field. Studies have shown that oil film pressure is one of the key operating parameters describing the operating conditions in hydrodynamic journal bearings. Measuring the oil film pressure in bearings has been a demanding task and therefore the subject has been studied mainly by mathematical means. The oil film pressure has been calculated by trying various approaches by many researchers. The aim of this paper is to analyze the oil film pressure in hydrodynamic journal bearings for an irregular geometry of shaft (elliptical) by applying Finite Difference Method approach.

In the journal bearing studied in this work, there is a small deviation in the geometry of the shaft from the regular circular shape. Crosby [5] investigated the performance of a journal bearing having a finite length, a bush with an elliptical bore and subjected to a set of prescribed bearing loads. Mishra et al. [6] assumed the non-circularity in bearing bore to be elliptical and carried out the numerical solution of Reynolds equation and Energy equation to outline the temperature profile. We have assumed the geometry of the shaft as *elliptical* (an irregular shape) in the present work studied. The effect of varying the ellipticity of the bearing on non-dimensional bearing parameters (pressure and load) is investigated.

1.1 Assumptions [7]

1. Body forces are neglected, i.e. there are no extra fields of forces acting on the fluid. This is true except magnetohydrodynamics.

2. The pressure is constant through the thickness of the film. As the film is only one or two thousands of an inch thick it is always true. With elastic fluids there may be exceptions.
3. The curvature of surfaces is large compared with film thickness. Surface velocities need not be considered as varying in direction.
4. There is no slip at the boundaries. The velocity of the oil layer adjacent to the boundary is the same as that of the boundary.
5. The lubricant is Newtonian.
6. Flow is laminar, in big turbine bearings this is not true and the theory is being slowly developed.
7. Fluid inertia is neglected. Several studies have shown that even if Reynolds number is 1,000 the pressures are only modified by about 5 %.
8. The viscosity is constant through the film thickness. This is certainly not true but leads to great complexity if it is not assumed.

1.2 Elliptical Shaft Geometry: Oil Film Thickness

As we know, oil film thickness for a journal bearing (circular shaft) is given by

$$h = c(1 + \varepsilon \cos \theta)$$

In case the shaft is elliptical, the oil film thickness (as shown in Fig. 1) will be

$$h = c(1 + \varepsilon \cos \theta) + \boldsymbol{diff} \tag{1}$$

where diff $= MN = CN - CM$.

Let's first find co-ordinates of point M. The point M can be said as intersection of ellipse and a line CA. The equation for ellipse

$$\frac{x^2}{a^2} + \frac{y^2}{b^2} = 1 \tag{2}$$

Equation for the line passing from C (0, 0) and at an angle θ

$$y = mx + c$$

As its passing from (0, 0), the intercept is 0 and m = tan θ

$$y = x \tan \theta \tag{3}$$

Putting value of y in Eq. (2)

$$\frac{x^2}{a^2} + \frac{(x \tan \theta)^2}{b^2} = 1$$

$$\therefore \frac{x^2}{a^2} + \frac{x^2 \tan^2 \theta}{b^2} = 1$$

Fig. 1 Geometry of elliptical shaft—bearing

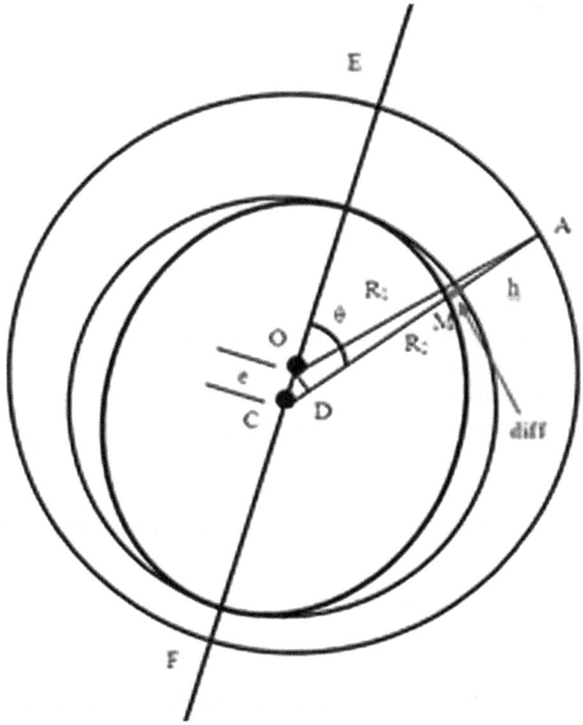

$$\therefore \frac{x^2}{a^2}\left\{1+\frac{\tan^2\theta}{b^2/a^2}\right\}=1$$

Taking b/a = e_e (which is less than unity)

$$\therefore \frac{x^2}{a^2}\left\{1+\frac{\tan^2\theta}{e_e^2}\right\}=1$$

and a = radius of shaft (R) as can be seen from Fig. 1

$$\therefore \frac{x^2}{R^2}\left\{1+\frac{\tan^2\theta}{e_e^2}\right\}=1$$

$$\therefore x^2=\frac{R^2}{\left\{1+\frac{\tan^2\theta}{e_e^2}\right\}}$$

$$\therefore x^2=\frac{R^2}{\left\{1+\frac{\sin^2\theta}{e_e^2\cos^2\theta}\right\}}$$

$$\therefore x^2 = \frac{R^2 e_e^2 \cos^2\theta}{\{e_e^2 \cos^2\theta + \sin^2\theta\}}$$

$$\therefore x = \frac{Re_e \cos\theta}{\sqrt{\{e_e^2 \cos^2\theta + \sin^2\theta\}}}$$

and from Eq. (3)

$$y = \frac{Re_e \cos\theta}{\sqrt{\{e_e^2 \cos^2\theta + \sin^2\theta\}}} \tan\theta$$

$$\therefore y = \frac{Re_e \sin\theta}{\sqrt{\{e_e^2 \cos^2\theta + \sin^2\theta\}}}$$

Co-ordinates of point M

$$(x, y) =$$

$$\left(\frac{Re_e \cos\theta}{\sqrt{\{e_e^2 \cos^2\theta + \sin^2\theta\}}}, \frac{Re_e \sin\theta}{\sqrt{\{e_e^2 \cos^2\theta + \sin^2\theta\}}} \right)$$

So

$$CM = \frac{Re_e}{\sqrt{\{e_e^2 \cos^2\theta + \sin^2\theta\}}}$$

Now diff $= MN = CN - CM$

$$= R\left\{ 1 - \frac{e_e}{\sqrt{\{e_e^2 \cos^2\theta + \sin^2\theta\}}} \right\}$$

From Eq. (1), oil film thickness

$$h = c\begin{bmatrix} (1 + \varepsilon\cos\theta) \\ + \\ \frac{R}{c}\left\{ 1 - \frac{e_e}{\sqrt{\{e_e^2 \cos^2\theta + \sin^2\theta\}}} \right\} \end{bmatrix}$$

$$h = c\begin{bmatrix} (1 + \varepsilon\cos\theta) \\ + \\ \frac{1}{c_r}\left\{ 1 - \frac{e_e}{\sqrt{\{e_e^2 \cos^2\theta + \sin^2\theta\}}} \right\} \end{bmatrix} \quad (4)$$

The oil film thickness in case of elliptical shaft can be given by above Eq. (4) which is function of c, R, ε, θ, e_e, c_r.

1.3 Mathematical Modeling

Reynolds equation

$$\frac{\partial}{\partial x}\left(h^3\frac{\partial p}{\partial x}\right)+\frac{\partial}{\partial y}\left(h^3\frac{\partial p}{\partial y}\right)=6U\eta\frac{dh}{dx}$$

$$\frac{\partial^2 p}{\partial x^2}+\left(\frac{3}{h}\frac{dh}{dx}\right)\frac{\partial p}{\partial x}+\frac{\partial^2 p}{\partial y^2}=\frac{6U\eta}{h^3}\frac{dh}{dx}$$

$$x=R\theta$$

$$\frac{\partial^2 p}{R^2\partial\theta^2}+\left(\frac{3}{h}\frac{dh}{Rd\theta}\right)\frac{\partial p}{R\partial\theta}+\frac{\partial^2 p}{\partial y^2}=\frac{6U\eta}{h^3}\frac{dh}{Rd\theta}$$

$$\frac{\partial^2 p}{\partial\theta^2}+\left(\frac{3}{h}\frac{dh}{d\theta}\right)\frac{\partial p}{\partial\theta}+\frac{\partial^2 p}{\frac{1}{R^2}\partial y^2}=\frac{6U\eta R}{h^3}\frac{dh}{d\theta}$$

$$y=RY$$

$$\frac{\partial^2 p}{\partial\theta^2}+\left(\frac{3}{h}\frac{dh}{d\theta}\right)\frac{\partial p}{\partial\theta}+\frac{\partial^2 p}{dY^2}=\frac{6U\eta R}{h^3}\frac{dh}{d\theta} \tag{5}$$

From Eq. (4)

$$\frac{dh}{d\theta}=c\left[-\varepsilon\sin\theta+\frac{e\sin\theta\cos\theta(1-e^2)}{c_r\left(e^2\cos^2\theta+\sin^2\theta\right)^{\frac{3}{2}}}\right]$$

From Eq. (5)

$$\frac{\partial^2 p}{\partial\theta^2}+\left(3\frac{1}{c(\boldsymbol{h}_{val})}c(\boldsymbol{h}_{der})\right)\frac{\partial p}{\partial\theta}+\frac{\partial^2 p}{dY^2}$$

$$=\frac{6U\eta R}{c^3(\boldsymbol{h\_val})^3}c(\boldsymbol{h\_der})$$

$$\frac{\partial^2 p}{\partial\theta^2}+\left(3\frac{\boldsymbol{h\_der}}{\boldsymbol{h\_val}}\right)\frac{\partial p}{\partial\theta}+\frac{\partial^2 p}{dY^2}=\frac{6U\eta R}{c^2}\frac{\boldsymbol{h\_der}}{(\boldsymbol{h\_val})^3}$$

Taking non dimensional pressure $p^* = \frac{c^2}{6U\eta R}p$

$$\frac{\partial^2 p^*}{\partial\theta^2} + \left(3\frac{\boldsymbol{h\_der}}{\boldsymbol{h\_val}}\right)\frac{\partial p^*}{\partial\theta} + \frac{\partial^2 p^*}{dY^2} = \frac{\boldsymbol{h_{der}}}{(\boldsymbol{h_{val}})^3} \tag{6}$$

The Eq. (6) can be used for Elliptical shaft bearing—Two dimensional analysis.

1.4 Finite Difference Approach

For two dimensions [8],

$$\frac{\partial^2 p^*}{\partial\theta^2} = \frac{p^*_{i+1,j} - 2p^*_{i,j} + p^*_{i-1,j}}{\Delta\theta^2}$$

$$\frac{\partial p^*}{\partial\theta} = \frac{p^*_{i+1,j} - p^*_{i-1,j}}{2\Delta\theta}$$

$$\frac{\partial^2 p^*}{\partial Y^2} = \frac{p^*_{i,j+1} - 2p^*_{i,j} + p^*_{i,j-1}}{\Delta Y^2}$$

Equation for Elliptical shaft bearing—Two dimensional analysis

$$\frac{\partial^2 p^*}{\partial\theta^2} + \left(3\frac{\boldsymbol{h\_der}}{\boldsymbol{h\_val}}\right)\frac{\partial p^*}{\partial\theta} + \frac{\partial^2 p^*}{dY^2} = \frac{\boldsymbol{h_{der}}}{(\boldsymbol{h_{val}})^3}$$

Replacing values of $\frac{\partial^2 p^*}{\partial\theta^2}$, $\frac{\partial p^*}{\partial\theta}$ and $\frac{\partial^2 p^*}{\partial Y^2}$

$$\frac{p^*_{i+1,j} - 2p^*_{i,j} + p^*_{i-1,j}}{\Delta\theta^2}$$

$$+\left(3\frac{\boldsymbol{h_{der_i}}}{h_{val_i}}\right)\left\{\frac{p^*_{i+1,j} - p^*_{i-1,j}}{2\Delta\theta}\right\}$$

$$+\frac{p^*_{i,j+1} - 2p^*_{i,j} + p^*_{i,j-1}}{\Delta Y^2} - \left(\frac{\boldsymbol{h_{der_i}}}{(\boldsymbol{h_{val_i}})^3}\right) = 0$$

$$\frac{1}{\Delta\theta^2}\left\{p^*_{i+1,j} - 2p^*_{i,j} + p^*_{i-1,j}\right\}$$

$$+\left(3\frac{\boldsymbol{h\_der_i}}{\boldsymbol{h\_val_i}}\right)\left\{p^*_{i+1,j} - p^*_{i-1,j}\right\}$$

$$+\frac{1}{\Delta Y^2}\left\{p^*_{i,j+1}-2p^*_{i,j}+p^*_{i,j-1}\right\}-\left(\frac{\boldsymbol{h}_{der_i}}{(\boldsymbol{h}_{val_i})^3}\right)=0$$

$$p^*_{i-1,j}\left\{\frac{1}{\Delta\theta^2}-3\frac{\boldsymbol{h}_{der_i}}{\boldsymbol{h}_{val_i}}\right\}+p^*_{i,j}\left\{\frac{-2}{\Delta\theta^2}\right\}$$

$$+p^*_{i+1,j}\left\{\frac{1}{\Delta\theta^2}+3\frac{\boldsymbol{h}_{der_i}}{\boldsymbol{h}_{val_i}}\right\}$$

$$+p^*_{i,j-1}\left\{\frac{1}{\Delta Y^2}\right\}+p^*_{i,j+1}\left\{\frac{1}{\Delta Y^2}\right\}-\left(\frac{\boldsymbol{h}_{der_i}}{(\boldsymbol{h}_{val_i})^3}\right)=0$$

$$p^*_{i,j}\left\{\frac{-2}{\Delta\theta^2}\right\}=\left(\frac{\boldsymbol{h}_{der_i}}{(\boldsymbol{h}_{val_i})^3}\right)-p^*_{i-1,j}\left\{\frac{1}{\Delta\theta^2}-3\frac{\boldsymbol{h\_der_i}}{\boldsymbol{h\_val_i}}\right\}$$

$$-p^*_{i+1,j}\left\{\frac{1}{\Delta\theta^2}+3\frac{\boldsymbol{h}_{der_i}}{\boldsymbol{h}_{val_i}}\right\}$$

$$-p^*_{i,j-1}\left\{\frac{1}{\Delta Y^2}\right\}-p^*_{i,j+1}\left\{\frac{1}{\Delta Y^2}\right\}$$

$$p^*_{i,j}=\left\{\frac{\Delta\theta^2}{-2}\right\}\begin{bmatrix}\left\{\frac{\boldsymbol{h}_{der_i}}{(\boldsymbol{h}_{val_i})^3}\right\}\\ -p^*_{i-1,j}\left\{\frac{1}{\Delta\theta^2}-3\frac{\boldsymbol{h}_{der_i}}{\boldsymbol{h}_{val_i}}\right\}\\ -p^*_{i+1,j}\left\{\frac{1}{\Delta\theta^2}+3\frac{\boldsymbol{h}_{der_i}}{\boldsymbol{h}_{val_i}}\right\}\\ -p^*_{i,j-1}\left\{\frac{1}{\Delta Y^2}\right\}-p^*_{i,j+1}\left\{\frac{1}{\Delta Y^2}\right\}\end{bmatrix} \tag{7}$$

Above equation is used for finding values non-dimensional pressure at various locations for the elliptical shaft geometry.

1.5 Liebmann's Method

Consider 3 intermediate points in each direction, i.e. 3 intermediate points in x-direction and 3 intermediate points in y-direction. So there will be 9 intermediate points total. Take eccentricity ratio = 0.5, L/D = 1 and for ellipse b/a ratio = 0.995 [8].

$$\Delta\theta=\frac{\theta_{max}-\theta_{min}}{(totalpoints-1)}=\frac{2\pi-0}{(5-1)}=\frac{2\pi}{4}=\frac{\pi}{2}$$

$$\Delta Y = \frac{Y_{max} - Y_{min}}{(totalpoints - 1)} = \frac{\frac{L}{D} - 0}{(5 - 1)} = \frac{\frac{L}{D}}{4} = \frac{0.25L}{D}$$

$$= \frac{0.5L}{R} = 0.5(as\frac{L}{R} = 1)$$

Initially, the values of non-dimensional pressure at all locations are taken zero. At boundary nodes also, the values are zero. But the values of non-dimensional pressure at boundary nodes will not alter, so not considering them.

In the next iteration, using Eq. (7) for i = 2, j = 2.

For reducing the number of iterations to achieve the final pressure values, we use over-relaxation.

$$\lambda = 1.5$$

$$\left(p_{2,2}^*\right)_{new} = \lambda\left(p_{2,2}^*\right)_{new} + (1 - \lambda)\left(p_{2,2}^*\right)_{old}$$

Thus calculated new value of $\left(p_{2,2}^*\right)_{new}$ is considered for the next calculation.

Then the next calculation of non-dimensional pressure for next node i = 2, j = 3 is done from Eq. (7).

Then again the over-relaxation is applied to the recently found $p_{2,3}^*$ value and then new value of $p_{2,3}^*$ is considered for the next all calculations.

Then error is calculated for each i, j using following equation.

$$|(\in_a)|_{i,j} = \left| \frac{\left(p_{i,j}^*\right)_{new} - \left(p_{i,j}^*\right)_{old}}{\left(p_{i,j}^*\right)_{new}} \right| \times 100\,\%$$

And then the iterations are repeated until the absolute values of all the percent relative errors $(\in_a)_{i,j}$ fall below a prescribed stopping criterion $\in_s$. For this example, stopping criterion $\in_s$ is taken 0.001.

Non-dimensional load W* can be calculated by integrating [8] p* over θ and over Y (Fig. 2).

$$W^* = \int_{\theta_{min}}^{\theta_{max}} \int_{Y_{min}}^{Y_{max}} p^* dY d\theta$$

1.6 Equations Used in One Dimensional Analysis

As seen in the above sections, similar analysis of elliptical shaft-bearing also can be done in 1-D. The equations derived to be used for the computations are as follows

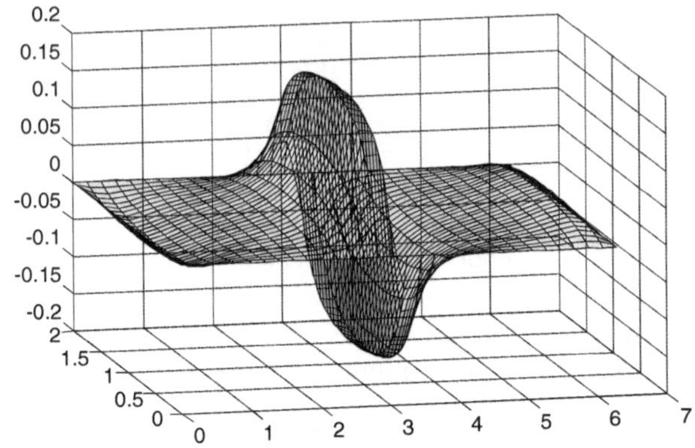

Fig. 2 Non-dimensional pressure profile for elliptical shaft-bearing

$$\frac{\partial^2 p^*}{\partial \theta^2} + \left(3\frac{\boldsymbol{h\_der}}{\boldsymbol{h\_val}}\right)\frac{\partial p^*}{\partial \theta} = \frac{\boldsymbol{h\_der}}{(\boldsymbol{h\_val})^3}$$

And to use Finite Difference approach, following equations can be used.

$$\frac{\partial^2 p^*}{\partial \theta^2} = \frac{p_{i+1}^* - 2p_i^* + p_{i-1}^*}{\Delta\theta^2}$$

$$\frac{\partial p^*}{\partial \theta} = \frac{p_{i+1}^* - p_{i-1}^*}{2\Delta\theta}$$

2 Results and Discussion

Computations have been carried out for the elliptical shaft bearing (1-D analysis) having b/a ratio = 0.995, clearance ratio = 0.003 and taking 80 intermediate points and the non-dimensional load W* is compared in elliptical shaft and in circular shaft.

Then keeping eccentricity ratio (0.5) same, the different values of b/a ratio have been taken and calculations are done for Non-dimensional load for elliptical shaft bearing (1-D analysis). The results computed are shown in Table 1. The same has been extended to 2-D analysis and the results (Table 2) achieved are as expected (as in 2-D analysis of circular shaft geometry, similar deviation is noticed from 1-D analysis of the circular shaft geometry), which confirms the attained results for elliptical shaft geometry.

As seen from the Fig. 3 Plot of W^*versus ε Journal bearing (circular vs. elliptical shaft), the non-dimensional load for elliptical shaft is less than the same for circular shaft for any value of eccentricity ratio. From Table 1, it is observed that the load carrying capacity is affected considerably when the b/a ratio of elliptical shaft exceeds 0.995 and the load carrying capacity is noticeable only after value of b/a ratio exceeds 0.99.

Table 1 Journal bearing (elliptical shaft)– for different values of b/a ratio

b/a ratio	Non–dimensional load W^*
0.1	3.3412×10^{-4}
0.2	0.0051
0.3	0.0209
0.4	0.2323
0.5	0.2149
0.6	0.3995
0.7	0.6029
0.8	0.2819
0.9	0.2003
0.95	0.4060
0.98	0.7117
0.99	1.0298
0.995	1.4018
0.999	2.0072
1 (circular)	2.1658

Table 2 Liebmann—over relaxation approach applied for clearance ratio = 0.003; eccentricity ratio = 0.5; and intermediate points taken for calculation = 30

b/a ratio	Non–dimensional load W^*	
	One dimensional	Two dimensional
0.95	0.4060	0.2411
0.98	0.7117	0.2885
0.99	1.0298	0.4288
0.995	1.4018	0.6140
0.999	2.0072	1.0708
1 (circular)	2.1658	1.3059

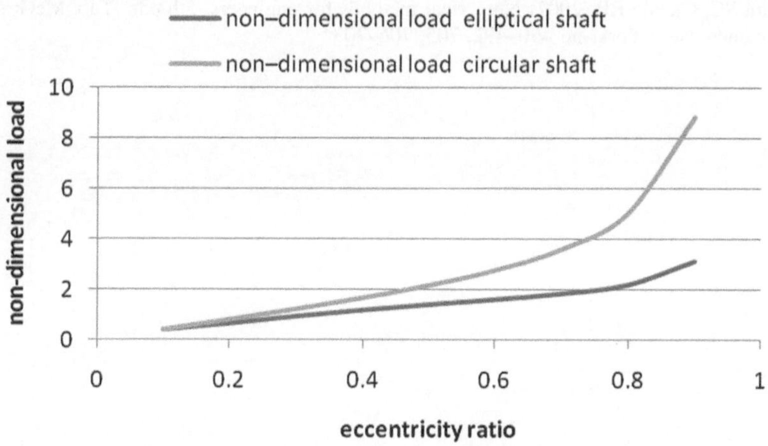

Fig. 3 Plot of $W^* vs \varepsilon$—journal bearing (circular vs. elliptical shaft)

3 Conclusion

Sometimes the shaft geometry is not exactly circular because of manufacturing defect or might be due to deformation. The shaft geometry is considered as elliptical in this study and the effect of the ellipticity of the shaft geometry on the non-dimensional pressure and load for a journal bearing with an elliptical shaft is studied. As b/a ratio (1 for circular) decreases, the non-dimensional load also decreases. It means if the shaft is deformed it can be allowed to a certain extent only, as once b/a ratio descends than the limit then the load-carrying capacity of the shaft (elliptical) may go beyond the lowest value of load-carrying capacity of the exactly circular shaft which is not desired scenario in any case. Thus, in working conditions, the b/a ratio can be considered as a measure to take judgment on a bearing life or the load-carrying capacity of a deformed shaft.

References

1. Theyse FH (1964) Fundamentals of hydrodynamic lubrication and their consequences in design engineering—part I. Wear 7:419–434
2. Kvitnitsky YI, Kirkatch NF, Poltavsky YD (1976) The solution of reynolds equation under natural boundary conditions for hydrodynamic journal bearings. Wear 37:217–231
3. Jakeman RW (1984) A numerical analysis method based on flow continuity for hydrodynamic journal bearings. Tribol Int 17(6):325–333
4. Chandrawat HN, Sinhasan R (1987) A comparison between two numerical techniques for hydrodynamic journal bearing problems. Wear 119:77–87
5. Crosby WA (1992) An investigation of the performance of a journal bearing with a slightly irregular bore. Tribol Int 25(3):199–204
6. Mishra PC, Pandey RK, Athre K (2007) Temperature profile of an elliptic bore journal bearing. Tribol Int 40:453–458
7. Cameron A (1981) Basic lubrication theory. Longman Green & Co. Ltd., London, pp 18–27
8. Chapra SC, Canale RP (2007) Numerical methods for engineers, 5th edn. Tata McGraw Hill Companies, New York, pp 490–492, 705, 706–708

Effect of Localized Defect on the Vibration Behavior of Cylindrical Roller Bearing-Rotor System

U. A. Patel and S. H. Upadhyay

Abstract This paper presents a mathematical model to investigate the vibration behavior of a rotor-bearing system due to localized defects of inner race and outer race for cylindrical roller bearing. In the formulation, the contacts between rolling elements and inner/outer races are considered as nonlinear springs as well as nonlinear damping. The governing equations of motion are formulated using energy approach, nonlinear equations are solved by Newton Rap son method for n-unknown nonlinear simultaneous equation. Equations of motions are solved by implicit type numerical integration method -New mark- β method. Here results obtained in the form of Fast Fourier Transformation and Phase plot/Poincare map. The validity of the proposed model verify by comparison of frequency components of the system response with those obtain from experiments.

Keywords Cylindrical roller · Defects · Nonlinear vibrations

Nomenclature

Q	Total contact force
δ	Deformation
W	Width of laminate
A_1	Const. clearance
R_o	Outer race radius
ω_c	Cage angular velocity
Z	Number of roller
D_h	Depth of defect

U. A. Patel (✉)
Department of Mechanical Engineering, L. D. College of Engineering, Ahmadabad, Gujarat 380015, India
e-mail: utkarsh12345@rediffmail.com

S. H. Upadhyay
Department of Mechanical and Industrial Engineering, Indian Institute of Technology, Roorkee, India

H. C. Patel et al. (eds.), *Proceedings of International Conference on Advances in Tribology and Engineering Systems*, Lecture Notes in Mechanical Engineering, DOI: 10.1007/978-81-322-1656-8_27, © Springer India 2014

D_L Depth of defect
P_o Amplitude of outer race defect
Pi Amplitude of inner race defect
ζ_j Roller tilts due to thrust load
P_d Internal radial clearance
k_{eq} Equivalent contact stiffness
Δ_j Contact deformation due to ideal normal loading
E Modulus of elasticity for steel
μ Poisson's ratio
B Radial internal clearance
ψi Initial position of defect
α Length of defect in degrees
L Length of roller
k No. of laminas
q Contact force per unit length
n_i Roller speed
R_o Inner race radius
ω_2 Angular velo. (inner race)
C Viscous damping coefficient
$\dot{\delta}$ Deformation rate
Q_{ry} Dissipative force in y direction
Q_{rz} Dissipative force in z direction
R_{r1} Radius of outer race
R_{r2} Radius of inner race
R Roller radius
R_{x1} Equivalent roller outer race curvature
R_{x2} Equivalent roller inner race curvature
X1 Displacement in y direction
Xd1 Velocity in y direction
X2 Displacement in Z direction
Xd2 Velocity in Z direction
H Height of local defect

1 Introduction

The defect in rolling element bearing can be categorised as local and distributed. Local defects include cracks, pits and spalls on the rolling surfaces. The distributed defect in rolling element bearing includes surface roughness, waviness, and misaligned races and off sized rolling element. Modelling and simulation method provide an accurate approach for predicting the dynamic behaviour for a cylindrical roller bearing-rotor system.

The number of rolling elements and their positions in load zone change with bearing rotation, give a periodic vibration, variation in stiffness generates vibration commonly known as varying compliance vibrations [1]. A lot of research work have been published, on the detection and diagnosis of bearing defects [2, 3]. Most of proposed model investigated the distributed defects particularly waviness of race way [4–6] just few works present mathematical models for simulation of local bearing defect. Tondon et al. [7] proposed an analytical model for predicting the vibration frequencies of rolling bearings and amplitude of frequency component due to localised defect under radial and axial load. Mayer et al. [8] studied effect on vibration due to distributed defect of bearing on rotor bearing system. Patil et al. [9]. studied localised defect on ball bearing and solved the nonlinear equation using state space variable method. Gafari et al. [10] studied the effect of localised defect for rolling element bearing. Wang Liqin et al. [11] predicted result for Rotor Roller Bearing System with Radial Clearances and Waviness. Upadhyay et al. [12] analysis the Nonlinear Phenomenon in high speed Ball Bearing System due to Radial clearance and unbalanced rotor effects. Cao et al. [13] made a dynamic model for double-row spherical roller bearing considering surface defects, pre-loads, and radial clearance, local as well as distributed defect in his model.

In this paper, a mathematical model was developed in which the contact between rolling elements and roller are modeled as nonlinear springs according to Hertz an contact theory and nonlinear damping fora cylindrical roller bearing-rotor system. Localized defect on inner race and outer race is modeled considering half sinusoidal wave.

2 Problem Formulation

A schematic diagram of rolling element bearing is shown in Fig. 2. For investigating the structural vibration characteristics of a cylindrical roller bearing, bearing assembly can be considered as a spring mass damper system. Contact deformation between races and roller gives a nonlinear force deformation relation, which is derived using Hertzian Theory. In modeling as shown in Fig. 3 the rolling element bearing is considered as a spring mass damper system having nonlinear spring and nonlinear damping.

2.1 Contact Force for Roller Bearings

In Fig. 1, *D*m is a pitch diameter of the bearing, *D*r1 and *D*r2 are diameters of the outer race and inner race respectively, $P_d/4$ is a radial clearance of the bearing. On inner race and at outer race localized defect is inserted with nonconventional machining processes as shown in Fig. 4. Shaft is inserted in the bearing by press fit. Palmgren [14] developed empirical relation from laboratory test data which

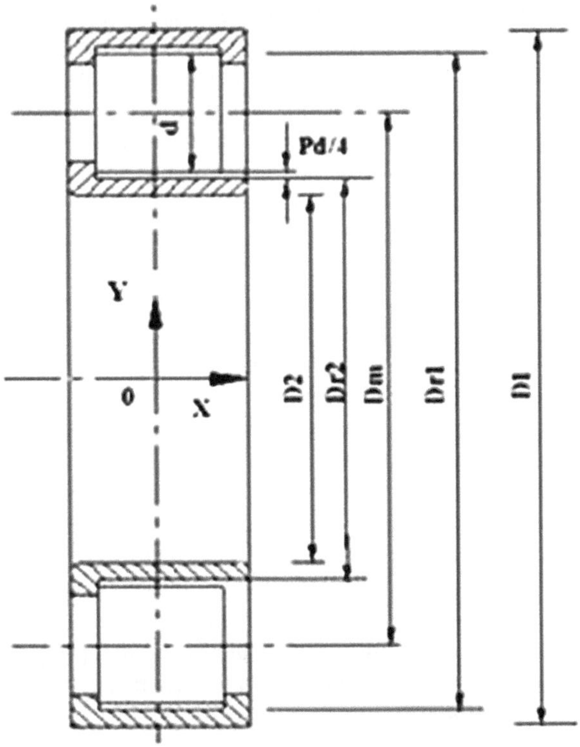

Fig. 1 Displacement Model of inner race outer race and roller

define relationship between contact force and deformation for line contact for roller bearing as,

$$\delta = 3.84 \times 10^{-5} \times \frac{Q^{0.9}}{l^{0.8}} \tag{1}$$

Considering that the contact length is divided into k lamina, each lamina of width w, the Eq. (1) can be written as,

$$q = \frac{\delta^{1.11}}{1.24 \times 10^{-5} \times (kw)^{0.11}} \tag{2}$$

The localized defect is considered as a circumferential half sinusoidal wave, as shown in Fig. 4, amplitude of outer race defect at the contact angle can be given by,

$$P_0 = A_1 + \left(D_h \times \sin\left(\pi \times \frac{R_0}{D_L} \times ((\omega_c)t + \frac{2\Pi(j-1)}{z} \right)\right)\right) \tag{3}$$

Amplitude of inner race defect at the contact angle can be given by,

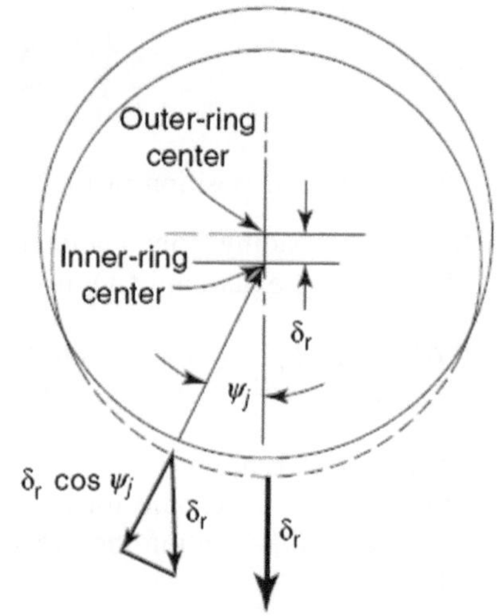

Fig. 2 Displacement Model of inner race outer race and roller

$$P_i = A_1 + \left(D_h \times \sin\left(\pi \times \frac{R_i}{D_L} \times ((\omega_c - \omega_2)t + \frac{2\Pi(j-1)}{z}\right)\right) \quad (4)$$

Total roller raceway deformation considering Contact deformation due to ideal normal loading, Radial defection due to thrust loading, radial internal clearance and localized defect can be given by,

$$\delta_{\lambda j} = \Delta_j + w\left(\lambda - \frac{1}{2}\right)\zeta_j - \frac{P_d}{2} - P_o - P_i \quad (5)$$

For k number of lamina Eq. (6) become as follow,

$$q_{jk} = \sum_{\lambda=1}^{\lambda=k} \frac{\left[\Delta_j + w\left(\lambda - \frac{1}{2}\right)\zeta_j - \frac{P_d}{2} - P\right]^{1.11}}{1.24 \times 10^{-5} \times (kw)^{0.11}}$$

By multiplying above equation with w which give total roller race way loading for length (l) for jth roller in radial direction and is given by,

Equilibrium condition for total loading is given by,

$$Q_j = \sum\nolimits_{\lambda=1}^{\lambda=k} \frac{\left[\Delta_j + w\left(\lambda - \frac{1}{2}\right)\zeta_j - \frac{P_d}{2} - P\right]^{1.11}}{1.24 \times 10^{-5} \times (k)^{0.11}} \times w^{0.89} \times 2 = 0 \quad (6)$$

In above Eq. 2 accounts for inner and outer race ways.

$$\frac{F_r}{2} - \sum_{j=1}^{j=\frac{Z}{2}+1} \tau_j Q_j \cos \Psi_j = 0$$

$$\Psi_j = \text{Angular position of the } j\text{th roller} = \frac{2\Pi(j-1)}{z} + \omega_c t$$

where, τ_j = Loading Zone parameter for jth rolling element

Substituting equation of Q_j in above equation we get,

$$\frac{0.62 \times 10^{-5} F_r}{w^{0.89}} - \sum_{j=1}^{j=\frac{z}{2}+1} \frac{\tau_j \cos \Psi_j}{k^{0.11}} \sum_{\lambda=1}^{\lambda=k} \left[\Delta_j + w\left(\lambda - \frac{1}{2}\right)\zeta_j - \frac{P_d}{2} - P\right]^{1.11} \quad (7)$$

Thrust load Equilibrium:

In above Eq. 2 account for inner and outer race ways.

At each roller location the thrust couple is balanced by a radial load couple.

Therefore,

$$\frac{h}{2} \times Q_{aj} = Q_j e_j \quad (8)$$

where, h = roller thrust couple moment arm

Therefore equation become as,

$$\frac{F_a}{2} - \frac{2}{h} \sum_{j=1}^{j=\frac{Z}{2}+1} \tau_j Q_j e_j = 0 \quad (9)$$

where, e_j is the eccentricity of the loading for jth roller and given by,

$$e_j = \frac{\sum_{\lambda=1}^{\lambda=k} q_{\lambda j}\left(\lambda - \frac{1}{2}\right) w}{\sum_{\lambda=1}^{\lambda=k} q_{\lambda j}} - \frac{l}{2}$$

Substitute the Qj and e_j in (7), we get

$$\frac{0.31 \times 10^{-5} F_a}{w^{0.89}} - \sum_{j=1}^{j=\frac{z}{2}+1} \frac{\tau_j}{k^{0.11}} \left\{ \sum_{\lambda=1}^{\lambda=k} \left[\delta_{j\lambda}\right]^{1.11} \left(\lambda - \frac{1}{2}\right) w - \frac{l}{2} \sum_{\lambda=1}^{\lambda=k} \left[\delta_{j\lambda}\right]^{1.11} \right\} \quad (10)$$

Equilibrium of deflection:

The shift of the inner-ring centre relative to the outer-ring centre owing to radial loading as shown in Fig. 2

where, δ_r = Radial deflection of the inner ring centre relative to outer ring centre due to radial loading

$$\left[\delta_a \times \frac{l}{D}\right] + \left[\delta_r \cos \Psi_j\right] - \frac{P_d}{2} - 2 \sum_{\lambda=1}^{\lambda=k} \left[\Delta_j + w\left(\lambda - \frac{1}{2}\right)\zeta_j - \frac{P_d}{2} - P\right] = 0 \quad (11)$$

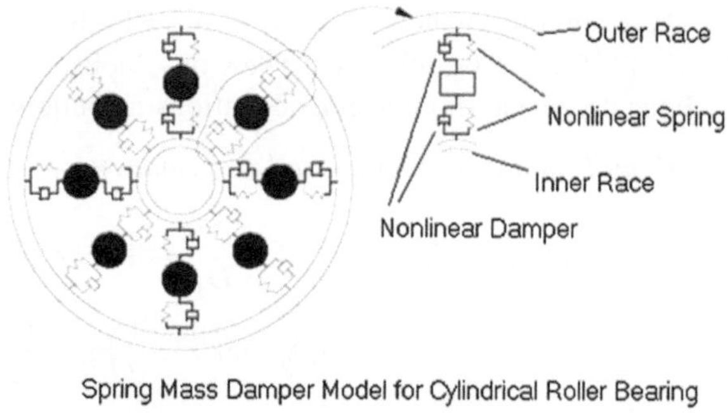

Fig. 3 Spring mass damper model for bearing

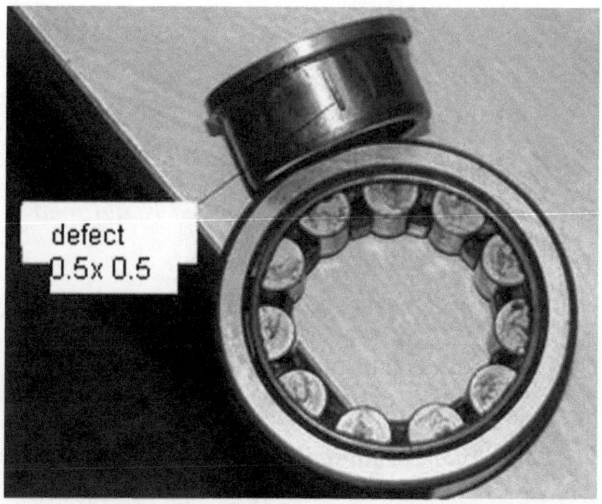

Fig. 4 Localised defect

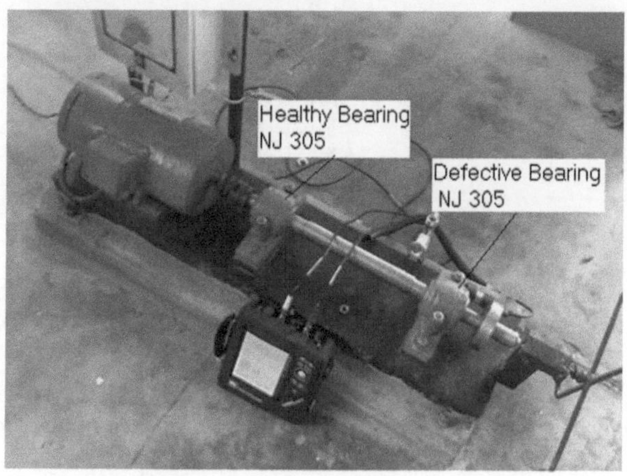

Fig. 5 View of experiment setup with Vibration Analyser, Experiment.rig

There are four unknown $(\Delta_j, \zeta_j, \delta_a, \delta_r)$ with four equilibrium equation (function f1, f2, f3, f4) can be solve with Newton–Raphson method (Figs. 3, 4).

Four nonlinear simultaneous equations as bellow,

(1) *Equilibrium of deflection and deformation*

$$f_1(\Delta_j, \zeta_j, \delta_a, \delta_r) = \left[\delta_a \times \frac{l}{D}\right] + [\delta_r \cos\Psi_j] - \frac{P_d}{2} - 2\sum_{\lambda=1}^{\lambda=k} \left[\Delta_j + w\left(\lambda - \frac{1}{2}\right)\zeta_j - \frac{P_d}{2} - P\right] = 0$$

(2) *Total roller loading*

$$f_2(\Delta_j, \zeta_j, \delta_a, \delta_r) = Q_j = \frac{w^{0.89}}{1.24 \times 10^{-5} k^{0.11}} \sum_{\lambda=1}^{\lambda=11} \left[\Delta_j + w\left(\lambda - \frac{1}{2}\right)\zeta_j - \frac{P_d}{2} - P\right]^{1.11} \times 2$$

(3) *Equilibrium condition for applied radial load*

$$f_3(\Delta_j, \zeta_j, \delta_a, \delta_r) = \frac{0.62 \times 10^{-5} F_r}{w^{0.89}} - \sum_{j=1}^{j=\frac{z}{2}+1} \frac{\tau_j \cos\Psi_j}{k^{0.11}} \sum_{\lambda=1}^{\lambda=k} \left[\Delta_j + w\left(\lambda - \frac{1}{2}\right)\zeta_j - \frac{P_d}{2} - P\right] = 0$$

(5) *Equilibrium condition for applied thrust load*

$$f_4(\Delta_j, \zeta_j, \delta_a, \delta_r) = \frac{0.31 \times 10^{-5} F_a \times h}{w^{0.89}} - \sum_{j=1}^{j=\frac{z}{2}+1} \frac{\tau_j}{k^{0.11}} \left\{ \sum_{\lambda=1}^{\lambda=k} \left[\Delta_j + w\left(\lambda - \frac{1}{2}\right)\zeta_j - \frac{P_d}{2} - P\right]^{1.11} \left(\lambda - \frac{1}{2}\right) w - \frac{l}{2} \sum_{\lambda=1}^{\lambda=k} \left[\Delta_j + w\left(\lambda - \frac{1}{2}\right)\zeta_j - \frac{P_d}{2} - P\right]^{1.11} \right\} = 0$$

From the theory of Newton–Raphson method, Objective function and Jacobean matrix for nonlinear stiffness can be define as follow.

$$F(\Delta_j, \zeta_j, \delta_a, \delta_r) = \begin{bmatrix} f_1(\Delta_j, \zeta_j, \delta_a, \delta_r) \\ f_2(\Delta_j, \zeta_j, \delta_a, \delta_r) \\ f_3(\Delta_j, \zeta_j, \delta_a, \delta_r) \\ f_4(\Delta_j, \zeta_j, \delta_a, \delta_r) \end{bmatrix} \tag{12}$$

Radial contact stiffness and axial stiffness is defined as follow,

$$K = Q_j/\Delta_j = \frac{\frac{w^{0.89}}{1.24\times10^{-5}\times(k)^{0.11}}\sum_{\lambda=1}^{\lambda=k}\left[\delta_j + w\left(\lambda-\frac{1}{2}\right)\zeta_j - \frac{P_d}{2} - P\right]^{1.11}}{\sum_{\lambda=1}^{\lambda=k}\left[\Delta_j + w\left(\lambda-\frac{1}{2}\right)\zeta_j - \frac{P_d}{2} - P\right]} \quad (13a)$$

$$k_a = \frac{w^{0.89}}{1.24\times10^{-5}\times(k)^{0.11}}\sum_{\lambda=1}^{\lambda=k}\left[\Delta_j + w\left(\lambda-\frac{1}{2}\right)\zeta_j - \frac{P_d}{2} - P\right]^{0.11}\times\left(\left(\lambda-\frac{1}{2}\right)-\frac{l}{2}\right)\times\frac{2}{h} \quad (13b)$$

2.2 Algorithm for n-Nonlinear Simultaneous Equation for Contact Force Calculation

Step 1: [Input] $X_i = (x_{1i}, x_{2i}, x_{3i}\ldots x_{ni})$

Step 2: [Initialization] Set Maximum no. of iteration and tolerance

Step 3: [Function evaluation at X_i] Compute $\mathrm{F}(X_i) = \begin{bmatrix} f_1(x_{1i}, x_{2i}, x_{3i}\ldots x_{ni}) \\ f_2(x_{1i}, x_{2i}, x_{3i}\ldots x_{ni}) \\ \vdots \\ f_n(x_{1i}, x_{2i}, x_{3i}\ldots x_{ni}) \end{bmatrix}$

Step 4: [Evaluate the Jacobean at X_i]

$$J(X_i) = \begin{bmatrix} \frac{\partial f_1}{\partial x_1} & \frac{\partial f_1}{\partial x_2} & \cdots & \frac{\partial f_1}{\partial x_n} \\ \frac{\partial f_2}{\partial x_1} & \frac{\partial f_2}{\partial x_2} & \cdots & \frac{\partial f_2}{\partial x_n} \\ & & \vdots & \\ \frac{\partial f_n}{\partial x_1} & \frac{\partial f_n}{\partial x_2} & \cdots & \frac{\partial f_n}{\partial x_n} \end{bmatrix}$$

Step 5: [Solve $\Delta_x \Delta_x = \frac{F(X_i)}{J(X_i)}$]

Step 6: [Update] $X_{i+1} = X_i + \Delta_X$

Step 7: [Increment loop] I = I + 1 If I < Maximum no. of iteration, Then go to Step 3

Step 8: [Output]

Step 9: [Termination]

Stop

After solving above nonlinear simultaneous equation iteratively with Newton–Raphson method contact forces can be calculated in radial vertical and axial direction as bellow.

$$Q_{ry} = (1/k)\sum_{j=1}^{j=z} Qj\sin\Psi j,\ Q_{rz} = (1/k)\sum_{j=1}^{j=z} Qj\cos\Psi j$$
$$Q_a = (1/k)\sum_{j=1}^{j=z} Qaj$$

2.3 Dissipative Force for Roller Bearing

According to Upadhyay and Harsha [12], a nonlinear damping formula, correlating the contact damping force with the equivalent contact stiffness and contact deformation rate is given by,

$$F_d = \left(ck_{eq}\dot{\delta}\right) \tag{14}$$

$$k_{eq} = \int_0^{\delta} k\delta^{N_c} d\delta \frac{1}{N_c + 1} k\delta^{N_c+1}, N_c = \frac{10}{9}$$

corresponding damping force for jth roller in radial direction is given by,

$$F_{djr} = \frac{9}{19} ck_{jr}\delta^{19/9}\dot{\delta} \tag{15}$$

Total damping force in radial vertical and axial direction is given by,

$$F_{dry} = \frac{1}{11} \times \sum_{j=1}^{j=z} F_{djr} \times \sin\left[\frac{2\Pi(j-1)}{z} + \omega_c t\right]$$

From the theory of Newton–Raphson method, function and Jacobean matrix for nonlinear Damping can be defined as bellow.

Function Matrix for Damping:

$$F\left(\dot{\Delta}_j, \dot{\zeta}_j, \dot{\delta}_a, \dot{\delta}_r\right) = \begin{bmatrix} \dot{f}_1\left(\dot{\Delta}_j, \dot{\zeta}_j, \dot{\delta}_a, \dot{\delta}_r\right) \\ \dot{f}_2\left(\dot{\Delta}_j, \dot{\zeta}_j, \dot{\delta}_a, \dot{\delta}_r\right) \\ \dot{f}_3\left(\dot{\Delta}_j, \dot{\zeta}_j, \dot{\delta}_a, \dot{\delta}_r\right) \\ \dot{f}_4\left(\dot{\Delta}_j, \dot{\zeta}_j, \dot{\delta}_a, \dot{\delta}_r\right) \end{bmatrix} \tag{16}$$

2.4 Dynamic Model of a Rigid Rotor System

The equations of the motion of a rigid rotor roller bearing system may be written as,

$$[M]\ddot{X} + [C]\dot{X} + [K]X = f(t) \tag{17}$$

where $[M]$, $[C]$ and $[K]$ are the mass, damping and stiffness matrices for the system and $\ddot{X}$, $\dot{X}, X$ refer to the acceleration, velocity and displacement vectors, respectively. $f(t)$ is a force vector.

$$m\ddot{y} + Fdry + \mathrm{Qry} = F_{ry} \tag{18}$$

$$m\ddot{z} + Fdrz + \mathrm{Qrz} = F_{rz} \tag{19}$$

$$m\ddot{x} + Fda + \mathrm{Qa} = F_{a} \tag{20}$$

where, F_{ry} is the load acting on the bearing in y direction and F_{rz} is the load acting in Z direction F_a is the load acting on the bearing in X direction, m is a mass of the rotor.

Fdry and *Fda*is nonlinear damping force in Y, Z, and X direction respectively. Qry, Qrz and Qaare the nonlinear stiffness force in Y, Z and X direction respectively. Using new mark-β method, the differential equations of motion can be solved and the transient responses at every time increment are obtained (as per algorithm given below).

3 Results and Discussion

In this work, in modeling contact between the roller and raceway is considered nonlinear stiffness and nonlinear damping, mass of the rotor bearing system is 1.3 kg, radial load of Y direction is 13 N and assuming the rotor mass eccentricity e is 0. In order to investigate the dynamic behavior, the geometric parameters of the bearing are given in Table 1. Here results are obtained using central difference method and new mark beta method for the phase plot.

Table 1 Geometric parameter of the bearing

Parameters	Value
Pitch (mm)	43.5
Diameter of roller (mm)	10
Length of roller (mm)	11
No. of rollers (Z)	11
Radial clearance (mm)	0.025
Length of defect in direction of rolling (mm)	0.5
Depth of defect (mm)	0.5

3.1 Initial Condition

For numerical solution, the initial condition and step size are very important for successive and economic computational solution. The lager the time step, Δt, faster the computation. On the other hand time step should be small enough to achieve an adequate accuracy, the time step for the investigation is taken as $\Delta t = 10^{-4}$s at time $t = 0$ the following assumptions are made. The rotor is given initial displacements and velocities. For fast convergence the initial displacements are set to the following values:

$x_0 = 1 \text{micrometer} y_0 = 1 \text{micrometer}$

Initial velocities are assumed to be zero:

$\dot{x}_0 = 0 \text{ micrometer/s} \dot{y}_0 = 0 \text{ micrometer/s}$

For shaft speed 1,000 rpm, Shaft frequency (fs) = 16.67 Hz, Ball pass frequency outer (fco) = 72 Hz, Ball pass frequency inner (fci) = 112.7 Hz, Roller defect frequency (fr) = 68.67 Hz, VC = Variable compliance frequency.

3.2 System Under Investigation

As shown in Figs. (5 and 6) shaft having diameter 25 mm is supported by two bearings (NJ 305). One end of the shaft is connected to motor with the help of coupling and other end of the shaft is free to placed rotor mass. As shown in Fig. (6) at right side power screw arrangement is made to applied centric axial load. Bearings are placed in the adjustable pedestal.

In the present study, the analysis applied to a NJ-305 Radial cylindrical roller bearing. There is also another bearing (NJ-305) at the motor side of the shaft, which was not studied in this work. To assure the validity of the proposed model, a number of simulated results compared with experimental data.

3.3 Experimental Validation

Effect of inner race defective bearing: Theoretical Results are plotted in Figs. 7, 8, 9, 10 and Experimental results are in Figs. 15, 16, 17, and 18. Effect of outer race

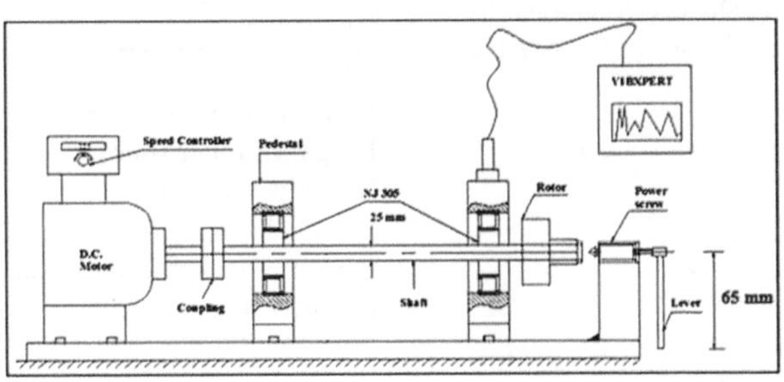

Fig. 6 View of experiment setup with Vibration Analyser, Experiment.rig

Fig. 7 FFT Freq.-x1

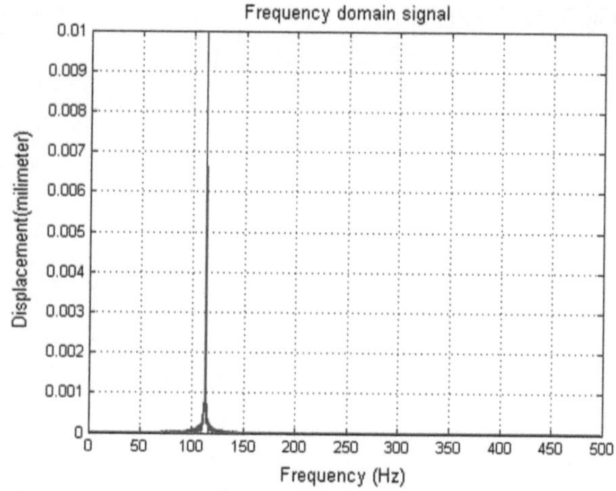

Fig. 8 FFT Freq-xd1

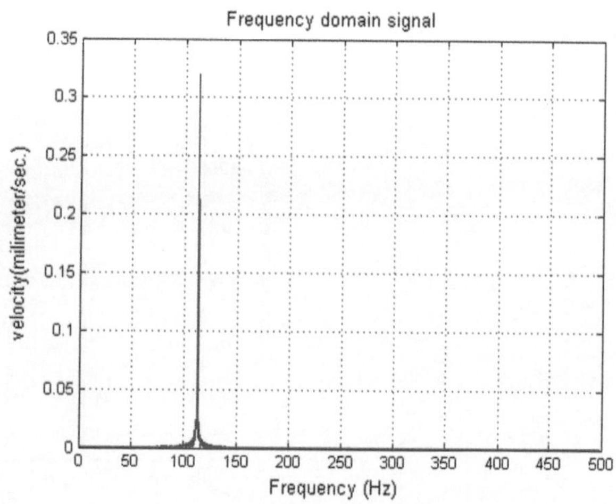

defective bearing: Theoretical Results are plotted in Figs. 11, 12, 13, 14 and Experimental results are in Figs. 19, 20, 21, and 22.

For inner race defective bearing displacement in Y direction from Fig. 7 maximum displacement 10 μm is theoretically and experimentally from Fig. 15 is 8 μm at variable compliance frequency (fci = 112.7 Hz) another peaks are at frequency(fci−fs = 96 Hz) and at frequency (fci + fs = 129.66 Hz). For inner race defective bearing velocity in Y direction from Fig. 8 maximum velocity 0.32 mm/sec, experimentally from Fig. 16 is 0.3 mm/sec. at variable compliance frequency (fci = 112.7 Hz) another peak observed at frequency (fci−fs = 96 Hz). For inner race defective bearing displacement in Z direction from Fig. 9 maximum displacement 1.1 μm theoretically, and experimentally from Fig. 17 is 1.3 μm at

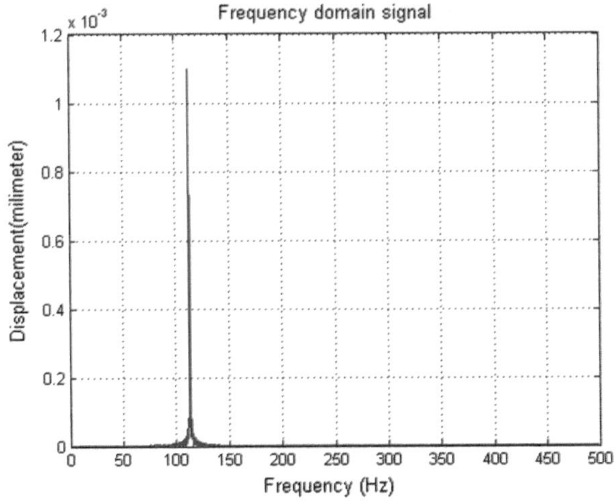

Fig. 9 FFT Freq-x2

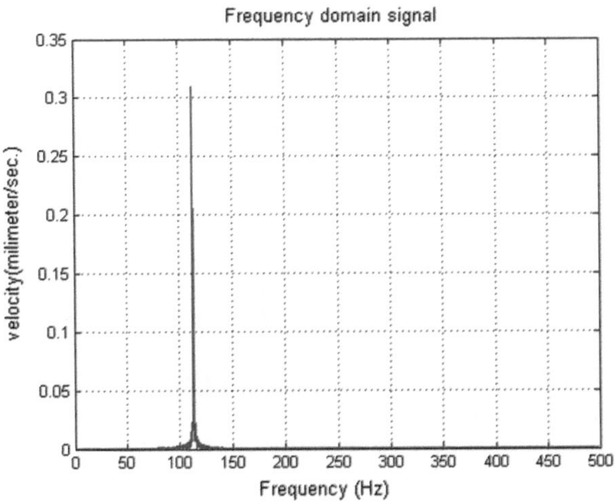

Fig. 10 FFT Freq-xd1

variable compliance frequency (fci = 112.7 Hz) another peaks are at frequency(2fci = 225 Hz) and at frequency (3fci = 338 Hz).For inner race defective bearing velocity in Z direction from Fig. 10 maximum velocity 0.31 mm/sec, experimentally from Fig. 18 is 0.29 mm/sec at variable compliance frequency (fci = 112.7 Hz) another peak are at frequency(fci + fs = 129.66 Hz). For outer race defective bearing displacement in Y direction from Fig. 11 maximum displacement 12 μm theoretically, and experimentally from Fig. 19 is 10.5 μm at variable compliance frequency (fco = 72 Hz) another peaks are at frequency(fco−fs = 55 Hz) and at frequency (fco + fs = 88 Hz). For outer race

Fig. 11 FFT Freq-x1

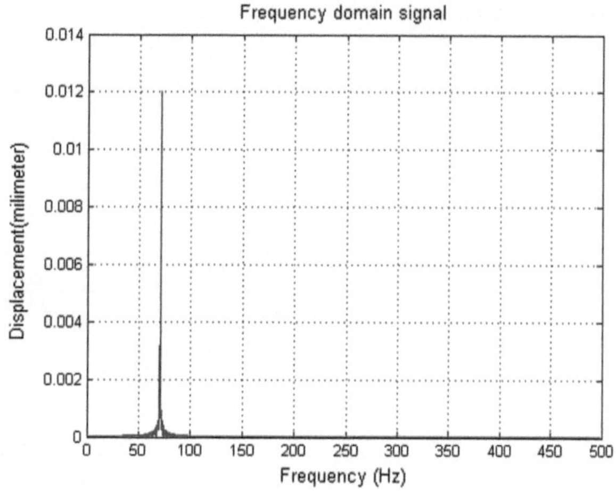

Fig. 12 FFT Freq-xd1

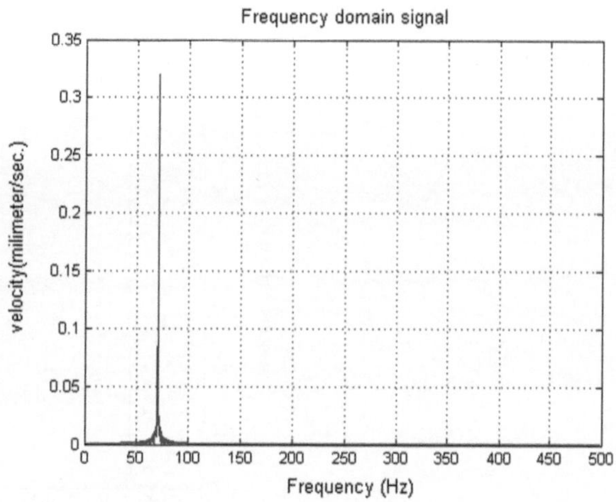

defective bearing velocity in Y direction from Fig. 12 maximum velocity 0.32 mm/sec, experimentally from Fig. 20 is 0.35 mm/sec variable compliance frequency (fco = 72 Hz) another peaks are at frequency (fco−fs = 55 Hz) and at frequency (fco + fs = 88 Hz). For outer race defective bearing displacement in Z direction from Fig. 13 maximum displacement 0.13 μm theoretically, and experimentally from Fig. 21 is 0.15 μm at variable compliance frequency (fco = 72 Hz) another peaks are at frequency (fco−fs = 55 Hz) and at frequency (fco + fs = 88 Hz) and at frequency (2fco Hz). For outer race defective bearing

Fig. 13 FFT Freq-x2

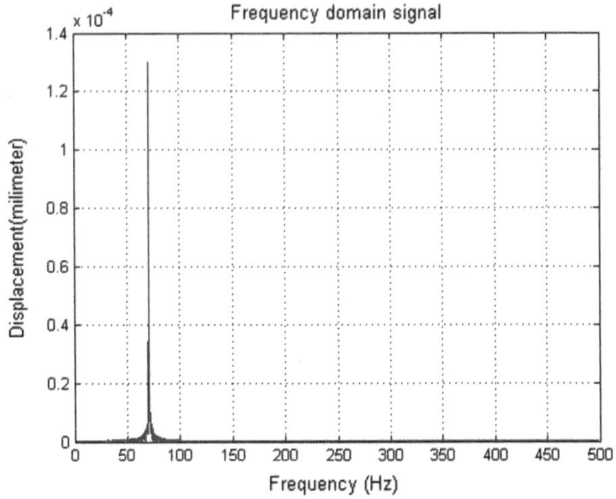

Fig. 14 FFT Freq-xd2

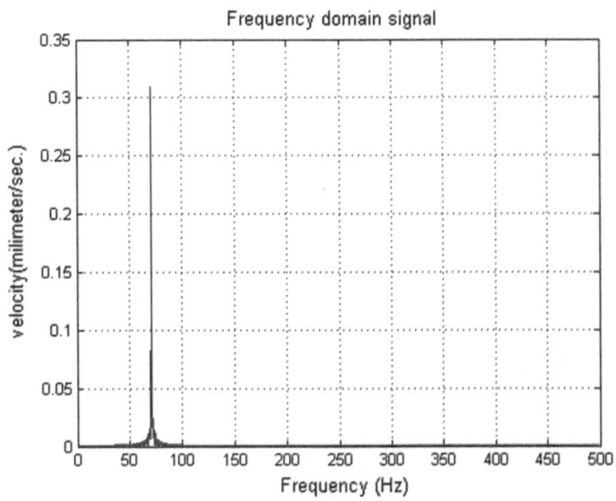

velocity in Z direction from Fig. 14 maximum velocity 0.31 mm/sec, experimentally from Fig. 22 is 0.30 mm/sec at variable compliance frequency (fco = 72 Hz) another peaks are at frequency (fco−fs = 55 Hz) and at frequency (fco + fs = 88 Hz) and also at frequency (2fco Hz) and (2fco-fs) Hz. Phase plot Figs. 23 and 24 for displacement velocity plot for Y and Z direction having nonlinear periodic vibration for inner race defect bearing, Figs. 25 and 26 displacement velocity plot for Y and Z direction nonlinear periodic vibration for outer race defective bearing, Poincare map fractal like structure is observed.

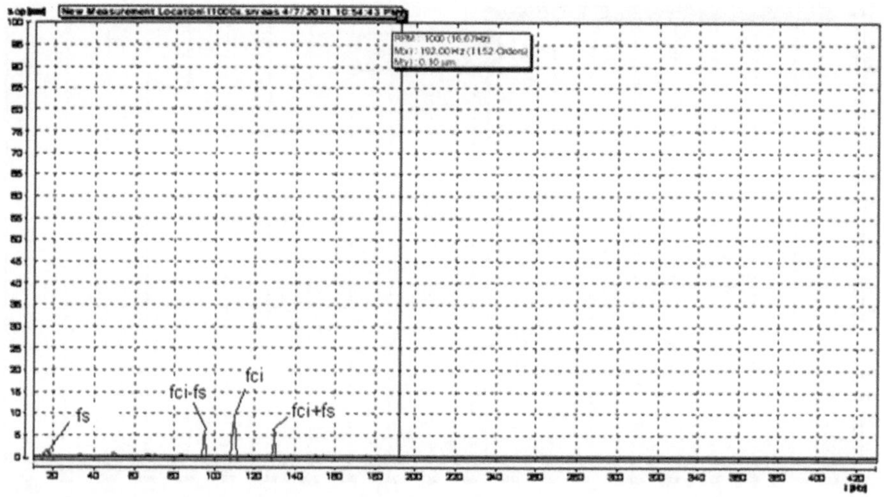

Fig. 15 Experimental FFT Freq-x1

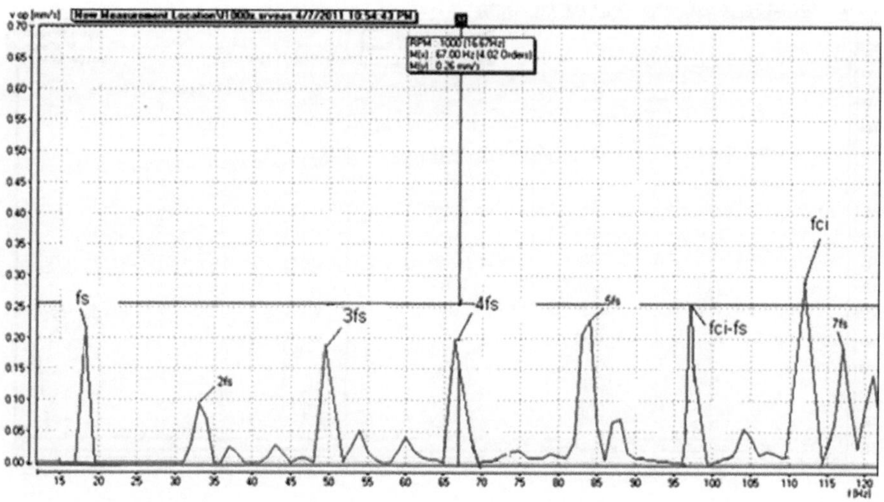

Fig. 16 Experimental FFT Freq-xd1

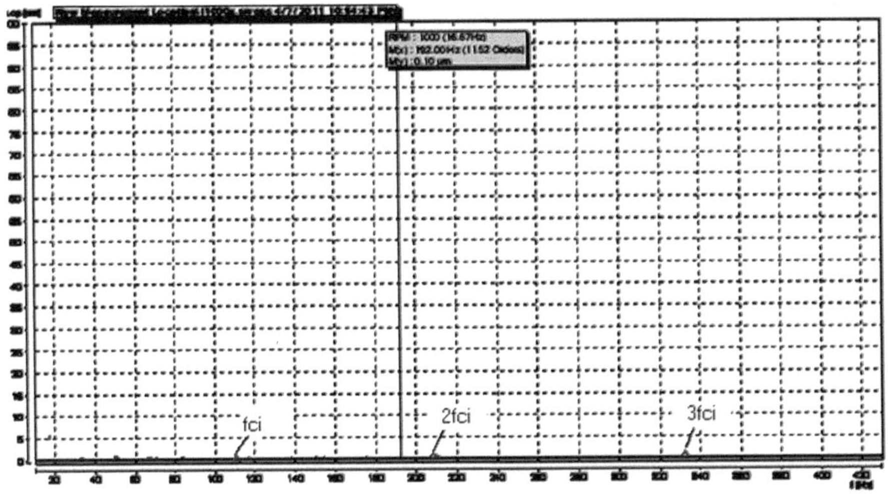

Fig. 17 Experimental FFT Freq-x2

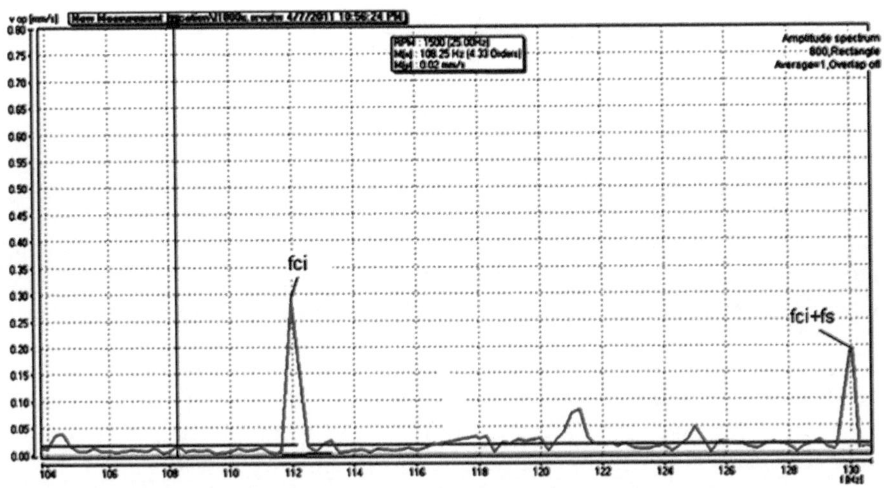

Fig. 18 Experimental FFT Freq-xd2

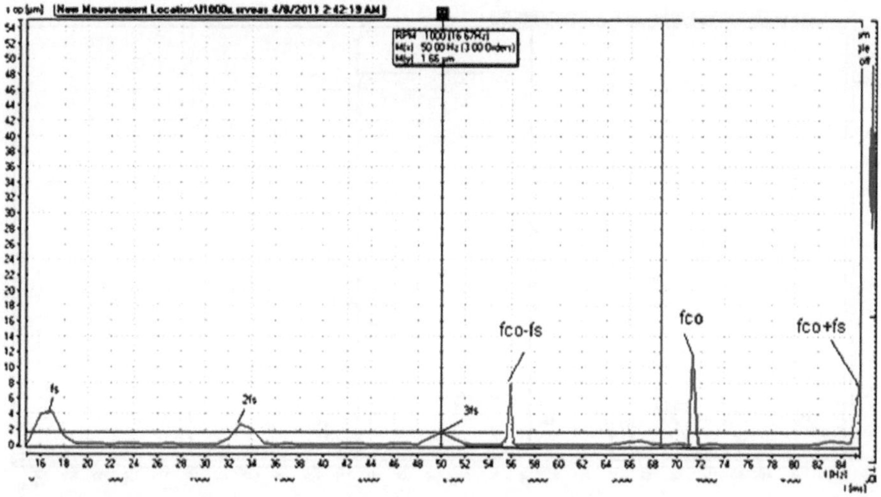

Fig. 19 Experimental FFT Freq-x1(outer race)

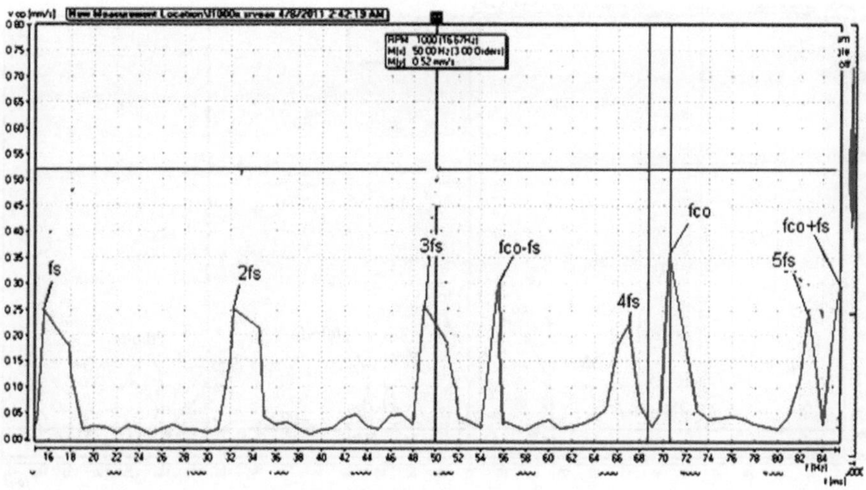

Fig. 20 Experimental FFT Freq-xd1(outer race)

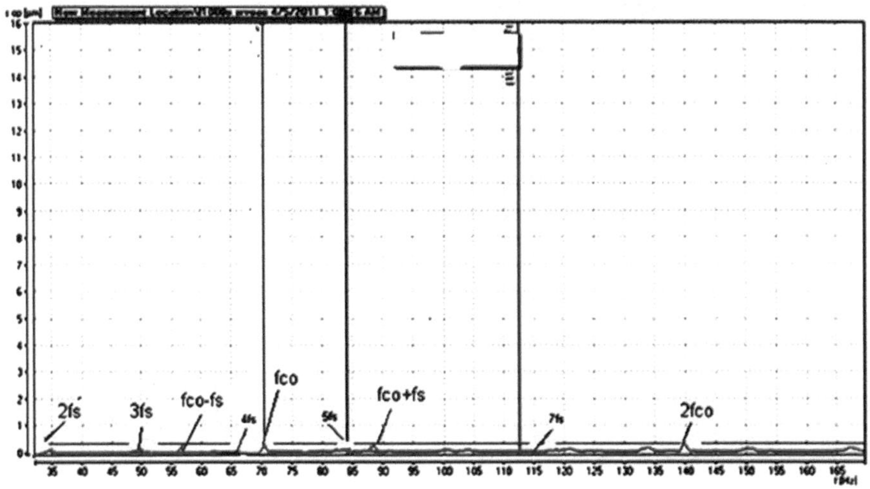

Fig. 21 Experimental FFT Freq-x2(outer race)

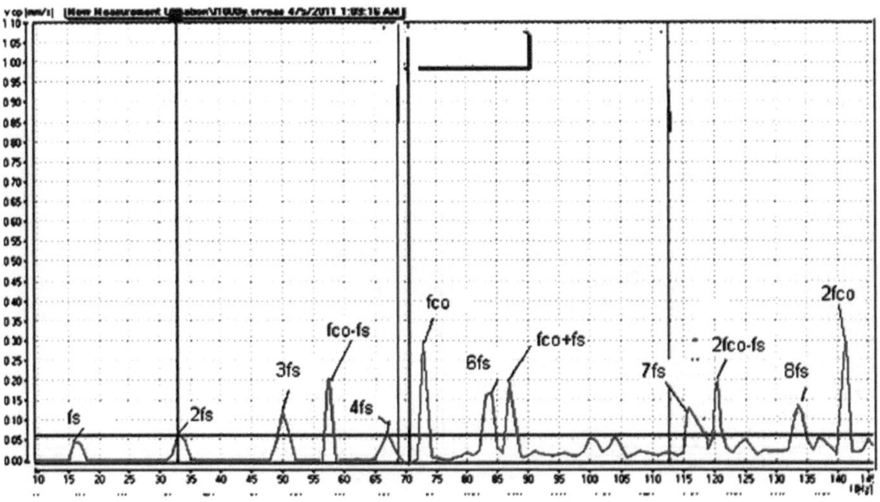

Fig. 22 Experimental FFT Freq-xd2(outer race)

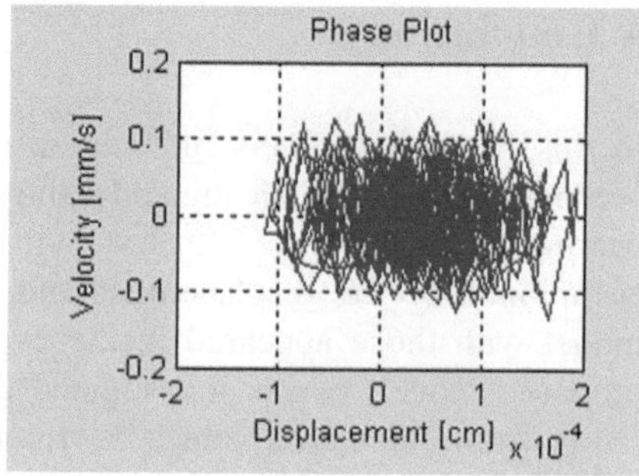

Fig. 23 Phase plot in Y direction for inner race defect

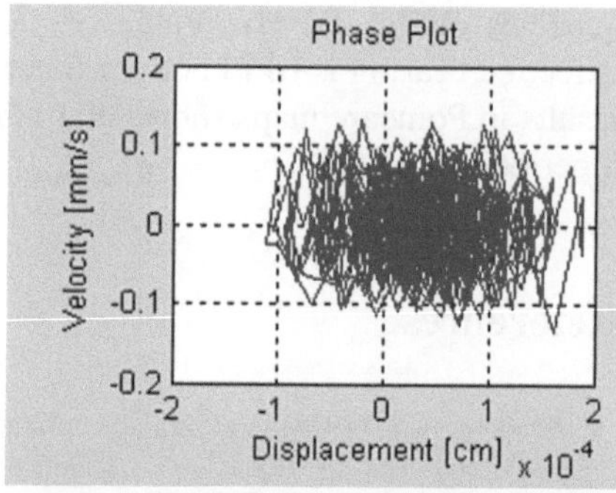

Fig. 24 Phase plot in Z direction for inner race defect

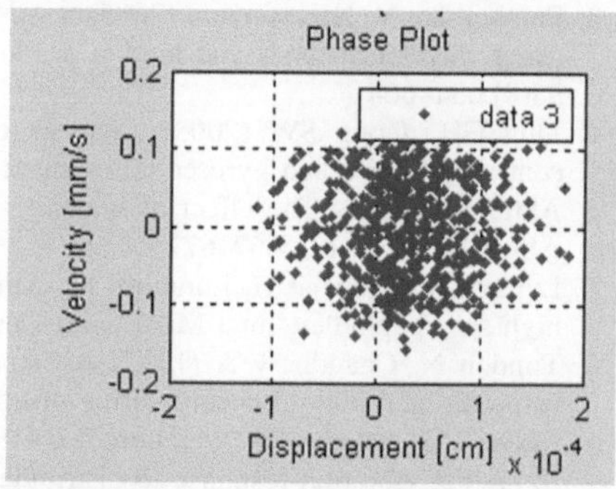

Fig. 25 Phase plot in Y direction for outer race defect

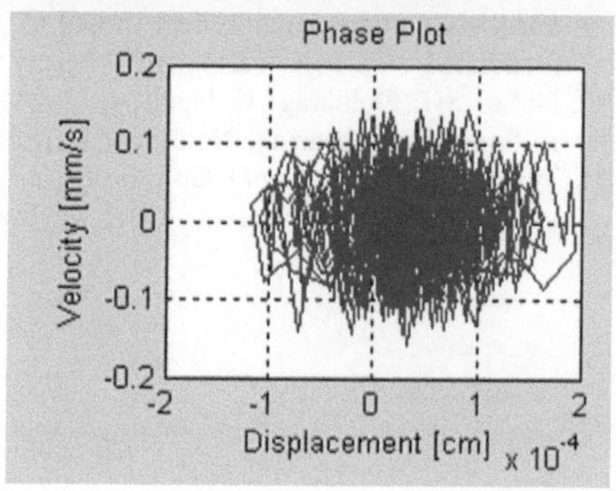

Fig. 26 Phase plot in Z direction for outer race defect

4 Conclusion

In this paper the effects of localize surface defects on a rolling element rotor bearing system are investigated using an analytical model. The response of the bearing system was compared with those obtained from experiments. The accordance of the displacement and velocity obtained from the mathematical model with those appeared in the experimental data verifies the validity of the proposed model. It was investigated that the defect frequencies are slightly different from calculated values as a consequence of slipping and skidding in the rolling element bearings. Frequency spectra display multiples of 1 fs Hz and VC Hz (Variable Compliance frequency) and the linear combination of two frequencies (VC $\pm$ fs) Hz. Value of variable compliance frequency for innerrace defective bearing is BPFI and for outer race defective bearing is BPFO. Phase plot results in Poincare maps shows that vibratory motion is a nonlinear motion having a periodic orbit.

References

1. Sunersjo CS (1978) Varying compliance vibrations of rolling bearings. J Sound Vibr 58(3):363–373
2. Jack LB, Nandi AK (2001) Support vector machine for detection and characterization of rolling element bearing faults. Proc Inst Mech Eng C J Mech Eng Sci 215:1065–1074
3. Purshotham V, Narayanan S (2005) S.A.N. Prashad Multi-fault diagnosis of rolling elements using wavelet analysis and hidden Markov model based fault recognisation. N.D.T&E Int 38(8):654–664
4. Jang GH, Jeong SW (2003) Analysis of a ball bearing with waviness considering the centrifugal force and gyroscopic moment of the ball. ASME J Tribol 125:487–498
5. Akturk N (1999) The effect of waviness on vibration associated with ball bearings. Trans ASME-J Tribol 121:667–677
6. Lynagh N, Rahnejat H, Ebrahimi M, Aini R (2000) Bearing induced vibration in precision high speed spindles. Int J Mach Tool Des Res 40:561–577
7. Tondon N, Chaudhary A (1997) An Analytical model for the prediction of the vibration response of rolling element bearing due to a localized defect. J Sound Vib 205(3):275–292
8. Mayer LD, Ahlgren FF, Weichbrodt B (1980) An Analytical model for ball bearing vibration to predict vibration response to distributed defects. J Mech Des-Trans ASME 102:205–210
9. Patil MS (2010) A theoretical model to predict the effect of localized defect on vibrations associated with ball bearing. Int J Mech Sci 52:1193–1201
10. Gafari SH, Golnaragi F, Ismail F (2008) Effect of localized faults on chaotic vibration of rolling contact bearing. Nonlinear Dyn 53:287–301
11. Wang L, Cui L, Zheng D, Gu L (2008) Nonlinear dynamics behaviors of a rotor roller bearing system with radial clearances and waviness considered. Chin J Aeronaut 21:86–96

12. Upadhyay SH, Harsha SP, Jain SC (2010) Analysis of nonlinear phenomenon in high speed ball bearing system due to radial clearance and unbalanced rotor effects. J Vibr Contr 16(1):65–88
13. Cao M, Xiao J (2008) A comprehensive dynamic model of double-row spherical roller bearing, model development and case studies on surface defects, preloads, and radial clearance. Mech Syst Signal Process 22:467–489
14. Palmgren A (1959) Grundlagen der Walzlagertechnik, Franckh'scheVerlagshandlung, 2nd edn. W.Keller &Co

Experimental Investigation on Life Cycle Analysis of the Moly (Mo) Coated Piston Ring in C. I. Engine

P. D. Patel, R. N. Patel, H. C. Patel and Pradip M. Patel

Abstract To increase service life of Internal Combustion engine component and improve efficiency of engine various surface treatment methods are used on cylinder piston group (CPG) components. The piston ring is the very critical component and for the improvement coating is done by the thermal spraying methods. Among thermal spraying methods, plasma spraying is most easily commonly applicable method in the automotive industries because it has a high spray rate and deposition, the process consumes fuel gases which are inexpensive and easily obtainable, the process requires minimum preheating and cooling during spraying, the technical reliability of plasma systems is well established in industrial applications, and spraying conditions can be easily controlled upon various applications. In particular, molybdenum coatings fabricated by atmospheric plasma spraying have enhanced resistance to wear and heat, and thus this coating technology was commercialized for application to the automotive industry. The wear resistance of plasma sprayed molybdenum coatings applicable to piston rings was investigated in this study. The performance parameters of the C I engine (as per IS 10000: Part 8) and wear condition (Endurance Test) of the piston and cylinder assembly (as per IS 10000 Part 5) has been done with and without using moly coated piston rings.

Keywords C I Engine · Plasma coating · Moly coating · Performance parameters · Wear condition

P. D. Patel (✉) · R. N. Patel
Department of Mechanical Engineering, Nirma University, Ahmedabad, Gujarat 380015, India
e-mail: pdpatel.auto@gmail.com

H. C. Patel
Department of Mathematics, L. D. College of Engineering, Ahmedabad, Gujarat 380015, India

P. M. Patel
Department of Mechanical Engineering, L. D. College of Engineering, Ahmedabad, Gujarat 380015, India

H. C. Patel et al. (eds.), *Proceedings of International Conference on Advances in Tribology and Engineering Systems*, Lecture Notes in Mechanical Engineering, DOI: 10.1007/978-81-322-1656-8_28, © Springer India 2014

1 Introduction

The requirement placed on the internal combustion engine is continuously change throughout the history of its development. To reduce the exhaust emission of the engine one way is to to improve the wear resistant of the piston ring [1]. In engine 30 % energy of total energy is lost due to the friction between piston ring and cylinder liner [2]. For this purpose material use for these components is more important. Today, the most commonly used piston rings are coated with electroplated chromium layers. In recent years, much development work has been performed to replace chromium plating by chromium nitride and molybdenum alternatives [3, 4, 5]. Among thermal spraying methods, plasma spraying is most widely applied in the automotive industries because (a) it has a high spray rate and deposition, (b) the process consumes fuel gases which are inexpensive and easily obtainable, (c) the process requires minimum preheating and cooling during spraying, (d) the technical reliability of plasma systems is well established in industrial applications, and (e) spraying conditions can be easily controlled upon various applications [6, 7]. In particular, molybdenum coatings fabricated by atmospheric plasma spraying have enhanced resistance to wear and heat and thus this coating technology was commercialized for application to the automotive industry [8].

2 Significance of Piston Rings

Piston rings for current internal combustion engines have to meet all the requirements of a dynamic seal for linear motion that operates under demanding thermal and chemical conditions. In short, the following requirements for piston rings can be identified [9]:

- Low friction, for supporting a high power efficiency rate.
- Low wear of the ring, for ensuring a long operational lifetime.
- Low wear of the cylinder liner, for retaining the desired surface texture of the liner.
- Emission suppression, by limiting the flow of engine oil to the combustion chamber.
- Good sealing capability and low blow-by for supporting the power efficiency rate.
- Good resistance against mechano-thermal fatigue, chemical attacks and hot erosion.
- Reliable operation and cost effectiveness for a significantly long time.

2.1 Plasma Spraying

In plasma spraying process, the material to be deposited (feedstock) is typically as a powder, sometimes as a liquid, wire is introduced into the plasma jet, emanating from a plasma torch.

In the jet, where the temperature is on the order of 10,000 K, the material is melted and propelled towards a substrate. There, the molten droplets flatten, rapidly solidify and form a deposit. Commonly, the deposits remain adherent to the substrate as coatings; free-standing parts can also be produced by removing the substrate. There are a large number of technological parameters that influence the interaction of the particles with the plasma jet and the substrate and therefore the deposit properties. These parameters include feedstock type, plasma gas composition and flow rate, energy input, torch offset distance, substrate cooling, etc [6].

2.2 Plasma Sprayed Molybdenum Coating

Molybdenum is coated on the ordinary cast-iron piston ring by the plasma spray technique. After coating honing process is done on the ring to improve surface finish of the ring to improve the friction properties [6]. Ordinary and plasma sprayed molybdenum coated piston rings are shown in Figs. 1, 2.

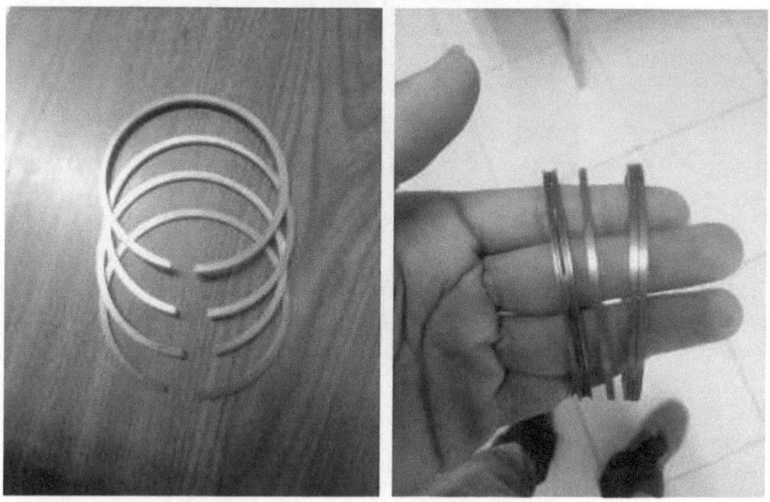

Fig. 1 Conventional piston rings

Fig. 2 Moly coated piston ring after honing

3 Experimental Methodology

Experimental set up with specification.

Make	Capton diesel
Type	Single cylinder, four stroke, vertical, water-cooled, cold start, compression ignition, high speed diesel engine
Bore	80 mm
Stroke	110 mm
Speed	1,500 rpm
Power	5 HP or 3.7 kW
Compression ratio	16.5:1
Fuel	Diesel

4 Gas Analyzer with Specification

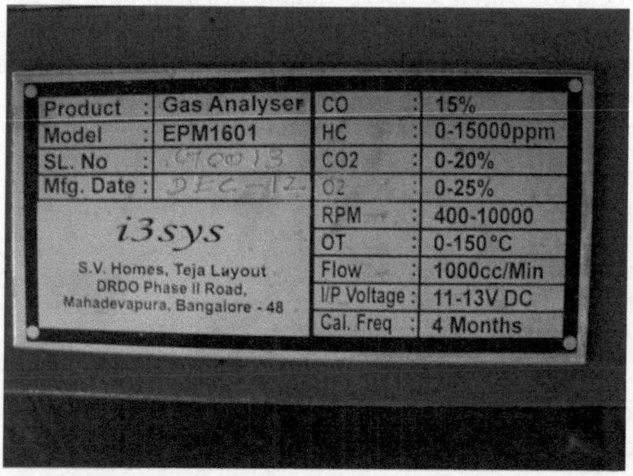

4.1 Performance Analysis

Experiments are carried out at constant engine speed of 1,500 rpm. Load is varied by rope brake dynamometer. Starting from no load observations are taken for five different loads [10]. Observations are taken at time when exhaust gas temperature remains steady. Various performance and emission parameters are measured at each load and test fuel are mentioned below. Using measured data, brake power, brake thermal efficiency are calculated for with and without Moly coated.

4.2 Parameter Under Observation

- Time taken for 25 ml fuel consumption
- Engine rpm
- Exhaust gas temperature
- Load on dynamometer
- Carbon monoxide
- Oxides of nitrogen
- Unburnt hydrocarbon
- Carbon dioxide

4.3 Life Cycle Analysis for Variable Speed Engine

Wear measurement for Variable Speed Engine has been done as per IS: 1000 (Part 5). The test shall last for a total of 100 h running time and shall consist of non-stop running periods of 10 h duration with not less than 2 h stoppage between consecutive running periods. Each running period shall consist of five cycles of 2 h duration at different loading [10].

After test wear has been measured as per IS: 1000 (Part 5) in Govt. Recognized & NABL Accredited laboratory.

4.4 Uncertainty Analysis

The errors involved with various measurements and in calculations of performance parameters are computed in this section using the method proposed by Jack P Holman (1994). The uncertainty in computed result of measured variables is given as,

$$\Delta R = \sqrt{\left[\left(\frac{\partial R}{\partial x_1} \times \Delta x_1\right)^2 + \left(\frac{\partial R}{\partial x_2} \times \Delta x_2\right)^2 + \cdots + \left(\frac{\partial R}{\partial x_n} \times \Delta x_n\right)^2\right]}$$

Computed result of the measured variables x_1, x_2, x_n (i.e. R is a function of x_1, x_2, and x_n), Δx_1, Δx_2, Δx_n are the uncertainties in the independent measured variables and ΔR = uncertainty of the computed result.

The uncertainty involved in brake power is,

$$\Delta BP = \sqrt{\left[\left(\frac{\partial BP}{\partial N} \times \Delta N\right)^2 + \left(\frac{\partial BP}{\partial F} \times \Delta F\right)^2\right]}$$

where,

Brake power $= \frac{2\Pi N(F \times R)}{60000}$
N = Speed
F = load
R = Radius of drum

Now, differentiating rake power with respect to N,

$$\frac{\partial BP}{\partial N} = \frac{2\Pi(F \times R)}{60000}$$

$$\frac{\partial BP}{\partial N} = \frac{BP}{N}$$

Now, differentiating rake power with respect to F,

$$\frac{\partial BP}{\partial F} = \frac{2\Pi N(R)}{60000}$$

$$\frac{\partial BP}{\partial F} = \frac{BP}{F}$$

Replacing above value in main equation we will get

$$\Delta BP = \sqrt{\left[\left(\frac{BP}{N} \times \Delta N\right)^2 + \left(\frac{BP}{F} \times \Delta F\right)^2\right]}$$

$$= \sqrt{\left[\left(3.7158 \times \frac{1.2}{1504}\right)^2 + \left(3.7158 \times \frac{0}{13}\right)^2\right]}$$

$$= \pm 0.0030 \text{ KW}$$

Uncertainty in Brake power = 0.08 %
The uncertainty involved in total fuel consumption (TFC) is,

$$\Delta TFC = \sqrt{\left[\left(\frac{\partial TFC}{\partial m_f} \times \Delta m_f\right)^2\right]}$$

$$= \sqrt{\left[\left(\frac{0.9618}{25} \times 0.1\right)^2\right]}$$

$$= \pm 0.0038 \text{ kg/hr}$$

Uncertainty in total fuel consumption = 0.3999 %

where,

m_f = the uncertainty involved in fuel flow rate.

The uncertainty involved in brake thermal efficiency is,

$$\eta_{bth} = \sqrt{\left[\left(\frac{\partial\eta_{bth}}{\partial BP} \times \Delta BP\right)^2 + \left(\frac{\partial\eta_{bth}}{\partial TFC} \times \Delta TFC\right)^2\right]}$$

$$= \sqrt{\left[\left(31.60 \times \frac{0.0030}{3.7158}\right)^2 + \left(31.60 \times \frac{0.0038}{0.9618}\right)^2\right]}$$

$$= \pm\, 0.1279$$

Uncertainty in brake thermal efficiency = 0.4048 %

Following table shows the uncertainty associated with brake power, fuel consumption and brake thermal efficiency:

BP (KW)	Uncertainty in BP	TFC (Kg/hr)	Uncertainty in TFC	BTE (%)	Uncertainty in BTE
0.9414	±0.0008	0.7059	±0.0030	10.91	±0.1170
1.8912	±0.0014	0.8186	±0.0034	18.90	±0.1191
2.7929	±0.0022	0.9053	±0.0037	25.24	±0.1210
3.7158	±0.0030	0.9618	±0.0038	31.60	±0.1279
3.9570	±0.0032	1.0260	±0.0038	31.55	±0.1285

Average uncertainty in brake power = ±0.00227 KW
Average uncertainty in total fuel consumption = ±0.003475 kg/hr
Average uncertainty in brake thermal efficiency = ±0.52 %

5 Conclusion

Experimental results shows that, by using Moly coated piston ring in place of the cast- iron piston ring for 100 h in the 4-stroke single cylinder C.I.Engine, 2.6 times less wear of compression ring and 2.24 times less wear of oil control ring. By using moly coated piston ring exhaust emission can be reduced up to some extent. In the exhaust gases UBHC reduces by 7.7 %, NO_x reduces by 2.29 %, CO reduces by 3.73 % and CO_2 reduces by 7.73 %. Also using moly coated ring engine oil property con be improve. Engine oil lubricity is improve and also reduction in the sediments of engine oil. It is therefore concluded that development of Moly coated piston ring is feasible since it gave satisfactory results for given operating conditions and reduction of HC, NO_x, CO and CO_2 emissions.

References

1. Tanaka M, Kitajima Y, Endoh Y, Watanabe M, Nagita Y (1993) Ceramic-metal composite coated piston ring and cylinder liner of marine low speed diesel engine. J MESJ, vol 12:77–88
2. Radil Kevin C (1996) Test method to evaluate cylinder liner-piston ring coatings for advanced heat engines. Technical memorendam NASA, 107526, 14:1–14
3. Wang Yucong, Tung Simon C (1999) Scuffing and wear behavior of aluminum piston skirt coatings against aluminum cylinder bore wear. 225–229:1100–1108
4. Bindumadhavan PN, Makesh S, Gowrishankar N, Keng Wah H, Prabhakar O (2008) Aluminizing and subsequent nitriding of plain carbon low alloy steels for piston ring applications. Surface and Coating Technology 127:251–258 may 2000
5. Bryskin B (2012) Galvanic chrome plating of piston rings is the most common way to improve engine performance. Advance material and processes, ASM International
6. Ahn Jeehoon, Hwang Byoungchul, Lee Sunghak (2005) Improvement of wear resistance of plasma-sprayed molybdenum blend coatings. Surface and Coating Technology 194:256–264
7. Lu S-P, Kwon O-Y, Guo Y (2003) Wear behavior of brazed WC/NiCrBSi(Co) composite coatings wear 254:421–428
8. Karamis MB, Yıldızlı K, Çakırer H (2004) An evaluation of surface properties and frictional forces generated from Al–Mo–Ni coating on piston ring. Applied Surface Science 230:191–201
9. Friedrich C, Berg G, Broszeit E, Rick F, Holland J (1997) PVD Cr N coatings for tribological application on piston rings. Surface and Coating Technology 97:661–668
10. Bureau of Indian Standard 10000: Part (V 2001 & VIII 1996) — Preparation od tests and measurement of wear performance

Load Capacity Analysis of Gas Foil Bearing (GFB) for Different Foil Materials

T. M. Jamir and S. K. Kakoty

Abstract There is a significant attention on GFBs in recent times owing to its oil free technology and suitability over a wide array of operating conditions. However GFBs have considerably low load carrying capacity compared to normal GFBs. The present paper investigated the load capacity of the bump-type GFBs for a simple model considering only deflection of bump foils. The load capacity of GFBs with different foil materials has been inspected in view of comparing the load capacity. Also the variations of GFBs load performance with respect to the compliance of the foil have been reviewed.

Keywords Gas foil bearings · Load capacity · Foil compliance coefficient

Nomenclature

C	Bearing radial clearance (m)
D	Diameter of journal (m)
e	Bearing eccentricity (m)
E_b	Young's modulus of bump foil
E_t	Young's modulus of top foil (N/m^2)
F_X, F_Y	Vertical and horizontal components of hydrodynamic forces (N)
$\bar{F}_X, \bar{F}_Y$	Non-dimensional vertical and horizontal components of hydrodynamic forces: $\frac{F_X}{p_a R^2}, \frac{F_Y}{p_a R^2}$
$F_{X0}; F_{Y0}$	Vertical and horizontal steady state components of hydrodynamic forces (N)
$\bar{F}_{X0}, \bar{F}_{Y0}$	Non-dimensional vertical and horizontal steady state components of hydrodynamic forces: $\frac{F_{X0}}{p_a R^2}; \frac{F_{Y0}}{p_a R^2}$

T. M. Jamir (✉) · S. K. Kakoty
Department of Mechanical Engineering, Indian Institute of Technology Guwahati, Guwahati 781039, India
e-mail: t.jamir@iitg.ernet.in

S. K. Kakoty
e-mail: sashin@iitg.ernet.in

H. C. Patel et al. (eds.), *Proceedings of International Conference on Advances in Tribology and Engineering Systems*, Lecture Notes in Mechanical Engineering, DOI: 10.1007/978-81-322-1656-8_29,

h	Film thickness (m)
h_o	Initial bump height (m)
$h_{\min}$	Minimum film thickness (m)
H	Non-dimensional film thickness
$H_{\min}$	Non-dimensional minimum film thickness
i, j	Grid location in circumferential and axial directions of FDM mesh
K_f	Bump foil structural stiffness per unit area (N/m^2)
l_o	Half bump length (m)
l_s	Length of segment between the bumps (m)
L	Bearing length (m)
m	Number of divisions along j direction of FDM mesh
n	Number of divisions along i direction of FDM mesh
O	Centre of bearing
O'	Centre of journal
p	Hydrodynamic pressure in gas film (N/m^2)
p_a	Atmospheric pressure in gas film (N/m^2)
$\bar{P}$	Arithmetic mean pressure along bearing length (N/m^2)
P	Non-dimensional hydrodynamic pressure
$\bar{p}$	Non-dimensional arithmetic mean pressure along bearing length
R	Radius of journal (m)
s	Bump foil pitch (m)
S	Compliance number: $\frac{p_a}{CK_f}$
t	Time (s)
t_b	Bump foil thickness (m)
t_t	Top foil thickness (m)
w_t	Top foil transverse deflection (m)
W	Non-dimensional top foil transverse deflection
W_0	Steady state load carrying capacity (N)
$\bar{W}_0$	Non-dimensional steady state load carrying capacity
x, y, z	Coordinate system on the plane of bearing
Z	Non-dimensional axial coordinate of bearing: $\frac{z}{R}$
α	Compliance of the bump foil $(\text{m}^3/\text{N}) : \frac{1}{K_f}$
ε	Eccentricity ratio
Λ	Bearing number: $\frac{6\mu\omega}{p_a}\left(\frac{R}{C}\right)^2$
μ	Gas viscosity (N - s/m^2)
ϕ	Attitude angle (rad)
θ	Angular coordinate of bearing (rad): $\frac{x}{R}$
θ_o	Half bump angle (rad)
υ	Poisson's ratio
τ	Non-dimensional time: ωt
ω	Rotor angular velocity (rad/s)
$\Delta\theta, \Delta Z$	Non-dimensional mesh size of FDM mesh

1 Introduction

Gas foil bearings (GFBs) are compliant surface, self-acting hydrodynamic bearings typically constructed from several layers of sheet metal foils which support a rotor by any combination of bending membrane or elastic foundation effects. Gas foil bearing technology has made significant progress over the last 40 years and fulfills most of the requirements of novel oil-free turbo-machinery by increasing tenfold their reliability in comparison to rolling elements bearings [1, 2]. GFBs compliant surface provides bearing structural stiffness and comes in several configurations such as the leaf type (Fig. 1a), bump type (Fig. 1b), and tape type, among others.

The underlying compliant structure (bumps) provides a tenable structural stiffness source [3, 4] resulting in a larger film thickness than the rigid wall bearings [5, 6] enabling high-speed operation and larger load capacity including tolerance for shaft misalignment [2]. Also damping of coulomb type arises due to the relative motion between the bumps and the top foil, and between the bumps and the bearing wall [7]. Foil Bearings generally operates with ambient air. However, some specific applications use other fluids such as helium, xenon, liquid nitrogen, and liquid oxygen among others [8]. Significant improvements in high-temperature limits Fig. 2 are obtained by using coatings (solid lubricants) [9, 10]. Since late 1960s, gas foil bearings are common in air cycle machines, the heart of the environment control system in aircraft. Current applications of foil bearings include oil-free cryogenic turbo-expanders for gas separation plants, auxiliary power units for various aerospace and ground vehicles, automotive gas turbine engines, vapour-cycle centrifugal compressors, and commercial air/gas compressors [7].

Peng and Carpino [6] developed a finite difference formulation, coupling hydrodynamic and elastic foundation effects to calculate stiffness and damping force coefficients in foil bearings. The analytical procedure results shows that the bearing direct stiffness increases with rotor speed and generally decreases with increased bump compliance. Peng and Khonsari [11] determined the ultimate load capacity of ideal foil bearings, showing its dependency on the stiffness (compliance) of the support structure beneath the top foil. Kim and San Andŕes [12]

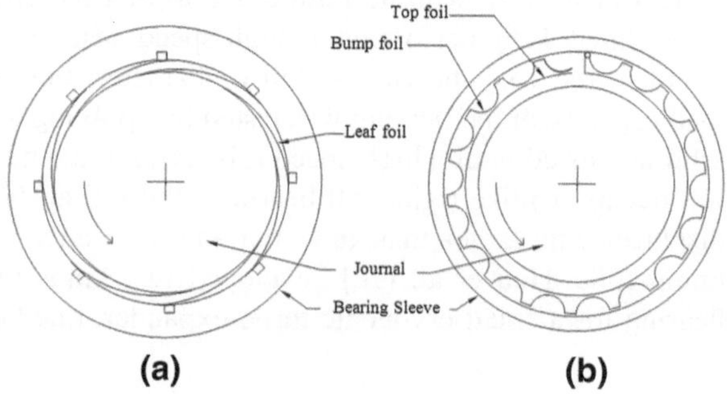

Fig. 1 Schematic views of two typical GFBs **a** Leaf type GFB, **b** Bump type GFB [10]

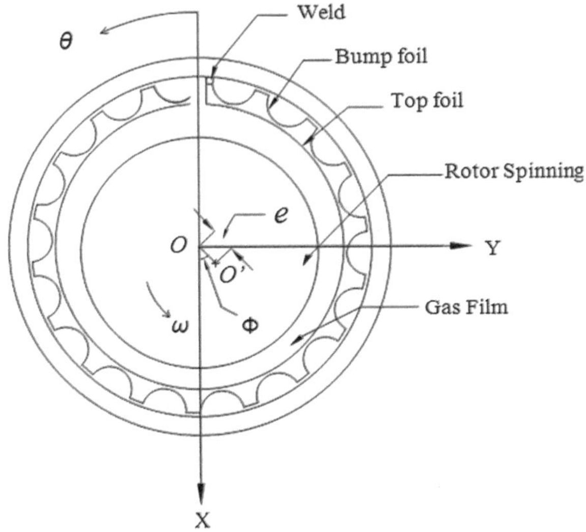

Fig. 2 Configuration of a typical bump type GFB [12]

further advanced the analysis to determine the effect of assembly preload on the ultimate load capacity for gas foil bearing operation at ultra-high operating speeds. Foil bearings are typically assembled with a slight preload, i.e., with the journal diameter being larger than the foil's. An initial assembly preload (interference between shaft and foil) increases the GFB static stiffness at both null and infinite rotor speeds. DellaCorte and Valco [13] introduced an empirical 'rule of thumb' relating the bearing load capacity to the bearing size and rotor speed of Generation I, II and III bump type foil bearings. Advanced foil bearings, with compliance tailored elastic support structures, exhibit load capacities up to five times that of the initial simple designs. Radil et al. [14] conducted tests to determine the effect of radial clearance on bearing load capacity. The authors compared results of two generation III bump type FBs with different initial clearances. The bearing with tighter initial clearance has a lower load capacity than the bearing with larger clearance. Each bearing has an optimum clearance which evidences maximum load capacity. Decreasing the clearance heavily preloads the shaft, which may lead to rotor-bearing system failure due to shaft thermal expansion and seizure. When the bearing clearance is greater than optimal, the bearing load capacity decreases by less than 20 % with no failure due to thermal effects.

Heshmat [15] performed a high-speed test using a journal foil bearing to establish the rotor-bearing stability characteristics and speed performance. The load capacity of the bearing was tested by applying a load to a centre bearing at an arbitrary speed until a high-speed rub between the mating surfaces of the shaft and foil occurs. Unlike rigid wall bearings, the foil air bearings exhibited eccentricity displacements larger than their nominal clearance due to the compliance of the bump foils. Hou et al. [16] developed two kinds of new compliant foil journal bearing for a small cryogenic turbo-expander, one bearing supported with copper wires and the other with elastic material (fluoride rubber) respectively. The performances of the bearings are studied under the same experiment conditions. The results indicate that the bearing supported by elastic material has excellent damping capacity which enhances the bearing tolerance to impact and vibration also the circumferential distribution of the stiffness in foil bearing supported with elastic material is more uniform, which contributes to improving the stability of the foil bearing.

It is noteworthy to observe from the available literatures over the years that the foils used in GFBs are mostly structural steel, which is mainly because of its favorable characteristics. This paper aims at exploring the possibility of using an alternative material for the bump foil. Further, an attempt has also been made to see the effects of compliance coefficients on the load capacity keeping the geometry and the material property same.

2 Gas Foil Bearing Formulation

The governing Reynolds equation for an isothermal, isoviscous ideal gas is given by [17]

$$\frac{\partial}{\partial x}\left(ph^3\frac{\partial p}{\partial x}\right)+\frac{\partial}{\partial z}\left(ph^3\frac{\partial p}{\partial z}\right)=6\mu\omega R\frac{\partial(ph)}{\partial x}+12\mu\frac{\partial(ph)}{\partial t} \tag{1}$$

where h is the film thickness;

$$h = C + e\cos(\theta - \phi) + w_t \tag{2}$$

Non-dimensional Reynolds equation is given by

$$\frac{\partial}{\partial\theta}(PH^3\frac{\partial P}{\partial\theta})+\frac{\partial}{\partial Z}(PH^3\frac{\partial P}{\partial Z})=\Lambda\frac{\partial(PH)}{\partial\theta}+2\Lambda\frac{\partial(PH)}{\partial\tau} \tag{3}$$

For steady state condition

$$\frac{\partial}{\partial\theta}(PH^3\frac{\partial P}{\partial\theta})+\frac{\partial}{\partial Z}(PH^3\frac{\partial P}{\partial Z})=\Lambda\frac{\partial(PH)}{\partial\theta} \tag{4}$$

where Λ is the bearing number

Non dimensional film thickness H becomes;

$$H = 1 + \varepsilon\cos(\theta - \phi) + W \tag{5}$$

The non-dimensional Reynolds equation is discretized in a Finite Difference Grid using central difference quotients by dividing the area into a number of meshes of size $\Delta\theta \times \Delta Z$ and a mesh of m nodes along circumferential direction and n nodes along axial direction.

For steady state, Eq. 4 is simplified as,

$$(P_{i,j})^2 K_1 + (P_{i,j}) K_2 + K_3 = 0 \tag{6}$$

where $P_{i,j}$ is the pressure at any point (i, j) and K_1, K_2 and K_3 are given in appendix A

Equation 6 is a nonlinear system of the form

$$F(P) = 0 \tag{7}$$

Using Newton–Raphson method for its solution,

$$P_{n+1} = P_n - \frac{F(P_n)}{F'(P_n)} \tag{8}$$

where, P_n is the pressures obtained after *nth* iteration and $F'(P)$ is first derivative of $F(P)$ with respect to P. Newton–Raphson iterative process is repeated until the following convergence criterion is satisfied.

$$\frac{\left|(\sum P_{i,j})_{n-1} - (\sum P_{i,j})_n\right|}{\left|(\sum P_{i,j})_n\right|} \leq 10^{-6} \tag{9}$$

2.1 Elastic Foundation Model for Foil Structure

A simple elastic foundation model which is the original work of Heshmat et al. [5] is considered for this work as shown in Fig. 3. The local deflection of the foil structure (w_t) depends on the bump compliance (α) and the average pressure across the bearing length,

$$w_t = \alpha(\bar{p} - p_a) \tag{10}$$

The compliance of the bump foil, with the geometry specified in Fig. 3 is determined by Heshmat et al. [5] as:

$$\alpha = \frac{2s}{E_b}\left(\frac{l_o}{t_b}\right)^3 (1 - \upsilon^2) \tag{11}$$

where;

s = Bump pitch (m)
l_o = Half of the bump length (m)
t_b = Bump foil thickness (m)
E_b = Young's modulus (Modulus of elasticity) of bump foil material
v = Poisson's ratio of bump foil material

By using following substitutions: $\bar{P} = \frac{\bar{p}}{p_a}$, $W = \frac{w_t}{C}$
Non-dimensional foil deflection equation is given by

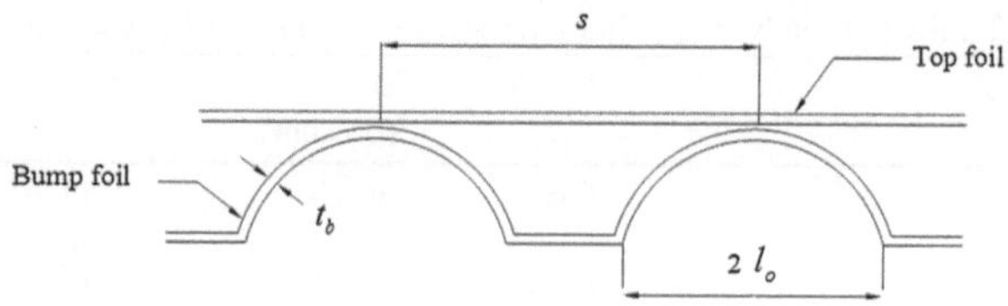

Fig. 3 Schematic of a foil bearing and detailed configuration of the foil [5]

$$W = S(\bar{P} - 1) \tag{12}$$

where S is so-called the Compliance number.

2.2 Steady State Characteristics

To obtain steady state characteristics of GFB, it is required to obtain pressure distribution by solving Eq. 6. The non-dimensional horizontal and vertical steady state load components can be obtained by following equations.

$$\overline{W}_X = -\int_{-L/D}^{L/D}\int_{0}^{2\pi} P\cos\theta d\theta dZ$$

$$\overline{W}_Y = -\int_{-L/D}^{L/D}\int_{0}^{2\pi} P\sin\theta d\theta dZ \tag{13}$$

The non-dimensional steady state load is obtained by;

$$\overline{W}_O = \sqrt{\overline{W}_X^2 + \overline{W}_Y^2} \tag{14}$$

3 Validation of Static Performance Analysis

The validity of the present analysis and computational program is assessed by comparison of steady state results with published results. Table 1 compares the attitude angle and the load capacity with the published results of Heshmat et al. [5] and Peng and Carpino [6] for L/D = 1.0 and S = 0 and S = 1. GFB reduces to ordinary gas bearing for S = 0.

From the above comparisons in Table 1, it has been observed that the present results are in good agreement with those from references validating the computational model and the present analysis. Table 2 shows a comparison of different

Table 1 Steady state characteristics for L/D = 1.0, Λ = 1.0

S	ε	ϕ (present)	ϕ	$\overline{W}$ (present)	$\overline{W}$ [5]
0	0.6	36.13	[a] 35.70 [b] 36.50	0.898	[a] 0.951 [b] 0.961
	0.75	24.12	[a] 24.10 [b] 24.70	1.847	[a] 1.894 [b] 1.922
1	0.6	35.55	[a] 32.10 [b] 34.00	0.515	[a] 0.568 [b] 0.567
	0.75	28.53	[a] 26.30 [b] 27.70	0.728	[a] 0.783 [b] 0.778

[a] Heshmat et al. [5]
[b] Peng and Carpino [6]

geometries adopted by different authors, thereby the variation of compliance number for different bearing geometries and designs are listed in the table.

In course of the investigation carried out for the comparison of bearing geometries, vis-a-vis, compliance numbers as shown in Table 2, it has been noticed that the materials which were considered by different authors are mainly structural steel. It has been, therefore, proposed to investigate the suitability of other material for bump foil of GFBs.

4 Load Capacity of GFBs with Different Foil Material

The load carrying capacity for bump type GFBs with different foil materials are calculated (shown in Table 3) considering a common geometry for bearing configuration as describes by Lez et al. [18].

Bearing length, L (m)	38.1
Bearing radius, R = D/2 (mm)	19.05
Bearing radial Clearance C (μm)	31.8
Bump foil thickness, t_b (μm)	102
Bump pitch, s (mm)	4.572
Bump half length, l_o (mm)	1.778
Length of segment between the bumps, l_s (mm)	1.016
Initial bump height, h_o (mm)	0.508
Atmospheric pressure, p_a $\mathrm{N/m^2}$	101.325×10^3
Gas viscosity, μ $\mathrm{N-s/m^2}$	2.98×10^{-5}

The load carrying capacity and altitude angle for different eccentricity ratio of the bearing has been evaluated for copper, structural steel and carbon fiber

Table 2 Variation of compliance number for different GFB geometries

Ref	s (mm)	l (mm)	C (μm)	t_b (μm)	υ	S
[18]	4.572	1.778	31.8	102	0.29	0.738
[19]	4.064	1.717	100	76.2	0.3	0.414
[20]	4.572	1.778	31.8	101.6	0.29	0.663
[21]	4.572	1.778	31.8	102	0.29	0.656
[22]	4.572	1.778	31.8	101.6	0.29	0.663
[23]	4.572	2.032	35.5	102	0.29	0.878
[24]	4.572	2.032	15	101.6	0.29	2.104
[25]	4.572	2.032	35.5	102	0.29	0.881
[26]	4.064	1.778	20	127	0.31	0.514
[27]	4.064	1.778	20	127	0.31	0.514
[28]	4.572	1.778	20	101.6	0.29	1.058

Table 3 GFBs load carrying capacity and attitude angle for different foil materials at $\Lambda = 1$ and L/D = 1

Material	S	ε	ϕ	$\bar{W}$
Copper	1.15	0.1	64.5896	0.0738
		0.2	58.6461	0.1496
		0.3	52.8208	0.2305
		0.4	47.2429	0.3193
		0.5	41.9937	0.4183
		0.6	36.9022	0.5289
		0.7	31.8603	0.6523
		0.8	27.9552	0.7887
Structural steel	0.66	0.1	68.0938	0.0766
		0.2	61.4502	0.1585
		0.3	55.0259	0.2499
		0.4	48.7889	0.3557
		0.5	42.7252	0.4809
		0.6	36.8591	0.6293
		0.7	31.2960	0.8040
		0.8	26.7532	1.0069
FRP carbon (pitch)	0.32	0.1	70.8995	0.0786
		0.2	63.5650	0.1651
		0.3	56.8998	0.2658
		0.4	50.1636	0.3890
		0.5	43.3878	0.5449
		0.6	36.7039	0.7450
		0.7	30.4782	1.0008
		0.8	25.1711	1.3224

reinforced polymer [29, 30]. The result has been plotted as shown in Fig. 5. Also the load capacity of plain gas bearing versus gas foil bearing has been plotted as shown in Fig. 4.

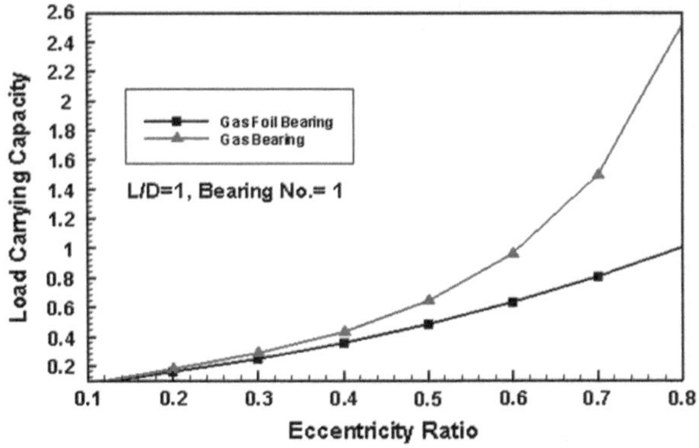

Fig. 4 Load carrying capacity with respect to eccentricity ratio for normal gas bearing and gas foil bearing

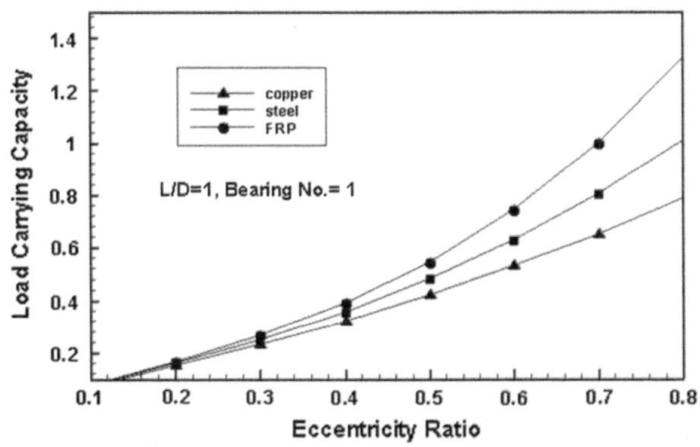

Fig. 5 GFB load capacity with respect to eccentricity ratio for different foil materials $\Lambda = 1$ and L/D = 1

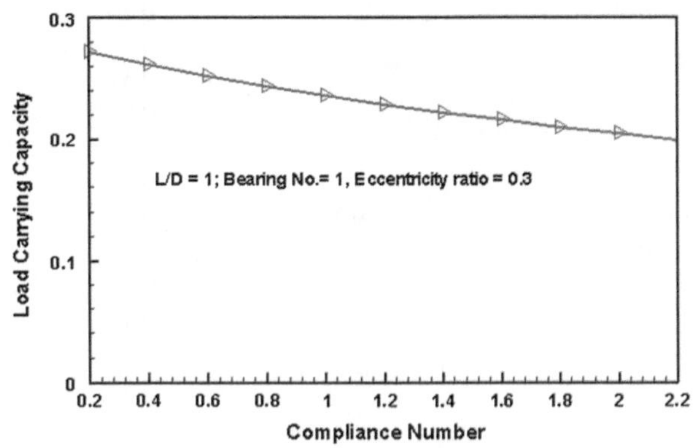

Fig. 6 Dependence of load carrying capacity on compliance number (S)

Figure 5 shows the GFB load carrying capacity for different bump foil material. It has been observe that there is a significant increase in the load capacity of the GFB with FRP carbon composite as foil material with respect to structural steel and copper amounting to 31.33 and 67.66 % respectively. Also Fig. 6 shows the dependency of load capacity on the compliance number of the bump foils and it has been observe that as the compliance number increases the load capacity decreases, hence lower compliance number would yield more load capacity but should limit to the value of normal gas bearing which is zero. Hence, the material properties and the structural geometry of the foil significantly influence the load capacity of the foil bearing.

5 Conclusion

This paper presents a simple model of bump-type foil bearings wherein the load capacity of the foil bearing with different foil material has been investigated. The results were encouraging and from the load capacity performance analysis, carbon FRP composite material as compared to structural steel and copper can be recommended particularly for bump type GFBs as the foil material. Experimentalists may consider varifying the model by experimental evidences, which is beyond the scope of the present work. Thermohydrodynamic study of GFBs with FRP composites as the foil material, the bonding of the FRP composite with steel and the anisotropic behavior of the composite materials also needs attention.

Acknowledgments This work has been carried out in the Department of Mechanical Engineering, Indian Institute of Technology, Guwahati, India, under the supervision of Prof. S. K. Kakoty.

Appendix A

$$K_1 = H_i^3(C_1 + C_2)$$

$$K_2 = H_i^3(C_3 + C_4)$$

$$K_3 = C_5 + C_6 - C_7$$

$$C_1 = \frac{-2}{(\Delta\theta)^2}$$

$$C_2 = \frac{-2}{(\Delta Z)^2}$$

$$C_3 = \frac{P_{i+1,j} + P_{i-1,j}}{(\Delta\theta)^2}$$

$$C_4 = \frac{P_{i,j+1} + P_{i,j-1}}{(\Delta Z)^2}$$

$$C_5 = \frac{(P_{i+1,j} + P_{i-1,j})(P_{i+1,j}H_{i+1}^3 + P_{i-1,j}H_{i-1}^3)}{4(\Delta\theta)^2}$$

$$C_6 = \frac{(P_{i,j+1} - P_{i,j-1})(P_{i,j+1}H_i^3 - P_{i,j-1}H_i^3)}{4(\Delta Z)^2}$$

$$C_7 = \frac{\Lambda(P_{i+1,j}H_{i+1} + P_{i-1,j}H_{i-1})}{2(\Delta\theta)}$$

References

1. Carpino M (1994) Foil bearing research at Penn state. Department of mechanical engineering the Pennsylvania state University park. N94–23058:148–151
2. Agrawal GL.1997. Foil air/gas bearing technology—an overview. International Gas Turbine and Aeroengine Congress and Exhibition, Orlando, ASME Paper No. 97-GT-347
3. Ku CP, Heshmat H (1992) Complaint foil bearing structural stiffness analysis part I: theoretical model—including strip and variable bump foil geometry. ASME J Tribol 114:394–400
4. Ku CP, Heshmat H (1993) Complaint foil bearing structural stiffness analysis part II: experimental investigation. ASME J Tribol 113:364–369
5. Heshmat H, Walomit J, Pinkus O (1983) Analysis of gas-lubricated compliant journal bearings. ASME J Lubr Technol 105:647–655
6. Peng JP, Carpino M (1993) Calculation of stiffness and damping coefficient for elastically supported gas foil bearings. ASME J Tribol 115:20–27
7. Ku CP, Heshmat H (1994) Structural stiffness and coulomb damping in compliant foil journal bearing: theoretical considerations. STLE Tribol Trans 37:525–533
8. Agrawal GL (1998) Foil air bearings cleared to land. ASME Mech Eng 120(7): 78–80
9. DellaCorte C, Fellenstein JA, Benoy P (1999) Evaluation of advanced solid lubricant coatings for foil air bearings operating at 25°C and 500°C. STLE Tribol Trans 42:338–342
10. DellaCorte C, Zaldana AR, Radil KC (2004) A systems approach to the solid lubrication of foil air bearings for oil-free turbomachinery. ASME J Tribol 126:200–207
11. Peng Z-C, Khonsari MM (2004) On the limiting load-carrying capacity of foil bearings. ASME J Tribol 126:817–818
12. Kim TH, San Andrés L (2006) Limits for high speed operation of gas foil bearings. ASME J Tribol 128:670–673
13. DellaCorte C, Valco MJ (2000) Load capacity estimation of foil air journal bearings for oil-free turbomachinery applications. NASA/TM-2000- 209782, ARL-TR-2334
14. Radil K, Howard S, Dykas B (2002) The role of radial clearance on the performance of foil air bearings. STLE Tribol Trans 45(4):485–490
15. Heshmat H (1994) Advancements in the performance of aerodynamic foil journal bearings: high speed and load capacity. ASME J Tribol 116:287–295
16. Hou Y, Zhu ZH, Chen CZ (2004) Comparative test on two kinds of new compliant foil bearing for small cryogenic turbo-expander. Cryogenics 44:69–72
17. Majumdar BC (2008) Introduction to tribology of bearings. S. Chand & Company Ltd, New Delhi
18. Lez SL, Arghir M, Frene J (2007) A new bump-type foil bearing structure analytical model. ASME J Eng Gas Turb Power 129:1047–1057

19. Peng Z-C, Khonsari MM (2006) A thermohydrodynamic analysis of foil journal bearings. ASME J Tribol 128:534–541
20. Kim TH, San Andrés L (2008) Heavily loaded gas foil bearings: a model anchored to test data. ASME J Eng Gas Turb Power 130:012–504
21. Lee DH (2010) The effect of coulomb friction on the static performance of foil journal bearings. Elsevier Tribol Int 43:1065–1072
22. Feng K, Kaneko S (2010) Analytical model of bump-type foil bearings using a link-spring structure and a finite-element shell model. ASME J Tribol 132:021–706
23. Rubio D, San Andrés L (2006) Bump-type foil bearing structural stiffness: experiments and predictions. ASME J Eng Gas Turb Power 129:494–502
24. Kim TH, Breedlove AW, San Andrés L (2009) Characterization of a foil bearing structure at increasing temperatures: static load and dynamic force performance. ASME J Tribol 131:041–703
25. San Andrés L, Rubio D, Kim TH (2007) Rotordynamic performance of a rotor supported on bump type foil gas bearings: experiments and predictions. ASME J Eng Gas Turb Power 129:850–857
26. San Andrés L, Kim TH (2010) Thermohydrodynamic analysis of bump type gas foil bearings: a model anchored to test data. ASME J Eng Gas Turb Power 132:042–504
27. San Andrés L, Kim TH (2010) Thermohydrodynamic model predictions and performance measurements of bump-type foil bearing for oil-free turboshaft engines in rotorcraft propulsion systems. ASME J Tribol 132:011–701
28. Ruscitto D, McCormick J, Gray S (1978) Hydrodynamic air lubricated compliant surface bearing for an automotive gas turbine engine. I-Journal Bearing Performance CONS/9427-1, NASA CR-135368
29. Tuakta C (2005) Use of fiber reinforced polymer composite in bridge structures. Thesis, Master, Massachusetts Institute of Technology
30. Chaban V (2011) Strength and structure of carbon–carbon reinforced composite. Department of Chemistry, University of Rochester, Rochester

Nonlinear Dynamic Behavior of Balanced Rotor Bearing System Due to Various Localized Defects

D. H. Pandya, S. H. Upadhyay and S. P. Harsha

Abstract In this paper, Improved dynamic model is propose to study the nonlinear dynamic behavior of a high speed balanced rotor supported on deep groove ball bearings (6205) with localized defect such as spall of micron level on races. Complex mathematical model simulates nonlinear vibrations due to both nonlinear contact stiffness and damping at the contact of rollers with both cage and races. The contact of rollers with cage and races are treated as nonlinear springs with contact damping where stiffness is obtained by using Hertzian elastic contact deformation theory. The explicit type numerical integration technique Runge–Kutta Fourth Order method is used to solve the coupled nonlinear differential equations. Computed results from the model are validated with experimental results, which are generated using healthy and defective deep groove ball bearings. Characteristic defect frequencies and its harmonics are broadly investigated using both theoretical and experimental results. The responses are presented in the form of bifurcation diagrams, Time domain displacement, Fast Fourier Transformation, Poincaré maps and orbit plot. The numerical simulations of the model agree with the experimental evidence and provide in-sight into the bearings chaotic response in a wide range of rotational speeds.

Keywords Non linear dynamics · Localized defect · Rolling element bearing · Bifurcation · Poincarèmaps

---

D. H. Pandya (✉) · S. H. Upadhyay · S. P. Harsha
Vibration and Noise Control Laboratory, Mechanical and Industrial Engineering Department, Indian Institute of Technology-Roorkee, Roorkee, Uttarakhand 247667, India
e-mail: veddhrumi@gmail.com

H. C. Patel et al. (eds.), *Proceedings of International Conference on Advances in Tribology and Engineering Systems*, Lecture Notes in Mechanical Engineering, DOI: 10.1007/978-81-322-1656-8_30, © Springer India 2014

Nomenclature

k_{in}	Equivalent non linear contact stiffness of the roller-inner race contact
k_{out}	Equivalent non linear contact stiffness of the roller-outer race contact
$k_{in\text{-}contact}$	Contact stiffness of the roller-inner race contact
$k_{out\text{-}contact}$	Contact stiffness of the roller-outer race contact
k_c	Contact stiffness of the roller-cage contact
r_{in}	Position of mass centre of inner race
r_{out}	Position of mass centre of outer race
x_{in}, y_{in}	Centre coordinates of inner race
x_{out}, y_{out}	Centre coordinates of outer race
δ_{in+}	Contact deformation of the roller-inner race
δ_{out+}	Contact deformation of the roller-outer race
m_{in}	Mass of the inner race, kg
m_j	Mass of the rolling elements, kg
m_{out}	Mass of the outer race, kg
m_{rotor}	Mass of the rotor, kg
N_b	Number of balls
R	Radius of outer race, mm
r	Radius of inner race, mm
c	Contact damping
c_{in}, c_{out}	Equivalent contact damping factor of roller-inner and outer race contact
d	Ball diameter, mm
D	Pitch diameter of bearing, mm
$\mathrm{F_u}$	unbalanced rotor force, N
W	Radial preload, N
β	Constant angular separation between rolling elements
γ	Internal radial clearance
Δ	Additional deformation due to defects
$\sum\rho$	Sum of curvature of the contacting bodies
$\left(\dot{\phi}\right)_{in}$	Angular velocity of inner race
$\left(\dot{\phi}\right)_{out}$	Angular velocity of outer race
χ_j	Radial position of *j*th rolling element from center of inner race
ρ_j	Radial position of *j*th rolling element from center of outer race
ρ_r	Radius of each rolling element, mm
ω_{cage}	Angular velocity of cage
ω_{inner}	Angular velocity of inner race
θ_j	Angular position of rolling element
VC	Varying compliance frequency
FFT	Fast fourier transformations
ORD	Outer race defect

IRD	Inner race defect
BPFO	Ball passage frequency on outer race
BPFI	Ball passage frequency on inner race

1 Introduction

The critical role of the rolling element bearings played in the operation and performance of machine systems have rendered them vitally important. The high-speed rotor bearing systems often show the unpredictable dynamic response with defect free bearing in balanced rotor condition. Study on dynamic behavior of ball bearing due to localized and distributed defects has been done. But, few papers have reported diagnosis of localized defects and their effect on chaotic behavior of high speed ball bearing. McFadden and Smith studied vibration response considering single and multiple point defects on the inner race of the rolling element bearing [1, 2]. Mevel and Guyader [3] have developed a theoretical model of a ball bearing supporting on a balanced horizontal rigid rotor, with a constant vertical radial force. This is similar to the work done by Fukata et al. [4] but more results have been reported for parametric studies undertaken and routes to chaos traced out. Chaos in this model of bearing has been reported to come out of the sub-harmonic route and the quasi-periodic route. Igarashi and Kato [5] studied the effects of multiple local defects on inner and outer races through the frequency spectrum. Where, the authors reported the presence of a frequency due to time delay between impulses produced by two defects and concluded that amplitude of vibration at this frequency changes as function of the angle between two defects. Prabhakar et al. [6] have considered single and multiple point defects on inner race, outer race, and the combination faults and used discrete wavelet transform (DWT) to detect these faults on bearings. Kiral and Karagülle [7] studied the influences of defect position/location and number of defects of races on vibration. Choy et al. [8] utilized the modified Poincarè map for the damage quantification in tapered and roller bearings. Their work showed that the modified Poincarè map, based on the cage speed, provides information on the damage type. Arslan and Aktürk [9] developed a shaft-bearing system with bearing defects. The model has 3 DOF, 2 DOF for the radial displacement and 1 DOF for the axial displacement. The balls in the bearing have additional DOF since they can vibrate in a radial direction. The effect of the centrifugal load on the balls is neglected. Mevel et al. [10] found experimental routes to chaos, which was confirmed with proposed theoretical predictions of phenomena. Harsha et al. [11] analyzed the nonlinear behavior of a high speed horizontal balanced rotor supported by a ball bearing and concluded that the most severe vibrations occur when the varying compliance frequency (VC) and its harmonics coincide with natural frequency. Gafari [12] have performed experimental confirmation of existence of strange attractor in Poincarè maps and infinite number of open orbits in the phase plane in healthy bearing. Rafsanjani et al. [13] have presented an analytical model to study the non linear dynamic behavior of rolling element bearing systems including surface defects. Various surface defects due to local imperfections on race ways and rolling elements are introduced to the proposed model.

So far, small amount of research has been reported in the literature regarding the possible relation between chaotic parameters and bearing faults. The localized defect has been modeled and concern characteristic defect frequency along with it's harmonics in frequency spectrum are reported by earlier researchers. But, non linear behavior of the system will change in presence of localized defect was not attended in detail. In the present paper, the change in non linear behavior of the system with localized defect on outer race and inner race have been explained with the help of Poincare maps, orbit plots, peak to peak responses, Time domain displacement, and FFT plots. Experimental validation is also shows good agreement with numerical simulation.

2 Problem Formulation and Numerical Simulation

A schematic diagram of rolling element bearing is shown in Fig. 1. For investigating the structural vibration characteristics of rolling element bearing, a model of bearing assembly can be considered as a spring mass damper system. Elastic deformation between races and balls and cage and ball gives a non linear force deformation relation, which is obtained by Hertzian theory. In the mathematical modeling, the rolling element bearing is considered as spring mass damper system and rolling elements act as non linear contact spring as shown in Fig. 2. Since, the Hertzian forces arise only when there is contact deformation, the springs are

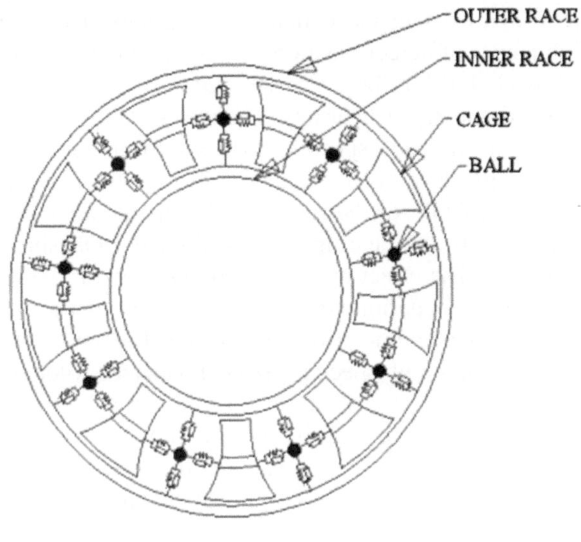

Fig. 1 The flexibility of the rolling contacts in a rolling element bearing is represented by non linear springs and non linear damping

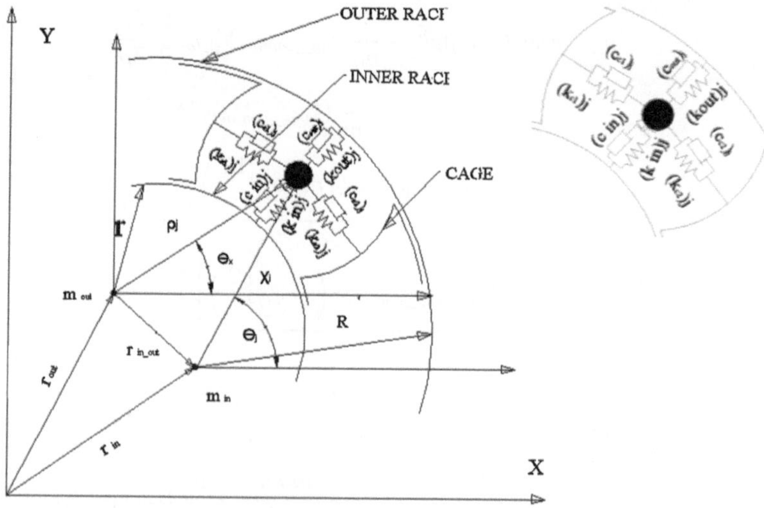

Fig. 2 Mass-spring-damper of rolling element bearing

required to act only in compression. In other words, the respective spring force comes into play when the instantaneous spring length is shorter than its unstressed length, otherwise the separation between balls and the races takes place and the resultant force is set to zero. The excitation is because of the varying compliance vibrations of the bearing which arise because of the geometric and elastic characteristics of the bearing assembly varying according to the cage position.

A real rotor bearing system is generally very complicated and difficult to model. First the expression for energies of the individual components of the bearing is formulated with assumptions as described by harsha et al. [11]. Using these energies, the equations of motion are derived with the help of Lagrange's equation.

Lagrange's equation for a set of independent generalized coordinates, as:

$$\frac{d}{dt}\frac{\partial T}{\partial\{\dot{p}\}} - \frac{\partial T}{\partial\{p\}} + \frac{\partial V}{\partial\{p\}} + \frac{\partial P_d}{\partial\{\dot{p}\}} = \{f\} \tag{1}$$

where T, V, p and f are kinetic energy, strain energy, vector with generalized degree-of-freedom (DOF) coordinate and vector with generalized contact forces respectively and P_d represent the dissipation energy due to damping. The kinetic and potential energies can be subdivided into the contributions from the various components i.e. from the rolling elements, the inner race, the outer race and the rotor.

The kinetic energy and strain energy contributed by the inner race, outer race, balls, rotor and springs, can be differentiated with respect to the generalized coordinates ρ_j ($j = 1,2,\ldots\ldots, N_b$), x_{in}, and y_{in} to obtain the equations of motion. For the generalized coordinates $\boldsymbol{\rho_j}$, where $j = 1,2,\ldots\ldots, N_b$, the equations are:

$$\begin{aligned}
&\ddot{\rho}_j + g\sin\theta_j + \rho_j\dot{\theta}^2 - \frac{1}{m_j}(k_{in_contact})\,[\delta_{in}]_+^{3/2}\frac{\partial\chi_j}{\partial\rho_j} \\
&\quad + \frac{1}{m_j}(k_{out_contact})\,[\delta_{out}]_+^{3/2} \\
&\quad + \frac{1}{2\,m_j}\frac{\partial\left[(k_{in_contact})\left([\delta_{in}]_+^{1/2}\right)\right]}{\partial\rho_j}[\delta_{in}]_+^2 \\
&\quad + \frac{1}{2\,m_j}\frac{\partial\left[(k_{out_contact})\left([\delta_{out}]_+^{1/2}\right)\right]}{\partial\rho_j}[\delta_{out}]_+^2 \\
&\quad + \frac{2kc\rho_r}{m_j}\left[\frac{2\pi}{Nb}(j-1) + \frac{\omega inner}{2}\left(1-\frac{\rho_r}{\rho_j}\right)\times t - 1\right] \\
&\quad * \left[\frac{\omega inner}{2}\left(\frac{\rho_r}{\rho_j^2}\right)\times t\right] \\
&\quad + \frac{3}{2\,m_j}\sum_{j=1}^{N_{r.e.}}\left\{C_{in}\,(K_{in_contact})\,\delta_{in_+}^{3/2}\;(-\chi_j)^q\frac{\partial\dot{\chi}_j}{\partial\dot{\rho}_j}\right\} \\
&\quad + \frac{3}{2\,m_j}\sum_{j=1}^{N_{r.e.}}C_{out}\,(K_{out_contact})\,\delta_{out_+}^{3/2}\;(-\dot{\rho}_j)^q \\
&\quad + \frac{3}{m_j}\times\sum_{j=1}^{N_{r.e.}}\left[\left\{C_c\,K_c\,\delta_{c_+}^{3/2}\;\dot{\delta}_{c+}^q\right\}\right]\frac{\partial\dot{\delta}_c}{\partial\dot{\rho}_j} = 0 \\
&\qquad j = 1, 2, \ldots, N_b
\end{aligned} \tag{2}$$

For the generalized coordinate $\boldsymbol{x_{in}}$ the equation is:

$$\begin{aligned}
&\ddot{x}_{in} - \frac{1}{m_{rotation}}\sum_{j=1}^{r.e}(k_{in_contact})\;[\delta_{in}]_+^{3/2}\frac{\partial\chi_j}{\partial x_{in}} \\
&\quad + \frac{3}{2\,m_{rotation}}\sum_{j=1}^{N_{r.e.}}\left\{C_{in}\,(K_{in_contact})\,\delta_{in_+}^{3/2}\,(-\dot{\chi}_j)^q\frac{\partial\dot{\chi}_j}{\partial\dot{x}_{in}}\right\} = \frac{F_u\sin(\omega_s\,t)}{m_{rotation}}
\end{aligned} \tag{3}$$

For the generalized coordinate $\boldsymbol{y_{in}}$ the equation is:

$$\begin{aligned}
&\ddot{y}_{in} + g - \frac{1}{m_{rotation}}\sum_{j=1}^{N_b}(k_{in_contact})\;[\delta_{in}]_+^{3/2}\frac{\partial\chi_j}{\partial y_{in}} + \\
&\quad + \frac{3}{2\,m_{rotation}}\sum_{j=1}^{N_{r.e.}}\left\{C_{in}\,(K_{in_contact})\,\delta_{in_+}^{3/2}\,(-\dot{\chi}_j)^q\frac{\partial\,\dot{\chi}_j}{\partial\dot{y}_{in}}\right\} = \frac{(W + F_u\cos(\omega_s\,t))}{m_{rotation}}
\end{aligned} \tag{4}$$

where $m_{rotation} = (m_{inner} + m_{rotor})$.

This is a system of ($N_b + 2$) second order, non linear differential equations. No external radial force is allowed to act on the bearing system and no external mass is attached to the outer race. The "+" sign as subscript in these equations signifies that if the expression inside the bracket is greater than zero, then the rolling element at angular location 'θ_j' is loaded giving rise to restoring force and if the expression inside bracket is negative or zero, then the rolling element is not in the load zone, and restoring force is set to zero. For the balanced rotor condition, the unbalanced rotor force (F_u) is set to be zero.

Equations (2–4) are modified to consider the effect of the defect on the bearing surface. The total deflection of the *j*th ball of bearing is the sum of the characteristic of the outer race defect and that of the static deflection considered. Radial displacement is obtained by considering the resulting distortion. Considering the internal radial clearance (γ) and additional deformation due to defects in bearings (Δ_j), the contact deformation considering ORD becomes,

$$\delta_{out} = \gamma + \Delta_{j(ord)}$$

The inner race is rotating at the shaft speed and the ball center is rotating at the speed of cage, so the contact angle 'θ_i' is

$$\theta_j = \frac{2\pi}{N_b}(j-1) + \left(\omega_{\text{cage}} - \omega_{\text{inner}}\right) \times t$$

Defect is on the surface of the inner race, so it rotates at speed ω_{inner}. Additional deflection 'Δ_j' in contact deformation is obtained when defect angle 'α' coincides with the contact angle 'θj' of *j*th rolling element, the contact deformation considering IRD becomes,

$$\delta_{in} = \gamma - \Delta_{j(ird)}$$

The coupled non linear second order differential equations are solved by numerical integration technique which is a time domain approach. The non-analytic nature of the stiffness term renders the system equations difficult for analytical solution. The equations of motion (2–4) are solved by using the explicit type numerical integration technique Runge–Kutta Fourth-Order method to obtain radial displacement, velocity, and acceleration of the rolling elements.

For numerical solutions, the initial conditions and step size Δt are very important for successive and economic computational solutions. Any non linear system is very sensitive to initial condition and for the very small perturbation of the initial condition gives a totally different behavior. So, improper selection of initial conditions can lead to a larger computation time or sometimes unexpected results.

Table 1 Geometrical and physical properties of the DGB bearing 6205

Rotor bearing specification	6205
Mass of the rotor (M_r)	5 kg
Diameter of inner race with point of contact with the rolling element (d)	31.12 mm
Diameter of outer race with point of contact with the rolling element (D)	47 mm
Radius of each rolling element (ρ_r)	3.965 mm
Radial load (W)	50 N
Outside diameter	52 mm
Number of rolling elements (N_b)	9
Angular separation between elements ($\beta = 2\pi/N_b$)	40 deg
Spall size (μm)	300 × 300 × 20
Bearing clearance (μm)	14

3 Experimental Setup

In the paper, experimentation is carried out on a rotor-bearing test rig. The rotor is mounted on the shaft, driven by DC motor whose operating speed can be controlled. A spring type flexible coupling has been used to connect rotor shaft and motor to compensate any misalignment. The SKF 6205 bearings have been taken for the study. The horizontal rigid rotor of weight 2.5 kg is used and rig is connected to a data acquisition system through proper instrumentation. Two piezoelectric accelerometers (bi-axial) are used for picking up the vibration signals from various stations on the test rig. These special piezoelectric pickup type sensors are used with frequency range of 1–30 kHz, measurement range ±500 g peak, resolution of 0.005 g and resonant frequency of 70 kHz. The sensitivity of sensor 1.02 mv/(m/s2) with IEPE (Integral Electronics piezoelectric accelerometer) input mode of sensor. The horizontal rotor is used, which is supported by radial contact ball bearings with dedicated localized defect. Bearing geometry is described as per Table 1. The localized defects are introduced as spall of 300 × 300 × 20 microns on outer race, inner race and rolling elements with laser machining process as shown in Fig. 3.

4 Result and Discussion

To get the satisfactory post transient steady state condition, numerical simulation is run for 2 s. To save the computational time, artificial damping, c = 50 N-s/m is used. In the present study, radial bearing SKF 6205, class 3 type is used, in which internal radial clearance is 14 μm.The other geometrical and physical properties of bearing is listed in Table 1.

Here, weight of balanced rigid rotor (50 N) is taken as constant radial load (W) and different types of nonlinear behavior is identified for the speed range of

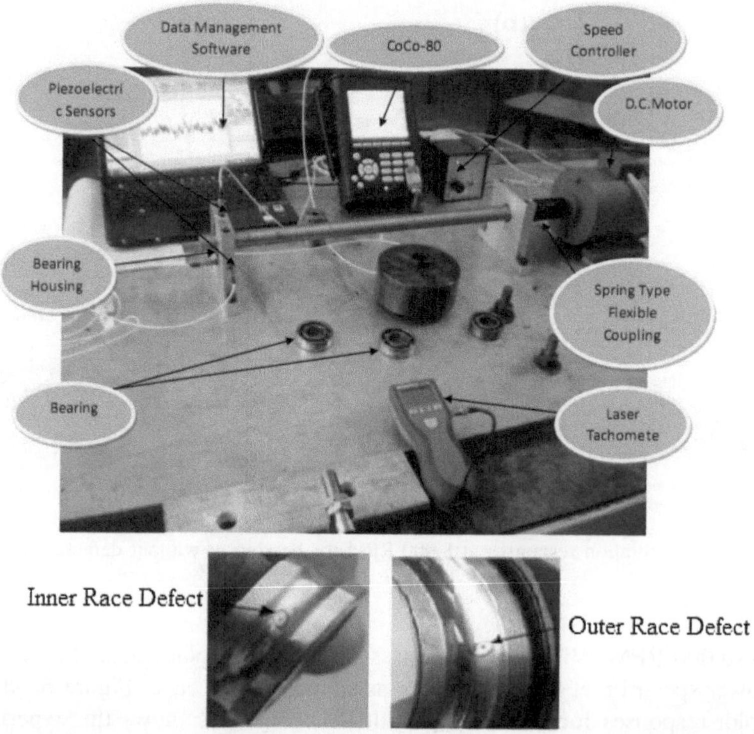

Fig. 3 Experimental setup

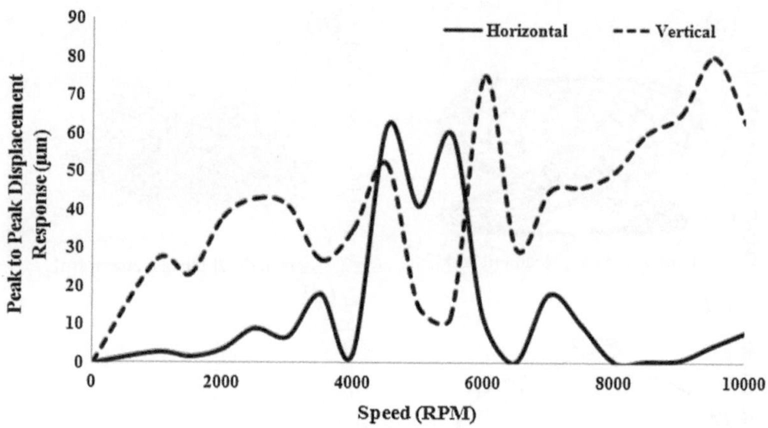

Fig. 4 Response plot for bearing with inner race defect at w = 5 0 n, $y_0 = 5\ \mu m$, c = 50 N-s/m

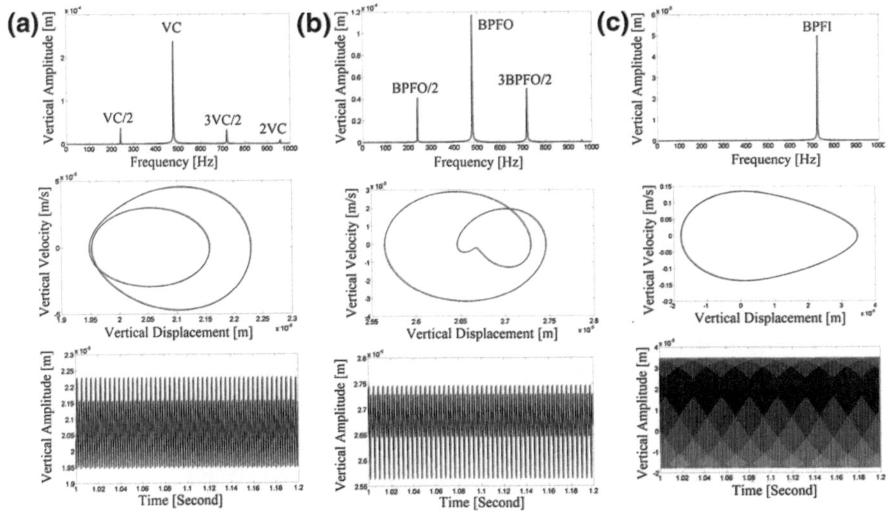

Fig. 5 Vertical vibration responses at 8,000 RPM for bearing, **a** without defect, **b** with ORD, **c** with IRD

1,000–10,000 RPM. Vibration responses like Time-displacement, Poincaré map and power spectrum at 8,000 RPM are as shown in Fig. 5a–c. Figure 6a–d shows orbit plot responses for the bearing with IRD. Figure 7 shows the experimental responses.

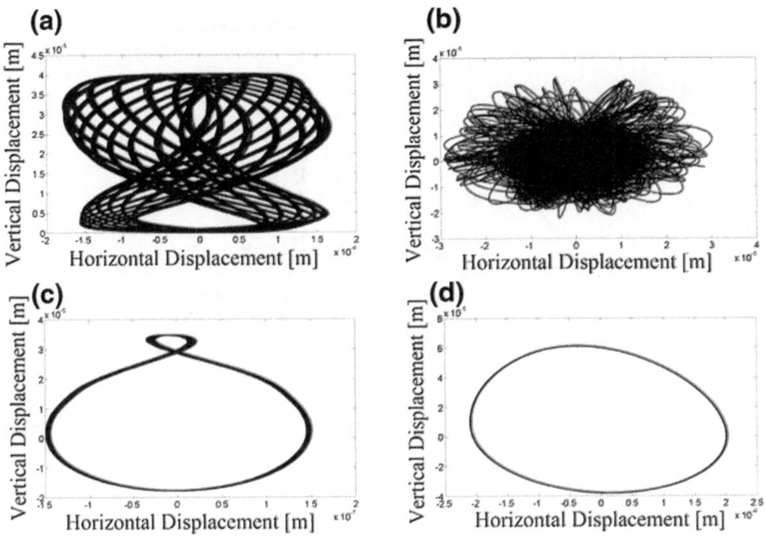

Fig. 6 Orbit plots for bearing with IRD at **a** 2,000, **b** 4,500, **c** 8,000 and **d** 9,500 RPM

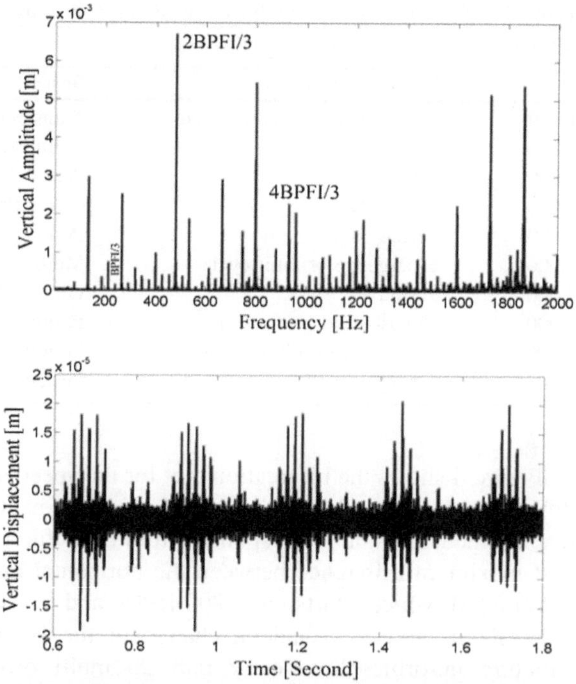

Fig. 7 Experimental vibration responses at 8,000 RPM for bearing with IRD

Here, during the simulation it is observed that vertical displacement takes more time (2.5–5 s) as compared to horizontal displacement (2–4 s) to die out. Peak to peak displacement in vertical is less as compared to horizontal in case of bearing without defect and bearing with ORD as reported by the authors [14]. But, in case of bearing with IRD, Peak to peak displacement in vertical is dominating as compared to horizontal as shown in bifurcation plot at Fig. 4.

In this paper, non linear dynamic behavior of bearing with localized defect on inner race defect is discuss as compare to defect on outer race. Vertical vibration responses of deep groove ball bearing without defect, with ORD and with IRD are as shown in Fig. 5a–c. Major peaks are reported VC and BPFO in case of healthy bearing and bearing with ORD respectively, other peaks are reported at sub harmonic and super harmonic peak of characteristic defect frequencies. Dynamic behavior of bearing without defect and with ORD is reported as Period 2T-non linear periodic motion at 8,000 RPM and more frequency amplitude is reported with ORD as compare to bearing without defect. As shown in Fig. 5c, power spectrum of IRD indicates the period 1T-non linear periodic motion. But, as observed in orbit plot a dense period 2T-non linear periodic motion is reported as shown in Fig. 6c. This dense orbit path is strongly supported by the experimental results as shown in Fig. 7. Experimental results show the major peak at 2BPFI/3 and other sub harmonic and super harmonic peaks are reported in FFT spectrum. Time response clearly shows the type-II intermittency, which indicates the reverse periodic doubling bifurcation.

Table 2 Behavior of the system for the speed as control parameter

RPM	Behavior of the system	
	Bearing with ORD	Bearing with IRD
1,000	Transient chaos-multiperiodic	Transient chaos-multiperiodic
2,000	Period 4T	Multi orbit periodic
3,000	Period 3T	Period 1T
4,000	Period 2T	Chaos through reverse periodic bifurcation
5,000	Period 1T	Period 4T
6,000	Emergence of limit cycle	Multi orbit periodic
7,000	Period 1T	Period 2T
8,000	Chaos—fold bifurcation	Period 2T-non linear periodic
9,000	Chaos—Hopf bifurcation	Period 1T-non linear periodic
10,000	Emergence of limit cycle	Period 1T-non linear periodic

Figure 4 shows the bifurcation plot for the speed as a control parameter and the difference between the peak to peak displacements of horizontal and vertical displacement is as output response. From the bifurcation plot, it can infer that the first maximum difference between the horizontal and vertical displacement is at 2,200 RPM which starts at 1,800 RPM and ends at 3,800 RPM. So, at the 2,000 RPM there is a sudden change in the behavior of the system and net structure in orbit plot clearly indicate multi orbit quasi periodic motion at 2,000 RPM as shown in Fig. 6a. As speed increase from 4,000 RPM reverse periodic bifurcation chaos is reported up to 5,500 RPM. Clear chaotic motion is as shown at 4,500 RPM in Fig. 6b. the sudden changes are also observed in peak to peak displacement plot. From 5,850 RPM, again vertical responses are dominant then horizontal. Further increasing speed the system will tends to become stable, as shown in Fig. 6c period 2T at 8,000 RPM and then period 1T at 9,500 RPM. The dense orbit path at Fig. 6c is indicated clear Modulation may also result from the changes in the length of the transmission path from the impact to the accelerometer. These Modulations of the transients will occur at the rotational frequency of the inner race. Experimental time displacement responses are validating intermittency type-I, which leads to cyclic fold bifurcation and the system has Periodic 1T behavior at 9,500 RPM. As the speed is changing, the dynamical behavior of the system is changed and that is listed in Table 2 for bearing with ORD and IRD.

5 Conclusion

The nonlinear responses of a balance rotor bearing system with ORD and IRD have been demonstrated to be chaotic with rotational speed as control parameter combined with ball-cage interaction. For cases, which are stable to free motion and not close to the neutral stability line, a limited range of chaos can be detected. The system has stable periodic 2T behavior at 2,000 RPM and transient type chaotic nature at 4,500 RPM. The system has chaos at 4,500 RPM through reverse periodic bifurcation. Further, the system shows quasi periodic stable 2T behavior with type-II intermittency at 8,000 RPM and stable periodic 1T behavior with type-I intermittency at 9,500 RPM.

Based on the characteristics of the dynamic behaviors of the system, the nonlinear dynamic responses are categorized as: First the system responses are periodic and are not sensitive to initial conditions. The predicted impact force also increases smoothly. This well-behaved region permits the designer to predict the trends accurately and without ambiguity. Second, the system responses show mixed intermittent behavior. For these responses, the hidden danger is periodicity. The periodic response may lead designers to overlook its large sensitivity to small variations of system parameters or operating conditions.

This study helps designer to predict non linear behavior of system accurately and without ambiguity.

References

1. Harsha SP, Nataraj C (2007) Non-linear dynamic response of a complex rotor bearing system. Int J Adv Vib Eng 12(1):379–399
2. McFadden PD, Smith JD (1984) Model for the vibration produced by a single point defect in a rolling element bearing. J Sound Vib 96:69–82
3. Mevel B, Guyader JL (1993) Routes to chaos in ball bearings. J Sound Vib 162:471–487
4. Fukata S, Gad EH, Kondou T, Ayabe T, Tamura H (1985) On the radial vibrations of ball bearings (computer simulation). Bull Jpn Soc Mech Eng 28:899–904
5. Igarashi T, Kato J (1985) Studies on the vibration and sound of defective roller bearings (3rd report: sound of ball bearing with multiple defects). Bull Japan Soc Mech Eng 28:492–499
6. Prabhakar S, Mohanty AR, Sekhar AS (2002) Application of discrete wavelet transform for detection of ball bearing race faults. Tribol Int 35:793–800
7. Kiral Z, Karagülle H (2003) Simulation and analysis of vibration signals generated by rolling element bearing with defects. Tribol Int 36:667–678
8. Choy FK, Zhou J, Braun MJ, Wang L (2005) Vibration monitoring and damage quantification of faulty ball bearings. ASME Trans J Tribol 127(4):776–783
9. Arslan H, Aktürk N (2008) An investigation of rolling element vibrations caused by local defects. ASME J Tribol 130:041101–041112
10. Mevel B, Guyader JL (2008) Experiments on routes to chaos in ball bearings. J Sound Vib 318:549–564
11. Harsha SP, Sandeep K, Prakash R (2003) The effect of speed of balanced rotor on non-linear vibrations associated with ball bearings. Int J Mech Sci 45(4):725–740
12. Ghafari SH, Golnaraghi F, Ismail F (2008) Effect of localized faults on chaotic vibration of rolling element bearings. Non-linear Dyn 53:287–301
13. Rafsanjani A, Abbasion S, Farshidianfar A, Moeenfard H (2009) Non-linear dynamic modeling of surface defects in rolling element bearing systems. J Sound Vib 319:1150–1174
14. Pandya DH, Upadhyay SH, Harsha SP (2013) The effect of localized defect on nonlinear dynamic behavior of balanced rotor supported by deep groove ball bearing. In: Proceedings of the ASME 2013 IDETC/CIE, Oregon, USA, 4–7 Aug 2013 (Accepted)

Performance of a Magnetic Fluid-Based Longitudinally Rough Short Bearing

P. I. Andharia, G. M. Deheri and Shruti Mehta

Abstract Efforts have been made to analyze the performance of a longitudinally rough short bearing formed by a magnetic fluid. Bearing surfaces are assumed to be longitudinally rough. The roughness of the bearing surfaces is characterized by a stochastic random variable with non-zero mean, variance and skewness. While modeling the roughness it has been taken into consideration that the standard deviation associated with the characterization of roughness plays a predominant role as compared to that of mean and skewness. The associated Reynolds equation is solved with appropriate boundary conditions to obtain the pressure distribution which is further used to calculate load carrying capacity. The results are presented numerically as well as graphically. It is observed that the performance of the bearing system gets improved due to the magnetization. It is seen that the roughness affects the system adversely. The study reveals that the negative effect of the roughness can be minimized by the positive effect of the magnetization. Thus, while designing the bearing system, the roughness must be given due consideration.

Keywords Short bearing · Surface roughness · Magnetic fluid · Reynolds equation · Longitudinal roughness

P. I. Andharia (✉)
Department of Mathematics, M.K. Bhavnagar University, Bhavnagar, Gujarat 364002, India
e-mail: pareshandharia@yahoo.com

G. M. Deheri · S. Mehta
Department of Mathematics, Sardar Patel University, Vallabh Vidyanagar, Gujarat 388120, India

H. C. Patel et al. (eds.), *Proceedings of International Conference on Advances in Tribology and Engineering Systems*, Lecture Notes in Mechanical Engineering, DOI: 10.1007/978-81-322-1656-8_31, © Springer India 2014

1 Introduction

The slider bearing are designed to support the axial loads. Exact solutions of Reynolds' equation for slider bearings with various simple film geometries are presented in a number of books and research papers (Cameron [1], Archibald [2], Lord Rayleigh [3]). Prakash and Vij [4] analyzed the hydrodynamic lubrication of a plane inclined slider bearing taking different geometries into consideration. Patel and Gupta [5] extended the above analysis of Prakash and Vij [4] by incorporating slip velocity.

However, bearing surfaces could be roughened through the manufacturing process, the wear and the impulsive damage. In order to account for the effect of surface roughness Christensen [6, 7] utilized a stochastic concept and introduced an averaging film model to lubricated surfaces with striated roughness. Several investigators have adopted a stochastic approach to model the random roughness (Tzeng and Saibel [8], Christensen and Tonder [9–11]). Christensen and Tonder presented a comprehensive general analysis for both transverse and longitudinal surface roughness based on a general probability density function by modifying and developing the approach of Tzeng and Saibel [8]. Subsequently many researchers have followed this model of Christensen and Tonder to study the effect of surface roughness such as the works in the hydrodynamic journal bearings by Guha [12] and Taranga et al. [13], the hydrodynamic slider bearing by Christensen and Tonder [14] and the squeeze film spherical bearings by Andharia et al. [15].

In all the above studies conventional lubricant were used. It has now been recognized that the use of magnetic fluid as a lubricant modifies the performance of the bearing. Agrawal [16] considered the configuration of Prakash and Vij [4] in the presence of a magnetic fluid lubricant and found its performance better than the one with conventional lubricant. Bhat and Deheri [17] extended the analysis of Agrawal [16] by investigating a magnetic fluid based porous composite slider bearing. Bhat and Deheri [18] discussed a general porous slider bearing with squeeze film formed by a magnetic fluid. Here it has been proposed to study the performance of a longitudinally rough short bearing in the presence of a magnetic fluid.

2 Analysis

The configuration of the bearing system which is infinite in Z-direction is shown in Fig. 1. The slider moves with the uniform velocity u in X-direction. The length of the bearing is L and the breadth B is in Z-direction where $B \ll L$. The dimension B being very small, the pressure gradient $\partial p/\partial z$ is much larger than the pressure gradient $\partial p/\partial x$ and hence latter can be neglected. The maximum and minimum film thicknesses are h_1 and h_2 respectively.

Fig. 1 Configuration of the bearing

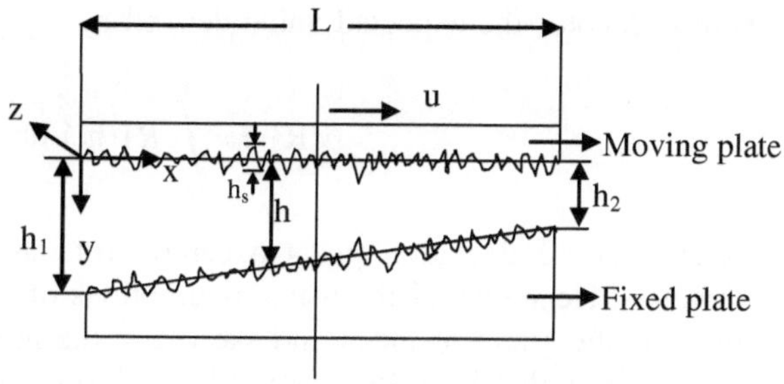

The lubricant film is considered to be isoviscous and incompressible and the flow is laminar. The magnetic field is oblique to the stator as in Agrawal [16]. Following the discussions carried out by Prajapati [19] regarding the effect of various forms of magnitude of the magnetic field, here it is expressed as

$$M^2 = KB^2\left\{\left(\frac{1}{2}+\frac{z}{B}\right)\sin\left(\frac{1}{2}-\frac{z}{B}\right)+\left(\frac{1}{2}-\frac{z}{B}\right)\sin\left(\frac{1}{2}+\frac{z}{B}\right)\right\} \tag{1}$$

where K is a suitably chosen constant from dimensionless point of view (Bhat and Deheri [18]).

The bearing surfaces are assumed to be longitudinally rough. The thickness h(x) of the lubricant film is

$$\mathrm{h(x)} = \bar{h}(\mathrm{x}) + \mathrm{h_s} \tag{2}$$

where $\bar{h}(x)$ is the mean film thickness and h_s is the deviation from the mean film thickness characterizing the random roughness of the bearing surfaces. h_s is considered to be stochastic in nature and governed by the probability density function

$$f(h_s) = \begin{cases} \frac{35}{32}\left(1-\frac{h_s^2}{c^2}\right)^3, & -c \le h_s \le c \\ 0 & \text{elsewhere} \end{cases} \tag{3}$$

where c is the maximum deviation from the mean film thickness. The mean α, the standard deviation σ and the parameter ε which is the measure of symmetry, of the random variable h_s, are defined by the relationships:

$$\alpha = \mathrm{E(h_s)} \tag{4}$$

$$\sigma^2 = \mathrm{E}\left[(\mathrm{h_s}-\alpha)^2\right] \tag{5}$$

and

$$\varepsilon = \mathrm{E}\left[(\mathrm{h_s}-\alpha)^3\right] \tag{6}$$

where E denotes the expected value defined by

$$\mathrm{E(R)} = \int_{-c}^{c} R\,\mathrm{f(h_s)\ dh_s} \tag{7}$$

The standard deviation plays a central role as compared to the other two parameters. Hence under the usual assumptions of hydrodynamic lubrication and neglecting the effect of mean and skewness, the modified Reynolds equation for film pressure (Bhat [20], Prajapati [19], Deheri et al. [21]) is given by

$$\frac{d^2}{dz^2}\left(p - 0.5\mu_0\overline{\mu}M^2\right) = 6\mu\,u\,l(h)\,\tfrac{d}{dx}\left(\tfrac{1}{n(h)}\right) \tag{8}$$

where $h = h_2\left\{1 + m\left(1 - \frac{x}{L}\right)\right\}$

$$\mathrm{l(h)} = \mathrm{h}^{-3}\left[1 + 6\mathrm{h}^{-2}\sigma^2\right]$$

and $\mathrm{n(h)} = \mathrm{h}^{-1}\left[1 + \mathrm{h}^{-2}\sigma^2\right]$
where in $m = \frac{h_1 - h_2}{h_2}$

while μ_0 is the magnetic susceptibility, $\overline{\mu}$ is the free space permeability, and μ is the lubricant viscosity.

The associated boundary conditions are

$$\mathrm{p} = 0;\ \mathrm{z} = \pm\frac{B}{2} \quad \text{and} \quad \frac{dp}{dz} = 0;\ \mathrm{z} = 0 \tag{9}$$

By integrating Eq. (8) twice and introducing the following dimensionless quantities

$$Z = \frac{z}{B},\ X = \frac{x}{L},\ P = \frac{h_2^3 p}{\mu\,u\,B^2},\ \mu^* = \frac{h_2^3 K\,\mu_0\,\overline{\mu}}{\mu\,u} \tag{10}$$

one obtains the pressure distribution in dimensionless form as

$$\begin{aligned} P = \frac{\mu^*}{2}&\left\{\left(\frac{1}{2}+Z\right)\sin\left(\frac{1}{2}-Z\right)+\left(\frac{1}{2}-Z\right)\sin\left(\frac{1}{2}+Z\right)\right\} \\ &+\frac{3mh_2}{LQ^3}\left[\frac{1}{4}-Z^2\right]\left[1+\frac{6\sigma^2}{h_2^2Q^2}\right]\left[\frac{3}{1+\frac{\sigma^2}{h_2^2Q^2}}-\frac{2}{1+\frac{\sigma^2}{h_2^2Q^2}}\right] \end{aligned} \tag{11}$$

where $\mathrm{Q} = \{1 + \mathrm{m}(1 - - \mathrm{X})\}^2$.

The load carrying capacity of the bearing is given by

$$w = \int_{-B/2}^{B/2}\int_{0}^{L} p(x,z)\,dx\,dz \tag{12}$$

Dimensionless load carrying capacity is obtained from

$$W = \frac{h_2^3 w}{\mu u\, B^4} = \frac{L}{B} \int_{-1/2}^{1/2} \int_{0}^{1} P dX dZ \tag{13}$$

3 Results and Discussions

Equation (11) presents the expression for the pressure distribution in dimensionless form while the dimensionless load carrying capacity is determined by Eq. (13). These performance characteristics depend on several parameters such as μ^*, σ, m, L/h_2 and B/h_2. Setting the roughness parameter σ to be zero this study reduces to the performance of a magnetic fluid based short bearing. Further taking the magnetization parameter μ^* to be zero, this study reduces to the performance analysis of short bearing as discussed in Basu et al. [22].

Tables 1 and 2 presents the effect of μ^* on the load carrying capacity for various values of σ/h_2 and m respectively. It is noticed that the effect of σ/h_2 and aspect ratio m is negligible with respect to μ^*. Tables 3 and 4 shows the effect of L/h_2 on the load carrying capacity for various values of σ/h_2 and m respectively. From these figures it is clear that the load carrying capacity increases considerably due to L/h_2. However, the effects of σ/h_2 and aspect ratio m are negligible with respect to L/h_2. Tables 5 and 6 suggest that the the effect of σ/h_2 and m on load carrying capacity is negligible with respect to B/h_2.

Table 1 Variation of load carrying capacity with respect to μ* and σ/h2

σ/h_2	Load				
	$\mu^* = 0$	$\mu^* = 0.025$	$\mu^* = 0.05$	$\mu^* = 0.075$	$\mu^* = 0.1$
0.03	0.005629	0.824412	1.643196	2.461979	3.280762
0.06	0.005679	0.824462	1.643246	2.462029	3.280812
0.09	0.005762	0.824545	1.643329	2.462112	3.280895
0.12	0.005878	0.824661	1.643444	2.462228	3.281011
0.15	0.006024	0.824808	1.643591	2.462374	3.281158

Table 2 Variation of load carrying capacity with respect to μ* and m

M	Load				
	$\mu^* = 0$	$\mu^* = 0.025$	$\mu^* = 0.05$	$\mu^* = 0.075$	$\mu^* = 0.1$
0.25	0.003175	0.821958	1.640742	2.459525	3.278308
0.5	0.004868	0.823651	1.642434	2.461218	3.280001
0.75	0.005878	0.824661	1.643444	2.462228	3.281011
1	0.006528	0.825312	1.644095	2.462878	3.281662
2	0.007701	0.826484	1.645268	2.464051	3.282834

Table 3 Variation of load carrying capacity with respect to L/h_2 and σ/h_2

σ/h_2	Load				
	$L/h_2 = 500$	$L/h_2 = 1000$	$L/h_2 = 1500$	$L/h_2 = 2000$	$L/h_2 = 2500$
0.03	0.824412	1.643196	2.461979	3.280762	4.099546
0.06	0.824462	1.643246	2.462029	3.280812	4.099596
0.09	0.824545	1.643329	2.462112	3.280895	4.099679
0.12	0.824661	1.643444	2.462228	3.281011	4.099794
0.15	0.824808	1.643591	2.462374	3.281158	4.099941

Table 4 Variation of load carrying capacity with respect to L/h_2 and m

m	Load				
	$L/h_2 = 500$	$L/h_2 = 1000$	$L/h_2 = 1500$	$L/h_2 = 2000$	$L/h_2 = 2500$
0.25	0.821958	1.640742	2.459525	3.278308	4.097092
0.5	0.823651	1.642434	2.461218	3.280001	4.098784
0.75	0.824661	1.643444	2.462228	3.281011	4.099794
1	0.825312	1.644095	2.462878	3.281662	4.100445
2	0.826484	1.645268	2.464051	3.282834	4.101618

Table 5 Variation of load carrying capacity with respect to B/h_2 and σ/h_2

σ/h_2	Load				
	$B/h_2 = 10$	$B/h_2 = 20$	$B/h_2 = 30$	$B/h_2 = 40$	$B/h_2 = 50$
0.03	4.929587	2.464793	1.643196	1.232397	0.985917
0.06	4.929737	2.464869	1.643246	1.232434	0.985947
0.09	4.929986	2.464993	1.643329	1.232497	0.985997
0.12	4.930333	2.465166	1.643444	1.232583	0.986067
0.15	4.930773	2.465387	1.643591	1.232693	0.986155

Table 6 Variation of load carrying capacity with respect to B/h_2 and m

m	Load				
	$B/h_2 = 10$	$B/h_2 = 20$	$B/h_2 = 30$	$B/h_2 = 40$	$B/h_2 = 50$
0.25	4.922225	2.461113	1.640742	1.230556	0.984445
0.5	4.927303	2.463652	1.642434	1.231826	0.985461
0.75	4.930333	2.465166	1.643444	1.232583	0.986067
1	4.932285	2.466143	1.644095	1.233071	0.986457
2	4.935803	2.467901	1.645268	1.233951	0.987161

Figures 2 and 3 presents the distribution of non-dimensional load carrying capacity with respect to the magnetization μ^* for various values of L/h_2 and B/h_2 respectively. It is clear that the load carrying capacity increases sharply due to magnetization.

Fig. 2 Variation of load carrying capacity with respect to μ^* and L/h_2

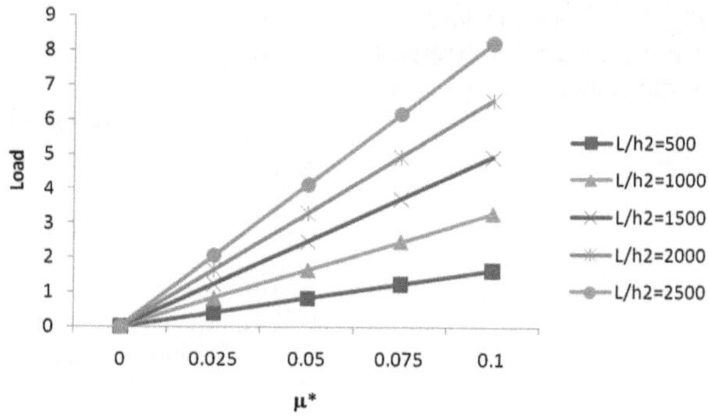

Fig. 3 Variation of load carrying capacity with respect to μ^* and B/h_2

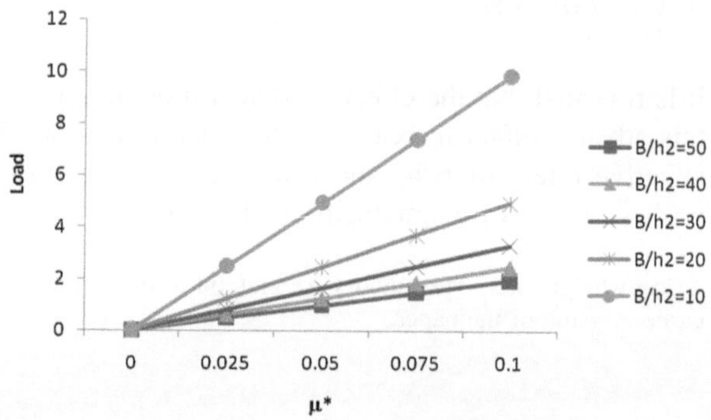

Figure 4 shows the effect of L/h_2 on the load carrying capacity for various values of B/h_2. From this figure it is clear that the load carrying capacity increases considerably due to L/h_2. Lastly, Fig. 5 suggests that the effect of the standard deviation is almost negligible so far as the load capacity is concerned. Furthermore, the aspect ratio has a strong positive effect in the sense that the load capacity increases sharply.

Fig. 4 Variation of load carrying capacity with respect to L/h_2 and B/h_2

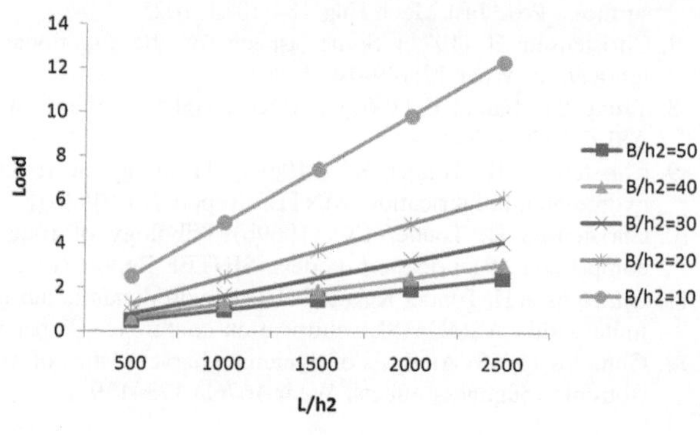

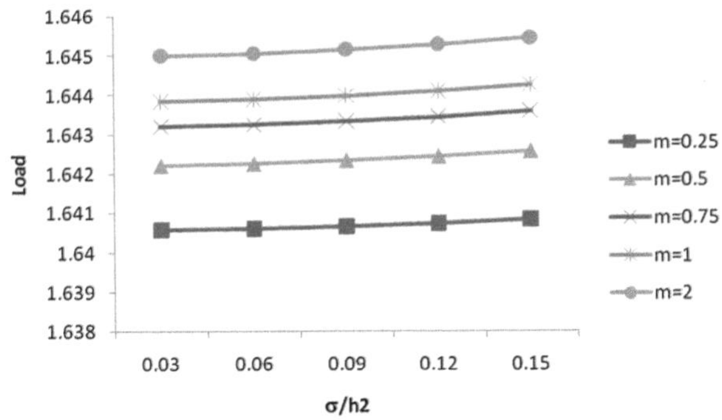

Fig. 5 Variation of load carrying capacity with respect to σ/h2 and m

4 Conclusion

It is revealed that the effect of standard deviation is negligibly adverse. However, this adverse effect increases with larger values of σ/h_2. The results show that the negative effect of B/h_2 and σ/h_2 can be compensated to a larger extent by the positive effect of magnetization and L/h_2.

Acknowledgments The authors are grateful to the referees for their valuable comments on the earlier version of the paper.

References

1. Cameron A (1966) Principles of lubrication. Longmans, London
2. Archibald FR (1956) Load capacity and time relations for squeeze films. J Basic Eng Trans ASME Ser D78:231–245
3. Lord Rayleigh (1918) Notes on the theory of lubrication. Philos Magaz J Sci 53:1
4. Prakash J, Vij SK (1973) Hydrodynamic lubrication of porous slider. J Mech Eng Sci 15:232–234
5. Patel KC, Gupta JL (1983) Hydrodynamic lubrication of a porous slider bearing with slip velocity. Wear 85:309–317
6. Christensen H (1969–1970) Stochastic model for hydrodynamic lubrication of rough surfaces. Proc Inst Mech Eng 184:1013–1025
7. Christensen H (1971) Some aspects of the functional influence of rough surfaces in lubrication. Wear 17:149–162
8. Tzeng ST, Saibel E (1967) Surface roughness effect on slider bearing lubrication. Trans. ASLE 10:334–340
9. Christensen H, Tonder KC (1969a) Tribology of rough surfaces: stochastic models of hydrodynamic lubrication. SINTEF Report No.10/69-18
10. Christensen H, Tonder KC (1969b) Tribology of rough surfaces: parametric study and comparison of lubrication models. SINTEF Report No.22/69-18
11. Christensen H, Tonder KC (1970) The hydrodynamic lubrication of rough bearing surfaces of finite width. ASME-ASLE lubrication conference, Paper no. 70-lub-7
12. Guha SK (1993) Analysis of dynamic characteristics of hydrodynamic journal bearings with isotropic roughness effects. Wear 167(1):173–179

13. Taranga R, Sekhar AS, Majumdar BC (1999) The effect of roughness parameter on the performance of hydrodynamic journal bearings with rough effects. Tribol Int 32:231–236
14. Christensen H, Tonder KC (1971) The hydrodynamic lubrication of rough bearing surfaces of finite width. ASME J Lubr Technol 93:324–330
15. Andharia PI, Gupta JL, Deheri GM (2001) Effect of longitudinal surface roughness on the behavior of squeeze film in a spherical bearings. Int J Appl Mech Eng 6:885–897
16. Agrawal VK (1986) Magnetic fluid based porous inclined slider bearing. Wear 107:133–139
17. Bhat MV, Deheri GM (1991) Porous composite slider bearing lubricated with magnetic fluid. Jpn J Appl Phys 30:2513–2514
18. Bhat MV, Deheri GM (1995) Porous slider bearing with squeeze film formed bya magnetic fluid. Pure Appl Math Sci 39(1–2):39–43
19. Prajapati BL (1995) On certain theoretical studies in hydrodynamic and electro-magneto hydrodynamic lubrication. Ph.D. thesis, S.P. University, Vallabh Vidyanagar
20. Bhat MV (2003) Lubrication with a magnetic fluid. Team Spirit (India) Pvt. Ltd., New Delhi
21. Deheri GM, Andharia PI, Patel RM (2004) Longitudinally rough slider bearings with squeeze film formed by a magnetic fluid. Ind Lubr Tribol 56(3):177–187
22. Basu SK, Sengupta SN, Ahuja BB (2005) Fundamentals of tribology. Prentice-Hall of India Pvt. Ltd., New Delhi

Prediction of Useful Life of Rolling Contact Bearings Using Monte Carlo Simulation Technique

Eshan Singh, G. D. Thakre, P. K. Arya and B. M. Shukla

Abstract Antifriction bearings are the most critical components of mechanical systems. Unpredicted failures of these components incur huge production losses along with undue downtime of the machinery. In context to this, the present paper deals with the prediction of useful life of antifriction bearings using the Monte Carlo Simulation method. The study was carried out on 6,205 deep grove ball bearings. The geometric factors and the stresses influencing the life of these bearing were selected using random number generator. Prediction based on these inputs was modeled using a MATLAB code followed by statistical analysis of the data. The useful life of the bearings, in terms of cycles to failure, was determined. The failure probability corresponding to the L_{10}, L_{50} and L_{90} life of the bearing was investigated. The method utilized and the model developed is very simple and can be efficiently used by practicing engineers for the life predictions of the antifriction bearings.

Keywords Antifriction bearing · Monte carlo simulation · Useful life

1 Introduction

Ball bearings have varied applications ranging from kids' yo-yos to horological and aerospace applications. These bearings, in particular are extremely critical to machinery as it may lead to shutdown of the equipments in case of an unpredicted

E. Singh (✉)
PGRPE—AAT, CSIR-IIP, Dehradun 248005, India
e-mail: eshan@iip.res.in

G. D. Thakre · P. K. Arya · B. M. Shukla
Tribology and Combustion Division, CSIR-IIP, Dehradun 248005, India

H. C. Patel et al. (eds.), *Proceedings of International Conference on Advances in Tribology and Engineering Systems*, Lecture Notes in Mechanical Engineering,
DOI: 10.1007/978-81-322-1656-8_32, © Springer India 2014

failure. Hence prediction of useful life of these bearings becomes highly demanding in industrial applications [1].

Various researchers have put forward new prognostics based solution schemes to investigate bearing failures. Damage mechanics, Wavelet Packet decomposition—HMM and neural networks are a few steps in the direction. Monte Carlo simulation technique too can be used in the prediction of failures and useful life of the bearings [2]. The present paper tries to estimate useful life of an antifriction bearing using a much proven technique of Monte Carlo Simulation that works on random numbers.

The parameters included for the calculation of useful life are the geometric, environmental and operational factors. These have duly been assigned random values based on the trends that they generally follow.

The paper also provides the reliability analysis of the bearings along with statistics involved in the more precise useful life prediction.

1.1 Failure Model

Failure model, based on stress and geometric factors of a ball bearing, was developed considering Paris' criterion for failure due to fatigue. The model considers two elastic bodies in the form of a ball on a concave race as shown in Fig. 1.

The bearing material selected was AISI E52100 and the geometry pertaining to the standard 6,205 bearing was adopted. The expressions used for model development are as given below;

Equivalent Radius:

$$\frac{1}{R} = \frac{1}{R_x} + \frac{1}{R_y} - \frac{1}{R_r} \tag{1}$$

Hertzian contact:

$$a = \sqrt[3]{\frac{3WR}{E}} \tag{2}$$

Fig. 1 Ball on a concave race

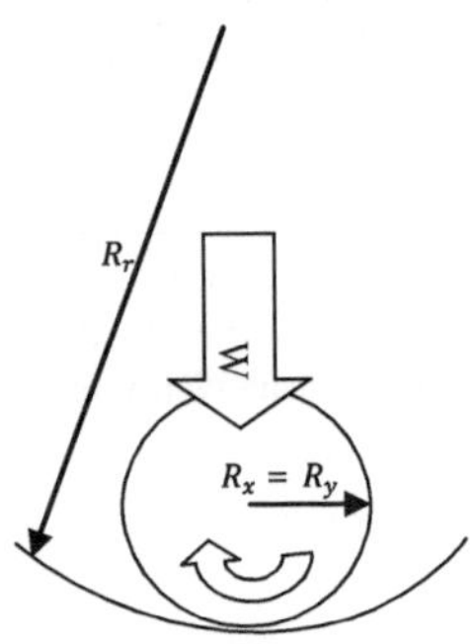

Maximum contact Pressure:

$$P_{\max} = \frac{3W}{2\pi a^2} \tag{3}$$

Maximum principal stress:

$$\sigma_{\max} = \frac{2P}{3} \tag{4}$$

Critical crack length:

$$a_c = \frac{1}{\pi}\left(\frac{K1_c}{\propto \sigma_{\max}}\right)^2 \tag{5}$$

Cycles to failure:

$$N_f = \frac{(a_0)^{1-m/2}}{C(\frac{m}{2}-1)\propto^m \pi^{m/2} \sigma_{\max}^m} \tag{6}$$

1.2 Solution Scheme

The solution process used involved the determination of maximum pressure within the contact given by Eq. (3). The max. Contact pressure is the function of equivalent radius of contact and the Hertzian contact area and is determined using Eqs. (1) and (2). Subsequently the maximum stress put forth by the bearing (Eq. 4) is calculated, which gives information about the number of cycles it would take for the bearing to fail. Geometric factors are generated randomly and reduced radius is calculated using Eq. (1). The pressure at contacting area determines the maximum stress generated in the bearings, which in turn aid in calculating the Critical crack length and subsequently the fatigue life of bearing (Eq. 6).

1.3 Results and Discussions

Investigations were made for 10,000 virtual bearings. Useful life of these bearings in terms of cycles to failure is determined and the most widely used representatives of failure life i.e. the L_{10}, L_{50} and L_{90} life determined are given in Table 1.

Figure 2 shows the co-relation between the Median Rank and the cycles. The trend followed typically represent that the bearing lives follow the Weibull distribution.

Figure 3 shows the survival probability for the 10,000 virtual bearings using the Monte Carlo simulation technique. The nature of the curve is in good agreement with the trends for the survival probability.

Table 1 Calculation of useful life

Life	Probability	Cycles	Reliability
L10	0.1	1560114439	0.9
L50	0.5	1461726621	0.5
L90	0.9	1319696281	0.1

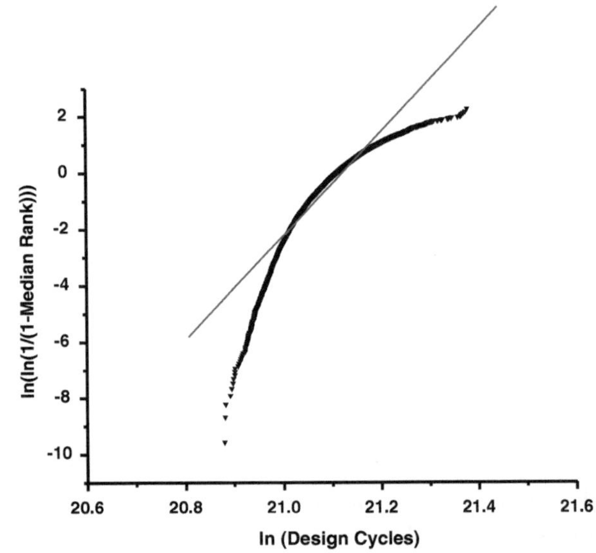

Fig. 2 Median Rank and Cycle co-relation

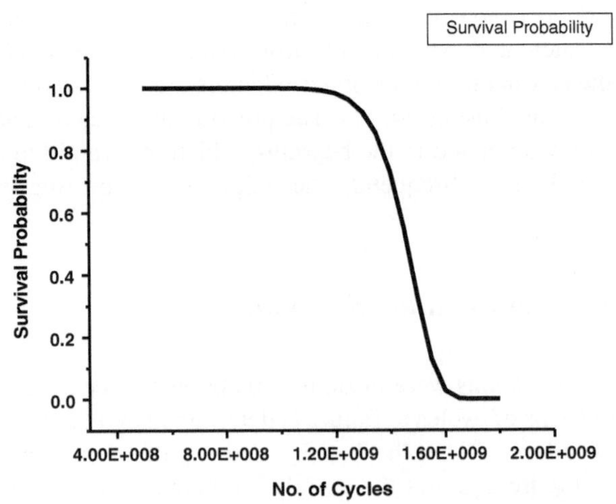

Fig. 3 Reliability Plot

The Weibull plot representing the cumulative failure probability as a function of cycles to failures is given in Fig. 4. From the Weibull plot it is determined that the present study on bearings has a shape factor (b) of 18.43 and the characteristic life (a) of 1491086592 cycles. A shape factor greater than 1 signifies increasing

Fig. 4 Weibull plot

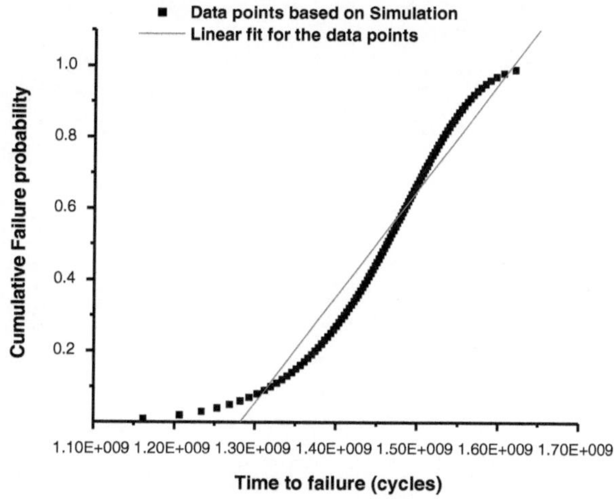

failure rate. An increase in failures is quite obvious for the bearings with increasing number of operating cycles. The characteristic life on the contrary represents the number of cycles for 63.2 % of bearing failures. Using these two values the MTTF of the bearings is given by;

$$MTTF = a * \gamma(1 + \frac{1}{b}) \tag{7}$$

The MTTF in the present study is reported to be 1.4485×10^9 cycles.

2 Conclusion

An estimation of the useful life prediction of bearings has been proposed in this work. The developed model can be used for useful life calculation of any bearing with changes in operating conditions in the MATLAB code. The L_{10}, L_{50} and L_{90} lives of the bearings have been calculated to be 1.5E9, 1.4E9 and 1.3E9 respectively. The reliability plots have been prepared and MTTF has been calculated to be 1.4E9 cycles.

References

1. Gwidon WS, Andrew WB (2005) Engineering tribology, 3rd edn. Elsevier Butterworth-Heinemann
2. Sayyad ZQ, Anwar KS, Tasneem P, Abul FMA (2011) Using Monte Carlo Simulation for prediction of tool life, Application Monte Carlo Method in Science and Engineering, Prof. Shaul Mondechai, ISBN: 978-953-307-691-1 InTech

Side Edge Effect on Elastic Contact Stress and Deformation

W. Wang, P. L. Wong and Z. M. Zhang

Abstract A convenient and fast solution for elastic quarter space contact problems is developed based on matrix representation. This new solution makes use of discretization into the form of matrices which conveniently facilitate the overlapping solution process for the elastic quarter space. Verification of the present method was accomplished by comparison with the existing quarter space results.

Keywords Contact mechanics · Solid mechanics · Elastic quarter space · Explicit solution · Matrix formulation

1 Introduction

Elastic half space model (half infinite body) has been widely accepted for the calculation of contact stress and deformation in practical mechanical systems such as the contact of rail-wheels, cam-followers, gears, bearings etc. In these practical contacts, the assumption of half infinite body model is not always satisfied. For example, the contact of gears and roller bearings, the length of the gear tooth and bearing roller are finite and the effect of free sides cannot be ignored. For these contact cases, elastic quarter space model is more accurate and preferred.

Figure 1 is a typical elastic quarter space problem. The free side surface is close to the loading or contact region. Thus, the free side surface (or edge) effects cannot be ignored. Hetenyi [1] is the first to investigate theoretically the elastic quarter

W. Wang (✉) · Z. M. Zhang
Department of Mechanical Engineering and Automation, Shanghai University, Shanghai 200072, China
e-mail: mewwang@shu.edu.cn

P. L. Wong
Department of Mechanical and Biomedical Engineering, City University of Hong Kong, Hong Kong, China

H. C. Patel et al. (eds.), *Proceedings of International Conference on Advances in Tribology and Engineering Systems*, Lecture Notes in Mechanical Engineering, DOI: 10.1007/978-81-322-1656-8_33, © Springer India 2014

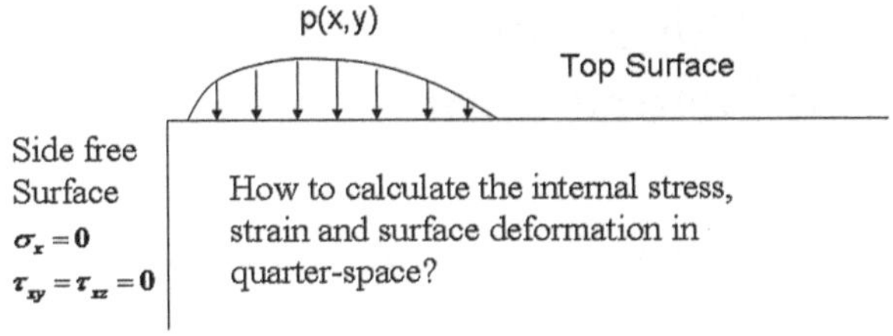

Fig. 1 Typical elastic quarter space problem

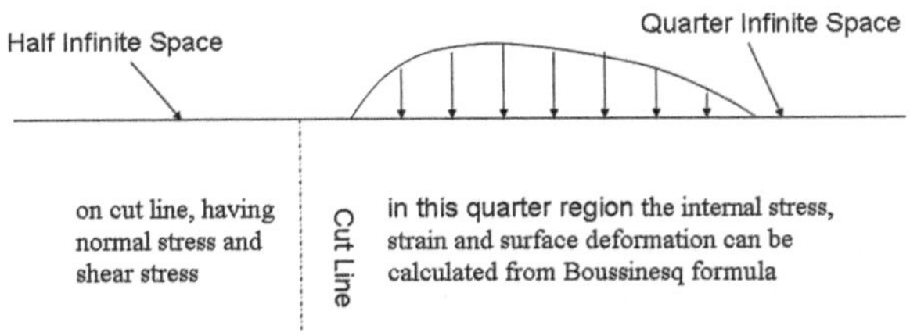

Fig. 2 Quarter space as cut half of half infinite space

space problem with an ingenious method of reflection and iteration of the reflected solution to obtain results of acceptable accuracy. Hetenyi infers the solution of the quarter space problem starting with a half infinite space. From the consideration of geometric structure, the quarter space can be taken as the cut half of a half infinite space as shown in Fig. 2. The stress and strain in the cut line can be calculated with Boussinesq formula. The difference of this cut quarter from an infinite model, as shown in Fig. 2, and a real quarter space is that the cut surface has normal and shear stresses, but the free surface of an elastic quarter space has none. Hence, the solution of an elastic quarter problem can be obtained if the conditions of zero normal and shear stresses on the cut surface are achieved.

Hetenyi proposed an ingenious reflection method with symmetric loading to eliminate the shear stresses on the cut line. Through iteratively solving the horizontal half infinite space and the vertical half infinite space, the normal stresses can also be eliminated. The process of Hetenyi's method is illustrated in Figs. 3 and 4. During the iteration of H style and V style half infinite space solutions with symmetric loadings, the normal loading on vertical cut line $\sigma_n(\eta, \zeta)$ is balanced

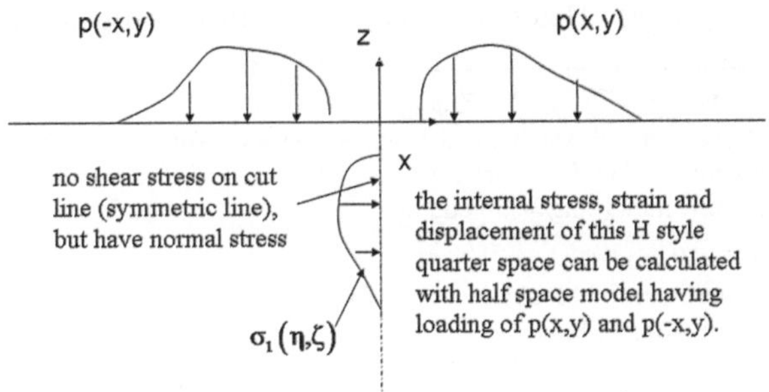

Fig. 3 Horizontal (H style) half infinite space with symmetric vertical loading

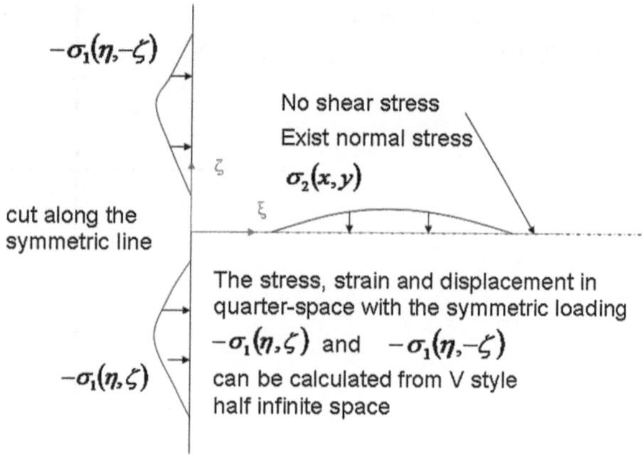

Fig. 4 Vertical (V style) half infinite space with symmetric horizontal loading

automatically. As to the excess loading on horizontal cut line, $\sigma_n(x, y)$ will approach to zero as iteration continues. This iteration convergence was proved by Hetenyi, because the Poisson's ratio is less than 0.5 and normally 10 iterations are enough for accurate results. The final quarter space solution is achieved from all the iteration of H and V style half infinite space solutions.

Following Hetenyi's work, Sneddon [2] derived the theoretical limit of the H style and V style iteration with two integrated equations as the final loading applied on horizontal and vertical half infinite space. The solution to this two coupled integration equations is still complex. Keer and his co-workers [3–6] put efforts to find methods to solve the integral equations. Decoupling of the two integral equations and special process of singularity were successfully achieved by Keer. Fourier transform method and direct numerical solving method were also used by Keer et al. to solve the integration equations. Through the work of Keer et al., a very strict and feasible solving method for quarter space problem was established successfully. However, the practical implement of Keer's method is still very complex.

2 Explicit Matrix Solution

It is known from Hetenyi [1] that the superposition of a horizontal and a vertical elastic half space can yield the solution of the elastic quarter space. For the convenience of description, the two bonding surfaces of the quarter space will be termed as 'horizontal' and 'vertical' respectively, with the first located at $z = 0$ and the second at $x = 0$, as shown in Fig. 5.

Let $\mathbf{H}$ and its x-symmetrical $\overline{\mathbf{H}}$ denote the normal loads on the horizontal half space, and $\mathbf{V}$ and its z-symmetrical $\overline{\mathbf{V}}$ be those on the vertical. $\mathbf{H}$ and $\overline{\mathbf{H}}$ form the

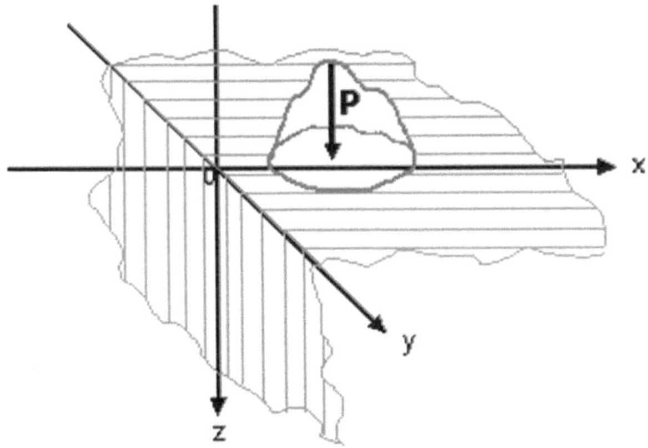

Fig. 5 Coordinate definition of quarter space

normal stress vector $\mathbf{T}_v = \mathbf{A} \cdot \mathbf{H}$ in the $x = 0$ vertical plane. Similarly, $\mathbf{V}$ and $\overline{\mathbf{V}}$ produce the normal stress vector $\mathbf{T}_h = \mathbf{B} \cdot \mathbf{V}$ in the horizontal half space.

The resulting pressure vector on the horizontal plane of the quarter space should be equal to the applied load vector $\mathbf{P}$, while that on the vertical plane should be zero. Thus,

$$\mathbf{H} - \mathbf{T}_h = \mathbf{P}; \mathbf{V} - \mathbf{T}_v = \mathbf{0} \text{ or}$$

$$\mathbf{H} - \mathbf{B} \cdot \mathbf{V} = \mathbf{P}; \mathbf{V} - \mathbf{A} \cdot \mathbf{H} = \mathbf{0}$$

These simultaneous equations are easily decoupled as

$$\mathbf{H} - \mathbf{B} \cdot (\mathbf{A} \cdot \mathbf{H}) = \mathbf{P}; \ \mathbf{V} - \mathbf{A} \cdot (\mathbf{B} \cdot \mathbf{V} + \mathbf{P}) = \mathbf{0}$$

The solution is obtained as

$$\mathbf{H} = (\mathbf{I} - \mathbf{B} \cdot \mathbf{A})^{-1} \cdot \mathbf{P} \tag{1}$$

$$\mathbf{V} = \mathbf{A} \cdot (\mathbf{I} - \mathbf{B} \cdot \mathbf{A})^{-1} \cdot \mathbf{P} \tag{2}$$

Therefore, the limits of Hetenyi-styled iteration are explicitly in Eqs. (1) and (2) in matrix form. The explicit stress and deformation solutions to elastic quarter space can be written as,

$$\boldsymbol{\sigma} = (\mathbf{C}_H + \mathbf{C}_H^*) \cdot \mathbf{H} + (\mathbf{C}_V + \mathbf{C}_V^*) \cdot \mathbf{V} = \mathbf{C}_\sigma \cdot \mathbf{P} \tag{3}$$

$$\boldsymbol{\delta} = (\mathbf{D}_H + \mathbf{D}_H^*) \cdot \mathbf{H} + (\mathbf{D}_V + \mathbf{D}_V^*) \cdot \mathbf{V} = \mathbf{D}_\delta \cdot \mathbf{P} \tag{4}$$

All the matrices in Eqs. 1–4, such as $\mathbf{A}$, $\mathbf{B}$, $\mathbf{C}_\sigma$ and $\mathbf{D}_\delta$ can be explicitly expressed in the formula of Bussinessq for point loading or Love's formula for uniform loading on a small patch.

3 Validation of Explicit Matrix Solution

A case is selected from Hanson and Keer [5], and the results obtained by the present solution are compared with those of Hanson and Keer. The very detailed loading condition and abundance of results provided by Hanson and Keer [5] are very suitable for verification of the present method. A half hemispherical loading is applied to the top surface of an elastic quarter-space, with the spherical centre located at the edge. The load can be described by the following formulae:

$$\begin{aligned} &p(x,y) = p_0 \frac{\sqrt{a^2 - x^2 - y^2}}{a}, \quad \sqrt{x^2 + y^2} \le a \\ &p(x,y) = 0, \quad \sqrt{x^2 + y^2} > a \end{aligned} \tag{5}$$

The stress distributions on the *x–z* plane were obtained with the present matrix formulation and are plotted in contour together with the results of Hanson and Keer [5] in Fig. 6. Comparing the two plots, it shows that the present results are in good correlation with Hanson and Keer's. The matrix solution is thus validated.

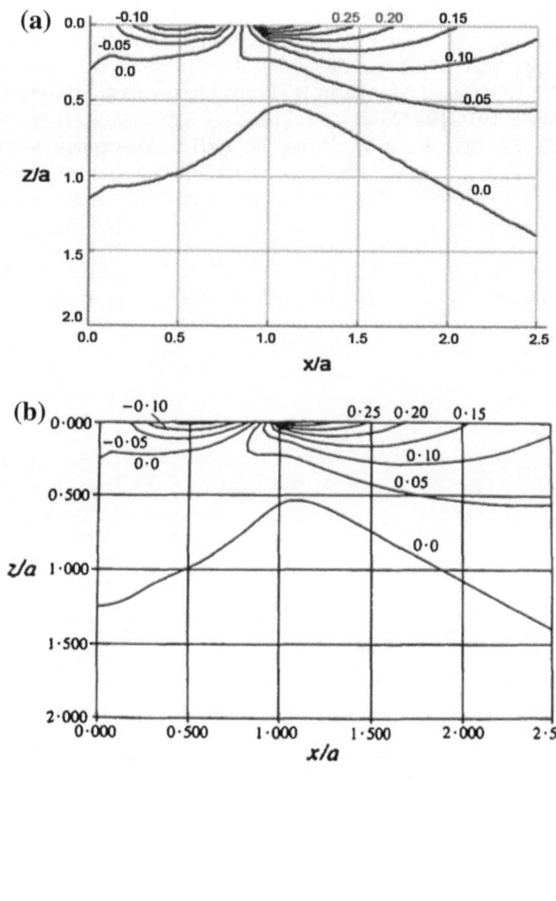

Fig. 6 Comparison of stress of σxx/p0 in the (x, z)-plane, **a** Results of present method, **b** Results of Hanson and Keer [5]

4 Conclusions

The current authors have proposed a new explicit solution for the elastic quarter-space problem using matrix formulation [7]. The good correlation with existing results from other investigators confirms the reliability of the proposed solution. The detail of this matrix solution and the advantages are introduced. The effects of the side free surface on the contact stress and deformation results are investigated. This work is much beneficial to solving practical mechanical contact problems in terms of higher accuracy and efficiency.

References

1. Hetenyi M (1970) A general solution for the elastic quarter space. J Appl Mech 70–76 (ASME)
2. Sneddon IN (1971) Fourier transformation solution of the quarter plane problem in elasticity, File PSR-99/6. Applied Mathematics Research Group, North Carolina State University
3. Keer LM, Lee JC, Mura T (1983) Hetenyi'selastic quarter space problem revisited. Int J Solid Struct 19(6):497–508
4. Keer LM, Lee JC, Mura T (1984) A contact problem for the elastic quarter space. Int J Solid Struct 20(5):513–524
5. Hanson MT, Keer LM (1990) A simplified analysis for an elastic quarter-space. Q J Mech Appl Math 43(4):561–587
6. Yu CC, Keer LM, Moran B (1996) Elastic-plastic rolling-sliding contact on a quarter space. Wear 191:219–225
7. Zhang ZM, Wang W, Wong PL (2013) An explicit solution for the elastic quarter-space problem in matrix formulation. Int J Solids Struct 50:976–980

Smart Journal Bearing Using Giant Magnetostrictive Actuators

Z. M. Fang, Z. Li and W. Wang

Abstract A smart hydrodynamic journal bearing with giant magnetostrictive actuators (GMA) is introduced, including the bearing design and potential performance improvement. The static and dynamic performance of this smart bearing prototype are evaluated experimentally. It is proved that the GMA smart bearing have good auto-centering and vibration suppressing capabilities.

Keywords Smart journal bearing · Hydrodynamic · Giant magnetostrictive actuator · Auto-centering · Vibration suppression

1 Introduction

In some rotary machines such as machine tools, turbines and compressors, hydrodynamic journal bearings are widely used due to the merits of high loading capacity and long service life. However, their applications are also limited by the relatively low high-speed performance and position precision. To improve these performances, many innovative ideas have been proposed. Most of the practical designs are developed using hydraulic systems to adjust the bearing clearance. Based on this concept, an active journal bearing with adjustable supporting flexible sleeve was proposed by Sun and Krodkiewski [1]. Also, similar idea was employed for active tilting-pad journal bearings by Wu and Queiroz [2]. However, the major problem of these designs is the very low response time of a hydraulic system. In contrast, active magnetic bearings [3] have a high response time and high-speed capacity, but their low loading capacity and high cost limit their practical application. On the other hand, bearings can be controlled to reduce their vibration by

Z. M. Fang · Z. Li · W. Wang (✉)
Department of Mechanical Engineering and Automation, Shanghai University, Shanghai 200072, China
e-mail: mewwang@shu.edu.cn

H. C. Patel et al. (eds.), *Proceedings of International Conference on Advances in Tribology and Engineering Systems*, Lecture Notes in Mechanical Engineering,
DOI: 10.1007/978-81-322-1656-8_34,

piezoelectric actuators [4]. However, the magnitude of output force and the displacement amplitude of the piezoelectric actuators are limited. Theoretical investigations for rotor dynamic of controllable journal bearings have been performed. Rho and Kim [5] studied the ability of an actively controlled hydrodynamic journal bearing to suppress whirl instability and reduce the unbalance response of a rotor-bearing system. Nikolakopoulos and Papadopoulos [6] investigated the stability of an active journal bearing with a controllable electro-rheological lubricant and the application of magnetorheological fluid for journal bearing control has been studied [7].

Compared with piezoelectric materials, giant magnetostrictive materials (GMM) have higher energy density and larger strain [8]. The actuators developed from GMM have a large loading capacity and large output displacement amplitude, which allow them to be successfully applied in industries such as high-power sonar system [9]. The applications of GMM in precision positioning were also reported [10]. Moreover, it was also proposed to use GMM for force sensing purposes [11].

In recent years, we have tried to use GMM for the journal bearing control and a prototype of an active controlled journal bearing using GMM actuators (named a smart GMA journal bearing) has been successfully built and tested [12, 13]. Figure 1 is a schematic sketch of the smart GMA journal bearing.

As illustrated in Fig. 1, the cylindrical journal bearing is excited by two GMAs, respectively, in vertical and horizontal directions. Springs are used to add preload on the GMAs to attain better performance. Two eddy current sensors are installed for providing feedback signals of the instantaneous rotor position. A GMM rod, which is 10 mm in diameter and 30 mm in length, is installed in the center of the GMA. The coil is composed of two windings of enamel insulated wire, supplying for the generation of the pre-excursion magnetic field and dynamic magnetic excitation field, respectively.

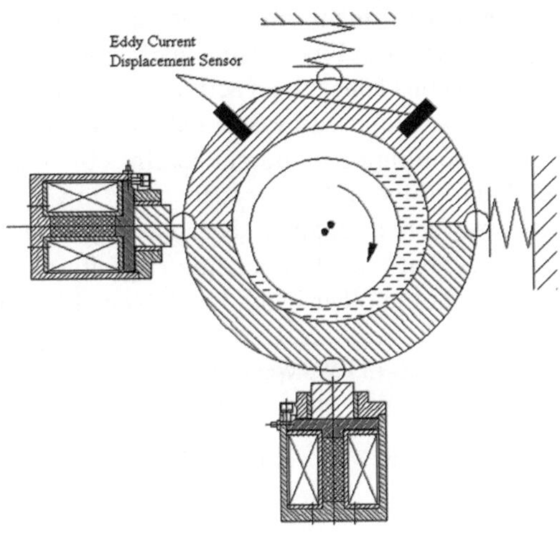

Fig. 1 Sketch of a smart GMA journal bearing

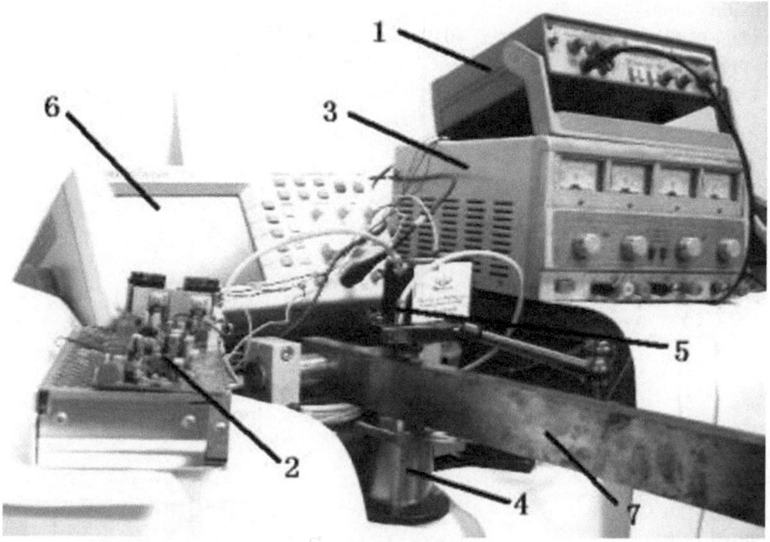

Fig.2 Test-rig for GMA performance *1* Signal generator *2* Power amplifier modular *3* Constant current source *4* GMA *5* Eddy displacement sensor *6* Ondoscope *7* Lever device *8* Load

2 Performance of GMA

A lever test-rig as shown in Fig. 2 is developed to measure the static and dynamic displacement output of the designed GMA. The GMA displacement output is measured by an eddy current sensor. The experimental results show that the maximum static displacement can reach 30 μm as shown in Fig. 3, which is the same order of magnitude as that of the oil film clearance of a typical small journal bearing with diameter under 100 mm. Dynamic tests of the GMA were also performed with the excitation current at different frequencies. The results also show that even an excitation frequency of 2,000 Hz, the GMA output displacement amplitude can reach 20 μm. The static and dynamic test results prove that the response frequency and vibration amplitude of the GMA can meet the requirements for use in vibration suppression of a journal bearing rotor system.

3 Performance of Smart Journal Bearing

A rotational test-rig is developed for the evaluation of GMA smart journal bearing. Figure 4 is the photo of the test-rig. The shaft is driven by an electric spindle. One side of the shaft is supported by a ball bearing and another side is supported by the smart journal bearing. The test loading is applied at the middle of the shaft. The dimensions of smart journal bearing are 20 mm in diameter and 20 mm in length.

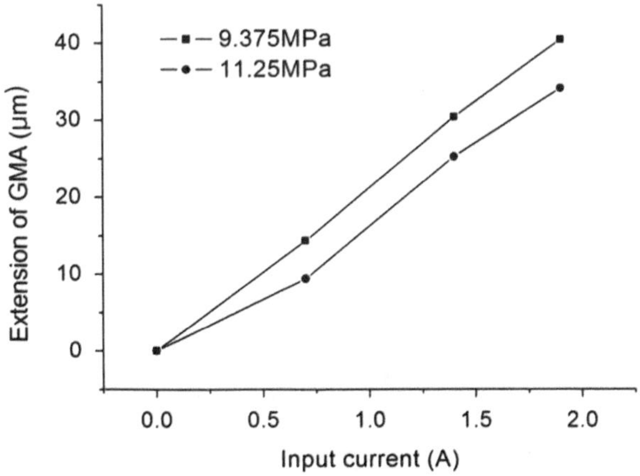

Fig. 3 Static output displacement of GMA under different pre-loading pressure

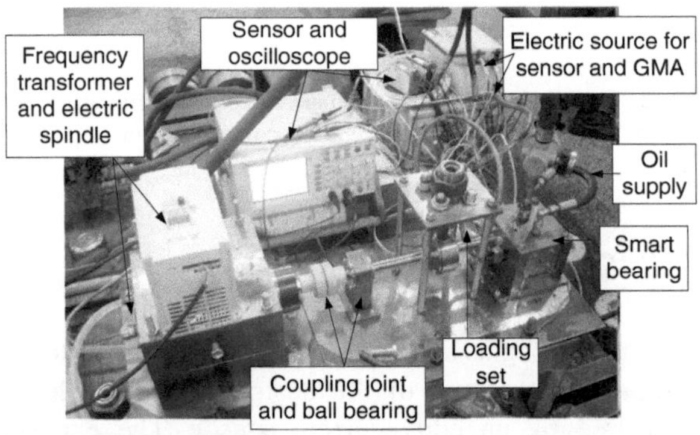

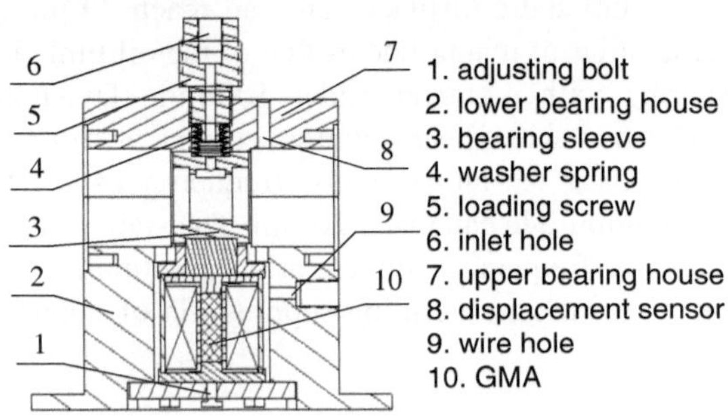

Fig. 4 Test-rig of GMA smart journal bearing

The radial clearance of the bearing is 20 μm. The structure of the smart bearing is also shown in Fig. 4.

The performance of the smart journal bearing has been evaluated on this test-rig. Table 1 illustrates the auto-centering capability of the GMA smart journal bearing with different loadings. Through increase or decrease the excitation current to the GMAs, the shaft center position can be almost adjusted to the initial

Table 1 Auto-centering of smart bearing with different loadings at speed of $\omega = 12$ Hz

Loading	0 kg		5 kg		10 kg	
Position of shaft center	x	y	x	y	x	y
Initial current (A)	1	1	1	1	1	1
Initial position (μm)	0.26	0.12	3.7	−4.4	3.9	−8.5
Excitation current to GMA (A)	–	–	−0.19	0.22	−0.20	0.43
Shaft position after controlled (μm)	–	–	0.24	0.07	0.32	0.98

position of loading zero. The position error after auto-centering by GMA is less than 1 μm.

The vibration suppression capability of the smart journal bearing also was investigated. A sinusoidal excitation current with amplitude of 1.5 A and same frequency of the rotation speed is exerted to the GMA. The phase difference is 180°. Figure 5 is the comparison of shaft locus with GMA control or without control. It can be seen that the locus is decreased, but it is not very obvious. Because the vibration locus is not only the unbalance vibration, it is more obvious when check the vibration FFT spectrum as in Fig. 6.

As shown in Fig. 6, the unbalance vibration at rotation speed decrease very obviously. The vibration of double rotation frequency decrease very less. If take out the unbalance vibration locus to compare, as shown in Fig. 7. The unbalance vibration is suppressed successfully.

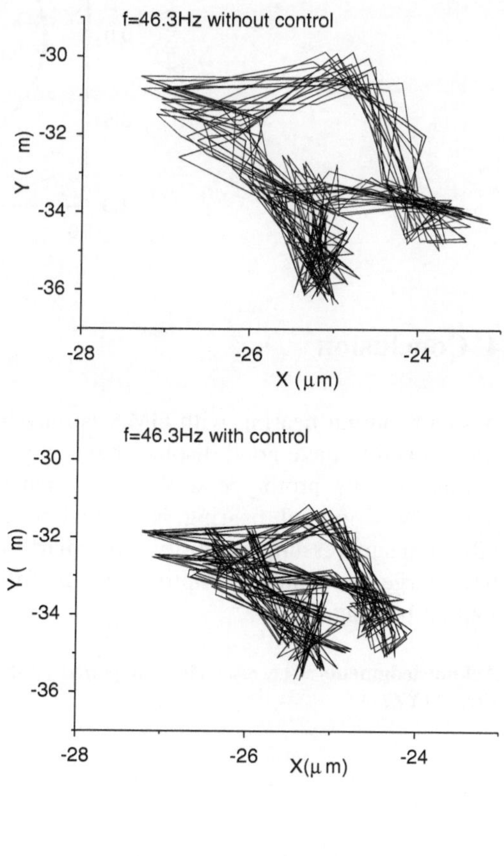

Fig. 5 Comparison of shaft locus with and without GMA control

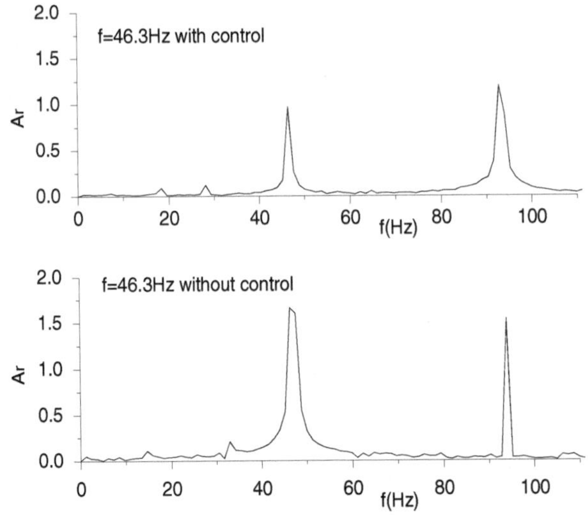

Fig. 6 Comparison of locus FFT spectrum with and without GMA control

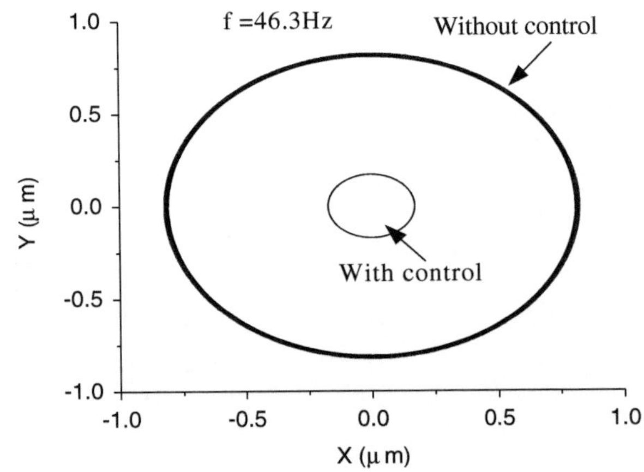

Fig. 7 Suppression of unbalance vibration with GMA smart journal bearing

4 Conclusion

A smart journal bearing with GMA is introduced. The performance of the GMA was proved to have good displacement output and frequency response. The smart journal bearing prototype was also evaluated experimentally. The newly developed GMA journal bearing is proved to have very good auto-centering and vibration suppressing capability. The smart journal bearing based on GMA should have large potential to improve the performance of traditional hydrodynamic journal bearing.

Acknowledgments The research is supported by the Shanghai Municipal Education Commission (11YZ21).

References

1. Sun L, Krodkiewski JM (2000) Experimental investigation of dynamic properties of an active journal bearing. J Sound Vib 230(5):1103–1117
2. Wu A, Cai Z, Queiroz MS (2007) Model-based control of active tilting-pad bearings. IEEE/ASME Trans Mechatron 12(6):689–695
3. Lum KY, Coppola VTD, Bernstein DS (1996) Adaptive autocentering control for an active magnetic bearing supporting a rotor with unknown mass imbalance. IEEE Trans Control Syst Technol 4(5):587–597
4. Palazzolo AB, Jagannathan S, Kascak A, Montague GT, Kiraly LJ (1993) Hybrid active vibration control of rotor bearing systems using piezoelectric actuators. J Vib Acoust Stress Reliab Des 115:111–119
5. Rho BH, Kim KW (2002) A study of the dynamic characteristics of synchronously controlled hydrodynamic journal bearings. Tribol Int 35:339–345
6. Nikolakopoulos PG, Papadopoulos CA (1998) Controllable high speed journal bearings, lubricated with electro-rheological fluids—an analytical and experimental approach. Tribol Int 31(5):225–234
7. Bompos DA, Nikolakopoulos PG (2011) CFD simulation of magnetorheological fluid journal bearings. Simul Model Pract Theory 19(4):1035–1060
8. Chung KS (2007) Actuator—application of giant magnetostrictive material. B.Sc. Final Year Project Report, MEEM Department, City University of Hong Kong, 2007
9. Dhilsha R, Rajeshwari PM, Rajendran V (2005) Advanced magnetostrictive materials for sonar applications. Def Sci J 55(1):13–20
10. Yang B, Meng G, Feng Z, Yang D (2011) Giant magnetostrictive clamping mechanism for heavy-load and precise positioning linear inchworm motors. Mechatronics 21:92–99
11. Jia Z, Liu H, Wang F, Ge C (2011) Research on a novel force sensor based on giant magnetostrictive material and its model. J Alloys Compd 509:1760–1767
12. Lau HY, Liu KP, Wang W, Wong PL (2009) Feasibility of using GMM based actuators in active control of journal bearing system. Proceeding of world congress on engineering, 2009, London, July 2009
13. Ma KD, Wu C, Fu YQ, Wang W, Li BF (2009) Experimental investigation of giant magnetostrictive actuators on actively controlled oil film bearing. Lubr Eng 1 (in Chinese)

Part V
Nano Technology/Tribology

Experimental Evaluation of the Tribological Properties of CuO Nano-Lubricants at Elevated Temperatures

Chacko Preno Koshy, P. K. Rajendrakumar and Manu V. Thottackkad

Abstract Tribofilms formed from Copper oxide (CuO) nanoparticles in the nano-lubricant under boundary/thin film lubrication conditions have been studied. The tribological performance is linked to the tribofilm properties and consequently to the lubricating conditions. In the present study, analysis of mineral oil (paraffin oil) and vegetable oils (coconut oil and mustard oil) as a nano-lubricant has been carried out in the perspective of its tribological behaviour using a pin-on-disc tribometer. Small quantities of CuO nanoparticles are added to the base lubricating oil and ultrasonic shaking is carried out on the mixture, to formulate the corresponding nano-lubricant. The variation of its friction-reduction and anti-wear properties have been analysed at ambient and elevated temperatures. Viscosity, flash-point and fire-point of the base lubricants and the nano-lubricants have been measured. From the dispersion analysis it is seen that the nano-lubricant is not suitable enough for long stationary applications. Surface structure of the worn surfaces obtained by Atomic Force Microscope (AFM) and Scanning Electron Microscope (SEM) shows that the average roughness value decreases drastically when nano-lubricants are used rather than the base lubricants. With the current interest in lubricants from renewable sources, studies on vegetable oil based nano-lubricants and means of predicting their performance are given predominance in this work.

Keywords Thin/boundary lubrication · Nano-lubricants · Vegetable oils

---

C. P. Koshy (✉) · P. K. Rajendrakumar · M. V. Thottackkad
Department of Mechanical Engineering, National Institute of Technology,
Calicut, Kerala 673601, India
e-mail: chackopreno@yahoo.co.in

H. C. Patel et al. (eds.), *Proceedings of International Conference on Advances in Tribology and Engineering Systems*, Lecture Notes in Mechanical Engineering,
DOI: 10.1007/978-81-322-1656-8_35,

1 Introduction

Lubrication is an art that has been practiced for thousands of years from the early days of human civilization. There has been growing concern on the use of mineral oils as lubricants because of the worldwide interest in environmental issues [1]. Nowadays mineral oil which is derived from petroleum and synthetic oil from petrochemicals are mainly used as base oil in lubricants. Recent environmental awareness has put aside mineral oils which has been the most widely used lubricant base fluid, and are being replaced by biodegradable fluids like vegetable oils and other synthetic fluids for new lubricant formulation. Vegetable oils are better options for replacing mineral oils, as they are wholly biodegradable, non-toxic, and 'Generally Regarded as Safe' (GRAS) products [2]. Today, vegetable oils are gaining popularity because of the significant advantages to environment as the vegetable oil is a renewability resource and possess adequate performance in a variety of applications. Indeed, vegetable oils possess many of the desirable properties of lubricants, such as good contact lubrication, high viscosity index and high flash-point. Bio-oils provide intrinsically strong lubricating films and as such possess higher lubrication properties than conventional mineral oils as metal working fluids [3]. Vegetable oils are particularly effective as boundary lubricants as the high polarity of the entire base oil allows strong interactions with the lubricated surfaces. However, the main weakness is in their relatively poor oxidation stability [4]. Also at extreme loads and relatively high temperatures, vegetable oil based lubricants become significantly less effective [5].

Basically, vegetable oil is made of two simple building blocks which are glycerol and fatty acids. Actually, there is only one type of glycerol but the fatty acids can vary widely in their structure. Glycerol consists of three alcohol groups where fatty acids can be attached and the resulting products are termed monoglycerides (one fatty acid), diglycerides (two fatty acids) or triglycerides (three fatty acids). Coconut oil finds mention as a lubricant among other widely used vegetable oils such as soybean oil, olive oil, rapeseed oil, etc. during the pre-mineral oil era. Coconut oil is being widely used as a two stroke engine lubricant in autorikshaws and scooters in Kerala, a southern state of India. There are many advantages obtained by the use of coconut oil including improved mileage, better pick up (acceleration), smoother engine operation, less smoke etc. [6]. Coconut oil shows good lubricant properties such as high viscosity index, good lubricity, high flash point and low evaporative loss. One of the drawbacks of coconut oil when used as a lubricant is that the wear rate in sliding in its presence is higher compared to that of mineral oil based lubricants [7]. Coconut oil has very high pour point because of the predominantly saturated nature of its fatty acid constituents precluding its use as base oil for lubricants in temperate and cold climatic conditions [8]. Table 1 shows the gas chromatography mass spectroscopy (GCMS) results, showing the constituents of coconut oil [6].

Tribology is based on the study of lubrication, friction and wear at macro level. Nanotribology deals with the variation in properties of lubricants and contacting

Table 1 Fatty acid constitutions of coconut oil

Components	Fraction (%)
Lauric acid ($CH_3(CH_2)_{10}COOH$)	51.0
Myristic acid ($CH_3(CH_2)_{12}COOH$)	18.5
Caprilic acid ($CH_3(CH_2)_6COOH$)	9.5
Palmitic acid ($CH_3(CH_2)_{14}COOH$)	7.5
Oleic acid ($CH_3(CH_2)_7CH{=}CH(CH_2)_7COOH$)	5.0
Capric acid ($CH_3(CH_2)_8COOH$)	4.5
Stearic acid ($CH_3(CH_2)_{16}COOH$)	3.0
Linoleic acid ($CH_3(CH_2)_4CH{=}CHCH_2CH{=}CH(CH_2)_7COOH$)	1.0

surfaces at micro/nano scales. The term "nano-lubricant" in the present context represents a lubricant which is used for nano applications or a lubricant obtained by the addition of nanoparticles. There have been several investigations on the tribological properties of mineral oil based lubricants with different nanoparticles added. Many investigators have reported that the addition of nano particles to the lubricants is an effective method to reduce friction and wear [9–13]. The friction-reduction and anti-wear behaviours are dependent on the characteristics of nano-particles such as size, shape and concentration. The sizes of nanoparticles used were mostly in the range of 20–150 nm. Experimental studies were carried out on the variation in properties of SAE-40 and coconut oil as base lubricants and boric acid solid lubricant of 50 nm particle size as suspensions [14]. The results showed a reduction in temperature at the contacting points when nanoparticles were used.

In the present study, property variations of different lubricating oils obtained by the addition of nanoparticles on weight percentage basis have been carried out at ambient temperature and at elevated temperatures (up to 100 °C). The oils selected as base lubricant for the analysis have been coconut oil, mustard oil and paraffin oil. Due to their eco-friendly characteristics, vegetable oils are given due importance as base oils in the present study. The tribological analysis of the base oils are carried out initially, using a pin-on-disc tribometer. Nanoparticle selected for addition to the base lubricants has been copper oxide (CuO), in its pure form. Table 2 shows the properties of CuO nanoparticles. Along with the tribological studies, thermo-physical analysis, dispersion analysis and surface morphology analysis have also been conducted. Table 3 shows some typical properties of base lubricants selected for this experimental study.

Table 2 Nanoparticle properties

Nanoparticle	Morphology	Purity (%)	Size (nm)
CuO	Nearly spherical	99.0	20–150

Table 3 Typical properties of base lubricants

Properties	Coconut oil	Mustard oil	Paraffin oil
Appearance	Clear and bright	Dark brown	Clear and bright
Specific gravity	0.917	0.926	0.859
Kinematic viscosity, at 40 °C (cSt)	22	82	95
Kinematic viscosity, at 100 °C (cSt)	9.5	10	38
Viscosity index	202	167	112
Pour point, °C	–	–	−12
Flash point, COC, °C	298	–	240
Sulphur, ppm	–	–	<50
Aromatics, %wt.	–	–	0.5
Saturates, %wt.	–	–	99.5
Acid value	4.3	–	–
Saponification value	250	–	–
Iodine value	13	–	–

2 Experimental Procedure

2.1 Preparation of Nano Lubricant

CuO nanoparticles are suspended in coconut oil, mustard oil and paraffin oil respectively in different concentrations to formulate the nano-lubricant which is applied in small quantities at the sliding interface between aluminium alloy (Al 98 %, Si 2 %) pin and steel disc on a pin-on-disc tribometer. 500 ml of base oil is used for making each nano-lubricant. Density of the oil is measured on weight to volume basis using a 25 ml jar and a precision balance. Commercially available nanoparticles in the range of 20–150 nm supplied by M/s Sigma Aldrich Ltd., USA, were used in the experimental study, for which the size and true density have been provided by the suppliers. The size distributions of the particles are verified using a Scanning Electron Microscope (SEM). These nanoparticles are added to the oil on weight percentage basis, such as 0.1, 0.2 %, etc. The oil is then agitated using ultrasonic shaker for 40 min to ensure uniform dispersion and good suspension stability. Temperature is maintained at 30 °C.

2.2 Tribological Study

The tribological studies have been conducted using a pin-on-disc tribometer in accordance with ASTM G-99 standards. The pin material used is aluminium alloy and the disc is of steel with a hardness of 60 HRC. The diameter of the pin has been selected to be 8 mm and length 27 mm. After the turning operation of the aluminium pin, its surface is polished using emery paper of 800 grit size; the same polishing method is adopted for steel disc also. The surfaces are then cleaned by acetone. Profilometer readings show that the roughness value (R_a) of pin surface varies from 0.29 to 0.32 μm and of disk surface, 0.13–0.14 μm. The sliding distance is taken as 1000 m and track diameter of the disc, 100 mm. The speed of rotation of the disc has been 300 rpm. Normal load applied at the interface is 2 MPa. Oil is supplied at the interface in a drop-wise manner in order to maintain boundary/thin film lubrication conditions. Frictional force is measured directly by using a load cell and an electronic display. Three trials have been conducted for each concentration. The average values of the three trials are taken for the calculation of coefficient of friction.

2.3 Viscosity Analysis

The viscosities of the base oils and nano-lubricants are measured using a Redwood Viscometer with ASTM D-445 standards. 125 ml of oil is used for each trial. Time required for emptying 50 ml of oil is measured and viscosity is calculated using Redwood formula. Three trials are conducted for each sample. The variation of kinematic viscosity has been obtained over a temperature range of 30–95 °C.

2.4 Flash and Fire Point Measurements

Flash and fire points are measured by standard methods using a Cleveland open cup (COC) flash and fire point apparatus (ASTM D-92).

2.5 Dispersion Analysis

The analysis of dispersion of nanoparticles in the nano-lubricants is done by Ultra-Violet (UV) Spectroscopy Analyzer. Base oil is filled in two cuvettes and base line correction is done. (Cuvette is a square shaped 3.5 ml capacity bottle made of quartz). In one cuvette nano-lubricant is added and the base oil in the other. The absorbance level of visible light, which is proportional to the dispersion of nanoparticles in the oil, is measured using visible spectroscopy over different time intervals.

3 Results and Discussions

The results of the thermo-physical tests carried out and the effect of temperature on the tribological properties of nano-lubricants are discussed here. Figure 1 shows the Scanning Electron Microscope (SEM) image of CuO nanoparticles. Some of

Fig. 1 SEM image of copper oxide nanoparticles

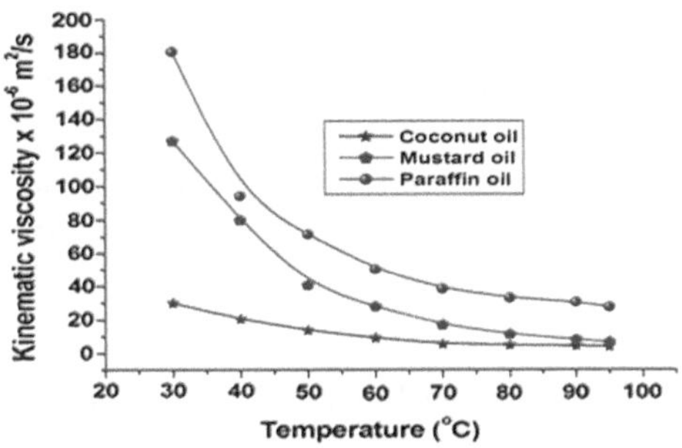

Fig. 2 Variation of kinematic viscosity with temperature for different lubricating oils

the nanoparticles which are seen to be agglomerated would get separated during agitation by ultrasonic shaker. The degree of agglomeration can be obtained using dispersion analysis.

3.1 Viscosity Analysis

The influence of temperature on kinematic viscosity of the base lubricants and the nano-lubricants is illustrated in Figs. 2 and 3 respectively. As evident from Fig. 2, paraffin oil exhibits higher viscosity than other vegetable oils in the temperature range investigated. However the variation in kinematic viscosity with temperature is found to be the least for coconut oil compared to mustard oil and paraffin oil. This aspect of coconut oil promotes its utilization in lubricating surfaces.

In the case of nano-lubricants, the kinematic viscosity follows the same trend as that of the base lubricant. However the numerical values of kinematic viscosity are found to be higher by 87.9, 63.6 and 42.6 % respectively for coconut oil, mustard oil and paraffin oil, in the range of temperature used.

As shown in Fig. 3, the decrement rate of kinematic viscosity with temperature is minimum (72.6 %) for coconut oil nano-lubricant. Thus it is inferred that

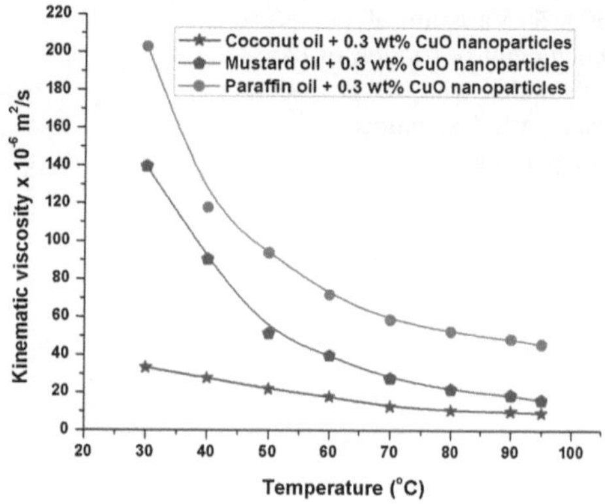

Fig. 3 Variation of kinematic viscosity with temperature for nano-lubricants

coconut oil nano-lubricant offers great prospect in the nano lubrication research which necessitates further studies on its tribological properties.

3.2 Frictional Force Analysis at Elevated Temperatures

In majority of applications (such as in engines), the lubricants are used at high temperatures, so it is relevant to conduct the effects of nano lubricant at elevated temperatures. The variation in tribological characteristics at higher temperatures has been obtained by incorporating a suitable attachment for heating the oil supplied to the pin-on-disc tribometer, which consists of a heating coil, blower and a temperature sensor. The experiment has been conducted up to 100 °C. When base oils are analysed, the coefficient of friction at the interface is found to increase with increase in temperature as shown in Fig. 4. The rate of increase is seen to be very high for mustard oil. This may be attributed to the larger reduction of viscosity with increasing temperature.

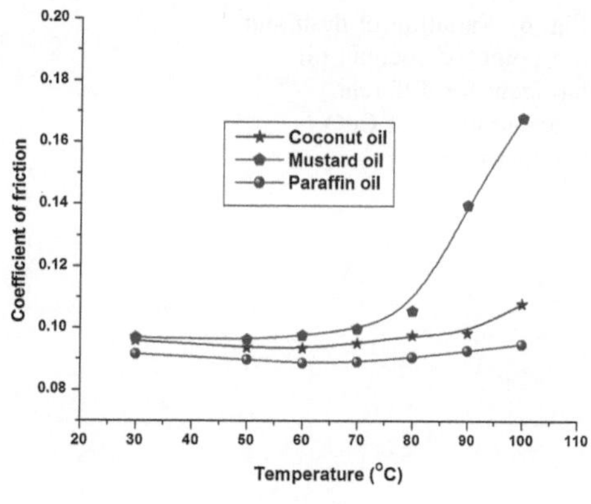

Fig. 4 Variation of coefficient of friction of base oils with temperature

Fig. 5 Variation of coefficient of friction of base oils added with 0.3 % CuO nanoparticles against temperature

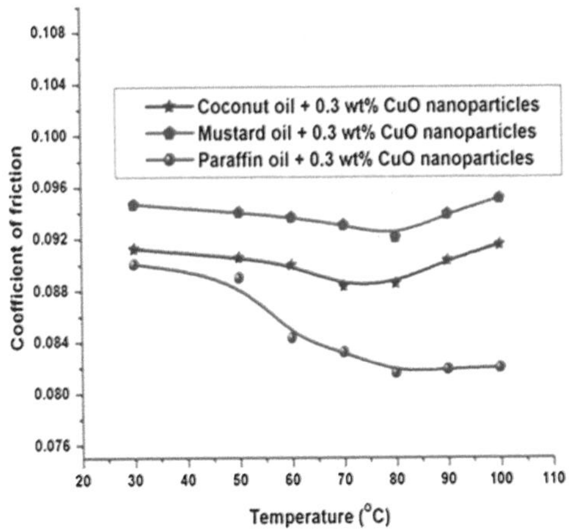

Furthermore, when CuO nanoparticles are added to the base oils, this increasing tendency is diminished. Up to a temperature of around 70 °C, mustard oil and coconut oil added with nanoparticles show a reduction in frictional force and thereafter it increases. In the case of paraffin oil added with nanoparticles, the frictional force shows a decreasing tendency even up to 100 °C as shown in Fig. 5. The thinning of vegetable based nano-lubricant at higher temperature is possibly causing the increase in friction.

3.3 Flash and Fire Point Analysis

As shown in Fig. 6, the effect of concentration of nanoparticles on flash point is insignificant, whereas the fire point shows an increase with increase in concentration of nanoparticles. It may be due to the variation in chain structure of the fatty acid at elevated temperatures.

Fig. 6 Variation of flash and fire points of coconut oil lubricant for different concentrations of CuO nanoparticles

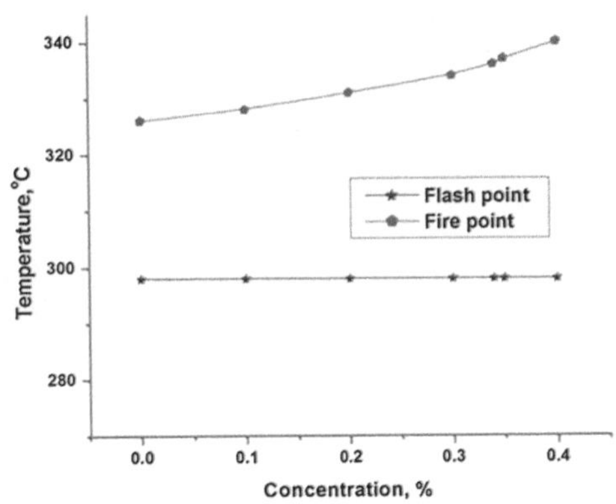

3.4 Dispersion Analysis

From the UV spectroscopy dispersion analysis of coconut oil nano-lubricant, it is evident that the nanoparticles in the nano-lubricant are likely to settle down after a period of time as in Fig. 7. Thus the nano-lubricant developed from coconut oil is applicable only for short term applications.

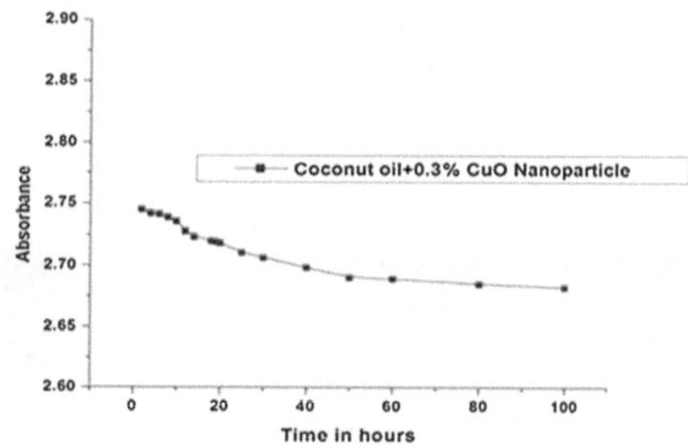

Fig. 7 Variation of absorbance levels of visible light at different time intervals for coconut oil nano-lubricant with 0.3 % concentration of nanoparticles at 450 nm wave length

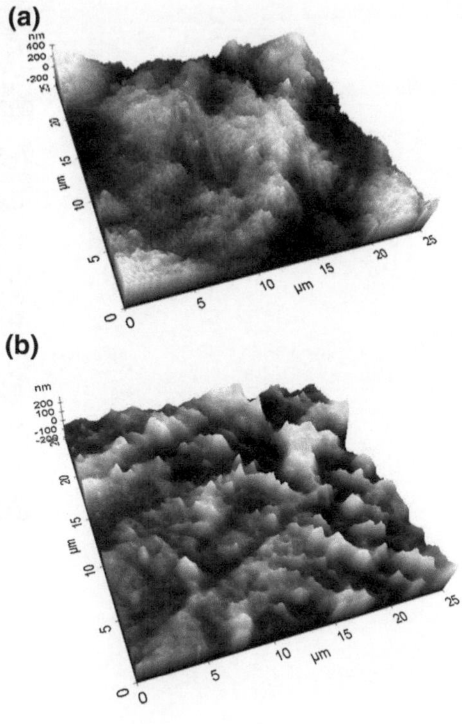

Fig. 8 AFM images of the friction surfaces of the pin before and after sliding, **a** Surface of pin before sliding, **b** Surface of pin after sliding in presence of coconut oil with 0.3 % CuO nanoparticles

3.5 Surface Roughness and Surface Topography Analysis

AFM and SEM have been used for surface topography analyses. AFM images provide three dimensional (3D) topographic features. Non-contact mode has been used for the present analyses.

From Fig. 8, it is observed that the average roughness of the pin surface before the test has been 107 nm. After the experiment using coconut oil with 0.3 % CuO nanoparticles, the average roughness is seen to reduce drastically to 66 nm. SEM images in Fig. 9a and b also illustrates the same effects.

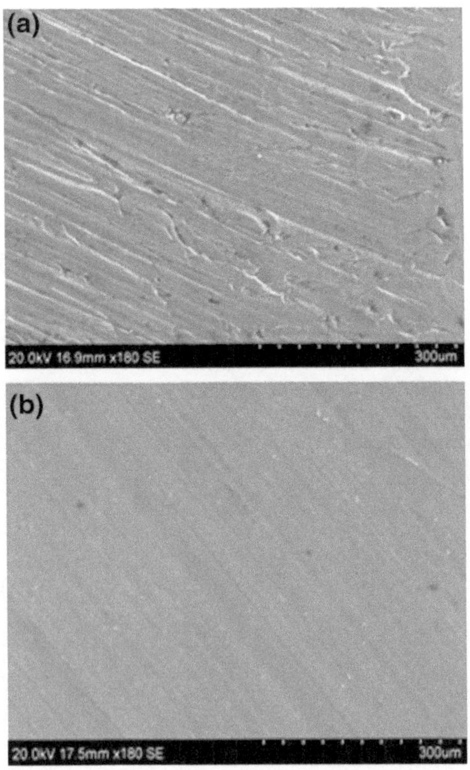

Fig. 9 SEM images of the pin surfaces before and after sliding, **a** Surface of pin before sliding, **b** Surface of pin after sliding in presence of 0.3 % CuO nanoparticle in coconut oil

4 Conclusions

The major conclusions that have been reported from the work are:

- The variation of kinematic viscosity with temperature is less for coconut oil when compared with mustard oil and paraffin oil.
- By the addition of CuO nanoparticles in paraffin oil, coefficient of friction is lower compared to coconut oil and mustard oil, when sliding is taken place in presence of nano-lubricants at elevated temperatures.
- Tribological properties of vegetable oils are further enhanced by the addition of CuO nanoparticles.
- The flash point of coconut oil does not change with the concentration of CuO nanoparticles, but the fire point increases.
- Vegetable oils find predominant significance from the environmental perspective in the lubrication scenario. Coconut oil nano-lubricant is a potential candidate for boundary/thin film lubrication as it offers better lubricity.
- Besides these factors, local availability and economic feasibility of coconut oil calls for comprehensive studies on coconut oil nano-lubricants.

References

1. Bartz Wilfried J (1998) Lubricants and the environment. Tribol Int 31(1–3):35–47
2. Erhan SZ, Sharma BK, Liu Z, Adhvaryu A (2008) Lubricant base stock potential of chemically modified vegetable oils. J Agric Food Chem 56:8919–8925
3. Erhan SZ, Asadaukas S (2000) Lubricant basestocks from vegetable oils. Ind Crops Prod 11(2–3):277–282
4. Fox NJ, Stachowiak GW (2007) Vegetable oil-based lubricants-A review of oxidation. Tribol Int 40(7):1035–1046
5. Jagadeesh KM, Satish VK (2009) Studies on boundary lubrication properties of oxidized coconut and soy bean oils. Lubr Sci 21(9):355–365
6. Jayadas NH, Prabhakaran Nair K, Ajithkumar G (2007) Tribological evaluation of coconut oil as an environment-friendly lubricant. Tribol Int 40:350–354
7. Maleque MA, Masjuki HH, Sapuan SM (2003) Vegetable-based biodegradable lubricating oil additives. Ind Lubr Tribol 55(3):137–143
8. Jayadas NH, Prabhakaran Nair K (2006) Coconut oil as base oil for industrial lubricants-evaluation and modification of thermal, oxidative and low temperature properties. Tribol Int 39(9):873–878
9. Thottackkad MV, Rajendrakumar PK, Prabhakaran Nair K (2012) Experimental Evaluation on the Tribological Properties of Coconut oil by the Addition of CuO Nanoparticles. Int J Precision Eng Manuf 13:111–116
10. Hsu Stephen M (2004) Nano-lubrication: concept and design. Tribol Int 37:537–545
11. Wu YY, Tsui WC, Liu TC (2007) Experimental analysis of tribological properties of lubricating oils with nanoparticle additives. Wear 262:819–825
12. Morina A, Neville A, Priest M, Green JH (2006) ZDDP and MoDTC interactions in boundary lubrication-the effect of temperature and ZDDP/MoDTC ratio. Tribol Int 39:1545–1557

13. Thottackkad MV, Rajendrakumar PK, Prabhakaran Nair K (2012) Tribological property variations of lubricating oils with nanoparticle additives-an experimental study. J Tribol Surf Eng 3:219–228
14. Vamsi Krishna P, Srikant RR, Nageswara Rao D (2010) Experimental investigation on the performance of nano boric acid suspensions in SAE-40 and coconut oil during turning of AISI 1040 steel. Int J Mach Tools Manuf 50(10):911–916

On Using Nanoporous Alumina Films as Tribological Coating

Arti Yadav, Prashant Pendyala and M. S. Bobji

Abstract Anodization of aluminum alloys is a common surface treatment procedure employed for the protection against corrosion. A thin amorphous layer of alumina is formed on the surface of alloy, which seals the alloy surface from the surrounding. This alumina layer being harder than the base aluminum alloy can be useful as a tribological coating. But since this alumina layer is randomly formed with disordered voids and pores, predicting the mechanical properties is difficult. Specific anodizing conditions can be used to form highly ordered anodic nanoporous alumina films [1] on the aluminum alloy surface. These nanoporous alumina layer can be effectively used as a tribological coating, because of the highly ordered controllable geometry and the empty pores which can be used as reservoirs for lubricant.

Keywords Nano porous alumina · Thin film · Nanoindentation · Nanotribology

1 Introduction

Ordered Nanoporous alumina is a self-organized nano-structured material containing a high density of uniform cylindrical pores that are aligned perpendicular to the surface of the metal. These types of films were first reported as early as 1950, following pioneering work in 1970 and 1980s by Moskovits in Canada and Masuda. A regular porous structure is formed when aluminum is electrochemically anodized in certain electrolytes [2–7]. These films are being used in various application such as sensors, Nanorods and Nanotube templates etc.,. In this study, we tried to study the usefulness of these nanoporous alumina films prepared on

A. Yadav (✉) · P. Pendyala · M. S. Bobji
Department of Mechanical Engineering, Indian Institute of Science, Bangalore 560012, India
e-mail: arti@mecheng.iisc.ernet.in

H. C. Patel et al. (eds.), *Proceedings of International Conference on Advances in Tribology and Engineering Systems*, Lecture Notes in Mechanical Engineering, DOI: 10.1007/978-81-322-1656-8_36, © Springer India 2014

Pure aluminum for tribological applications, by controlling the geometry of pores and film thickness.

The nanoindentation technique is an effective tool for probing mechanical properties of materials at very small scales. Its principal advantage is to continuously monitor with high precision and accuracy, both the load (P) and displacement (h) of an indenter during loading and unloading in the micro-Newton and nanometer ranges respectively [8–11].

2 Experiment

2.1 Sample Preparation

We have prepared ordered porous alumina samples in 0.3 M Oxalic acid electrolyte with a platinum as cathode and pure aluminum as anode as described elsewhere [12]. There are four stages of Nanoporous alumina preparation. First anodization, Etching, Second Anodization, Barrier layer reduction.

For all the samples in the study, First anodization and Etching were done as described elsewhere [13]. Using these conditions samples with varying time of second anodization have been obtained and after that barrier layer reduction was done with same conditions. Samples after the barrier layer reduction were cleaned in distilled water were stored in vacuum and sent for Nanoindentation studies.

2.2 Nanoindentation

All the samples were indented in Hysitron Nanoindenter in Load-controlled mode using a Berkovich tip. Loading rates were maintained constant for all the experiments at o.1 mN/sec. Hardness values were directly obtained from the Nanoindenter sytem software.

3 Results and Discussions

The pore-diameter of the ordered porous alumina film is observed to be linearly proportional to the anodizing potential with a proportional constant λ_p of approximately 1.29 nm/V^{-1} [14–17].

$$D_p = \lambda_p \cdot U$$

where, D_p is a pore diameter (nm) and U denotes an anodizing potential (V). The pore diameter is highly dependent on the voltage but is less sensitive to the electrolyte pH. So films with similar pore dimensions and of varying thickness can be produced to control the mechanical properties as shown in Figs. 1 and 2.

It is clear from the figures, pore diameter remain constant for 1 and 10 min anodization time at around 40 nm. But the film thickness increases (as seen in Scanning Electron Microscopy using a cross section of the sample) from 200 nm to 1.2 micron (Fig. 3).

The film thickness is observed to increase linearly with time of anodization till 10 min anodization (Fig. 4). This shows that the geometry of the ordered porous alumina layer can be tuned to the requirement and is helpful in creating a tribological coating with a highly ordered structure.

Figure 4 shows the Current vs time graph of 10 min second anodization and barrier layer reduction. Barrier layer (thin continuous oxide layer at the interface of porous film and aluminum) can be thinned by the various methods. We used the voltage reduction method as shown.

In Fig. 3 we observe that the thickness of pore depends on anodizing time. At 1 min of second anodization the depth of porous alumina was 200 nm and at 10 min it was 1,250 nm.

Figure 5 shows SEM image of indent from nanoindentation of porous alumina for 10 min of second anodization. It can be observed that though alumina is brittle, there is no clear breakage of alumina in the periphery of the indent and the deformation is due to compaction.

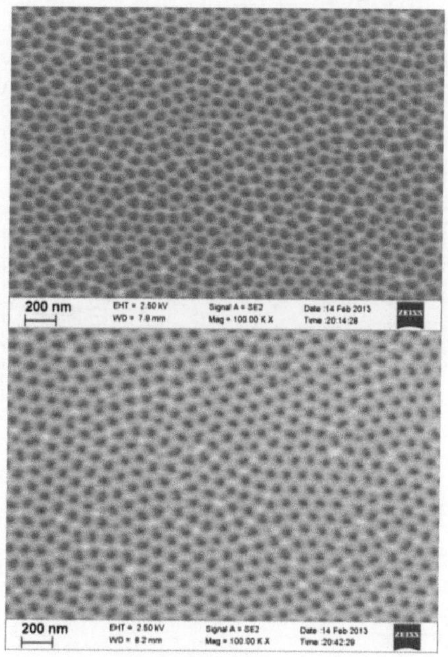

Fig. 1 *Cross-section* images of porous alumina. **a** 1 min of second anodization, **b** 10 min of second anodization

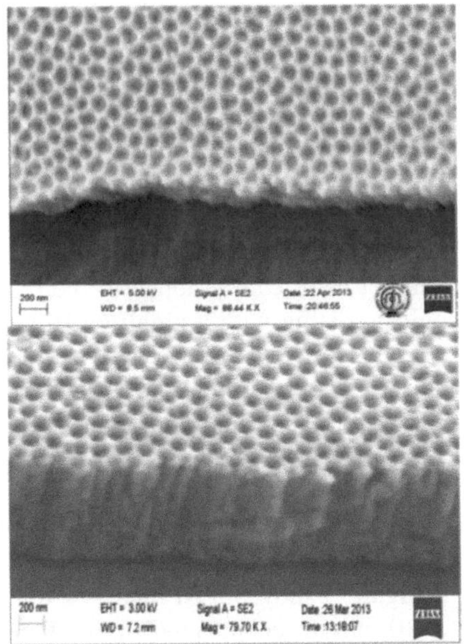

Fig. 2 *Cross-section* images of porous alumina. **a** 1 min of second anodization, **b** 10 min of second anodization

From the Fig. 6, the load-Depth of penetration curve shows the plastic behavior of the nanoporous alumina film.

Figure 7 Hardness versus Anodizing time for second anodization of porous alumina it is clear that the Hardness is increasing when the second anodization time is increasing. At 1–10 min of second anodizing time the hardness value is varying from 500 to 760 MPa respectively.

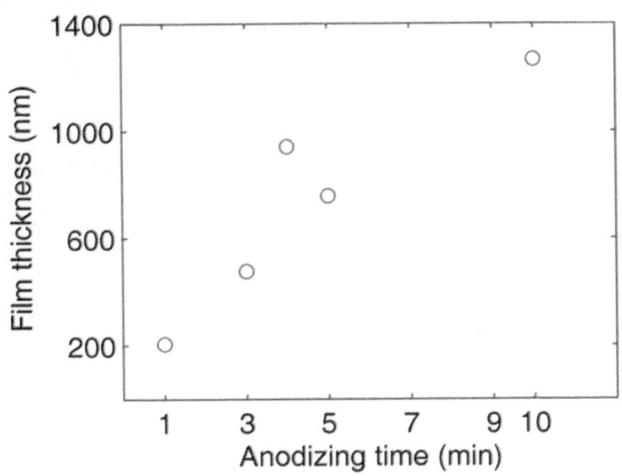

Fig. 3 Porous alumina film thickness after second anodization for various times

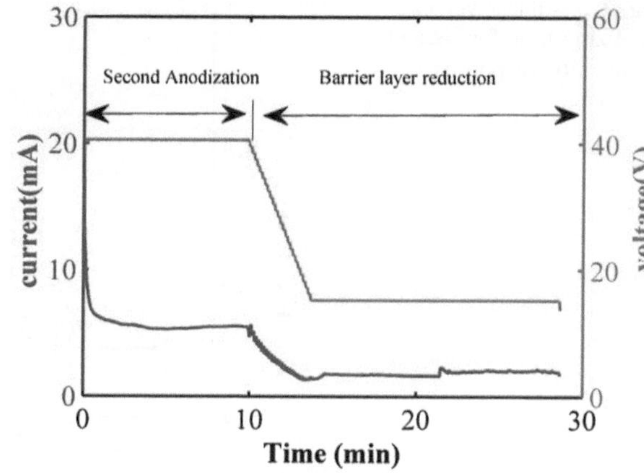

Fig. 4 Current versus time graph at constant voltage of 10 min of second anodization and barrier layer thinning

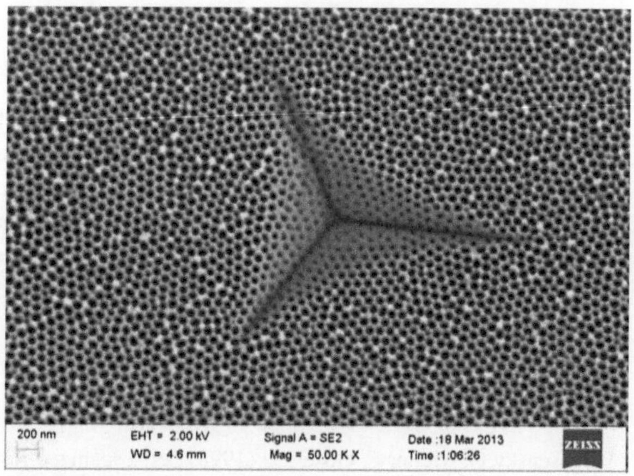

Fig. 5 SEM image of indent from nanoindentation of porous alumina for 10 min of second anodization

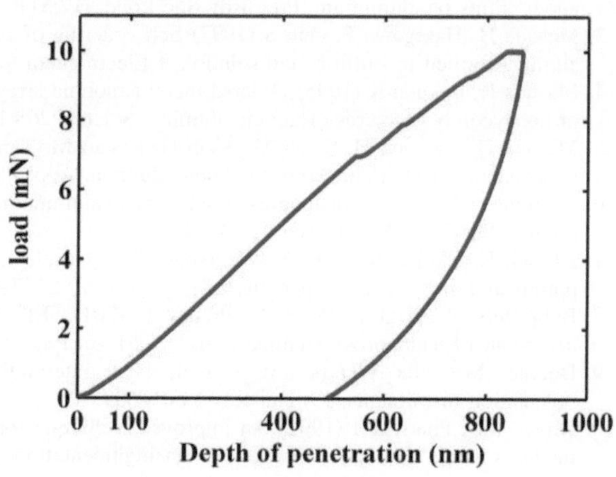

Fig. 6 Nanoindentation graph of porous alumina at 10 min second anodization

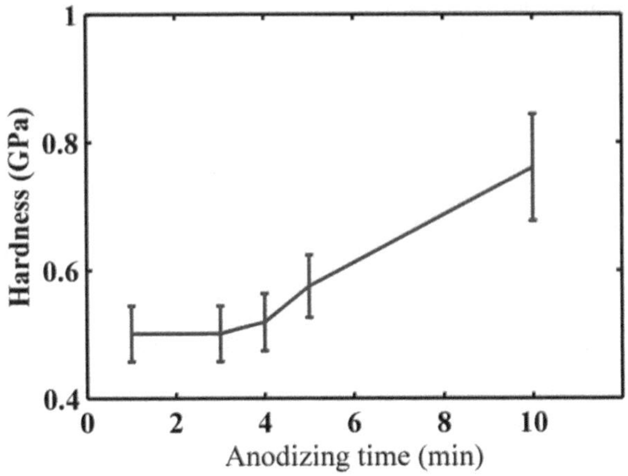

Fig. 7 Hardness versus anodizing time for second anodization of porous alumina

4 Conclusion

We tried to vary the film thickness of ordered porous alumina keeping the pore geometry identical. The effect ordered porous alumina film thickness on the hardness obtained from the nanoindentatio technique studied. We find that the hardness of this material can be varied by varying the time of second anodization of porous alumina.

References

1. Jessensky O, Müller F, Gösele U (1998) Self-organized formation of hexagonal pore arraysin anodic alumina. Appl Phys Lett 72:1173–1175
2. Sullivan JPO, Wood GC (1970) The morphology and mechanism of formation of porous anodic films on aluminium. Proc Roy Soc Lond 317:511–543
3. Masuda H, Hasegawa F, Ono S (1997) Self ordering of cell arrangement of anodic porous alumina formed in sulfuric acid solution. J Electrochem Soc 144:L127
4. Masuda H, Fukuda K (1995) Ordered metal nanohole arrays made by a two-step replication of honeycomb structures of anodic alumina. Science 268:1466–1468
5. Masuda H, Yamada H, Satoh M, Asoh H, Nakao M, Tamamura T (1997) Highly ordered nanochannel-array architecture in anodic alumina. Appl Phys Lett 71:2770–2772
6. Lohrengel M (1993) Thin anodic oxide layers on aluminum and other valve metals: high-field regime. Mater Sci Eng R-Rep 11:243–294
7. Nielsch K, Choi J, Schwirn K, Wehrspohn RB, Gosele U (2002) Self-ordering regimes of porous alumina: the 10 % porosity rule. Nano Lett 2:677–680
8. Bobji MS, Prashant P, Preeti G, Pranav K (2011) Effect of porosity on the indentation behaviour of nanoporous alumina films. J Surf Sci Eng 5:51–62
9. Doerner MF, Nix WD (1986) A method for interpreting the data from depth-sensing indentation instruments. J Mater Res 1:601–609
10. Oliver WC, Pharr GM (1992) An improved technique for determininghardness and elastic modulus using load and displacement sensingindentation. J Mater Res 7(6):1564–1583

11. Zheng CM, Zheng YT (1997) On the initial unloading slope in indentation of elastic–plastic solids by an indenter with an axisymmetric smooth profile. Appl Phys Lett 71:2623–2625
12. Nix WD (1997) Elastic and plastic properties of thin films on substrates: nano-indentation techniques. Mater Sci Eng A234–236:37–44
13. Giannakopouls AE, Suresh S (1999) Determination of elastoplastic properties by instrumented sharp indentation. Scripta Mater 40(10):1191–1198
14. Pharr GM (1998) Measurement of mechanical properties by ultra-low load indentation. Mater Sci Eng A253:151–159
15. Alcal AG et al (2002) Mechanical properties of amorphous anodic alumina and tantala films using nanoindentation. Nanotechnology 13:451–455
16. Krell A Schadlich S (2001) Nanoindentation hardness of submicormeter alumina ceramics. Mater Sci Eng A307:172–81
17. Surganov V, Morgen P, Nielsen JC, Gorokh G, Mozalev A (1987) Highly ordered anodic porous alumina formation by self-organized anodizing. Electrochim Acta 32:1125–1127

Solid Particle Erosion of *Luffa cylindrica* Fiber Reinforced Polymer Composite

Niharika Mohanta and S. K. Acharya

Abstract Experiments were carried out to study the effects of impingement angle and particle velocity on the solid particle erosion behavior of *Luffa cylindrica* fire reinforced epoxy composites. The erosive wear is evaluated at different impingement angles from 30 to 90° at three different velocities of 48, 70 and 82 m/s. The erodent used is silica sand with the size range 150–250 μm of irregular shape. The result shows semi ductile behaviour with maximum erosion rate at 45–60° impingement angles. The morphology of the eroded surfaces was examined by using Scanning Electron Microscopy (SEM).

Keywords *Luffa cylindrica* fiber (LCF) · Erosive wear · Erodent · Semi-ductile · SEM

1 Introduction

Increasing environmental awareness throughout the world is motivating the researchers to design material that are compatible with the environment. Synthetic fiber such as glass, carbon and aramid are being widely used in polymer composites because of their high stiffness and strength properties [1, 2]. Natural fiber such as jute, bagasse, pineapple, sisal, banana [3–10] etc. based polymer composite form a new class of materials which have good potential to be used in polymer composite and have established themselves as potential fillers materials for numerous applications. The attractive features of these natural fiber reinforced polymer composites such as light weight, high specific modulus, and nontoxity, friendly processing and absorbed CO_2 during their growth [11, 12] have shifted the

N. Mohanta (✉) · S. K. Acharya
Department of Mechanical Engineering, NIT, Rourkela 769008, India
e-mail: mohanta.niharika@gmail.com

H. C. Patel et al. (eds.), *Proceedings of International Conference on Advances in Tribology and Engineering Systems*, Lecture Notes in Mechanical Engineering,
DOI: 10.1007/978-81-322-1656-8_37,

focus of researchers from synthetic to natural fiber reinforced polymer composite. These composites after being widely accepted in aerospace applications heading their way towards tribo application such as bearing and gears etc. Being used in applications such as radomes, surfing boats, gas and steam turbine blades, gears of locomotive, conveyor belts, pump impellers in mineral slurry processing where components encounter impact of lot of abrasion like dust sand, splinters of material slurry of solid particle and consequently the material undergoes erosive wear [13]. Therefore the study of erosion characteristics of natural fiber reinforced polymer composites is of vital importance.

In the pursuit of visualizing the importance of natural fiber reinforced polymer composites in tribological applications, extensive research work has already been published on various types of polymers and fibers [14–17]. It is well established that the erosive resistance of polymer composites is low as compared to monolithic materials [18] and the erosive wear of reinforced polymer composite is higher than unreinforced polymer matrix [19]. Mishra and Acharya [20], Deo and Acharya [21] have reported erosive wear behaviour of bagasse and *Lantana camara* fiber reinforcement in polymer matrix. However there is no information available on the erosive wear behaviour of *Luffa cylindrica* fiber reinforced polymer composite. Various researchers have co related erosion rate of composite with some important factors such as properties of target materials, testing environment, operating parameters, and properties of erodent [22, 23]. Hence the priority of this work is to prepare Polymer Matrix Composites (PMC) with *Luffa cylindrica* as reinforcement to study the erosive behaviour. Experiments were carried out to study the effect of impingement angle and particle velocity on the solid particle erosion behaviour *Luffa cylindrica* reinforced epoxy resin composites.

2 Materials and Method

2.1 Raw Materials

2.1.1 Matrix Material

Epoxy LY 556 is the resin which is used as matrix material. Its common name is bisphenol-A-diglycidyl-ether and it chemically belongs to the epoxide family. The epoxy resin and the hardener are supplied by Ciba Geigy Ltd.

2.1.2 Fiber Material

Luffa cylindrica is a tropical plant belonging to the family of *Cucurbitacea*, with a fruit possessing netting like fibrous vascular system. The LC strut are characterized by a micro cellular architecture with continuous hollow micro channels which

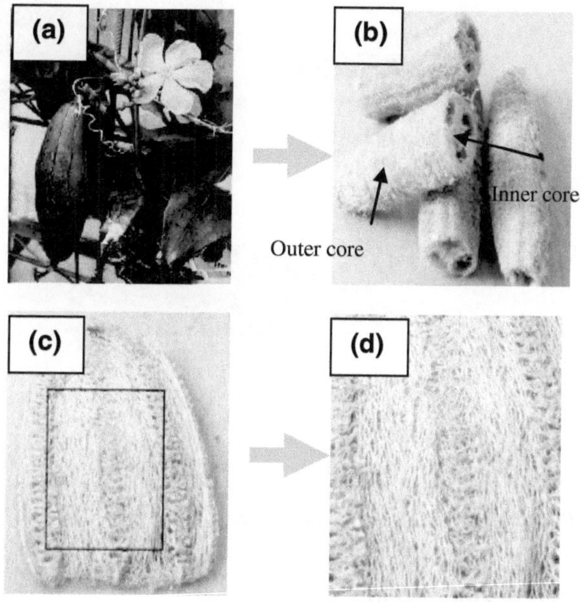

Fig. 1 **a** The *L. cylindrica* fruit **b** Sponge guard with hollow micro channels **c** *Outer core* open as mat **d** The *rectangular* portion used for making composite

forms a vascular bundles and yield a multimodal hierarchical pore system [24]. Figure 1b shows the sponge guard and the hollow micro channels. This specific morphology makes it possible to imagine a specific composition on crystallinity cellulose. In this work LC fibers were cut to rectangular mat like after opening the outer core and the micro channel portion as shown in Fig. 1c, d from the sponge guard neglecting the end portion to keep the thickness same for the mat and have been used for manufacturing the layered composite.

2.2 Composite Fabrication

A wooden mould of dimension 140 × 100 × 6 mm was used for casting the composite slab. The composites were manufactured with single, double and triple layer of *Luffa cylindrica* fiber in three different weight proportions (8, 13 and 19 wt%). For different wt% of fibers, a calculated amount of epoxy resin and hardener (ratio of 10:1 by weight) was thoroughly mixed with gentle stirring to minimize air entrapment. For easy removal of the composite from the mold after curing, a releasing agent (silicon spray) is used. Each ply of the LC fiber is of dimension 140 × 100 mm. The layered composites were made by conventional hand lay-up technique. The cast of each composite is cured under a load of 25 kg for 72 h before they are removed from the mold. Specimens of required dimensions were cut using a diamond cutter for physical characterization and mechanical testing. At most care has been taken to maintain uniformity of the composites. The schematic view of the layered composites is shown in Fig. 2.

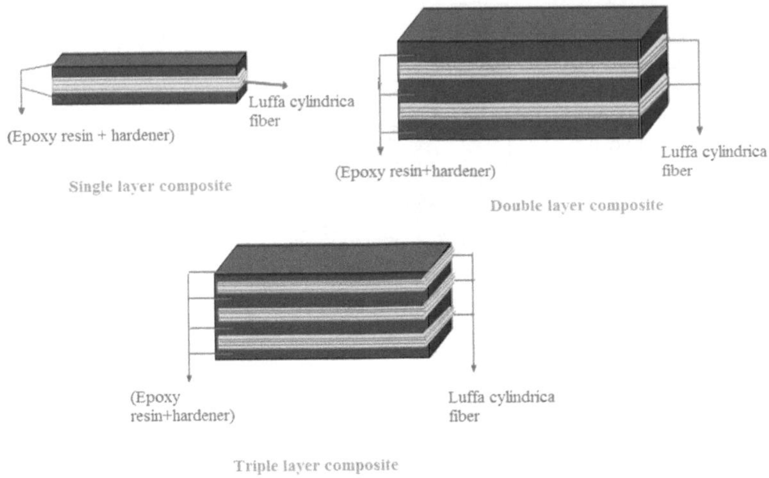

Fig. 2 Schematic view of the composites

2.3 *Experiment*

2.3.1 Erosion Wear Studies

The solid particle erosion experiments were carried out as per ASTM G76 standard on the erosion test rig shown in Fig. 3. The test rig consist of an air compressor, an air drying unit, a conveyor belt type particle feeder and a particle mixing and accelerating chamber. The dried and compressed air is mixed with silica sand (150–250 μm size) which was feed constantly by conveyor belt feeder into the mixing chamber. Samples of composite (30 × 30 × 3 mm) were held at selected angles (30, 45, 60 and 90°) with respect to flow of the impinging sand particles and eroded. The silica particles were accelerated by passing through a converging tungsten carbide nozzle of 4 mm diameter to bombard the target. The distances between the target material and nozzle was approximately 20 mm. Wear was measured by weight loss method Samples were cleaned with acetone before and after each test. Eroded samples were cleaned with a brush to remove fine sand particles attached to the surface and then wiped with a cotton plug dipped in acetone to avoid any air entrapment of wear debris in the samples. The wear rate was expressed in terms of ΔWc (g)/ΔWs (g); where ΔWc loss of weight of composites and ΔWs total weight of erodent used. ΔWc was determined by weighing the sample before and after each experiment on a weighing balance having an accuracy of 0.001 mg. The experimental detail is presented in Table 1.

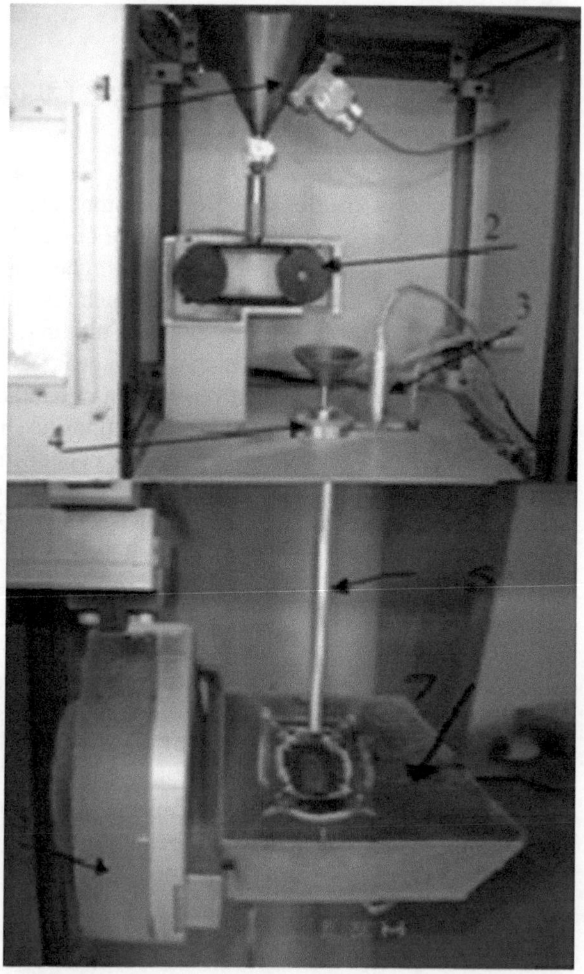

Fig. 3 Details of erosion test rig *(1)* Sand hoppe *(2)* Conveyor belt *(3)* Pressure transducer *(4)* Particle–air mixing chamber *(5)* Nozzle *(6)* X–Y assembly *(7)* Sample holder

Table 1 Test parameter

Erodent	Silica sand
Erodent size (μm)	200 ± 50
Erodent shape	Irregular
Impingement angle (°)	30, 45, 60, 90
Impact velocity (m/s)	48, 70, 82
Erodent feed rate (gm/min)	10
Test temperature (°C)	27
Nozzle to sample distance (mm)	20

3 Result and Discussion

3.1 *Influence of Impingement Angle (α) on Erosion Wear Behavior*

Figure 4a–c shows the result of the erosion rate for single, double and triple layered LCF epoxy composite as function of angle of impingement. The wear rates of all samples were remarkable higher at higher particle speed. Particles have a higher kinetic energy at higher velocity, which result in greater impingement effect and result in higher wear. It is evident from plot that the erosion rate increases with increase of impingement angle and attains a peak value (α_{max}) at 45° for single and double layered composite. However erosion rate attains a peak value (α_{max}) at 60° impingement angle for triple layered composite. In the literature materials as classified as ductile or brittle based on the dependence of their erosion rate with impingement angle. The ductile behavior is characterized by maximum erosion rate at low impingement angle typically $15° < \alpha < 30°$ On the other hand, if the maximum erosion rate occurs at normal impact ($\alpha = 90°$) the behavior of the material is brittle.

The present reinforced composite exhibit semi ductile behavior with maximum erosion occurring in the range of 45–60°. As evident from literature and pointed out by Rattan and Bijwe [25] there is no fixed trends correlating ductile and brittleness of material with α_{max} or α_{min}. Thermoplastic generally exhibit a more ductile response than the thermoset. Deo et al. [21] reported that lantana camara fiber reinforced polymer composite shows the semi ductile behavior with maximum erosion rate at 45°. Similar results were obtained in the present work. Therefore it can be concluded that behavior of these *Luffa cylindrica* fiber reinforced epoxy composites can be termed as semi-ductile in nature.

3.2 *Surface Morphology of Eroded Surface*

The eroded surfaces of the specimens are examined directly by scanning electron microscope JEOL JSM-6480 LV. The composite samples are mounted on the stubs with silver paste. To enhance the conductivity of the samples a thin film of platinum is vacuum evaporated on to them before the photomicrographs are taken. Figure 5a shows the single layered composite eroded at 45° impingement angle with velocity 70 m/s. It can be seen from the surface of the sample that the material removal is mainly due to micro cutting and micro ploughing. Figure 5b, c show the micrograph of eroded surface of double and triple layered composite at 45° impingement angle with particle velocity of 70 m/s. It is well known that the fiber in composite subjected to solid particle erosion, encountered intensive debonding and breakage of the fibers, which are not supported by enough matrix [26].

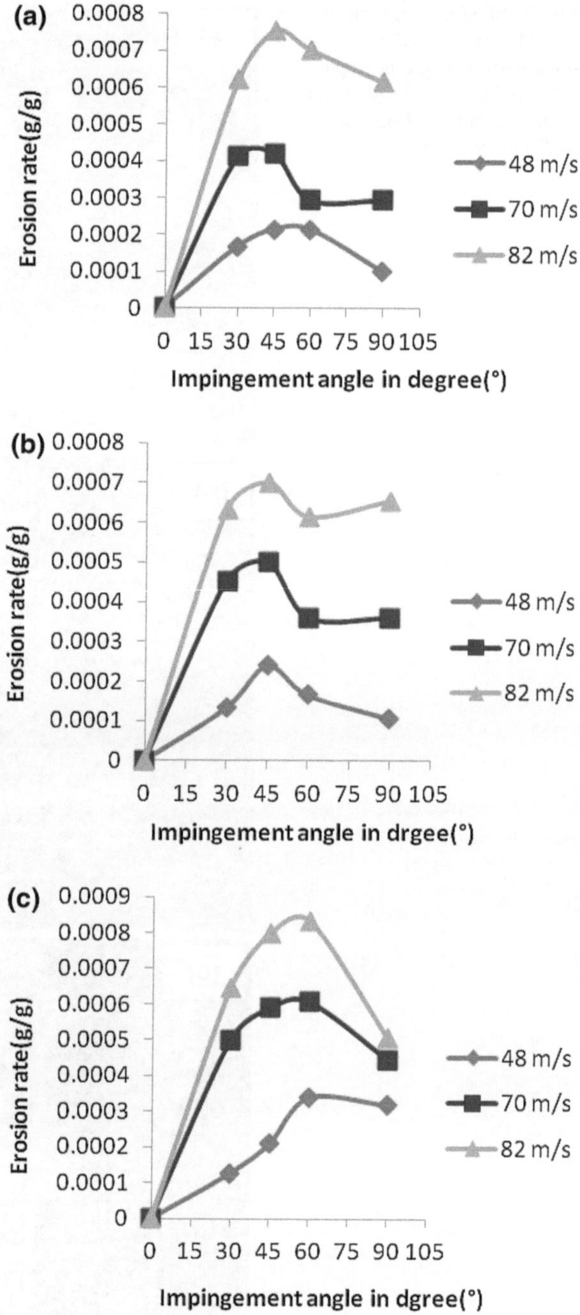

Fig. 4 Erosion as a function of impingement angle for different particle speeds of **a** Single layer **b** Double layer **c** Triple layer LCF epoxy composite

Fig. 5 SEM micrograph of eroded surface at 45° impingement angle at impact velocity 70 m/s **a** Single layer **b** Double layer **c** Triple layer LCF epoxy composite

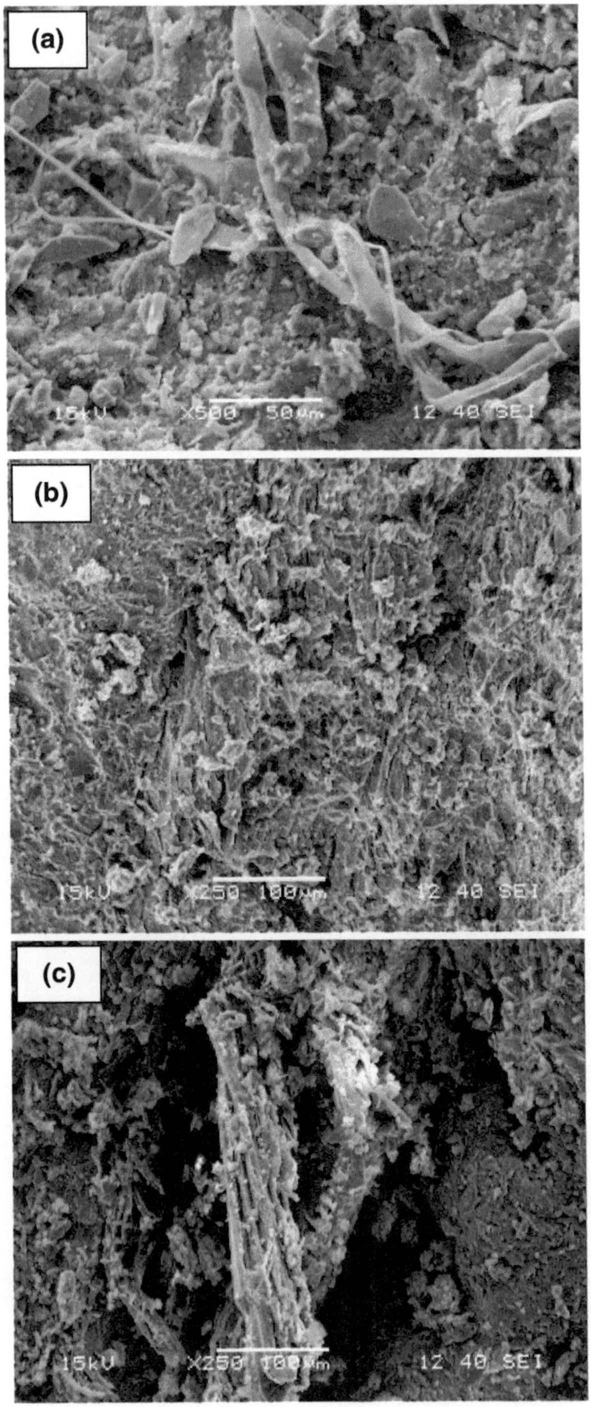

The same type of behavior observed in this present case. The maximum erosion rate occurs for triple layer composite may be due to the insufficient matrix material.

4 Conclusion

Experiments were carried out to study the solid particle erosion of *Luffa cylindrica* fiber reinforced in epoxy resin composites at various impingement angles and impact velocities with constant mass of erodent. Based on the studies the following conclusions can be drawn.

1. The influence of impingement angle on erosion wear of composite under consideration exhibits semi ductile behavior with maximum wear rate at 45° for single and double layer composite and maximum wear rate at 60° impingement angle for triple layer LCF epoxy composite.
2. The erosion rate increases with increase of velocity of impinging particle.
3. SEM studies or eroded surfaces support the evolved mechanisms and indicated micro-cracking, sand particle embedment, chip formation, exposure of fibers, fiber cracking and removal of fibers.

References

1. Rout J, Misra M, Tripathy S, Nayak SK, Mohanty AK (2001) The influence of the fibre treatment on the performance of coir-polyester composite. Compos Sci Tech 61:1303–1310
2. Li Q, Matuana LM (2003) Surface of cellulosic materials modified with functionalised polyethylene coupling agents. J Appl Polymer Sci 88:278–286
3. Roe PJ, Ansell MP (1985) Jute-reinforced polyester composites. Mater Sci 20:4015–4020
4. Sridhar MK, Basavarappa G, Kasturi SG, Balasubramaniam N (1984) Mechanical properties of jute/polyester composites. Indian J Tech 22:213–215
5. Kumar P (1986) Mechanical behaviour of jute fiber and their composites. Indian J Tech 24:29–32
6. Sha AN, Lakkad SC (1981) Mechanical properties of jute reinforced plastic. Fibre Sci Tech 15:41–46
7. Acharya SK, Mishra P, Meher SK, Dikshit V (1839) Weathering behaviour of bagasse fibres reinforced polymer composite. J Reinf Plast Compos 2008:27
8. Luo S, Netravali AN (1999) Interfacial and mechanical properties of environment- friendly 'Green' composites made from pineapple fibres and poly (Hydyoxybutyrate-co-valerate) resin. J Mater Sci Eng 34:3709–3719
9. Bisanda ETN, Ansell MP (1991) The effect of saline treatment on the mechanical and physical properties of sisal–epoxy composites. Comp Sci Tech 41:165–178
10. Pothan LA, Thomas S, Neelakantan NR (1997) Short banana fiber- reinforced polyester composites: mechanical, failure and aging characteristics. J Reinf Plast 16:744–765

11. Abdelmouleh M, Boufis S, Belgacem MN, Dufresne A (2007) Short natural fibre reinforced polyethylene and natural rubber composite: effect of silane coupling agents and fiber loading. Compos Sci Tech 67(7–8):1627–1639
12. Tserki V, Zfeiropoulo NE, Simon F and Panayiton CA (2005) Study of the effect of acetalation and propeionylation, surface treatment of natural fibers. Composites Part A 36(8):1110–1118
13. Rajesh JJ, Bijwe J, Tewari US and Venkataraman B (2001) Erosive wear behaviour of various polyamides. Wear 249:702–714
14. Tong J, Ma Y, Chen D, Sun J, Ren L (2005) Effects of vascular fibre content on abrasive wear of bamboo. Wear 259:37–46
15. Hornsby PR, Hinrichsen E, Tarverdi K (1997) Preparation and properties of polypropylene composites reinforced with wheat and flax straw fibres, part II. Analysis of composite microstructure and mechanical properties. J Mater Sci 32:1009–1015
16. Yousif BF, EL-Tayeb NSM (2006) Erosive wear behaviour of various polyamides, Wear. In: The Proceedings ICOMAST. GKH press, Melaka, pp 384–387
17. El Tayeb NSM (2008) A study on the potential of sugarcane fibres/polyester composite for tribological applications. Wear 265:223–235
18. Roy M, Vishwanathan B, Sundararajan G (1994) The solid particle erosion of polymer matrix composites. Wear 171:149–161
19. Hager A, Friedrich K, Dzenis YA, Paipetis SA (1995) Study of erosion wear of advanced polymer composites. In: Street K, Whistler BC (eds) Proceeding of ICCM 10. Wood Head Publishing, Cambridge, pp 155–162
20. Mishra P and Acharya SK (2010) Solid particle erosion of bagasse fiber reinforced epoxy composite. Int J Phy Sci 5(2):109–115
21. Deo C and Acharya SK (2009) Solid particle erosion of lantana camara fibre reinforced polymer matrix composite. Poly-Plas Technology Eng 48:1084–108
22. Tewari US, Harsha AP, Hager AM, Friedrich K (2002) Solid particle erosion of unidirectional carbon fibre reinforced polyetherketone composite. Wear 252:992–1000
23. Bhushan B (1999) Principle and application of tribology. Wiley, New York
24. Zampheri A, Mabande GTP, Selvam T, Schwieger W, Rudulph A, Hermann R, Sieber H, and Greil P (2006) Biotemplating of *Luffa cylindrica*, sponges to self supporting hierarchical zeolite macrostructures for bio-inspired structured catalytic reactors. Mater Sci Eng 26(1):130–135
25. Rattan R, Bijwe J (2007) Influence of impingement angle on solid particle erosion of carbon fabric reinforced polyethermide composite. Wear 262:568–574
26. Sari N, Sinmazcelik T (2007) Erosive wear behaviour of carbon fiber polyetherimide composites under low particle speed. Mater Des 28(1):351–355

Synthesis and Micro-Structural Characterization of CrAlN Coatings by Reactive Magnetron Sputtering

Hetal N. Shah and R. Jayaganthan

Abstract The metallic coating such as TiN, CrN and HfN, with the specific lattice structure may possess excellent tribological and chemical properties, which enable their use as wear resistant, corrosion resistant, and diffusion barrier applications. Further improvement in the mechanical and tribological properties, the addition of third element (Al, Si, Ti) in the binary system may possess higher thermal stability, higher hardness and improvement in wear and tear properties. In this context, the CrAlN coatings were syntheses by reactive magnetron sputtering and the influence of addition of Al content on structural and chemical properties were discussed. The microstructure and composition of the as-deposited coatings were systematically characterized by field emission scanning electron microscopy/EDS and atomic force microscopy, and the phase formation by x-ray diffraction (XRD).The coatings were deposited mainly on stainless steel SA304 and the results are analyzed.

Keywords CrAlN · Microstructure · Magnetron sputtering · Thin film

1 Introduction

Transition metal nitrides such as TiN, CrN and HfN, with the NaCl (B1) type lattice structure possess excellent tribological and chemical properties, which enable their use as wear resistant, corrosion resistant, and diffusion barrier coatings. The effect of additions of Al, Si, and, Va on the mechanical and tribological

H. N. Shah (✉)
Department of Mechanical Engineering, Venus International College of Engineering, Gandhinagar, Gujarat 382422, India
e-mail: amihetnadiad@gmail.com

R. Jayaganthan
Department of Metallurgical and Materials Engineering, Indian Institute of Technology-Roorkee, Roorkee, Uttrakhand 247667, India

H. C. Patel et al. (eds.), *Proceedings of International Conference on Advances in Tribology and Engineering Systems*, Lecture Notes in Mechanical Engineering,
DOI: 10.1007/978-81-322-1656-8_38,

properties of CrN coatings using cathodic arc ion plating [1, 2] cathodic arc evaporation [3] and DC/RF magnetron sputtering [4–9] has been reported in the literature. A significant increase in hardness value up to 27 GPa [4] to 35 GPa [10] or even higher values to 40 GPa [11], has been achieved for the CrAlN coatings when compared to CrN coatings. The coatings also exhibited higher thermal stability than pure CrN coatings.

Shah et al. [12] deposited CrAlN coatings with different Al contents on SA 304 substrates prepared by a magnetron sputtering technique and found that the coatings are highly textured with a dominating peak of CrN(111), with its position shifted to higher 2θ angle with increasing Al content deposited on SA 304. Sun et al. [13] investigated CrAlN coatings deposited on different substrate materials such as Si (100), commercial aluminum alloy, AA6061, and M42 high speed steel (HSS). They found that the substrate material influences phase evolution in a reactively co-sputtered CrAlN films with Al contents favorable for B1 phase formation. The films deposited on Si (100) substrate showed the formation of single B1 phase (CrAl)N, which correlated well with theoretical predictions. A clustering of Al atoms in the B1 CrN {200} planes seems to occur, which leads to the large contraction of inter-planar spacing for the films deposited on the HSS substrate. The coatings became more compact and denser, and the microhardness and fracture toughness of the coatings increased correspondingly with increasing substrate bias voltage [14]. $Cr_{1-x}Al_xN$ coatings deposited on Si and Stainless Steel substrates using RF magnetron sputtering with different atomic concentrations of aluminum ($0.51 < x < 0.69$) showed the evolution of (111), (200), and (102) crystallographic orientations associated to the cubic $Cr_{1-x}Al_xN$ and w-AlN phases, respectively [15]. The insertion of Al or AlN in the B1 type nitride coatings beyond solubility limits leads to phase transition from B1 to B4. Also, the solubility limit of AlN in B1 CrN has been calculated theoretically and found to be 77 % as reported in the literature [16]. The B4 Wurtzite phase showed low hardness and poor ductility, which are not desirable for many industrial applications [13].

The CrAlN coatings were characterized by using the techniques such as XRD, FE-SEM/EDS, and AFM. The mechanical properties were measured using microhardness tester while the wear rate and coefficient of friction of the coatings were measured by using Pin on disk tribo tester in the present work.

1.1 Experimental Procedure

CrAlN coatings were deposited on SA 304 steel (92 HRb) of 15 $\times$ 15 and 0.9 mm thick with a surface finish of 0.01 μm Ra by using DC/RF magnetron sputtering (Model: DCSS—12, Manufactured by Excel Industries, Mumbai). Chromium target (cathode power fixed at 60 W) and aluminum targets (cathode power varied from 60 to 110 W) in a mixed Ar/N_2 atmosphere were used for the deposition of coatings. The Ar/N_2 flow ratio was maintained constant at 1:1, while the argon and

Table 1 Deposition condition for CrAlN coatings

Parameters	Values
Target and size	Cr and Al (2″ dia.)
Substrate and size	Stainless steel SA304, 15 × 15 × 0.9 mm
Base pressure	1×10^{-4} Pa
Temperature	573 K
Sputtering pressure	0.66 Pa
Gas environment	Ar + N_2 (50:50)
Gas flow	5 sccm
Target to substrate distance	≈50 mm
Power density Cr target	3.1 W/cm^2
Al target	3.1–5.8 W/cm^2

nitrogen flows set to value 5 sccm. The thickness of as the deposited coatings varied from 4.0 to 4.5 μm. The SA304 samples were ground and mechanically mirror polished and then cleaned with acetone in an ultrasonic container for 15 min. The samples were mounted on the rotational substrate holder cum heater and rotated in forward and reverse manner. The rotation angle were set (30°) in such a way that it covers maximum plasma region. Prior to deposition, the targets were sputter cleaned in Ar gas (1.33 Pa) for 10 min. The deposition parameters are summarized in Table 1.

The coating thickness was measured by using its cross sectional scanning electron micrographs and stylus profile meter (XP-200 Ambios Technology Inc., USA). The microstructural and topographical analyses were made by field emission scanning electron microscopy (FESEM, Model: 200F, FEI Quanta) and atomic force microscopy (AFM, NTMDT). The films composition was determined by the energy-dispersive X-ray (EDS) technique. The coatings were also analyzed by X-ray diffraction (XRD, Model: D8 Advance, Brucker), to determine the phase composition and orientation using a Ni-filtered CuKα X-ray source.

1.2 Result and Discussion

1.2.1 Chemical Analysis

For the elemental compositions EDS is performed and the Fig. 1a shows the Al/Cr atomic ratio of as deposited CrAlN film as a function of the ratio of sputtering currents applied onto the Al targets and Cr targets (I_{Al}/I_{Cr}). The result shows that the Al/Cr atomic ratio in the coatings increased monotonously with the increase with the sputtering current applied to the Al targets. The sputtering current ratio I_{Al}/I_{Cr} is much higher than the Al/Cr atomic ratio for each sample, indicating the much lower sputtering yield of aluminium target in compare to that of chromium target. This fact could be ascribed to the formation of an insulating AlN layer on

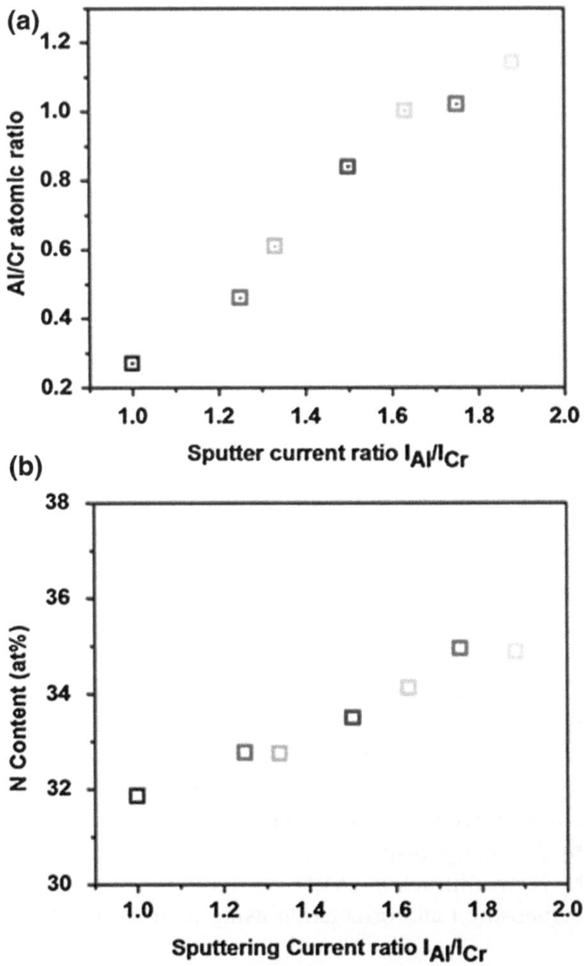

Fig. 1 **a** Al/Cr atomic ratio and **b** N content in the as-deposited CrAlN coatings as a function of the sputtering current ratio I_{Al}/I_{Cr}

the Al target surface during the reactive sputtering process, which is responsible for reduction in sputtering yield of Al. The nitrogen content in all of the as-deposited CrAlN coatings, as shown in Fig. 1b, was vary between 31 and 35 at %. It was found that the N content increased slightly with the increase of sputtering current applied on the Al targets, indicating that the aluminum incorporation is beneficial for the improvement of stoichiometric composition of the coatings [1].

1.2.2 Microstructural Analysis of CrAlN Coating Deposited on SA304 Substrate

With the addition of aluminum, the CrN coatings gradually crystallize, rocksalt-type cubic structure showing a CrAlN(111) as a preferential orientation along with

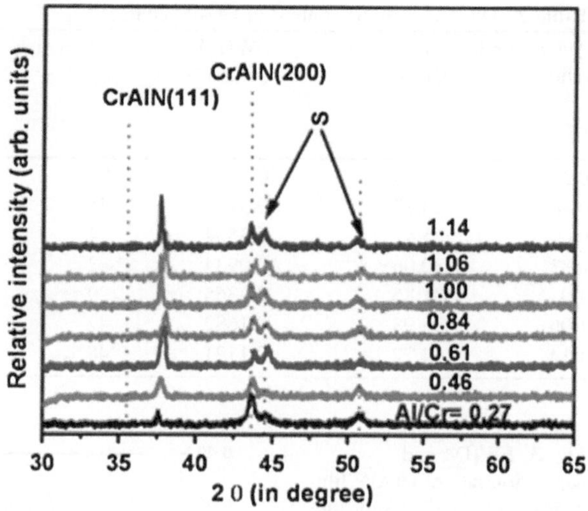

Fig. 2 The X-ray diffraction profiles of the as deposited CrAlN with different Al/Cr atomic ratio. The diffraction peaks marked with S comes from the stainless steel SA304 substrate

lower intense CrAlN(200) as shown in Fig. 2. The lower intensity of CrAlN(200) mainly due to higher nitrogen content in the chamber during deposition which favors (111) preferential growth due to comparatively lower strain energy. The Wurtzite-type (B4) structure phase was not observed as the AlN content in the coatings is below the critical composition for the phase transition from B1 to B4. It is reported that the solubility of AlN in cubic CrN has been predicted to go as high as 70 mol% [9, 10], the supersaturated CrAlN phase is thermodynamically metastable and hence the actual solubility limit finally depends on deposition condition [11].

The XRD result shows mainly two diffraction peaks (111) and (200). The intensity of both the peak increased and reach maximum then it is reduced to lower intensity value, which shows that the initially with the incorporation of Al content, the crystallinity increased or the crystallites in the coatings become increase. However after reaching maximum the peak intensity is reduced with further increase in Al content which leads to reduction in overall crystallinity of the film. The peak position of CrAlN(111) is shifted to higher angle side, indication a decrease in lattice constant gradually with increase in the aluminium content as shown in Table 2.

This shifting indicates the formation of the alloy nitride CrAlN and the fact could be attributed to lattice distortion caused by smaller aluminium atom incorporation which substitutes the chromium atom in the film. The incorporation of Al into the CrN coating, the full width at half maxima (FWHM) of the (111) diffraction peak increased to a maximum and then decreases with the increasing AlN content. But the abnormal result found for FWHM of CrAlN at 0.83 Al/Cr atomic ratio which has a lowest FWHM value among all. The abnormal result found may be due to formation of c-AlN. Generally the peak broadening of XRD profiles means the finer grain size and large residual micro stress. Also on the other

Table 2 The effect of Al content in CrAlN coating

Al/Cr atomic ratio	Al at % in CrAlN film	FWHM of (111) (°)	Grain size (nm)	Deposition rate (nm/min)	Surface roughness (nm)	
					RMS	Av
0.27	14.40	0.1495	88	35	7.42	5.73
0.46	20.77	0.3739	55	38	8.68	6.75
0.60	25.47	0.3504	24	40	8.81	6.76
0.83	30.06	0.0934	22	44	3.64	2.85
1.0	31.86	0.3365	29	42	8.02	6.45
1.06	33.98	0.1682	42	39	6.71	5.28
1.14	34.72	0.1121	48	37	9.34	7.54

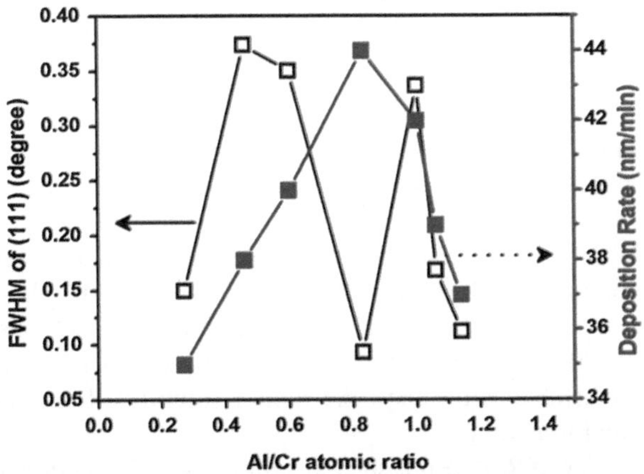

Fig. 3 FWHM and deposition rate of CrAlN film deposited on SA304 substrate as a finction of Al/Cr atomic ratio

side higher deposition rate will result in a finer grain size and a higher microstress [5, 6–7]. The deposition rate increased and touch at maximum at (44 nm/min) for the coating CrAl(30.06 at %)N having Al/Cr atomic ratio 0.86. Also the micro-stress is increased with deposition rate as mention in past literature [7], the higher kinetic energy of the incident ions distort the crystal lattice which leads to highest FWHM but in our case the CrAl(30.06 at %)N film having lowest FWHM. For clarification of above contra result, the grain size (Fig. 3) was calculated using Sherrrer formula.

The lattice constant 'a' is calculated from CrAlN(111) peaks for coating deposited on SA304 substrates and lattice microstrain 'ε' were calculated by using Eqs. 1 and 2 respectively [17–19].

$$\frac{1}{d^2} = \frac{h^2 + k^2 + l^2}{a^2} \tag{1}$$

where 'd' is the interplanar distance obtained from the position of the (111) peak using the Bragg condition, and

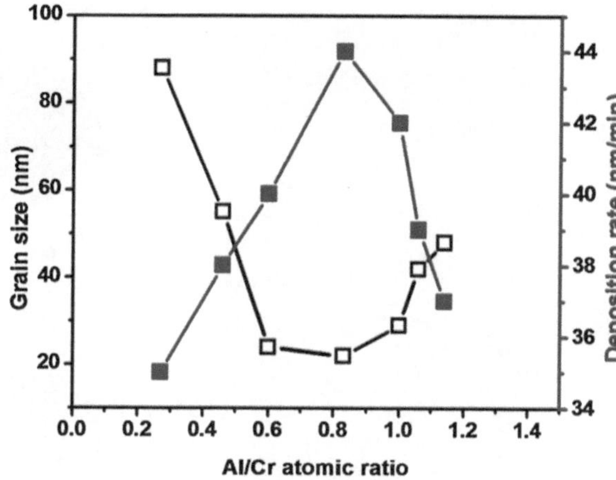

Fig. 4 Grain size and deposition rate of CrAlN film as a function of Al/Cr atomic ratio

$$\varepsilon = \frac{a - a_0}{a_0} \times 100 \quad (2)$$

where 'a' is the lattice constant in the strained condition and 'a_0' is the unstrained lattice constant (4.140 Å). The lattice constants were found to be decreasing with increasing Al content as shown in Table 2.

It has been reported in the literature [20] that the microstress increases with the deposition rate and hence coating thickness, because of the distortion of crystal lattice of the coating due to the comparatively high kinetic energy of the incident ions. The CrAlN coatings were deposited at the same temperature (573 K), and therefore thermal stresses of all the thin films were presumed to be the same in the present work. The grain size and deposition rate of CrAlN coating as a function of Al content is shown in Fig. 4. The grain size calculation of all the CrAlN film found that initially it is reducing and reach minimum for CrAl(30.06 at %)N film. The minimum grain size at 0.86 Al/Cr atomic ratio is due to higher deposition rate as shown in Fig. 4 at that atomic ratio. Also, the microstrain is found highest among all films as its deposition rate is highest.

Two dimensional surface topography of CrAlN coatings deposited at different Al content, were characterized by using atomic force microscopy (AFM) and the morphology of the films were obtained by using FE-SEM (Fig. 5). The coating becomes more compact and denser as seen in these micrographs. The overall roughness of the coating deposited on SA304 substrate was summarized in Table 2. The surface roughness is initially reducing up to 30.04 at % Al and then increasing with increasing Al content. Initially the reduction in surface roughness and high density of the coating may be due to increasing, applied voltage to Al target, which increase mobility of the atoms and results in higher nucleation density [21] during the formation of denser coating. The higher mobility of the adatoms can move into inter granular voids and diffuse in the films. Higher nucleation density helps to produce denser coating and atoms with high mobility diffuse into the intern granular voids [22, 23].

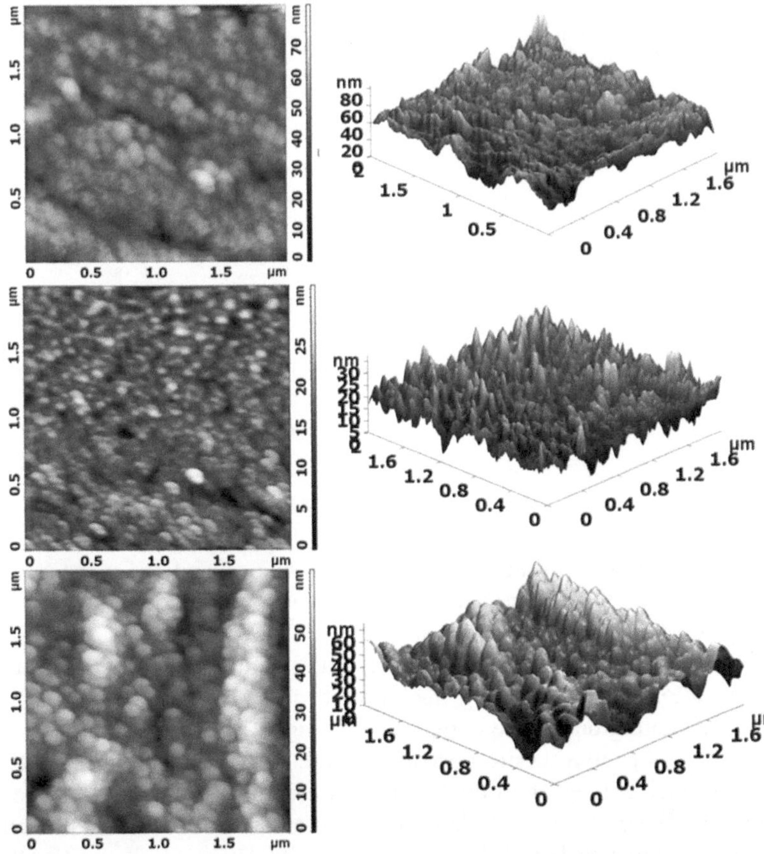

Fig. 5 2D and 3D AFM surface morphologies of CrAlN films deposited with different Al/Cr atomic ratio

2 Conclusion

CrAlN coatings with different Al content were deposited on SA304 substrates prepared by a magnetron sputtering technique. The deposited coatings are highly textured with dominating peak of CrN(111), with position being shifted to higher 2θ angle with increasing Al content deposited on SA304. The grain size is reduced initially up to 30.0 at % Al content and then grain size increases with further increase in Al content up to 7.15 %. The surface roughness is comparatively low, between 6 and 10 nm in all the coatings with varying Al content. The low surface roughness and high density of the coating may be due to increasing applied voltage to Al target, which increases mobility of the atoms and results in higher nucleation density leading to denser coating.

References

1. Kawate M, Hashimoto AK, Suzuki T (2003) Oxidation resistance of CrAlN and TiAlN films. Surf Coat Technol 165:163–167
2. Uchida M, Nihira N, Mitsuo A, Toyoda K, Kubota K, Aizawa T (2004) Friction and wear properties of CrAlN and CrVN films deposited by cathodic arc ion plating method. Surf Coat Technol 177–178:627–630
3. Romero J, Gómez MA, Esteve J, Montalà F, Carreras L, Grifol M, Lousa A (2006) CrAlN coatings deposited by cathodic arc evaporation at different substrate bias. Thin Solid Films 515:113–117
4. Shah HN, Jayaganthana R, Kaur D (2010) Influence of silicon content on the microstructure and hardness of CrN coatings deposited by reactive magnetron sputtering. Mater Chem Phys 121:567–571
5. Ding XZ, Zeng XT (2005) Structural, mechanical and tribological properties of CrAlN coatings deposited by reactive unbalanced magnetron sputtering. Surf Coat Technol 200:1372–1376
6. Barshilia HC, Selvakumar N, Deepthi B, Rajam KS (2006) A comparative study of reactive direct current magnetron sputtered CrAlN and CrN coatings. Surf Coat Technol 201:2193–2201
7. Pulugurtha SR, Bhat DG (2006) A study of AC reactive magnetron sputtering technique for the deposition of compositionally graded coating in the Cr–Al–N system. Surf Coat Technol 201:4411–4418
8. Li T, Li M, Zhou Y (2007) Phase segregation and its effect on the adhesion of Cr–Al–N coatings on K38G alloy prepared by magnetron sputtering method. Surf Coat Technol 201:7692–7698
9. Mayrhofer PH, Music D, Reeswinkel Th, Fuß H-G, Schneider JM (2008) Structure, elastic properties and phase stability of Cr1–xAlxN. Acta Mater 56:2469–2475
10. Brizuela M, Garcia-Luis A, Braceras I, Oñate JI, Sa′nchez-Lo′pez JC, Martı′nez-Martı′nez D, Lo′pez-Cartes C, Ferna′ndez A (2005) Corrosion resistance of CrAlN and TiAlN coatings deposited by lateral rotating cathode arc. Surf Coat Technol 200:192–197
11. Ding XZ, Tan ALK, Zeng XT, Wang C, Yue T, Sun CQ (2008) Corrosion resistance of CrAlN and TiAlN coatings deposited by lateral rotating cathode arc. Thin Solid Films 516:5716–5720
12. Shah HN, Jayaganthan R (2012) Influence of Al contents on the microstructure, mechanical, and wear properties of magnetron sputtered CrAlN coatings. J Mater Eng Perform 21(9):2002–2009
13. Sun Y, Wang YH, Seow HP (2004) Effect of substrate material on phase evolution in reactively sputtered Cr-Al-N films. J Mater Sci 39:7369–7371
14. Chunyan Y, Linhai T, Yinghui W, Shebin W, Tianbao L, Bingshe X (2009) The effect of substrate bias voltages on impact resistance of CrAlN coatings deposited by modified ion beam enhanced magnetron sputtering. Appl Surf Sci 255:4033–4038
15. Sanche′z JE, Sanche′z OM, Ipaz L, Aperador W, Caicedo JC, Amaya C, Herna′ndez Landaverde MA, Espinoza Beltran F, Muñoz-Saldaña J, Zambrano G (2010) Mechanical, tribological, and electrochemical behavior of CrAlN coatings deposited by r.f. reactive magnetron co-sputtering method. Appl Surf Sci 256:2380–2387
16. Makino Y, Nogi K (1998) Synthesis of pseudobinary CrAlN films with B1 structure by rf—assisted magnetron sputtering method. Surf Coat Technol 98:1008–1012
17. Gautier C, Machet J (1997) Study of the growth mechanisms of chromium nitride films deposited by vacuum ARC evaporation. Thin Solid Films 295:43–52
18. Pelleg J, Zevin LZ, Lungo S, Croitoru N (1991) Reactive sputter deposited TiN films on glass substrates. Thin Solid Films 197:117–128

19. Schell N, Petersen JH, Bøttiger J, Mucklich A, Chevallier J, Andr easen KP, Eichhorn F (2003) On the development of texture during growth of magnetron-sputtered CrN. Thin Solid Films 436:100–110
20. Lugscheider E, Bobzin K, Hornig Th, Maes M (2002) Investigation of the residual stresses and mechanical properties of (Cr,Al)N arc PVD coatings used for semi-solid metal (SSM) forming dies. Thin Solid Films 420–421:318–323
21. Cunha L, Andritschky M, Pischow K, Wang Z (1999) Microstructure of CrN coatings produced by PVD techniques. Thin Solid Films 355–356:465–471
22. Lin J, Moore JJ, Mishra B, Pinkas M, Sproul WD, Rees JA (2008) Effect of asynchronous pulsing parameters on the structure and properties of CrAlN films deposited by pulsed closed field unbalanced magnetron sputtering (P-CFUBMS). Surf Coat Technol 202:1418–1436
23. Cheikh Larbi AB, Tlili B (2006) Fretting wear of multilayered PVD TiAlCN/TiAlN/TiAl on AISI 4140 steel. Surf Coat Technol 201:1511–1518

Tribology Characteristics of Cold Rolled Spray Cast Al–6Si–20Pb Alloy

R. Mittal and D. Singh

Abstract Microstructural and tribological characteristics of cold rolled spray cast Al–6Si–20Pb alloy were studied for different percentage of thickness reduction. The length of aluminum grains and the width of grain boundary were observed to increase with the increase in percentage of thickness reduction. Wear rate was found to decrease with the increase in percentage of thickness reduction for all the applied loads.

Keywords Al–Si–Pb alloy · Spray casting · Rolling · Wear

1 Introduction

The use of Al–Pb alloys as bearing materials has been well known. The wearing properties of Al–Pb alloys can be further improved by adding alloying elements such as Cu, Mg and Si [1]. Due to very wide miscibility gap between Al and Pb, problem arises in the preparation of Al–Pb using the conventional casting technique. Spray forming technology has been widely used to produce preforms with homogenous microstructure. However some porosity is always there in the spray deposited perform which detroit the mechanical properties [2]. In order to eliminate porosity and to obtain high mechanical properties, the spray deposited preform must further be densified. Forging, extrusion, rolling, etc. are effective methods used to produce fully dense metal sheet as well as to achieve the high mechanical properties [3]. Out of these rolling is most effective method because of having both hydrostatic and deviatoric components. The application of these

R. Mittal (✉) · D. Singh
Department of Metallurgical and Materials Engineering, Indian Institute of Technology, Roorkee 247667, India
e-mail: rashmimittal3@gmail.com

H. C. Patel et al. (eds.), *Proceedings of International Conference on Advances in Tribology and Engineering Systems*, Lecture Notes in Mechanical Engineering, DOI: 10.1007/978-81-322-1656-8_39,

components on porous materials causes flattening of pores and brings about rapid densification by collapsing the pores at a faster rate [4].

The rolling behaviour of steel backed spray deposited Al–Sn strip has been studied by Tripathi et al. [5]. They found that during rolling densification in the Al–Sn deposit take place. Deshmukh et al. [4] have studied the analysis of cold densification rolling of a sintered porous metal strip. For 25 % thickness reduction the density was observed to decrease. Longitudinal cracks developed in the strip for a thickness reduction of 30 % or greater in a single pass. The wear behavior of cast and mushy state rolled Al–4.5Cu alloy and in situ Al4.5Cu–5TiB_2 composites have been studied by Mervin et al. [6] and they observed that both the mushy state rolled alloy and the composite have shown lower wear rates compared to that observed when those are in as cast condition.

In the present work Al–6Si–20Pb alloys were spray formed in the form of a disc and then cold rolled. The changes in microstructural and tribological characteristics are reported for different percentage reduction in thickness.

2 Experimental

A spray forming set-up is given elsewhere [7]. The base alloy composition used in the present work consisted of Al–6Si. The alloy was superheated to 200 °C in the graphite crucible using an induction furnace. A 20 wt% of lead was added in the Al–6Si alloy before atomization of melt. Nitrogen gas at a pressure of 10 bar was supplied for atomization prior to melt flow. The spray droplets were deposited over a copper substrate which was kept at a distance of 400 mm from the nozzle. The preform was taken out of the substrate after deposition and then samples were cut from the preform. After that these samples were cold rolled in a two-high mill. The speed of the rolls was 8 rpm and the diameter of the rolls was 110 mm.

Above samples were cut down for its microstructural study. These samples were polished using standard metallographic technique of polishing with an emery paper of 1/0, 2/0, 3/0 and 4/0 specification and then followed by wheel cloth polishing using an emulsion of alumina powder particles suspended in water. Afterwards, these samples were etched with keller's reagent and examined with Letiz optical microscope.

The cold rolled spray cast samples were machined into cylindrical pins of 6 mm diameter. Wear testing of cold rolled spray cast alloys was investigated using a pin on disc type wear testing machine. This machine consisted of a hardened EN-24 steel disc having surface roughness = 0.4–0.5 μm, diameter 12 cm, Rockwell hardness of 57 HRC and a specimen holder. Wear test samples was mounted in the specimen holder. Disc rotates at a constant speed of 2 m/s throughout the experiment. The load on the sample was applied by placing a load on the opposite side of a fulcrum of the lever attached to the specimen holder. The radius of rotation was fixed at 36.5 mm. Sliding distance for one run comes out to be around 3,000 m for this radius. The samples were run for a period of 25 min. Weight loss of the sample was determined which was used to find the resultant wear volume. The standard wear test procedure was followed for evaluating the wear rate for different load ranging from 10 to 50 N. The worn surfaces of the wear pins after the test and the debris generated during wear experiments were examined by using JEOL JXA840A scanning electron microscope (SEM).

3 Results and Discussion

3.1 Microstructure

Figure 1 shows the optical micrographs for (a) 0, (b) 20, (c) 40, (d) 60 and 80 % thickness reduction of spray deposited Al–6Si–20Pb alloy. The microstructure exhibits fine equiaxed grain morphology of the Al phase as shown in Fig. 1a.

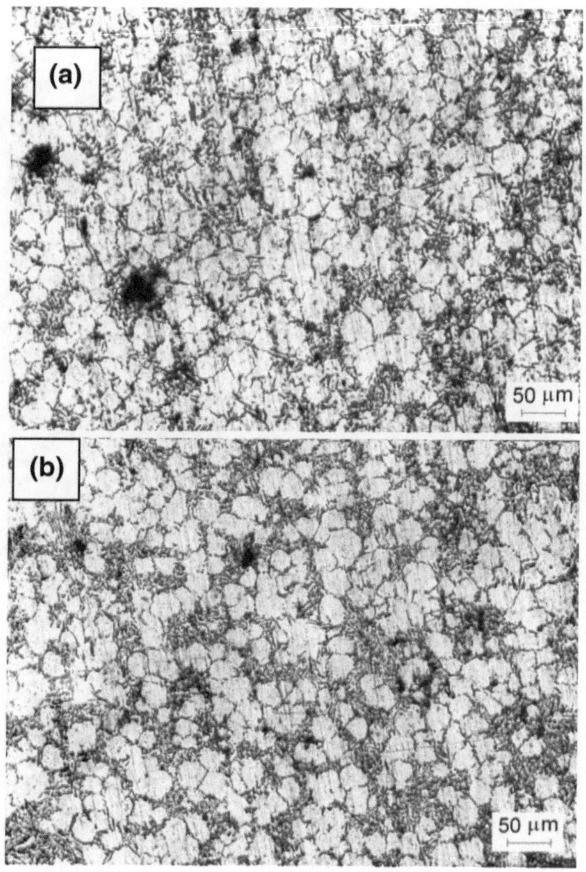

Fig. 1 Microstructure of spray deposited Al–6Si–20Pb alloy for **a** 0, **b** 20, **c** 40, **d** 60 and **e** 80 % thickness reduction

Fig. 1 continued

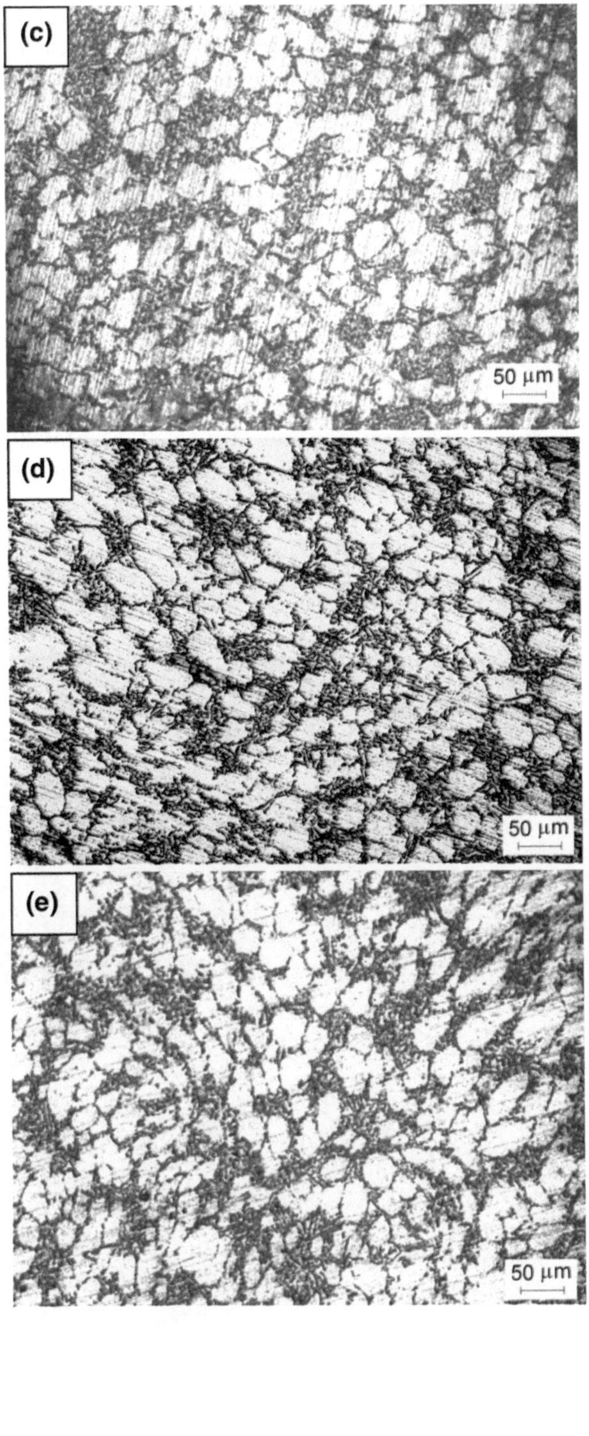

The identification of phases in spray deposited alloy was carried out by EDS analysis. The results indicate that the dark region corresponded to lead phase and the gray contrast region predominantly contained the Si phase whereas the bright region was of Al phase. A uniform distribution with a particulate morphology of primary Si phase has been shown in spray deposit Al–6Si–Pb alloy. The shape of the Pb phase is irregular in feature and it is located mainly at the grain boundaries. It can be seen that Pb distribution is almost uniform throughout the aluminum phase. The average Al grain size is 20–30 μm and the size of Si particles is submicron to 5 μm. Some irregular pores of 10–15 μm size can also be seen in the matrix.

Grains are elongated in the rolling direction as shown in Fig. 1b–e. Wherever there is lead phase along the grain boundary, the width of the grain boundary increases due to the spread of lead on rolling. The average length of Al-grains is also increases with the thickness reduction. In Fig. 1b and c, the average length of Al-grains is 25–35 μm. The average length of Al-grains is 35–45 μm in Fig. 1d and e.

The microstructural features of spray deposited preform are due to rapid solidification achieved during atomization of melt into fine droplets. The turbulent fluid flow condition and the high velocity gas jets give rise to the fragmentation of dendritic arms and thus homogeneity in the microstructure [8]. It is well known that there exists a certain amount of porosity in the spray deposited preform. The reduction in thickness is accompanied by an elongation of the grains in the rolling direction [9] as shown in Fig. 1b–e. In the initial stage of rolling, the metal flow is mainly in the thickness direction in the Al–Si–Pb deposit. Porosity is removed by restacking and rearrangement of spray deposited particles. Afterwards, plastic deformation becomes the predominant mechanism of densification during rolling [5]. As a result the porosity is eliminated by a process involving pore elongation followed by fragmentation into several smaller size pores in the direction of rolling. The opposite faces of the pores collapse when the width of the pores becomes very small and the porosity is almost eliminated. If there is an oxide layer in the pores that can also rapture by the deviatoric stresses but it would be difficult to have a proper bonding between the surfaces of the pores.

3.2 Wear Rate

Figure 2 shows the variation of wear rate of cold rolled spray deposit Al–6Si–20Pb alloy as a function of thickness reduction at different applied loads. It can be seen that wear rate decreases up to 40 % thickness reduction and then increases at 60 % thickness reduction at all the applied loads. Beyond 60 % thickness reduction wear rate again decreases. Obviously, wear rate is higher at higher loads.

Wear rate as a function of applied load for different thickness reduction in Al–6Si–20Pb alloy is shown in Fig. 3. Wear rate increases with the increase in applied load for all the thickness reduction. Wear rate versus load variation can be

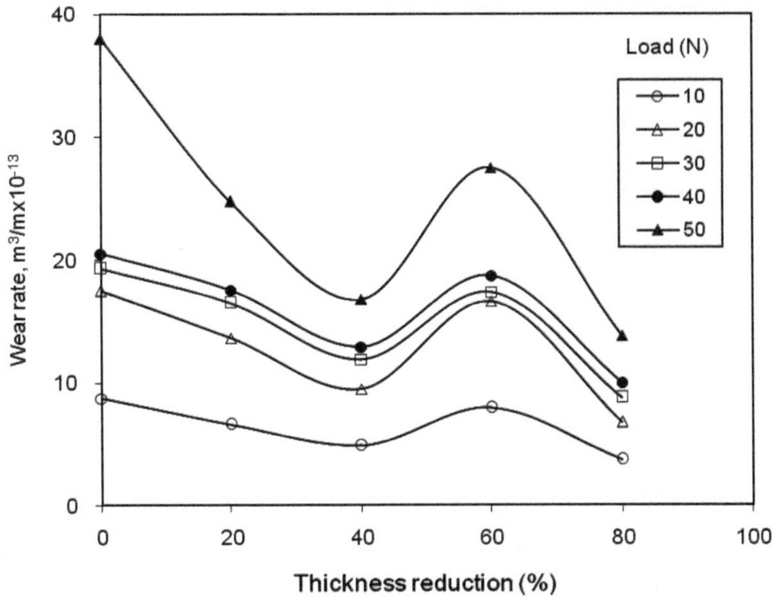

Fig. 2 Wear rate variation with thickness reduction for different applied loads

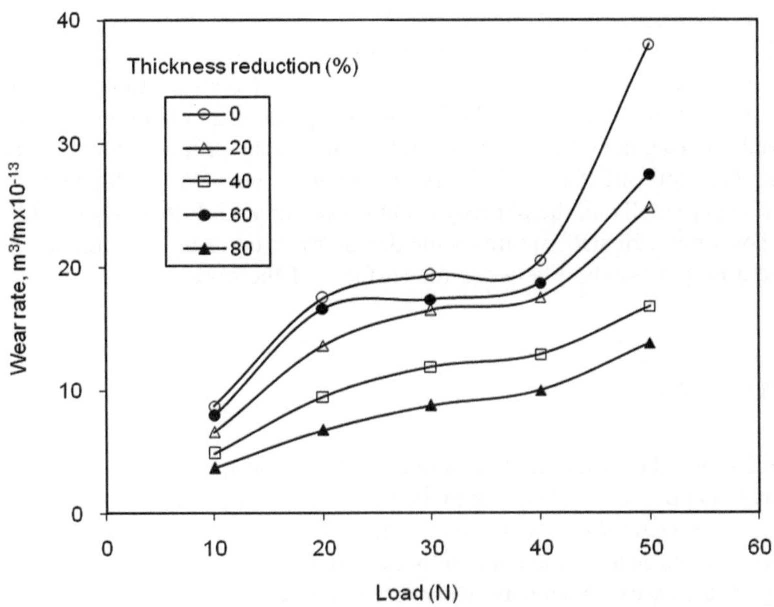

Fig. 3 Wear rate variation with applied loads for different thickness reductions

Fig. 4 SEM micrographs of worn surfaces and debris of Al–6Si–20Pb alloy for without (**a** and **b**) and with 80 % (**c** and **d**) thickness reduction at the load of 50 N

divided into three regions having three different rate of increase in wear rate with the increase in applied load.

In first region i.e. up to a load of 30 N, wear rate increases rapidly. In second region i.e. up to a load of 40 N, it is almost constant. In third region i.e. up to a load of 50 N, the wear rate again increases at a higher rate as compared to that of region one.

Porosity plays an important role on wear rate [10]. Porosity decreases with thickness reduction but at 60 % reduction, porosity increases due to the formation of cracks [11]. Therefore, wear rate decreases at all the percentage of thickness

Fig. 4 continued

reduction except 60 % reduction. At 60 % reduction, wear rate increases due to the increase in porosity. Beyond 60 % reduction, cracks annihilate and porosity decreases due to which wear rate decreases. The high applied promoted subsurface damage, eventually resulting in higher wear rate.

3.3 Worn Surface and Debris

Scanning electron micrographs of worn surfaces and debris of Al–6Si–20Pb alloy for without and with 80 % thickness reduction at the load of 50 N are shown in Fig. 4. In Fig. 4a, river like waves occur which is the indication of delamination wear mechanism. In this mechanism, a high amount of material transferred from pin to disk. While in Fig. 4c, smoothness of the worn surfaces shows the symptoms of oxidative wear. In this type of wear mechanism, a layer of oxide formed and covers the worn surface. Due to this oxide layer wear rate decreases.

Round shape and spherical debris formed in case of without rolled samples while flake type and platelets debris formed in case of 80 % rolled samples as shown in Fig. 4b and d respectively.

4 Conclusions

The grains are elongated and length of the Al grains increases with the thickness reduction as well as the width of the grain boundary increases due to the spread of lead on rolling. In general wear rate of spray cast Al–6Si–20Pb alloy decreases with the increase in thickness reduction and it is minimum for 80 % thickness reduction.

References

1. Zhu M, Gao Y, Chung CY, Che ZX, Luo KC, Li BL (2000) Improvement of the wear behavior of Al-Pb alloys by mechanical alloying. Wear 242:47–53
2. Hariprasad S, Sastry SML, Jerina KL, Lederich RJ (1993) Microstructure and mechanical properties of dispersion-strengthened high-temperature Al-8.5Fe1.3V-1.7Si alloys produced by atomized melt deposition process. Metall Trans A 24:865–875
3. Zhan MY, Chen ZH, Yan HG, Xia WJ (2007) Deformation behaviors of porous 4032 Al alloy preform prepared by spray deposition during hot rolling. J Mat Process Tech 182:174–180
4. Deshmukh AR, Sundararajan T, Dube RK, Bhargava S (1998) Analysis of cold densification rolling of a sintered porous metal strip. J Mat Process Tech 84:56–72
5. Tripathy MR, Dube RK, Koria SC (2007) Rolling behaviour of steel backed spray deposited Al-Sn strip. J Mat Process Tech 190:342–349
6. Mervin HA, Maiti R, Mitra R, Chakraborty M (2008) Wear behaviour of cast and mushy state rolled Al-4.5Cu-5TiB$_2$ composite. Wear 265:1606–1618
7. Singh D, Dangwal S (2006) Effects of process parameters on surface morphology of metal powders produced by free fall gas atomization. J Mat Sci 41:3853–3860
8. Anil M, Srivastava VC, Ghosh MK, Ojha SV (2010) Influence of tin content on tribological characteristics of spray formed Al-Si alloys. Wear 268:1250–1256
9. Rizzi P, Habib A, Castellero A, Battezzati L (2011) Constrained deformation of an Al based amorphous alloy by cold rolling. J Alloys Compd 5095:5275–5278
10. Hamid AA, Ghosh PK, Jain SC, Ray S (2006) Influence of particle content and porosity on the wear behaviour of cast in situ Al(Mn)-Al$_2$O$_3$(MnO$_2$) composite. Wear 260:368–378
11. Mittal R, Singh D (2012) Effect of cold rolling on the mechanical properties of spray cast Al-6Si-20Pb alloy. Adv Mat Res 585:402–406

Part VI
Skeleton and Structural Systems

An Innovative Approximate Method for Analysis of Continuous Beam

D. J. Varia and H. S. Patel

Abstract The safety and serviceability of a structure is dependent on how accurately the forces and the response associated with it are determined. Many precise methods of analysis have been well documented in the literature for structural analysis. Computer programs provide results with good cost and time efficiency. The main problem is that structural engineers are using these software's as black box and gross errors are left undetected. This problem can be overcome by using fast and efficient methods which yield results which are approximate and acceptable. These methods are called approximate methods and they have been used successfully for the analysis of the structures. An attempt to give overview of such approximate method for the solution of continuous beam is briefly done in this paper. This paper also intends to compare this approximate method of structural analysis to the values obtained from classical method. For practical purposes approximate solution is worked out using this approach, which is 90 % accurate, and this is to be justified on the grounds of resulting speed, clarity of calculation and feeling of structural behavior.

Keywords Approximate method · Continuous beam · Indeterminate structure · Classical methods

1 Introduction

Behr et al. [1] suggested some assumptions in selection of point of inflection in beam and column to obtain values closer to exact method. Potential errors in approximate methods of structural analysis have also brought out Behr and Goodspeed [2]. Wu [3] has carried out approximate analysis of building frames for vertical loads. Khuda

D. J. Varia (✉) · H. S. Patel
Department of Applied Mechanics, Government Engineering College, Patan, Gujarat 384265, India
e-mail: djvaria@gmail.com

H. C. Patel et al. (eds.), *Proceedings of International Conference on Advances in Tribology and Engineering Systems*, Lecture Notes in Mechanical Engineering, DOI: 10.1007/978-81-322-1656-8_40,

[4] developed moment coefficient to assist designers. Okonkwo et al. [5] developed the mathematical model for the evaluation of the internal support moments of a uniformly loaded continuous beam of equal span.

Numerous classical and matrix methods are available for analysis of continuous beam. Methods like Moment distribution and Kani's method are iterative type and accuracy of results will be obtained through more iteration [6]. Three moment equation, slope deflection method, energy method, consistent deformation method etc. are based on solution of simultaneous equations. While in stiffness and flexibility methods, linear simultaneous are to be formulated in matrix forms [7].

In all the above methods, one has to carry out iterations or to solve equations and that is most time and computer memory consuming part. To overcome these difficulties, an innovative approximate method is presented with steps for approximate solution of the problem.

1.1 Philosophy of Approximate Solution

In some circumstances, regardless of the particular computation method applied, prevent the carrying out of exact design calculations. At the same time it is to demonstrate that only reasonable accuracy of calculation is essential, since all assumptions do not always properly interpret the true structural action of the chosen system and building material.

It is quite usual to completely neglect the existing special continuity and the whole structure is replaced by a number of substitute structure, which are analyzed as two-dimensional problems. The effect of this plane continuity is often taken into account very accurately, despite small significance of the more remote span.

The resultant stresses are generally much more influenced by the stiffness factor, than by the number of neighboring spans. For this reason the difference in moments of inertia of individual structure-members should never be neglected, to achieve desired accuracy of results. The stiffness factor despite their decisive importance cannot be exactly evaluated, specially for RC structures.

The effect of normal forces, temperature, shrinkage, sinking of support, etc. are also neglected, at least for ordinary type of structures, despite the fact that in some cases these effects may change the calculated stresses by considerable percentage.

Looking to above factors, one can never get an exact solution, as exact mathematical modeling of a physical phenomenon is a formidable challenge. As a result one has to accept an approximate method giving numerically compatible results.

2 Steps for Approximate Solution of Continuous Beams

The method is demonstrated through benchmark problem and it is proved that the present approach gives exact result when compared numerically with the existing classical and matrix methods.

2.1 *Definition of Terminology*

2.1.1 Relative Deformation Coefficient (Cr)

Relative deformation coefficient is deformation of far end of a beam member if unit deformation is applied at near end of the member.

If unit rotation is applied to the near end of a fixed beam then Cr at far end is zero due to the fixed support. But in case of propped cantilever, if unit rotation is applied to fixed near end, then Cr at far end is 0.5 due to hinged support.

2.1.2 Fixity Coefficient (C_f)

Fixity coefficient gives the fixity provided by far end. The value of C_f at near end is always one and same for far end is dependent on relative deformation coefficient Cr at far end.

2.1.3 Actual Deformation (A_D)

Actual deformation of a joint is joint deformation due to unit deformation of any other joint. Actual deformation of a joint is computed by multiplying actual deformation of preceding joint with relative deformation coefficient of the joint.

2.2 *Procedural Steps for Continuous Beams*

1. Choose suitable sign conventions for forces and deformations.
2. Identify any one unknown action at a joint.
3. Fix up the number of spans adjoin to selected joint for contribution in calculation of unknown action. For joint at extreme support number of span can be one span/two spans/three spans/four spans/more on one side as per accuracy desired. For intermediate joint number of span can be one span/two span/three spans/more as per accuracy desired.
4. Compute relative deformation co-efficient Cr and fixity co-efficient C_f for all joints except at joint where unknown action is identified. Start computing from extreme support/s and move towards joint where unknown action is identified. If far ends of selected joint are not extreme support, than Cr and C_f will fall between ½ to 0 and ¾ to 1 respectively. As approximations Cr and C_f will be taken ¼ and ¼ respectively.
5. Take A_D equal to unity i.e. deformation corresponding to unknown force. Start computing Actual deformation A_D at each joint from joint where unit deformation is applied.

6. Compute fixed end actions corresponding to deformations.
7. Compute the sum of multiplication of actual deformations A_D with fixed end actions. The sum yields value of identified unknown action.

3 Analysis

To demonstrate the application of innovative approximate approach problem of multi-span continuous beam as per Fig. 1 is selected. Sample calculations are depicted for some joints and finally comparison with exact values are presented. This method is useful with any combination of loading, any no of spans and the secondary effects can also be incorporate in terms of fixity (Tables 1, 2, 3 and 4).

$M_A = \sum A_{D*}$ FEM $= 24.10$ kN m

Actual value of $M_A = 27.13$ kN m

Percentage Error $= -11.17$ % (Tables 2, 3 and 4)

Similar computation have been carried out for all other joints and results are compared in Table 5 (Figs. 2, 3, 4 and 5).

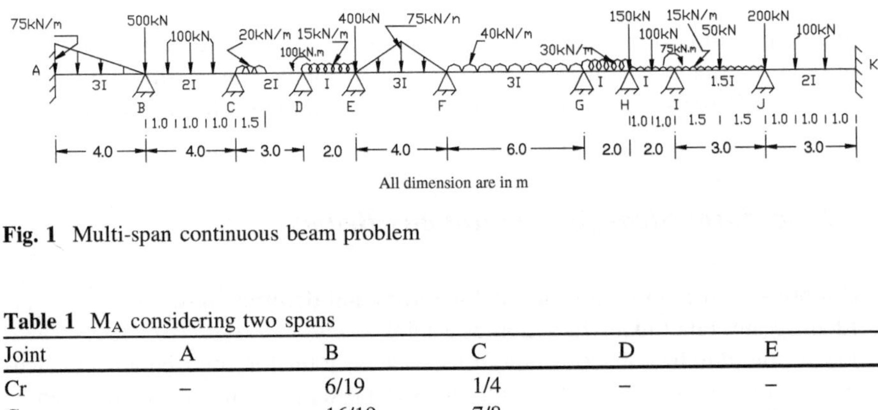

Fig. 1 Multi-span continuous beam problem

Table 1 M_A considering two spans

Joint	A	B	C	D	E
Cr	–	6/19	1/4	–	–
C_f	–	16/19	7/8	–	–
A_D	1	−6/19	3/38	–	–
FEM	60	85	−114.69	–	–
A_{D*} FEM	60	26.84	−9.05	–	–

Table 2 M_A considering four spans

Joint	A	B	C	D	E
Cr	–	0.31	53/226	16/53	1/4
C_f	–	0.84	399/452	45/53	7/8
A_D	1	−3148	0.07	−0.02	0.0056
FEM	60	85	−114.69	−99.68	57.50
A_{D*} FEM	60	−26.75	−8.46	2.22	0.32

Table 3 M_C considering two spans

Joint	A	B	C		D	E
Cr	–	1/4	–	–	1/4	–
C_f	–	7/8	–	–	7/8	–
A_D	–	−1/7	4/7	3/7	−3/28	–
FEM	–	−85	125	10.31	−99.68	–
A_{D*} FEM	–	12.14	71.42	4.41	5.24	–

Table 4 M_C considering four spans

Joint	A	B	C		D	E
Cr	0	1/5	–	–	16/53	1/4
C_f	1	9/10	–	–	45/53	7/8
A_D	0	−0.11	0.55	0.44	−0.13	0.03
FEM	60	−85	125	10.31	−99.68	57.50
A_{D*} FEM	0	9.46	69.63	4.56	13.32	1.92

Table 5 Table of moment comparison and % error

Joint	Actual moment in kN m	Computed moment considering two span in kN m	% error	Computed moment considering four span in kN m	% error
A	27.136	24.10	−11.17	27.32	0.68
B	105.72	111.79	5.73	105.33	−0.36
C	97.54	93.23	−4.41	98.92	1.41
D	−72.16(L)	−74.37	3.05	−72.71	0.76
	27.83(R)	25.63	−7.93	27.28	−1.97
E	8.95	8.95	−18.53	7.29	−10.00
F	124.04	119.25	−3.86	125.37	1.07
G	67.51	69.68	3.22	68.63	1.66
H	3.50	4.25	21.41	3.12	10.72
I	66.49	66.25	−7.88	65.97	−0.78
	−8.50	−10.75	20.89	9.02	6.10
J	67.52	69.05	2.27	67.68	0.24
K	116.23	115.47	−0.66	115.55	−0.58

4 Results and Discussion

1. The undetected errors of computer software's can be easily and quickly checked/detected accurately with this approximate method.
2. Tables 1 and 2 shows M_A calculated by approximate method considering two spans and four spans respectively.
3. Tables 3 and 4 shows M_C calculated by approximate method considering two spans and four spans respectively.

Fig. 2 Actual moment

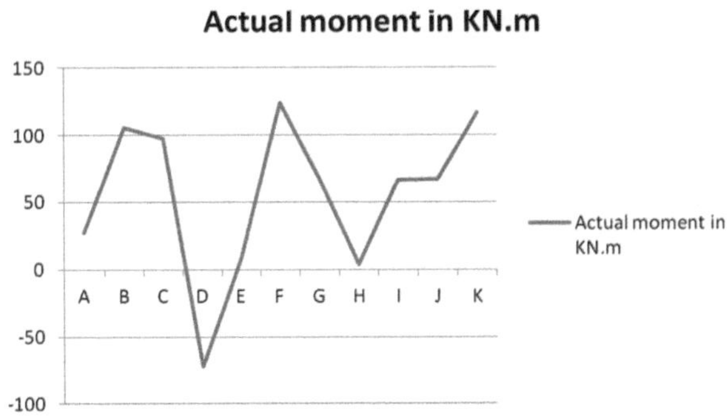

Fig. 3 Moment considering two spans

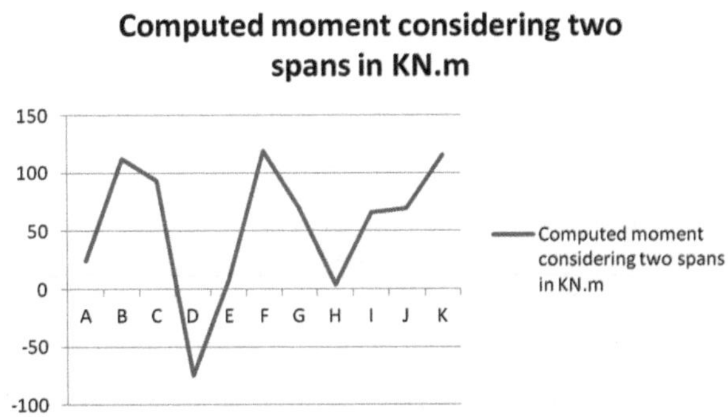

Fig. 4 Moment considering four spans

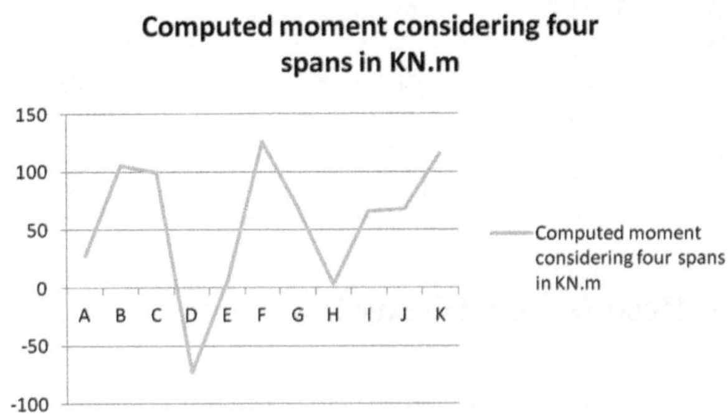

4. From the Table 5, it is observed that by the moment calculated by approximate method is very much near to moment calculated by classical method.
5. Computed moment considering two spans percentage error is approximately 21 % but same is 10 %, if four spans are considered (Table 5).

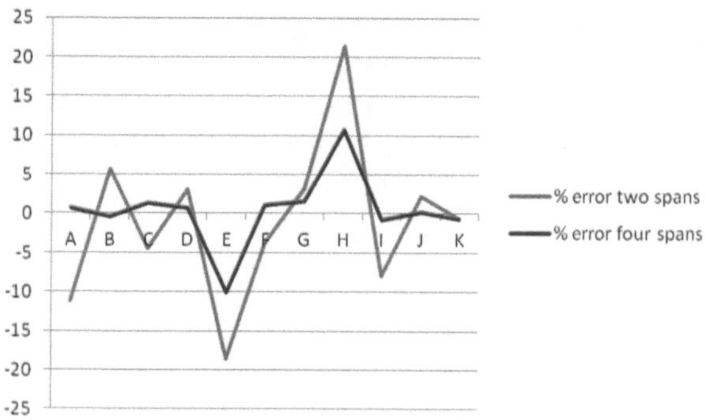

Fig. 5 % error in computed moment

6. Figure 2 shows actual moment calculated by classical method where Figs. 3 and 4 show moment calculated by approximate method considering two and four spans respectively.
7. Figure 5 revels that consideration of more spans yields highly accurate results.

5 Conclusion

From the above results and discussion it can be concluded that one has to accept that an approximate method for the solution of indeterminate structure like continuous beam giving numerically compatible results. This method is more helpful for design offices for the practical design application because one can quickly analyze the structure with 90 % accuracy.

References

1. Behr RA, Grotton EJ, Dwinal CA (1990) Revised method of approximate structural analysis. J Struct Div ASCE 116(11):3242–3248
2. Behr RA, Goodspeed CH (1990) Potential errors in structural analysis. J Struct Eng 116(11):3242–3247
3. Wu JH (1995) Approximate analysis of building frames for vertical loads. J Struct Eng 121(4):784–787
4. Khuda SN (2011) Design aid for continuous beam. Military Institute of Science and Technology, Dhaka
5. Okonkwo VO, Aginam CH, Chidolue CA (2010) Analysis of internal support moments of continuous beams of equal spans using simplified mathematical model approach. J Sci Multidiscip Res 2:72–80
6. Reddy CS (1987) Basic structural analysis. Tata Mc-Graw Hill, New Delhi
7. Wang CK (1983) Intermediate structural analysis. Mc-Grow Hill International Book Company, New York

Seismic Response of RC Elevated Water Tank Considering Site Specific Acceleration Time History

Chirag N. Patel and H. S. Patel

Abstract Elevated liquid storage tanks are extensively used to store water, inflammable liquids and other chemicals. Also, required to remain competent enough during and after earthquake to ensure availability of water supply in earthquake affected regions. A seismic analysis of such tanks must take into account considering the fluid–structure interaction and sub soil condition, which are highly responsible for the amplification of ground motion. In present paper time history analysis of elevated water tank is carried out for more realistic seismic analysis using finite element software SAP 2000 [1]. Site specific soil data has been collected for six different sites of Ahmedabad, metropolitan city of India and using one dimensional site specific ground response analysis software EERA [2] modified time history for acceleration data records of Bhuj earthquake 2001 is developed. Under modified time history analysis responses like base shear, overturning moment, top and sloshing displacement have been compared and contrasted. A study reflects the importance of site specific time history analysis and result shows that the structure responses are exceedingly influenced by water level and different subsoil condition.

Keywords Water tank · Seismic analysis · Site specific response

C. N. Patel (✉)
Faculty of Engineering, Pacific Academy of Higher Education and Research University, Udaipur, Rajasthan 313024, India
e-mail: cnpatel.693@gmail.com

H. S. Patel
Applied Mechanics Department, L. D. College of Engineering, Ahmedabad, Gujarat 380015, India

H. C. Patel et al. (eds.), *Proceedings of International Conference on Advances in Tribology and Engineering Systems*, Lecture Notes in Mechanical Engineering, DOI: 10.1007/978-81-322-1656-8_41,

1 Introduction

Water and its supply is very essential for humans and other life forms immediately after destructive earthquakes to reduce damages happens due to uncontrolled fires. So, the water supply system should be designed properly as seismic resistant, and seismic performance of elevated water tanks draws special importance in the structural design practice. In many water retaining structures distress has been observed even in 9–10 years of service life due to some problems related to structural aspects [3]. Also, from past earthquakes it has been seen that, the past designs of elevated water tank structures are tremendously vulnerable under earthquake forces and the Bhuj (India) earthquake provided another illustration when a great number of water tanks are damaged and few collapsed [4]. Unsuitable structural configuration, substandard materials and workmanship, reinforcement corrosion, wind and earthquake forces etc. are the reasons to deterioration of water tank [5, 6].

In current Indian code for seismic design IS 1893 (Part-1): 2002 [7], limited provisions on seismic design of elevated tanks are given. Those provisions are highly inadequate compared to present international practice [8]. The failure of most of the elevated water tanks in the Bhuj earthquake was also attributed to single mass idealisation of liquid. The revision of the current Indian code thus became inevitable. Indian Institute of Technology, Kanpur has proposed guidelines along with commentary and explanatory examples for seismic analysis of liquid storage tanks in association with Gujarat State Disaster Management Authority [9]. This publication cover seismic analysis and design of ground supported as well as elevated tanks including two degree of freedom idealization taking into account impulsive and convective hydrodynamic forces for convincing assessment of dynamic properties of tanks.

During an earthquake there is release of energy, which reaches to the ground surface by means of seismic waves travel through bed-rock and varied soil strata below the structure. The various soil parameters affect the properties of the seismic waves, which in turn affect the intensity of ground shaking or peak ground acceleration (PGA) at different sites. In order to determine the response of a soil deposition deposit to the motion of the bedrock immediate below the soil the effect of local soil condition on amplification of seismic waves on the intensity of ground shaking, site specific seismic analysis is performed. Estimation of seismic response on structures is an important aspect for earthquake resistant design of structures. The condition of the sub soil has an important contribution in influencing the response of ground surface under specified bed rock motion.

As per IS 1893 (Part-1): 2002 zone factor depends on location of building which represents peak ground acceleration of site and response spectrum which reflects the dynamic characteristics of building i.e. time period (natural period). Response spectrum is developed based on observation of time history i.e. records of displacement, velocity and acceleration at regular interval of time on the ground surface during the event of earthquake using accelerograph. But as specified in that, it is necessary to carry out site specific investigation for important structures which suggests that the standard response spectrum available from the Codal provisions is influenced by the local site conditions like soil profile, bedrock depth variations in the local characteristics of soil etc. However the response spectra of Codal provisions reflect only the general soil conditions and not the local sub soil characteristics which majorly influence the response of ground surface. A site specific ground response analysis is thus recommended for important structures rather than relying upon code specified response spectrum. The stimulus of the present study is to understand the seismic behaviour of reinforced concrete elevated water tank under different level of water inside the tank container, and soil strata below the structure. The responses like base shear, overturning moment, top and sloshing displacement have been compared and contrasted to reflect the importance of site specific time history analysis.

2 Model Provisions

A suitable spring mass analogue for two mass models to characterize basic dynamics of elevated tank was proposed by Housner [10] after the Chilean earthquake of 1960, which is more appropriate and is being commonly used in most of the international codes including GSDMA guideline.

The lateral shaking of the water tank divides the column of the water in the container in two different forms of mass i.e. convective mass and impulsive mass based on the type of hydrodynamic pressure induced by them. The upper portion of the water also termed as the "convective mass" moves in a long period causing the sloshing effect and induces convective type hydrodynamic pressure on the walls. The remaining portion of water termed as "impulsive mass" accelerates along with the wall and induces impulsive hydrodynamic pressure. The acceleration of this impulsive mass of water being same as that of the tank container, contributes predominantly to the base shear and overturning moment.

In spring mass model convective mass (m_c) is attached to the tank wall by the spring having stiffness (K_c), whereas impulsive mass (m_i) is rigidly attached to tank wall. For elevated tanks two-mass model is considered, which consists of two degrees of freedom system. Spring mass model can also be considered for elevated tanks, but two-mass model idealization is closer to reality as shown in Fig. 1. However, for most elevated tanks, one observes that the two time periods are well separated. Therefore, the idealization of two masses can be treated as two coupled single degree of freedom system. The stiffness (K_s) is lateral stiffness of staging and the structural mass (m_s) is considered, which include mass of tank container and one-third mass of staging, as staging will perform like a lateral spring. Elevated water tank staging follows the provisions given by Criteria for design of RCC staging for overhead water tanks [11].

Fig. 1 Two mass model

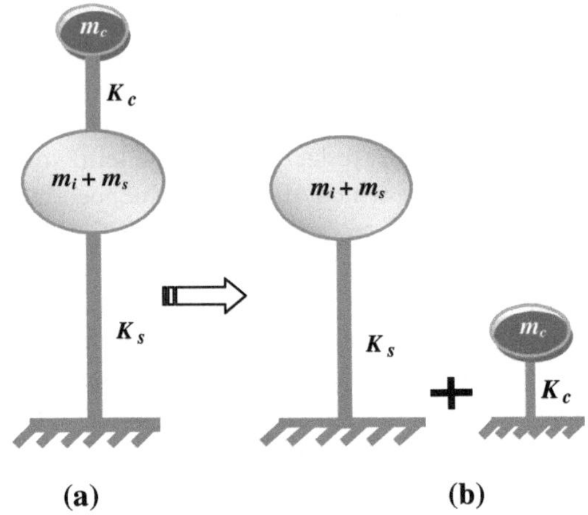

3 Fluid–Structure Interaction

Seismic ground acceleration triggers the lateral base excitation which causes hydrodynamic pressure on the tank walls. This hydrodynamic pressure depends on various parameters like geometry of the tank, height of the liquid, properties of liquid and fluid-tank interaction. A proper estimation of hydrodynamic pressure thus requires a rigorous fluid–structure interaction analysis. Tank-liquid system considered a mechanical analogue which divides liquid in two parts as, impulsive liquid and convective or sloshing liquid (Fig. 2). This mechanical model is quantified in terms of impulsive mass, convective mass, and flexibility of convective liquid. Expressions for mechanical analogue parameters of for circular and rectangular tanks have been developed by Housner. Some researchers have attempted numerical study for tanks of other shapes [12, 13]. Fluid–structure interaction problems can be investigated by various approaches such as added mass approach [14, 15], the Eulerian approach [16], the Lagrangian approach [17, 18] or the Eulerian–Lagrangian approach [19].

The simplest and optimised method of these is the added mass approach suggested by Westergaard (Fig. 3), can be examined by some conventional FEM software such as SAP2000, STAAD Pro and LUSAS.

Fig. 2 FEM model for fluid–structure interaction

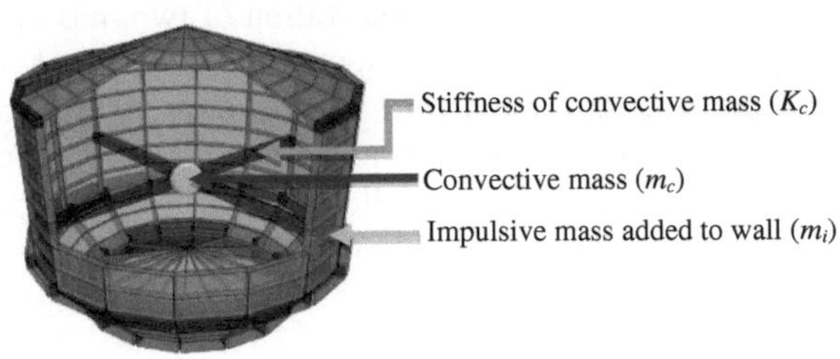

Fig. 3 Westergaard added mass concept with normal and cartesian directions journal bearing

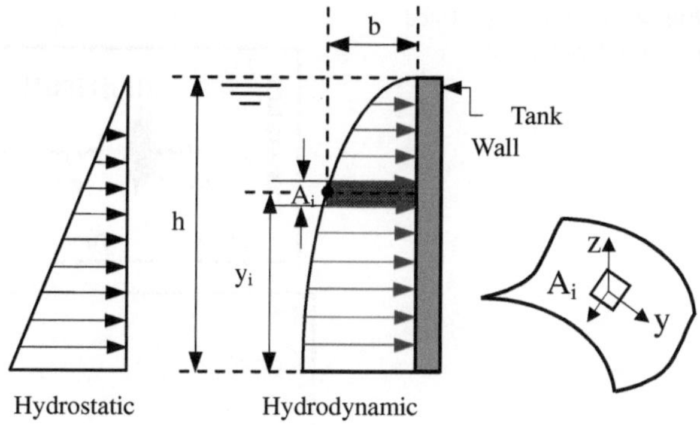

The general equation of motion for a system subjected to an earthquake excitation can be written as,

$$M\ddot{u} + C\dot{u} + Ku = -M\ddot{u}_g \tag{1}$$

In which M, C and K are mass, damping and stiffness matrices with $\ddot{u}$, $\dot{u}$ and u are the acceleration, velocity and displacement respectively. In the case of added mass approach the form of Eq. (1) become as above.

$$M * \ddot{u} + C\dot{u} + Ku = -M * \ddot{u}_g \tag{2}$$

In Eq. (2) $M*$ is a new mass matrix after addition of hydrodynamic mass to the structural mass, while the damping and stiffness matrices are same as given in Eq. (1). This method was originally developed for the dams but it can be applied to other hydraulic structure, under earthquake loads i.e. tank. In this paper the impulsive (m_i) and convective mass (m_c) along with spring stiffness (K_c) has been obtained according to GSDMA guideline equations and is added to the tank wall according to Westergaard approach as shown in Fig. 3 as per Eq. (3). Where, ρ is mass density, h is the depth of water and A_i is the curvilinear surface area.

$$m_{ai} = \left[\frac{7}{8}\rho\sqrt{h(h - y_i)}\right]A_i \tag{3}$$

4 Site Specific Response Analysis Using EERA

EERA software is a recent application of the equivalent-linear concept of earthquake site response analysis, which was formerly executed in the original and successive versions of SHAKE/SHAKE91. EERA is fully integrated with a spread sheet program EXCEL as shown in Fig. 4; Table 1.

Site specific response analysis studies have utilized data from pairs of nearby rock and soil recording. The ground motion recorded at Regional Passport Office

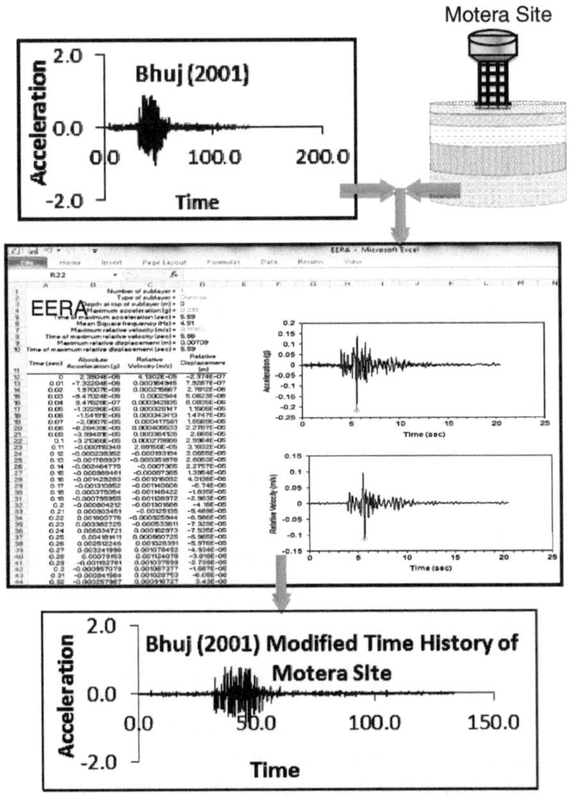

Fig. 4 Methodology Used by EERA Software

Table 1 Properties of earthquake record

Descriptions	Bhuj (2001) earthquake
Region	Gujarat, India
Date	26th January, 2001
Magnitude	7.7 M_w
Depth	16 km(10 mi)
PGA (m/s/s)	1.0382
PGV (cm/s)	111.3
PGD (cm)	8.821
Epicenter	23°25′08″N70°13′55″E23.419°N70.232°E
Casualties	19,727 believed dead, 166,001 injured
Max. intensity	X

(RPO) building during Bhuj earthquake on 26th January 2001 (Fig. 5 and Table 1) is used to develop the ground motion at the rock or hard soil level which is considered at 21 m below ground level. This artificial developed acceleration time history at 21 m depth at passport office is used as an input motion at other sites of Ahmedabad city (Fig. 6) for the development of acceleration time history. The acceleration time history developed at various site using one dimensional ground

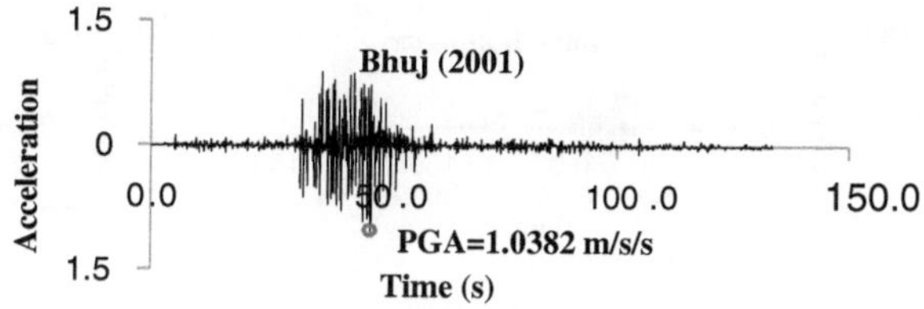

Fig. 5 Acceleration time history record of Bhuj (2001) earthquake

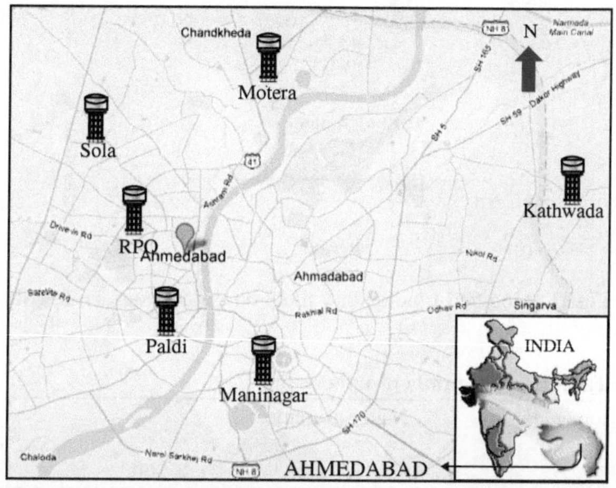

Fig. 6 Different site locations of Ahmedabad city

response analysis software EERA are shown in Fig. 7. For the development of artificial time history, longitudinal direction is considered because it is critical compared to other directions. Also, considered available soil data such as number of layers of soil, thickness of each layer, unit weight, and shear modulus are obtained from borehole data and geophysical testing. The soil profile of Motera site is tabulated in Table 2.

Nonlinearity of the shear modulus and damping are approximated by the use of equivalent linear soil properties and numerically obtaining values compatible with the effective strains in each layer. EERA work out the response in a horizontally layered soil-rock system subjected to transient and vertical traveling shear waves. EERA assumes that the cyclic soil behaviour can be replicated using an equivalent linear model, an assumption that has been extensively described in the geotechnical earthquake engineering literature [20, 21] and [22].

The soil amplification factor describes how the soil strata can amplify or attenuate the ground motion during an earthquake. Estimation of soil amplification factor is very important in any seismic microzonation study. In this study, the PGA at ground level has been computed using the EERA software by plotting the depth plot for variation of peak ground acceleration (PGA). This is obtained for site-specific input data for actual measured soil properties for each of the sites. The ratio between the actual time-history at the hypothetical engineering bed rock (at

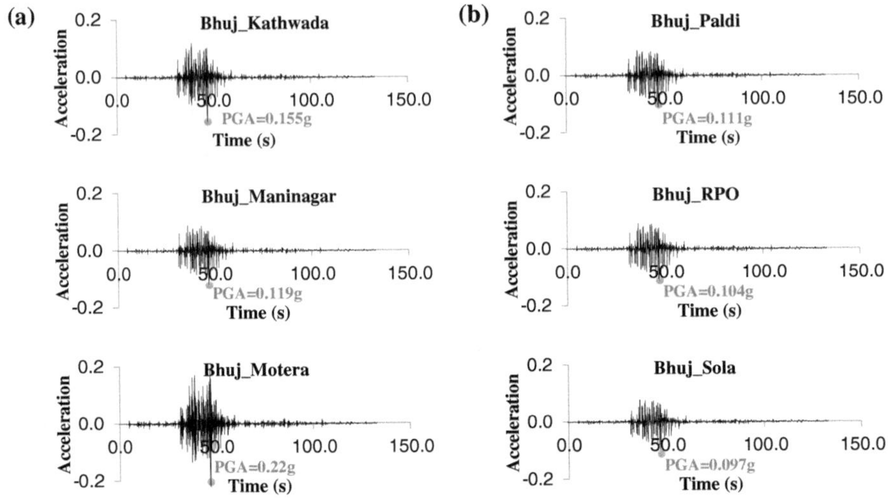

Fig. 7 Modified time history plots on ground surface for different site locations

Table 2 Soil profile of motera site

Layer no.	Material name	Thick -ness (m)	Vs (m/sec)
1	Medium brown clayey silt and sand	2.10	178.4
2	Loose to medium yellow silty fine sand	2.15	222.1
3	Medium yellow sand with kankar	0.30	226.5
4	Loose to medium yellow sand	0.75	236.9
5	Medium yellow clayey silty sand	0.70	245.9
6	Medium yellow silty sand	2.00	268.6
7	Medium yellow silty sand	2.00	288.1
8	Medium yellow silty sand	2.00	305.3
9	Medium yellow silty sand	3.00	325.3

21 m) at site to the amplified site response at the ground for soft soil sites give the amplification factor. This is estimated at various soil sites in Ahmadabad. The soil amplification factor (SAF) is ranging from 1.39 to 3.14 for selected six location of Ahmadabad region.

The site-specific acceleration depth plots are generated for all locations using EERA software, and the PGA values are obtained at ground surface. Figure 8 shows a few depth plots so obtained. Depth plots are useful to examine the variation of ground motion amplitudes with depth. These plots have been used to get the amplified peak ground acceleration (PGA) at the ground surface. In the six sites of Ahmadabad region, the areas like Kathwada, Maninagar, Motera, Paldi, RPO and sola have PGA range between 0.097 g and 0.22 g (Fig. 8).

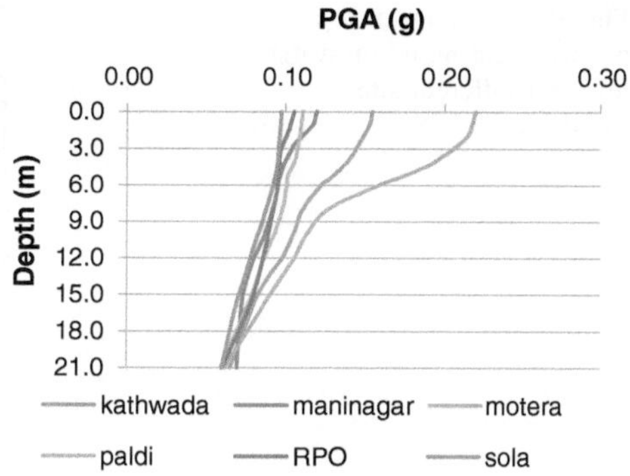

Fig. 8 Variation of PGA with depth for various sites

Fig. 9 SAP2000 FE model of elevated water tank

5 Problem Description

An intze type reinforced elevated water tank of 1000 m^3 storage capacity with their full, half and empty water level in the container has been considered for the present study. Columns are arranged on the periphery of staging and connected with three bracing levels. Other dimensions of the elevated tanks are illustrated in Table 3. Finite element model of elevated water tank is prepared and analysed in SAP2000 as shown in Fig. 9.

Columns and bracings in the frame type support system are modelled considering frame elements with six degrees of freedom per node. Conical part, bottom dome, top dome and container wall are modelled as thin shell elements with four nodes and six degrees of freedom per node.

In present study, time history analysis of elevated water tank is carried out for more realistic seismic analysis. Six sites of Ahmedabad city are chosen for the development of site specific acceleration time history analysis of water tank. The soil data has been obtained from borehole data and geophysical testing which are

Fig. 10 Variation of top displacement based on water level for different site locations

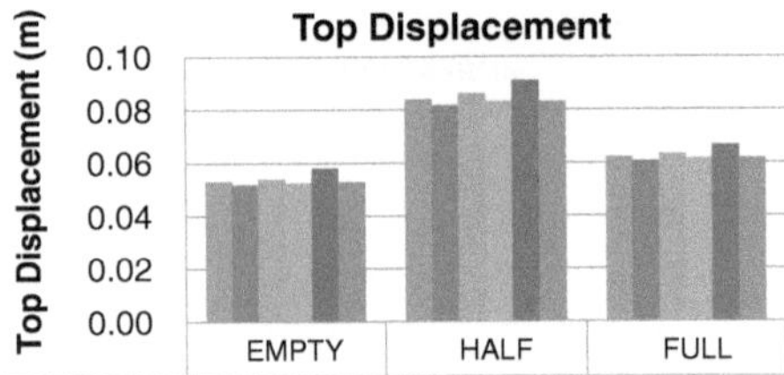

	EMPTY	HALF	FULL
Kathwada	0.053	0.084	0.062
Maninagar	0.052	0.082	0.061
Motera	0.054	0.086	0.063
Paldi	0.052	0.083	0.061
RPO	0.058	0.091	0.067
Sola	0.053	0.083	0.062

Fig. 11 Variation of sloshing displacement based on water level for different site locations

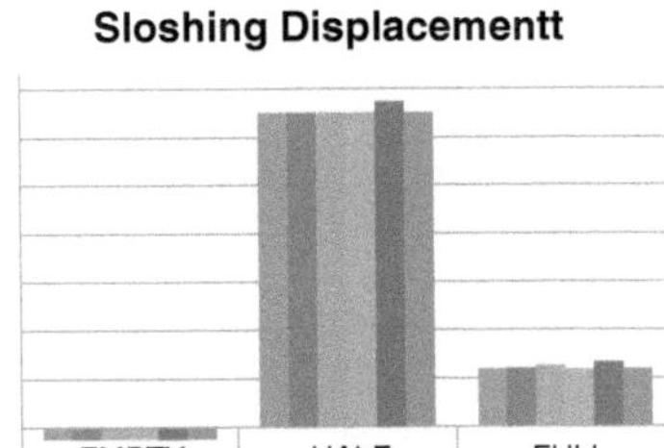

	EMPTY	HALF	FULL
Kathwada	0.000	0.295	0.136
Maninagar	0.000	0.295	0.136
Motera	0.000	0.295	0.138
Paldi	0.000	0.295	0.135
RPO	0.000	0.302	0.140
Sola	0.000	0.295	0.136

Fig. 12 Variation of base shear based on water level for different site locations

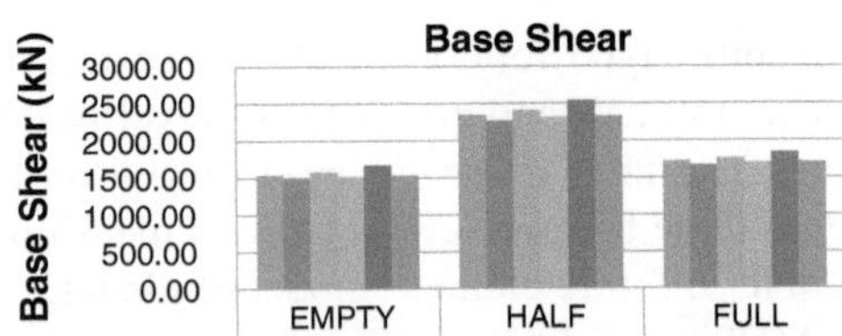

	EMPTY	HALF	FULL
Kathwada	1542	2346	1716
Maninagar	1484	2268	1667
Motera	1579	2408	1756
Paldi	1519	2314	1694
RPO	1674	2539	1839
Sola	1528	2328	1699

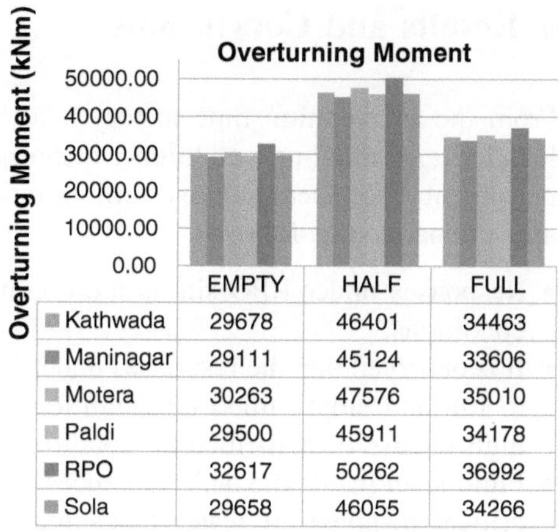

Fig. 13 Variation of overturning moment based on water level for different site locations

Table 3 Structural data for basic frame type staging

Parameter	Quantity
Capacity of the tank	1000 m^3
Unit weight of concrete	25 kN/m^3
Thickness of top dome	0.15 m
Rise of top dome	2.2 m
Size of top ring beam	0.35 m × 0.35 m
Diameter of tank	13.6 m
Height of cylindrical wall	6.8 m
Thickness of cylindrical wall	0.33 m
Size of middle ring beam	1.2 m × 0.6 m
Rise of conical dome	2.35 m
Thickness of conical shell	0.5 m
Rise of bottom dome	1.6 m
Thickness of bottom dome shell	0.2 m
Size of bottom circular girder	1 m × 1.2 m
Number of columns	8
Diameter of circular column	0.75 m
Size of bracing	0.5 m × 0.5 m
Distance between intermediate bracing	4 m
Height of staging above foundation	16 m

used to develop modified acceleration time history of various different sites of Ahmedabad city using software EERA, based on one dimensional ground response analysis. The results of RC elevated water tank are compared for different site locations. The above analysis is performed for elevated water tank for their full, half and empty water conditions. The responses include base shear, overturning moment, Sloshing displacement and Top displacement have been compared and contrasted. Above work reflects the importance of site specific time history analysis of elevated water tanks. The response results of the structure are exceedingly influenced by different subsoil and water filled up conditions (Figs. 10, 11, 12, 13; Table 3).

6 Results and Conclusions

From the acceleration time history analysis of elevated water tank for different location and filled-up conditions, responses like top displacement, sloshing displacement, base shear and overturning moment are compared and obtained results are summarized as follows:

- Responses under RPO site is more compared to other for all type of filled-up conditions.
- It is observed that the responses under half-filled up condition is more compared to full and empty filled up condition. Sloshing displacement is comparatively high in case of half-filled up condition.
- Fluid level inside the tank influences in the response of seismic parameters. The critical response of water tanks may not be always higher in full-filled up condition of tank.
- Acceleration time histories on ground are influenced by local sub soil characteristics.
- It is necessary to carryout site specific ground response analysis for important structures for realistic estimation of seismic force and better earthquake resistance.

Acknowledgments We acknowledge Mr. Shashi N. Vaghela, P.G. Student, Applied Mechanics Department, L. D. College of Engineering, Ahmedabad, Gujarat, India for helped us during the work. Also, to Mr. Deepak K. Jivani, Mr. Dilip Dharna and MAP Technical Consultancy and Civil Engineering Laboratory for providing soil data and related assistance.

References

1. Structural Analysis Program SAP (2000) User's manual. Computers and structures, Inc., Berkeley
2. Bardet JP, Ichii K, Lina CH (2000) EERA – A computer program for equivalent-linear earthquake site response analyses of layered soil deposits. University of Southern California, Department of civil engineering
3. Masood A, Ahmad T, Arif M, Mital VP (2008) Failure of overhead water tank in the state of Uttar Pradesh in India: a case study. In: ICCIDC-I, vol. 1, pp 185–191
4. Patel CN, Patel HS (2012) Former failure assessments of RC elevated water tanks: literature review. GIT J Eng Technol 5, Paper No. c12
5. Bhadauria SS, Gupta MC (2006) In-service durability performance of water tanks. J Perform Constr Fac, ASCE 2:136–145
6. Bhadauria SS, Gupta MC (2007) In situ performance testing of deteriorating water tanks for durability assessment. J Perform Constr Fac, ASCE 3:234–239
7. BIS: 1893-2002 (Part–1), Criteria for earthquake resistant design structures. BIS, New Delhi
8. Rai Durgesh C (2003) Performance of elevated tanks in Mω 7.7 Bhuj earthquake of January 26th, 2001. Indian Acad Sci (Earth Planet Sci) 112(3):421–429

9. IITK-GSDMA (2007) Guidelines for seismic design of liquid storage tanks provisions with commentary and explanatory examples
10. Housner George W (1963) The dynamic behaviour of water tanks. J Perform Constr Fac 53(2):381–387
11. BIS 11682:1985 (Preliminary Draft code) (2011) Criteria for design of RCC staging for overhead water tanks. Bureau of Indian Standards, New Delhi
12. Joshi SP (2000) Equivalent mechanical model for horizontal vibration of rigid intze tanks, J Earthquake Tech, Technical Note, 37:(1–3), (39–47)
13. Damatty AA, Korol RM and Tang LM (2002) Analytical and experimental investigation of the dynamic response of liquid-filled conical tanks, 12th World Conference on Earthquake Engineering, New Zealand, Paper 966, Topic 7
14. Westergaard HM (1931) Water pressures on dams during earthquakes. ASCE 57:1303
15. Barton DC and Parker JV (1987) Finite element analysis of the seismic response of anchored and unanchored liquid storage tanks, Earthquake Engineering and Structural Dynamics, 15:299–322
16. Zienkiewicz OC and Bettes P (1978) Fluid structure dynamic interaction and wave forces; an introduction to numerical treatment, Int. J Num. Methods Eng, 13:1–16
17. Wilson EL and Khalvati M (1983) Finite elements for the dynamic analysis of fluid-solid systems, Int. J Num. Methods Eng, 19:1657–1668
18. Olson LG and Bathe KJ (1983) A study of displacement-based fluid finite elements for calculating frequencies of fluid and fluid-structure systems, Nuclear Engineering and Design, 76:137–151
19. Donea et al. (1982) An arbitrary Lagrangian-Eulerian finite element method for transient dynamic fluid-structure interaction, Computer Methods in Applied Mechanics and Engineering, 33:689–723
20. Idriss IM, Seed HB (1968) Seismic response of horizontal soil layers. J Soil Mech Found Div, ASCE 94(Sm4):1003–1031
21. Seed HB, Idriss IM (1970) Soil moduli and damping factors for dynamic response analysis. Report no. EERC 70-10, University of California, Berkeley, December 1970
22. Kramer SL (1996) Geotechnical Earthquake Engineering, Prentice Hall, Upper Saddle River, New Jersey, 254–280

Soil-Foundation-Structure Interaction Effects in Seismic Behaviour of RC Elevated Water Tank

Chirag N. Patel and H. S. Patel

Abstract Presented here is an analysis of the effects of soil-foundation-structure interaction (SFSI) on seismic response of elevated water tanks. A particular importance is the issue of water motion inside the container, input motion with scaling and approaches for the SFSI models. Time history analysis is carried out for more realistic seismic analysis considering water as two- mass idealization and subsoil data of three different sites of Ahmedabad, metropolitan city of India using finite element software SAP2000. The analysis presented in this paper addresses the influence of inelastic behavior of both the soil and the structural components during seismic response evaluation of elevated water tank systems. Additionally, it is also demonstrated that that SFS interaction can sometimes have a beneficial effect on the superstructure response and sometimes produce detrimental effects on the system behavior and is dependent on the water filled-up in container and sub soil condition below the base. Finally, study reflects the importance of soil base parameters for the analysis and result shows that the structure responses are exceedingly influenced by water level inside the container of tank, different subsoil condition and approaches considered.

Keywords Water tank · Seismic soil structure interaction

C. N. Patel (✉)
Faculty of Engineering, Pacific Academy of Higher Education and Research University, Udaipur, Rajasthan 313024, India
e-mail: cnpatel.693@gmail.com

H. S. Patel
Applied Mechanics Department, L. D. College of Engineering, Ahmedabad, Gujarat 380015, India

H. C. Patel et al. (eds.), *Proceedings of International Conference on Advances in Tribology and Engineering Systems*, Lecture Notes in Mechanical Engineering, DOI: 10.1007/978-81-322-1656-8_42, © Springer India 2014

1 Introduction

Elevated water tanks are very important structure for community having basic purpose of to secure continuous water supply at the longer distance. It is also essential to ensure that, it does not hindered during an earthquake and should remain functional also after an earthquake. Even though, in current Indian code for seismic design IS 1893 (Part-1): 2002 [1], limited provisions on seismic design of elevated tanks are given. And those provisions are highly inadequate compared to present international practice [2]. Therefore, revision of the current Indian code thus became inevitable. Indian Institute of Technology, Kanpur has proposed guidelines along with commentary and explanatory examples for seismic analysis of liquid storage tanks in association with Gujarat State Disaster Management Authority [3]. It covers seismic analysis and design of elevated tanks including two degree of freedom idealization with impulsive and convective hydrodynamic forces for convincing assessment of dynamic properties of tanks.

Soil-structure interaction between a structure and the supporting soil is a phenomenon that influences many aspects in the design of a structure: safety, serviceability (crack propagation) and costs. Also, produce a significant effect on the response of the structure and the resulting localization of damage during seismic events. Especially, for structures like elevated tank with slenderness and heavy mass at the top.

The stimulus of the present study is to understand the seismic behaviour of reinforced concrete elevated water tank under different level of water inside the tank container and soil strata below the structure with different approach. The responses like base shear, overturning moment, top and sloshing displacement have been compared and contrasted to reflect the importance of soil structure interaction analysis and its approaches against consideration of fixed base assumption for structure foundation.

2 Fluid–Structure Interaction

A suitable spring mass analogue for two mass models to characterize basic dynamics of elevated tank was proposed by Housner [4] after the Chilean earthquake of 1960, which is more appropriate and is being commonly used in most of the international codes including GSDMA guideline. In spring mass model convective mass (m_c) is attached to the tank wall by the spring having stiffness (K_c), whereas impulsive mass (m_i) is rigidly attached to tank wall (Fig. 1). For elevated tanks two-mass model is considered, which consists of two degrees of freedom system. This hydrodynamic pressure depends on various parameters like geometry of the tank, height of the liquid, properties of liquid and fluid-tank interaction. A proper estimation of hydrodynamic pressure thus requires a rigorous fluid–structure interaction analysis.

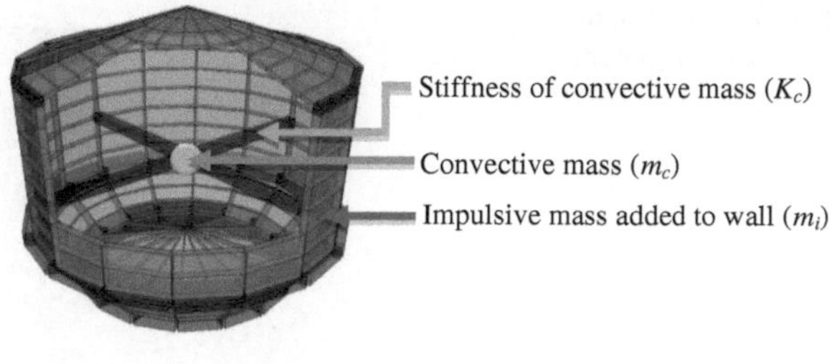

Fig. 1 FEM model for fluid-structure interaction

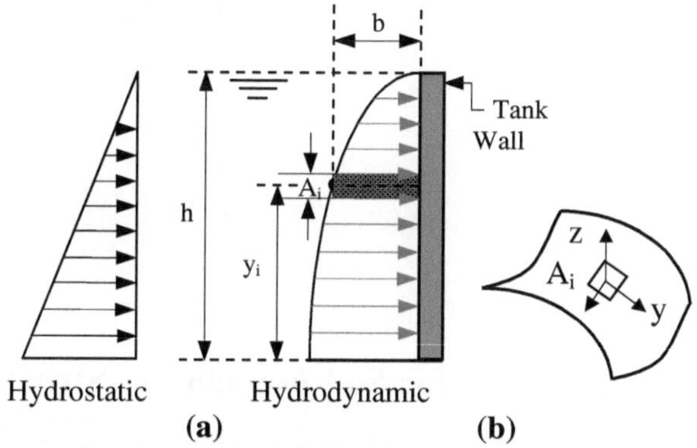

Fig. 2 Westergaard added mass concept with normal and cartesian directions journal bearing

The simplest and optimised method of these is the added mass approach suggested by Westergaard [5] as shown in Fig. 2, can be examined by some conventional FEM software such as SAP2000, STAAD Pro and LUSAS.

The general equation of motion for a system subjected to an earthquake excitation can be written as,

$$M\ddot{u} + C\dot{u} + Ku = -M\ddot{u}_g \tag{1}$$

In which M, C and K are mass, damping and stiffness matrices with $\ddot{u}$, $\dot{u}$ and u are the acceleration, velocity and displacement respectively. In the case of added mass approach the form of Eq. (1) become as above.

$$M * \ddot{u} + C\dot{u} + Ku = -M * \ddot{u}_g \tag{2}$$

In Eq. (2) M^* is a new mass matrix after addition of hydrodynamic mass to the structural mass, while the damping and stiffness matrices are same as given in Eq. (1). This method was originally developed for the dams but it can be applied to other hydraulic structure, under earthquake loads i.e. tank. In this paper the impulsive (m_i) and convective mass (m_c) along with spring stiffness (K_c) has been obtained according to GSDMA guideline equations and is added to the tank wall according to Westergaard approach as shown in Fig. 3 as per Eq. (3). Where, ρ is mass density, h is the depth of water and A_i is the curvilinear surface area.

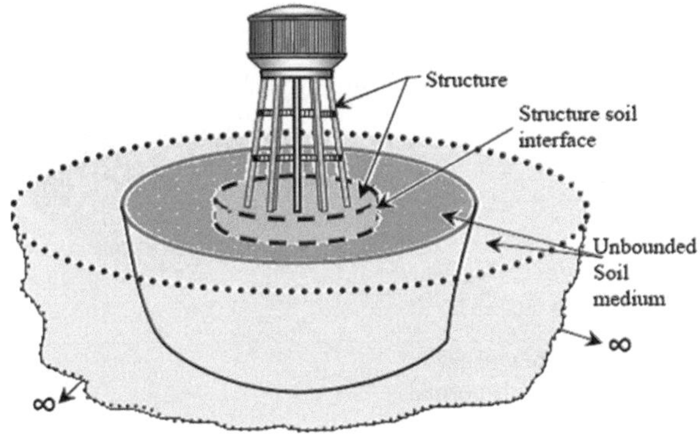

Fig. 3 Soil structure interaction problem

$$m_{ai} = \left[\frac{7}{8}\rho\sqrt{h(h - y_i)}\right]A_i \tag{3}$$

3 Approaches for Soil-Foundation-Structure Interaction

3.1 Spring Base or Cone Model (Sub Structure) Approach

In a dynamic soil–structure-interaction analysis a bounded structure (which may be linear or nonlinear), consisting of the actual structure and an adjacent irregular soil if present, will interact with the unbounded (infinite or semi-infinite) soil which is assumed to be linear elastic as shown in Fig. 4.

A cone model has been proposed by Meek and Wolf [6] for evaluating the dynamic stiffness and effective input motion of a foundation on the ground is used and static stiffness of this truncated cone (Fig. 4) for circular rigid foundation can be calculated with equations given in Table 1. Where, G: Shear modulus, r_0: Radius of circular foundation, υ: Poisson ratio, K_V, K_H, K_R and K_T are the vertical, horizontal, rocking and torsional stiffness, respectively.

In Fig. 5 k_x and k_θ represent the translation and rocking stiffness of foundation that can be modeled with spring. These are attached to the central point of the rigid circular foundation. Stiffness is calculated for individual layers and then equivalent stiffness calculated using Bowel's weighted approach as shown below Eqs. (4) and (5).

$$k_x = \frac{k_{x1}h_1 + k_{x2}h_2 + \cdots + k_{xn}h_n}{h_1 + h_2 + h_3 + \cdots + h_n} \tag{4}$$

$$k_\theta = \frac{k_{\theta 1}h_1 + k_{\theta 2}h_2 + \cdots + k_{\theta n}h_n}{h_1 + h_2 + h_3 + \cdots + h_n} \tag{5}$$

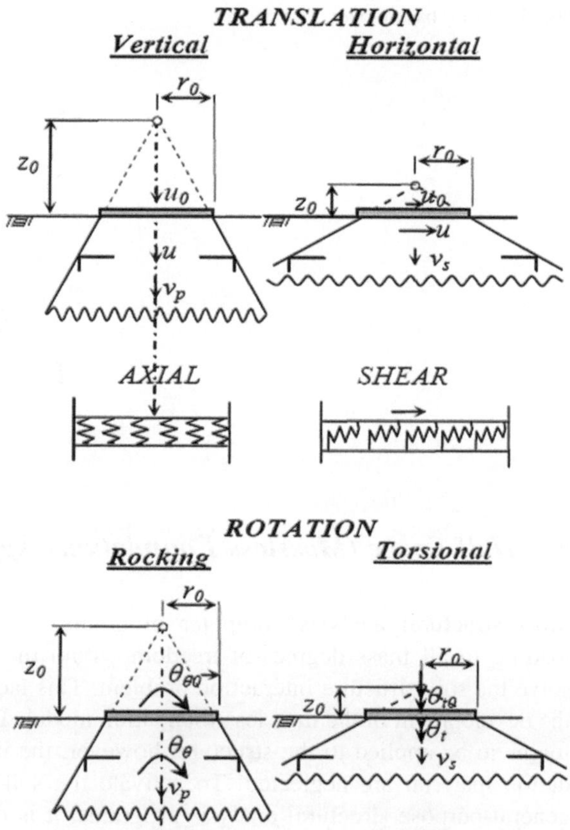

Fig. 4 Cone for various degree of freedom with corresponding apex ratio, wave propagation velocity and distortion

Table 1 Static stiffness value of rigid circular foundation

Stiffness	Foundation with no embedment
Vertical (K_V)	$\frac{4Gr_0}{1-\upsilon}$
Horizontal (K_H)	$\frac{8Gr_0}{2-\upsilon}$
Rocking (K_R)	$\frac{8Gr_0^3}{3(1-\upsilon)}$
Torsional (K_T)	$\frac{16Gr_0^3}{3}$

Fig. 5 Spring base model

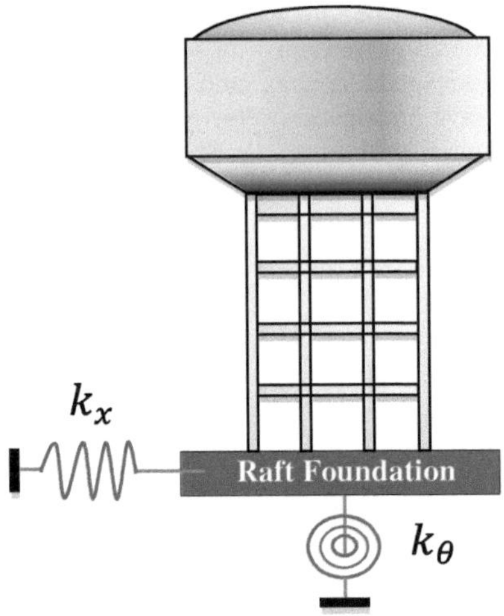

3.2 Half Space (Massless Foundation) Approach

Most structural analysis computer programs automatically apply the seismic loading to all mass degrees-of-freedom within the computer model and cannot solve the soil–structure interaction problem. This lack of capability has motivated the development of the massless foundation model. This allows the correct seismic forces to be applied to the structure; however, the inertia forces within the foundation material are neglected. To activate the soil–structure interactions within general-purpose structural analysis programs, it is only necessary to identify the foundation mass in order that the loading is not applied to that part of the structure. In this study, the SAP2000 [7] general purpose structural analysis program has been selected to consider the soil/foundation–structure interaction for the elevated tanks because it has this option and is capable of solving the SSI problem correctly. It allows the correct seismic forces to be applied to the structure; however, the inertia forces within the foundation material are neglected.

The model considered for the massless foundation approach may be seen in Fig. 6. In this model, the soil/foundation–structure model is divided into three sets of node points. The common nodes at the interface of the structure and the foundation are identified with ''*c*''; the other nodes within the structure are named ''*s*''; and the other nodes within the foundation are ''*f*'' nodes.

4 Subsoil and Earthquake Parameters

A study considered different three sites of Ahmedabad city as Kathwada, Paldi and, RPO (Regional Passport Office) as shown in Fig. 7. As shown in Fig. 8 and Table 2, properties of time history data records for Bhuj (2001) earthquake is used

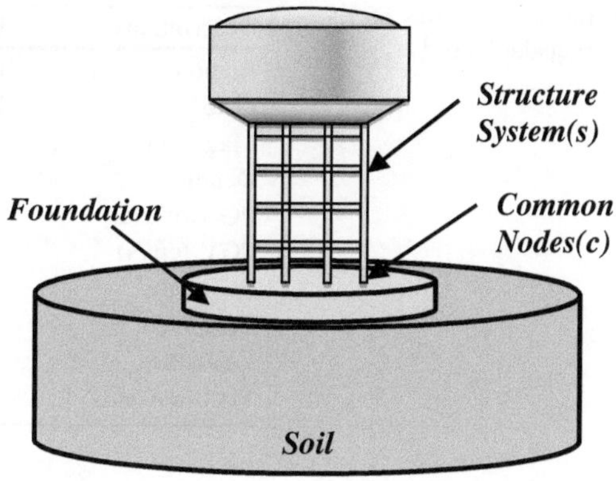

Fig. 6 Half space model

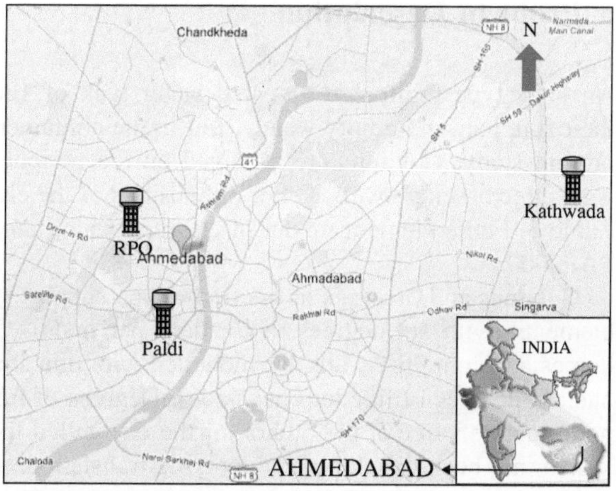

Fig. 7 Different site locations of Ahmedabad city

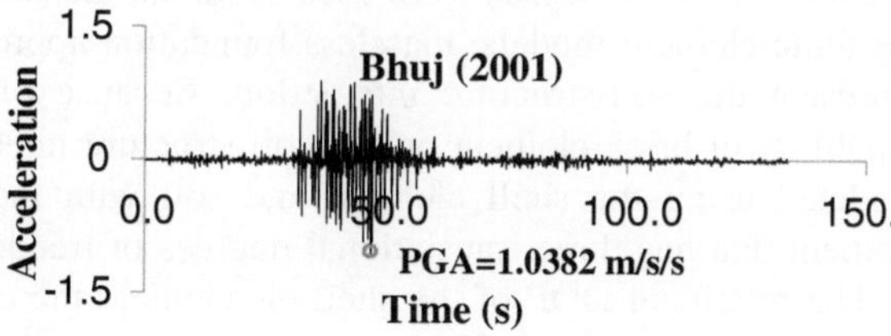

Fig. 8 Acceleration time history record of Bhuj (2001) earthquake

for the analysis and available soil data such as number of layers of soil, thickness of each layer, unit weight, shear modulus are obtained from borehole data and geophysical testing.

The soil characteristics can be calculated with well-known equations $G = E/2(1 + \upsilon)$ and $Vs = G/\rho$; where, υ: poisons ratio, E: Modulus of elasticity of soil layer, ρ: is the mass density, and G is the shear modulus, Vs: Shear wave velocity of soil layer.

Table 2 Properties of earthquake record

Descriptions	Bhuj (2001) earthquake
Region	Gujarat, India
Date	26th January, 2001
Magnitude	7.7 M_w
Depth	16 km (10 mi)
PGA (m/s/s)	1.0382
PGV (cm/s)	111.3
PGD (cm)	8.821
Epicenter	23°25′08″N70°13′55″E23.419°N70.232°E
Casualties	19,727 believed dead, 166,001 injured
Max. intensity	X

5 Problem Description

An intze type reinforced elevated water tank of 1000 m^3 storage capacity with their full, half and empty water level in the container has been considered for the present study. Columns are arranged on the periphery of staging and connected with three bracing levels. Other dimensions of the elevated tanks are illustrated in Table 3. Finite element model of elevated water tank is prepared and analysed in SAP2000.

Columns and bracings in the frame type support system are modelled as frame elements (with six degrees of freedom per node). Conical part, bottom and top domes and container walls are modelled with thin shell elements (with four nodes and six degrees of freedom per node). Degrees of freedom were fixed at the base nodes and left free at the others for the one called fixed-base system. Spring base system has been developed by applying translational and rocking spring as shown in Fig. 9.

Finite element meshes were used to model the subsoil for the other systems. In the finite element models, massless foundation approach (Fig. 10) was applied to represent the soil-structure interaction. Because of lack of rotational freedom capability of brick element on the soil-structure interaction surface, foundation is modeled using the shell element and soil with the isoperimetric 8 node-brick element that has three translational degrees of freedom per node.

The rotational DOF of the shell elements at the common nodes with the brick elements are neglected. 3D finite element meshes are generated and intended to model the influence of fluid–structure and soil-structure effects on the seismic behaviour of elevated tank. A parametric study is carried out to determine the distance of the boundary from the tank. For this purpose various different models with different mesh type of soil medium are used. A parametric study is performed by using transient analysis with direct integration technique to decide the dimension of the soil medium of the model and to select the mesh type of the soil medium.

Table 3 Structural data for basic frame type staging

Parameter	Quantity
Capacity of the tank	1000 m^3
Unit weight of concrete	25 kN/m^3
Thickness of top dome	0.15 m
Rise of top dome	2.2 m
Size of top ring beam	0.35 m × 0.35 m
Diameter of tank	13.6 m
Height of cylindrical wall	6.8 m
Thickness of cylindrical wall	0.33 m
Size of middle ring beam	1.2 m × 0.6 m
Rise of conical dome	2.35 m
Thickness of conical shell	0.5 m
Rise of bottom dome	1.6 m
Thickness of bottom dome shell	0.2 m
Size of bottom circular girder	1 m × 1.2 m
Number of columns	8
Diameter of circular column	0.75 m
Size of bracing	0.5 m × 0.5 m
Distance between intermediate bracing	4 m
Height of staging above foundation	16 m

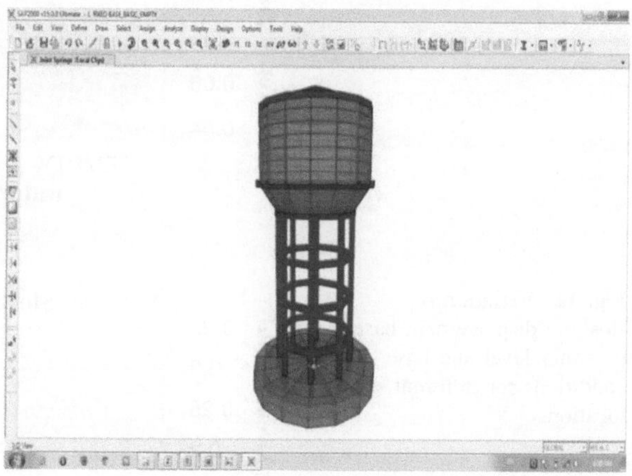

Fig. 9 Spring base model of elevated water tank

The analytical results of RC elevated water tank considering fixed base, spring base and half space are compared. Also, above analysis is performed for water tank for their full, half and empty water conditions. The responses include top displacement, sloshing displacement, base shear and overturning moment under the four different time- history data records have been compared and contrasted (Figs. 11, 12, 13 and 14).

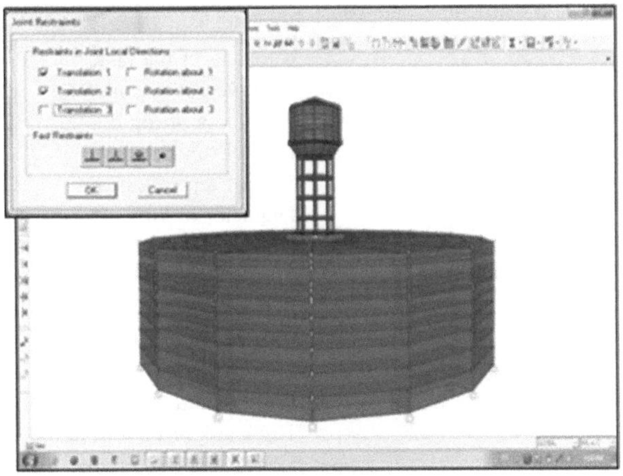

Fig. 10 Half space model of elevated water tank

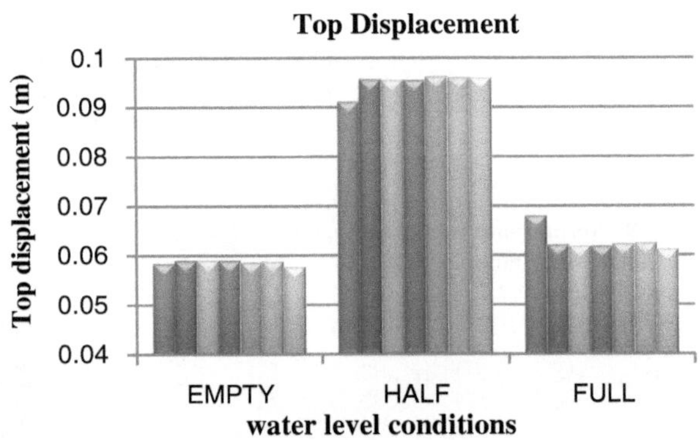

Fig. 11 Variation of top displacement based on water level and base conditions for different site locations

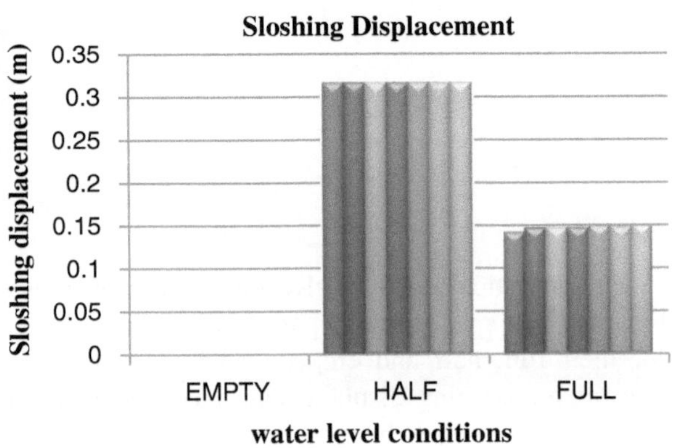

Fig. 12 Variation of sloshing displacement based on water level and base conditions for different site locations

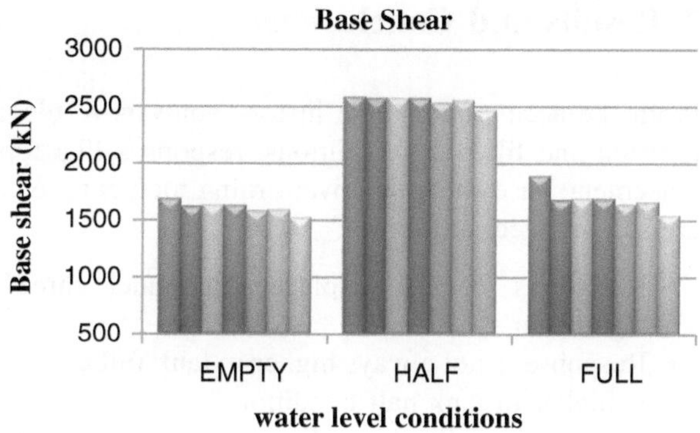

Fig. 13 Variation of base shear based on water level and base conditions for different site locations

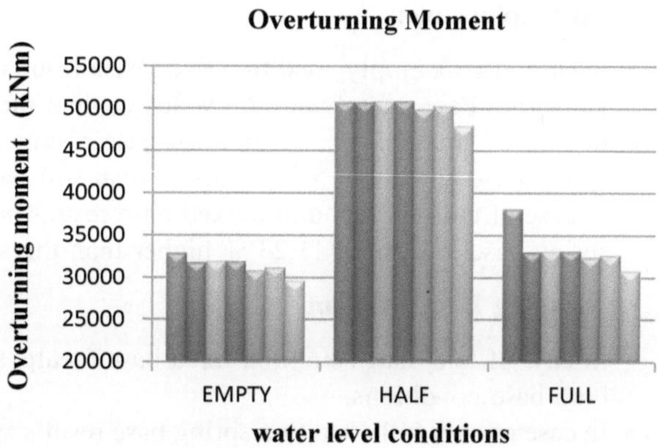

Fig. 14 Variation of overturning moment based on water level and base conditions for different site locations

6 Graphical Considerations

- FIXED BASE
- SPRING BASE KATHWADA
- SPRING BASE PALDI
- SPRING BASE PASSPORT OFFICE
- SOIL BASE KATHWADA
- SOIL BASE PALDI
- SOIL BASE PASSPORT OFFICE

7 Results and Conclusions

From the acceleration time history analysis of elevated water tank for different location and filled-up conditions, responses like top displacement, sloshing displacement, base shear and overturning moment are compared and obtained results are summarized as follows:

- Variations in top displacement under three different methods are not significant.
- Response is not always higher in tank full condition; in present study response is higher in tank half condition.
- Variations in Sloshing displacement under three different methods are not significant.

Top Displacement

- In case of tank empty condition fixed base results are about 0.85 % lower than the spring base and about 0.68 % lower than the soil base.
- In case of tank half condition fixed base results are about 4.51 % lower than spring base and about 5.20 % lower than soil base.
- In case of tank full condition fixed base results are about 10.03 % higher than spring base and about 11.23 % higher than the soil base.

Sloshing Displacement

- In case of tank half condition fixed base results are about 0.16 % higher than both base conditions.
- In case of tank full condition spring base results are about 3.39 % and soil base results are about 6.07 % higher than the fixed base results.

Base Shear

- In case of tank empty condition fixed base results are about 4.39 % higher than the spring base and about 11.04 % higher than the soil base.
- In tank half condition fixed base results are about 0.34 % higher than the spring base and about 4.43 higher than soil base.
- In tank full condition fixed base results are about 12.72 % higher than spring base and about 22.79 % higher than soil base.

Overturning Moment

- In case of tank empty condition fixed base results are about 3.84 % higher than the spring base and about 11.15 % higher than the soil base.
- In case of tank half condition fixed base results are about 0.34 % lower than the spring base and about 5.83 % higher than soil base results.
- In case of tank full condition fixed base results are about 15.42 % higher than spring base and about 23.72 % higher than soil base.
- The critical response of elevated tanks does not occur in the case of full tanks, but it may happen in a partially full or empty tank.

- Responses are varying in different site locations due to variation in base soil. The results shows that, sub soil parameters affect the responses which are very important for elevated tanks subjected to strong earthquakes.
- Approaches used for base soil also prove the effectiveness and provide deviation in the response of structure.
- Fixed base assumption may lead to uncertainty in seismic parameters and provide higher response amongst all.

Acknowledgments The authors would like to acknowledge Mr. Burhan K. Kanjetawala, P. G. Student, Applied Mechanics Department, L. D. College of Engineering, Ahmedabad, Gujarat, India for helped us during the work. Also, to Mr. Deepak K. Jivani and MAP Technical Consultancy & Civil Engineering Laboratory for providing soil data and related assistance.

References

1. BIS: 1893–2002 (Part: 1) Criteria for earthquake resistant design structures. Bureau of Indian Standards, New Delhi
2. Rai DC (2003) Performance of elevated tanks in Mω 7.7 Bhuj earthquake of January 26th, 2001. Indian Acad Sci (Earth Planet Sci) 112(3):421–429
3. IITK-GSDMA Guidelines for Seismic Design of Liquid Storage Tanks Provisions with commentary and explanatory examples, 2007
4. Housner GW (1963) The dynamic behaviour of water tanks. J Perform Constr Fac 53(2):381–387
5. Westergaard HM (1931) Water pressures on dams during earthquakes. Proc ASCE 57:1303
6. Wolf JP, Meek JW (1992) Cone models for homogeneous soil. J Geotech Eng ASCE 118:686–703
7. SAP (2000) Structural Analysis Program SAP2000. User's manual. Computers and Structures, Inc., Berkeley

Part VII
Engineering System

Nonlinear Dynamic Analysis of High Speed Unbalanced Rotor Supported on Deep Groove Ball Bearings Considering the Preload Effect

H. K. Yadav, S. H. Upadhyay and S. P. Harsha

Abstract The presented paper deals with the study of dynamic behavior of unbalanced rotor along with internal radial clearance (IRC-C3) as non-linearity. Even though the perfect alignment and precision manufacturing has been done, unbalance force cannot be eliminated completely. The dynamic behavior of a high speed unbalanced rotor supported on deep groove ball bearings (SKF-6205) has been studied along with internal radial clearance (IRC-C3) of 14 μm. Nonlinear dynamic response observed by keeping rotor speed constant and the preload as control parameter. The mathematical model has been developed considering both nonlinear contact stiffness and damping at the contact between rollers and races. The contact of rollers with races are treated as nonlinear springs with contact damping whose deformations are obtained by using Hertzian elastic contact deformation theory. The numerical integration technique Runge-Kutta-Fourth-Order method is used to solve the coupled nonlinear differential equations iteratively. Various techniques like Orbits plots, Poincaré map and power spectra are used to study the nature of response. It has been concluded that even though the IRC of bearing cannot be change, once the bearing is installed, the system can be made stable by giving optimized preload to the bearing.

Keywords Unbalanced rotor · Quasiperiodic · Internal radial clearance · Non linear dynamic response · Preload

Nomenclature

c	Contact damping
c_{in}	Equivalent contact damping factor of roller-inner race contact
c_{out}	Equivalent contact damping factor of roller-outer race contact
d	Ball diameter, mm
D	Pitch diameter of bearing, mm

H. K. Yadav (✉) · S. H. Upadhyay · S. P. Harsha
Mechanical and Industrial Engineering Department, Indian Institute of Technology Roorkee, Roorkee 247667, India
e-mail: himanshukyadav2@gmail.com

H. C. Patel et al. (eds.), *Proceedings of International Conference on Advances in Tribology and Engineering Systems*, Lecture Notes in Mechanical Engineering, DOI: 10.1007/978-81-322-1656-8_43,

FTF	Cage frequency
FFT	Fast fourier transformations
k_{in}	Equivalent non-linear contact stiffness of the roller-inner race contact
k_{out}	Equivalent non-linear contact stiffness of the roller-outer race contact
$k_{in\text{-}contact}$	Contact stiffness of the roller-inner race contact
$k_{out\text{-}contact}$	Contact stiffness of the roller-outer race contact
k_{in}	Equivalent non-linear contact stiffness of the roller-inner race contact
m_{in}	Mass of the inner race, kg
m_j	Mass of the rolling elements, kg
m_{out}	Mass of the outer race, kg
m_{rotor}	Mass of the rotor, kg
N_b	Number of balls
R	Radius of outer race, mm
r	Radius of inner race, mm
T	Kinetic energy of the bearing system
V	Potential energy of the bearing system
VC	Varying compliance frequency
W	Radial preload, N
X	Rotational frequency due to unbalanced weight

Greek symbols

$\delta_{in\,+}$	Contact deformation of the roller-inner race
$\delta_{out\,+}$	Contact deformation of the roller-outer race
γ_0	Internal radial clearance
β	Constant angular separation between rolling elements
χ_j	Radial position of jth rolling element from center of inner race
ρ_j	Radial position of jth rolling element from center of outer race
ρ_r	Radius of each rolling element, mm

1 Introduction

Ball bearing is one of the vital and crucial elements of small to large rotating machinery. It has very complicated dynamic behavior due to number of rolling element except the fixed outer race or inner race. Not only that, but its motion behavior is very sensitive to initial condition also. Further, each shaft is supported at the two ends by two ball bearings and different type of rotor like gear or pulley is mounted on that shaft. Now, this gear or pulley will create the unbalance force, because the centre of mass of rotor and shaft is not generally coincided. This unbalanced force is changed continuously with change in the speed of rotor and changing the dynamic behavior of the whole system continuously. So, in the present paper the authors' main intense is to study the effect of unbalance force on the dynamic behavior of the unbalanced rotor system. Also, how the optimized preload can make the system stable that is also studied.

The first study of the effect of preload has been done by Akturk [1]. Akturk [1] has showed that due to increase in the preload, there is a sharp decrease in the peak to peak amplitude at the resonant frequency (i.e., the BPF) and was experimentally done by Wardle and Poon [2]. This can be predicted because the balls get stiffer and they allow lower vibration amplitudes in the radial and axial directions. Also, the similar study was done by Purohit et al. [3]. Cao et al. [4] has studied the effect of axial preload for the double row spherical roller bearing. It has been shown that the preload does "clean" the dynamic response at the frequency domain (fewer frequency spikes). It reduces overall horizontal displacement response of the moving race, while increasing the displacement response at the axial direction. Therefore, preload might not be a good option if high position precision is required at the axial direction.

In the present paper, the change in non-linear behavior of the system with increase in preload has been explained in detail with the help of Poincare maps, orbit plot and FFT plots. The varying compliance effect studied theoretically by Perret [5] considering a deep groove ball bearing with the elastic deformation between race and balls modeled by the Hertzian theory and no bending of races. Meldau [6] studied theoretically the two-dimensional motion of shaft center. Both Perret and Meldau performed a quasi-static analysis since inertia and damping force were not taken into account.

Sunnersjo [7] studied the varying compliance vibrations theoretically and experimentally, taking inertia and damping forces into account. Fukata et al. [8] first took up the study of varying compliance vibrations and the nonlinear dynamic response for the ball bearing supporting a balanced horizontal rotor with a constant vertical force. It is a more detailed analysis compared with Sunnersjo's [7] works as regimes of super-harmonic, sub-harmonic and chaotic behavior are discovered.

Mevel and Guyader [8] have developed a theoretical model of a ball bearing supporting a balanced horizontal rigid rotor, with a constant vertical radial force. This is similar to the work done by Fukata et al. [9] but more results have been reported for parametric studies undertaken and routes to chaos traced out. Chaos in this model of bearing has been reported to come out of the sub-harmonic route and the quasi-periodic route. Tiwari et al. [10] has studied the effect of radial internal clearance and the appearance of sub-harmonics and Hopf bifurcation is seen theoretically whereas the shift in the peak response is also observed experimentally.

Harsha et al. [11] analyzed the nonlinear behavior of a high speed horizontal balanced rotor supported by a ball bearing. The conclusion of this work shows that the most severe vibrations occur when the varying compliance frequency (VC) and its harmonics coincide with natural frequency. Harsha [12] has studied the effects of radial internal clearance and rotor speed. The appearance of periodic, sub-harmonic, chaotic and Hopf bifurcation is seen theoretically. But Harsha considered only nonlinear stiffness. Harsha [13, 14] studied the effects of radial internal clearance for both balanced and unbalanced rotor speed. The appearance of periodic, sub-harmonic, chaotic and Hopf bifurcation is seen theoretically. But, only the nonlinear stiffness has been considered. The results are from a large number of numerical integrations and are mainly presented in the form of Poincaré maps and frequency spectra.

Upadhyay et al. [15] developed the mathematical model for the bearing by considering the non-linear spring along with contact damping at the ball-race contact. Effect of internal radial clearance along with unbalanced rigid rotor has been explained. Period doubling and mechanism of intermittency have been observed that lead to chaos.

So, after detailed literature survey, the authors have decided to study the unbalanced rotor along with the internal radial clearance as nonlinearity. The study has been done by numerically for deep groove ball bearing SKF 6205 having nine no. of balls. The bearing has internal radial clearance of class C3 (14 μm) and percentage of unbalanced is 10 %.

2 Problem Formulation

The equations of motion that describe the dynamic behavior of the complete model can be derived by using Lagrange's equation for a set of independent generalized coordinates, as:

$$\frac{d}{dt}\frac{\partial T}{\partial\left\{\dot{p}\right\}}-\frac{\partial T}{\partial\{p\}}+\frac{\partial V}{\partial\{p\}}+\frac{\partial P_d}{\partial\left\{\dot{p}\right\}}=\{f\} \tag{1}$$

where T, V, p and f are kinetic energy, potential energy, vector with generalized degree-of-freedom (DOF) coordinate and vector with generalized contact forces respectively and Pd represent the dissipation energy due to damping. The kinetic and potential energies can be subdivided into the contributions from the various components i.e. from the rolling elements, the inner race, the outer race and the rotor Fig. 1.

The kinetic energy and potential energy contributed by the inner race, outer race, balls, rotor and springs, can be differentiated with respect to the generalized coordinates ρ_j (j = 1, 2,……, N_b), x_{in}, and y_{in} to obtain the equations of motion. For the generalized coordinates ρ_j, where j = 1, 2,……, N_b, the equations are:

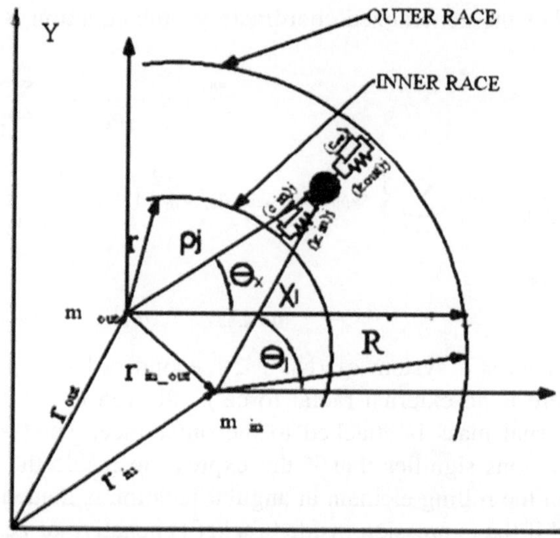

Fig. 1 Mass-spring-damper model of roller bearing

$$
\begin{aligned}
&\ddot{\rho}_j + g\sin\theta_j + \rho_j\dot{\theta}^2 - \frac{1}{m_j}(k_{in\_contact})\,[\delta_{in}]_+^{3/2}\frac{\partial\chi_j}{\partial\rho_j} \\
&+ \frac{1}{m_j}(k_{out\_contact})\,[\delta_{out}]_+^{3/2} \\
&+ \frac{1}{2\,m_j}\frac{\partial\left[(k_{in\_contact})\left([\delta_{in}]_+^{1/2}\right)\right]}{\partial\rho_j}[\delta_{in}]_+^2 \\
&+ \frac{1}{2\,m_j}\frac{\partial\left[(k_{out\_contact})\left([\delta_{out}]_+^{1/2}\right)\right]}{\partial\rho_j}[\delta_{out}]_+^2 \\
&+ \frac{3}{2\,m_j}\sum_{j=1}^{N_{r.e.}}\left\{C_{in}\,(K_{in\_contact})\,\delta_{in_+}^{3/2}\,(-\dot{\chi}_j)^q\frac{\partial\,\dot{\chi}_j}{\partial\,\dot{\rho}_j}\right\} \\
&+ \frac{3}{2\,m_j}\sum_{j=1}^{N_{r.e.}}C_{out}\,(K_{out\_contact})\,\delta_{out_+}^{3/2}\left(-\dot{\rho}_j\right)^q = 0 \\
&j = 1, 2, \ldots\ldots, N_b
\end{aligned}
\tag{2}
$$

For the generalized coordinate x_{in} the equation is:

$$
\begin{aligned}
&\ddot{x}_{in} - \frac{1}{m_{rotation}}\sum_{j=1}^{r.e}(k_{in\_contact})\,[\delta_{in}]_+^{3/2}\frac{\partial\chi_j}{\partial x_{in}} + \\
&\frac{3}{2\,m_{rotation}}\sum_{j=1}^{N_{r.e.}}\left\{C_{in}\,(K_{in\_contact})\,\delta_{in_+}^{3/2}\left(-\dot{\chi}_j\right)^q\frac{\partial\,\dot{\chi}_j}{\partial\,\dot{x}_{in}}\right\} = \frac{F_u\sin(\omega_s\,t)}{m_{rotation}}
\end{aligned}
\tag{3}
$$

For the generalized coordinate y_{in} the equation is:

$$\ddot{y}_{in} + g - \frac{1}{m_{rotation}} \sum_{j=1}^{N_b} (k_{in\_contact}) \, [\delta_{in}]_+^{3/2} \frac{\partial \chi_j}{\partial y_{in}} +$$

$$\frac{3}{2\, m_{rotation}} \sum_{j=1}^{N_{r.e.}} \left\{ C_{in} \left(K_{in\_contact}\right) \delta_{in_+}^{3/2} \left(-\dot{\chi}_j\right)^q \frac{\partial \dot{\chi}_j}{\partial \dot{y}_{in}} \right\} = \frac{(W + F_u \cos(\omega_s t))}{m_{rotation}}$$

where,

$$m_{rotation} = (m_{inner} + m_{rotor}) \tag{4}$$

This is a system of (N_b + 2) second order, non-linear differential equations. There is no external radial force is allowed to act on the bearing system and no external mass is attached to the outer race. The "+" sign as subscript in these equations signifies that if the expression inside the bracket is greater than zero, then the rolling element at angular location is loaded giving rise to restoring force and if the expression inside bracket is negative or zero, then the rolling element is not in the load zone, and restoring force is set to zero. For the balanced rotor condition, the unbalanced rotor force (Fu) is set to be zero.

In the present paper, the authors have written the equation of motion directly. The derivation of the equationof motion has been explained in detail in the Ref. Upadhyay et al. [16].

3 Methods of Solution

The coupled non-linear second order differential Eqs. (2–4) is solved by numerical integration technique which is a time domain approach. The non-analytic nature of the stiffness term renders the system equations difficult for analytical solution.

4 Results and Discussion:

To get the satisfactory post transient steady state condition, numerical simulation is run for 2 s. To save the computational time, artificial damping, c = 50 N-s/m is used. In the present study, DGB bearing SKF 6205, class three type is used, in which

Table 1 Behaviour of the system

Preload	Behavior of the system
5 N	Quasiperiodic
25 N	Quasiperiodic (3rd order)
50 N	Chaotic
75 N	Quasiperiodic (3rd order)
100 N	Quasiperiodic (2nd order)

internal radial clearance is 14 μm. The analysis has been done by keeping speed 9,000 RPM as constant and taking the value of preload as 5 N, 25 N, 50 N, 75 N and 100 N. The behavior of the system for different preload is listed in the Table 1

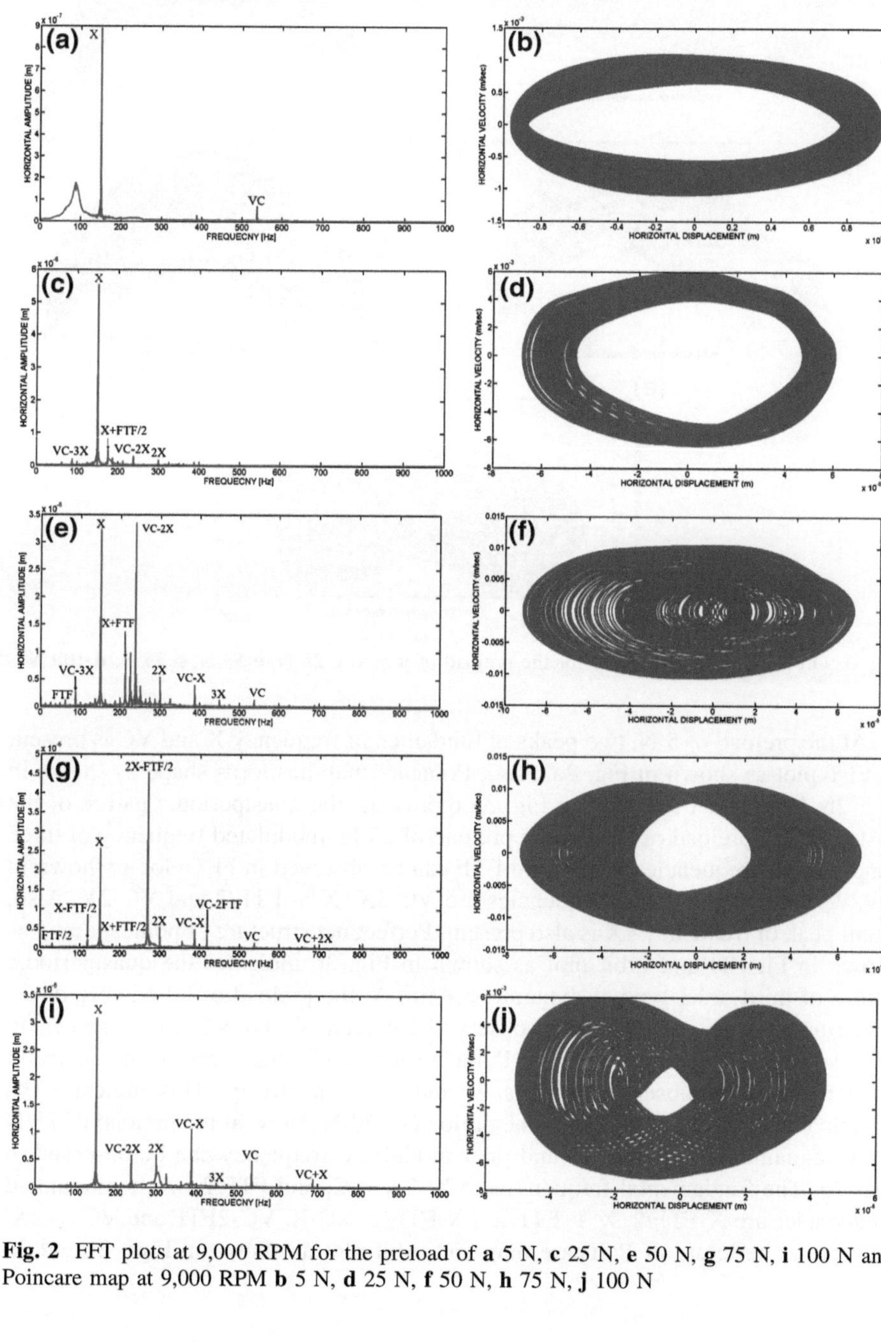

Fig. 2 FFT plots at 9,000 RPM for the preload of **a** 5 N, **c** 25 N, **e** 50 N, **g** 75 N, **i** 100 N and Poincare map at 9,000 RPM **b** 5 N, **d** 25 N, **f** 50 N, **h** 75 N, **j** 100 N

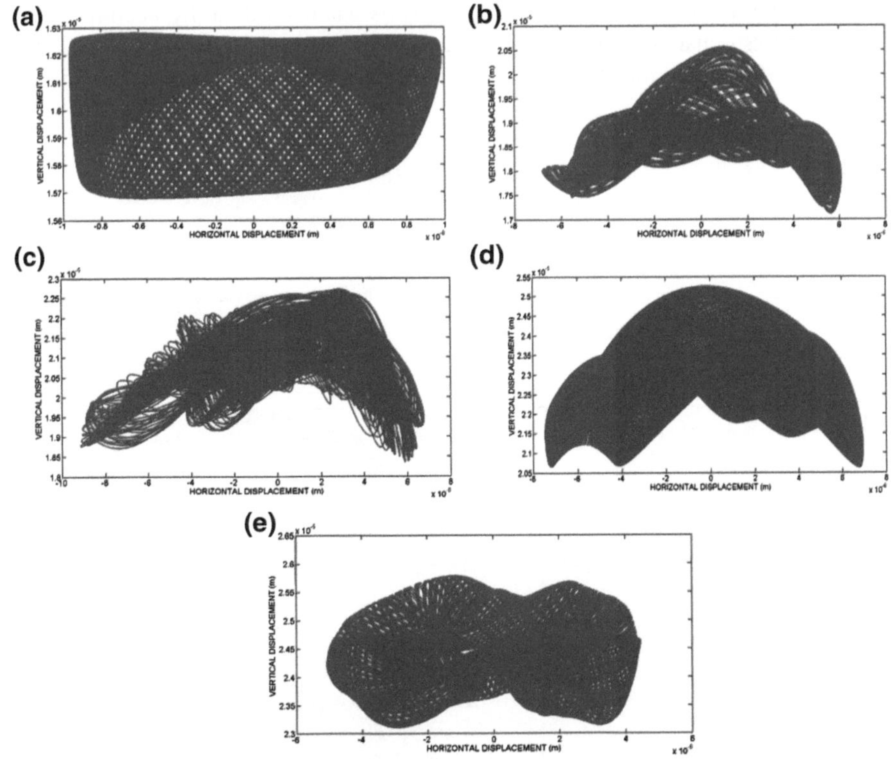

Fig. 3 Orbit plots at 9,000 RPM for the preload of **a** 5 N, **c** 25 N, **e** 50 N, **b** 75 N, **d** 100 N

At the preload of 5 N, two peaks of fundamental frequency X and VC is present in FFT plot as shown in Fig. 2a. Also, Poincare map has torus shape as shown in Fig. 2b. Perfect net structure in Fig. 3a represents the quasiperiodic nature of the system at the preload of 5 N.At the preload of 25 N, modulated frequency of three fundamental frequencies VC, X and FTF can be observed in FFT plot as shown in Fig. 2c. These modulated frequencies are VC-3X, X + FTF/2 and VC-2X. Also, small peak of frequency 2X is also present. Perfect net structure of poincare map as shown in Fig. 2d and orbit plot as shown in Fig. 3b indicates the quasiperiodic nature of third order type of dynamic system. At the preload of 50 N, very dense spectrum of frequency has been observed between X and VC-2X as shown in Fig. 2e. Further, aperiodic orbit in Poincare map and dense orbit at the centre in orbit plot can be observed in Figs. 2f and 3c, respectively. This indicates the chaotic behavior of the system at the preload of 50 N. Now, at the preload of 75 N three fundamental frequencies and their modulated frequency can be observed in Fig. 2g. The fundamental frequency are X, 2X, VC, and FTF/2, while modulated frequencies are X-FTF/2, X + FTF/2, 2X-FTF/2, VC-X, VC-2FTF and VC + 2X. Perfect net structure of Poincare map and orbit plot as shown in Figs. 2f and 3d respectively, indicates the quasiperiodic behavior (3rd order) of the system at the preload of 75 N. At 100 N, two fundamental frequencies X and VC and their modulated frequency can be observed. Fundamental frequencies are X, 2X, 3X and VC, while modulated frequencies are VC-2X, VC-X and VC + X. Also, perfect net structure of Poincare map and orbit plot as shown in Figs. 2j and 3e respectively indicates the quasiperiodic nature (2nd order) of the system. Here, at the preload of 100 N, the system behavior of the 2nd order because of only two fundamental frequencies X and VC.

5 Conclusion

Based on result and discussion following point can be concluded.

- At 50 N, the system has chaotic nature. More modulated frequencies present in FFT plot.
- At, 25 N and 75 N, the system has 3rd order quasiperiodic behavior, while at 100 N, the system has 2nd order quasiperiodic behavior.
- Perfect net structure of Poincare map and orbit plot indicates the quasiperiodic behavior of the system.
- As the preload increases the rate of increase of deflection decreases and therefore, preloading tends to reduce the bearing deflection under the additional loading. The reason for this is that the balls become stiffer with the increase in preload.
- Hence, it can be concluded that during the selection of the bearing, clearance should be decided carefully and preloading should be given to the bearing carefully when it is installed in the system, through which system can be stable. Preload should not be too small otherwise system has chaos and should not too high so the bearing will damage.

References

1. Aktürk N, Uneeb M, Gohar R (1997) The effects of number of balls and preload on vibrations associated with ball bearings. ASME J Tribol 119:747–753
2. Wardle FP, Poon SY (1983) Rolling bearing noise—cause and cure, chart. Mech eng 30:36–40
3. Purohit RK, Purohit K (2006) Dynamic analysis of ball bearings with effect of preload and number of balls. Int J Appl Mech Eng 11:77–91
4. Cao M, Xiao A (2008) A comprehensive dynamic model of double-row spherical roller bearing—model development and case studies on surface defects, preloads, and radial clearance. Mech Syst Sig Proc 22:467–489
5. Perret H (1953) Elastiche spielschwingungen konstant belaster walzlger, werstatt und betrieb 3:354–358

6. Meldau E (1951) Die Bewegung der achse von walzlagern bei geringen drehzahlen, werkstadt und betreib 7:308–313
7. Sunnersjo CS (1978) Varying compliance vibrations of rolling bearings. J sound vib 58:363–373
8. Mevel B, Guyader JL (1993) Routes to chaos in ball bearings. J Sound Vib 162:471–487
9. Fukata S, Gad EH, Kondou T, Ayabe T, Tamura H (1985) On the radial vibrations of ball bearings (computer simulation). Bull J Soc Mech Eng 28:899–904
10. Tiwari M, Prakash O, Gupta K (2000) Effect of radial internal clearance of a ball bearing on the dynamics of a balanced, horizontal rotor. J Sound Vib 238(5):723–756
11. Harsha SP, Sandeep K, Prakash R (2003) The effect of speed of balanced rotor on nonlinear vibrations associated with ball bearings. Int J Mech Sci 47(4):225–240
12. Harsha SP (2005a) Non-linear dynamic response of a balanced rotor supported on rolling element bearings. Mech Syst Sig Proc 19:551–578
13. Harsha SP (2005b) Non-linear dynamic analysis of an unbalanced rotor supported by roller bearings. Chaos, Solutions Fractals 26:47–66
14. Harsha SP (2006) Non-linear dynamic responses of a balanced rotor supported by rolling element bearings due to radial internal clearance effect. Mech Mach Theory 41:688–706
15. Upadhyay SH, Harsha SP, Jain SC (2010) Analysis of nonlinear phenomena in high speed ball bearings due to radial clearance and unbalanced rotor effects. J Vib Control 16(1):65
16. Upadhyay SH, Jain SC, Harsha SP (2009) Nonlinear vibration signature analysis of high speed rotating shaft due to ball size variations and varying number of balls. Proc IMechE Part K: J Multi-body Dyn 223:83–105

Part VIII
Bio-Tribology, Green Tribology and Biomimetics

Temperature Distribution in Living Tissue with Fractional Bioheat Model in Thermal Therapy

R. S. Damor, Sushil Kumar and A. K. Shukla

Abstract This paper deals the study of fractional bioheat equation for cancer treatment in thermal therapy with external EM (electromagnetic) heating. Numerical solution is obtained by implicit finite difference method. The effect of anomalous diffusion (super-diffusion) in tissue has been studied, the temperature profile over the entire affected region are obtained for different order of fractional bioheat equation.

Keywords Fractional bioheat equation · Hyperthermia · Finite difference method

1 Introduction

Hyperthermia treatment has been demonstrated effective as a cancer therapy in recent years. Its objective is to raise temperature of pathological tissue above cytotoxic temperature (41–45 °C) without overexposing healthy tissue. The success of hyperthermia treatment strongly depends on knowledge of the heat transfer process in blood perfused tissue [1]. Fractals and fractional calculus have been

---

AMS subject classification (2010): 35R11; 80M20; 65M06.

R. S. Damor (✉) · S. Kumar · A. K. Shukla
Department of Applied Mathematics and Humanities, S. V. National Institute of Technology, Surat 395007, India
e-mail: rameshsvnit2010@gmail.com

S. Kumar
e-mail: sushilk@ashd.svnit.ac.in

A. K. Shukla
e-mail: aks@ashd.svnit.ac.in

H. C. Patel et al. (eds.), *Proceedings of International Conference on Advances in Tribology and Engineering Systems*, Lecture Notes in Mechanical Engineering, DOI: 10.1007/978-81-322-1656-8_44,

used to improve the modelling accuracy of many phenomena in natural science. Numerical solution of fractional diffusion equation by finite difference method are studied by many researcher, Meerschaert et al. [2], gave a second order accurate numerical approximation for the fractional diffusion equation, Murio [3], discussed implicit finite difference approximation of time fractional diffusion equation.Many numerical and experimental methods have been developed in order to solve the bioheat equation. Tunc et al. [4] gave the temperature response of biological tissue with external EM heating and space dependent blood perfusion term.

Kumar and Katiyar [5], studied freezing and thawing process in tissue using porous media approach, recently Singh et al. [6] gave the solution of fractional bioheat equation by finite difference and homotopy perturbation.

In this paper, we consider space fractional derivative of order $\alpha \in (1, 2)$ which is in the form Riemann–Liouville fractional derivative and apply shifted Grunwald formula [2] to arrive at an approximation. We use implicit finite difference method to solve the fractional bioheat model. The temperature profiles are obtained for different values of α, to study the effect of α on temperature profile in tissue.

Riemann–Liouville Fractional Derivative [7]

$$\frac{\partial^{\alpha} u(x,t)}{\partial t^{\alpha}} = \begin{cases} \dfrac{1}{\Gamma(2-\alpha)} \dfrac{d^n}{dx^n} \displaystyle\int_0^x \frac{u(x,s)}{(x-s)^{-\alpha+1-n}} ds, & for\ 1 \le \alpha < 2 \\ \dfrac{\partial^{\alpha} u(x,t)}{\partial x^2} & for\ \alpha = 2 \end{cases} \tag{1}$$

2 Heat Transfer Model

In this study, we consider fractional form of Pennes [8] bioheat heat model by replacing space derivative by Riemann–Liouville fractional derivative.

$$\rho c \frac{\partial T}{\partial t} = K \frac{\partial^{\alpha} T}{\partial x^{\alpha}} + \rho_b w_b c_b (T_a - T) + q_{\text{met}} + q_{\text{ext}}, \tag{2}$$

where ρ, c, k, T, t, and x represents density, specific heat, thermal conductivity, temperature, time and distance respectively; the subscript t and b are for the tissue and blood respectively. T_a and w_b are considered as artillery temperature and blood perfusion rate respectively. q_{met} and q_{ext} are metabolic heat generation and external heat source in skin tissue respectively.

$$q_{\text{met}} = q_0 \left(1 + d \left(\frac{T - T_0}{T_0} \right) \right) \tag{3}$$

where $d = T_0/10$

$$q_{\text{ext}} = \rho SPe^{a(x-0.01)}. \tag{4}$$

where S and a are antenna constants, P transmitted power. x is the distance of tissue from outer surface. The initial and boundary conditions of the model are considered as,

$$T(x,0) = T_0 \tag{5}$$

$$\frac{\partial T}{\partial x}_{|x=0} = 0, T(x,t)_{|x=L} = T_w, \tag{6}$$

where T_w is the tissue wall temperature.

On making dimensionless variables,

$$\zeta = (x/L)^{(1/\alpha)}, \eta = \left(\frac{K}{\rho c L^2}\right)t, \theta = \left(\frac{T - T_0}{T_0}\right),$$

$$\Psi_m = \frac{q_0}{T_0 K}L^2, \Psi_f = \frac{W_b c_b}{K}L^2, \Psi_e = \frac{\rho SP}{T_0 K}L^2$$

Equation (2) reduces in the following form,

$$\frac{\partial \theta}{\partial \eta} = \frac{\partial^\alpha \theta}{\partial \zeta^\alpha} + A\theta + \Psi(\zeta), \tag{7}$$

where $A = (d\Psi_m - \Psi_f)$ and $\Psi_\zeta = (\Psi_m + \Psi_e e^{L(1-\zeta-0.01)})$

The initial and boundary conditions are expressed as,

$$\theta(\zeta, 0) = 0 \tag{8}$$

$$\frac{\partial \theta}{\partial \zeta}_{|\varsigma=0} = 0, \theta(\zeta,\eta)_{|\varsigma=1} = \theta_w. \tag{9}$$

3 Implicit Finite Difference Scheme

$$\frac{\partial^\alpha \theta}{\partial x^\alpha} = \frac{1}{h^{(\alpha)}} \sum_{l=0}^{i+1} g_l \theta^n_{i-l+1} \tag{10}$$

where,

$$g(l) = (-1)^l \frac{\alpha(\alpha-1)\cdots(\alpha-l+1)}{l!}, \textit{for } l = 1,2,3,\ldots$$

Now, the implicit finite difference formula for Eq. (7) is given as,

$$\frac{\theta_i^{n+1}-\theta_i^n}{k}=\frac{1}{h^\alpha}\left(\sum_{l=0}^{i+1}g_l\theta_{i-l+1}^n\right)+A\theta_i^n+\Psi(\zeta(i)) \tag{11}$$

On taking $\gamma=\frac{k}{h^2}$ and further simplification, gives

$$\begin{aligned}-\gamma g(2)\theta_{i-1}^n+(1-\gamma g(1)-kA)\theta_i^n-\gamma g(0)\theta_{i+1}^n-\gamma\sum_{l=3}^{i+1}g_l\theta_{i-l+1}^n\\=\theta_i^{n-1}\ +k*\Psi(\zeta(i))\end{aligned} \tag{12}$$

$$[M]\theta_i^n=\theta_i^{n-1}+k*\Psi(\zeta(i)) \tag{13}$$

where M is the lower triangular matrix with super diagonal of order $(n-1,n-1)$, and

$$M(i,i)=1+2\gamma-k*A,\ \ for\, i=2\, to n-2$$

$$M(i,i+1)=-\gamma,\ \text{if i<}l$$

$$M(i,l)=-\gamma g(i-l+1) for\, i=2\, to\, n-2$$

Initial condition and boundary conditions given by Eqs. (8) and (9) can be written as,

$$\theta_i^n|_{\eta=0}=0 \tag{14}$$

$$\theta_2^n=\theta_1^n;\theta_{m+1}^n=\theta_w; \tag{15}$$

4 Results

In present study we consider the following parameters [4].

$$\begin{aligned}x=0.05\text{ m},\rho=1050\,\text{kgm}^{-3},c=4180\,\text{Jkg}^{-1}C^{-1},\\c_b=3344\text{ Jkg}^{-1}\ C^{-1},\ K=0.5\text{ Wm}^{-1}C^{-1},\\W_b=8\text{ kgm}^{-3}s^{-1},q_0=1091\,\text{Wm}^{-3},S=12.5\text{ m}^{-1}\\P=10\text{ W},a=-127\text{ kg}^{-1},T_0=T_a=T_w=37^0\text{ C}\end{aligned}$$

To study the effect of α on temperature profile during hyperthermia, temperature profile for classic Pennes bioheat equation and fractional bioheat equation with $\alpha=1.9,1.8$ and 1.7 are plotted in Figs. 1, 2, 3 respectively, at different time.

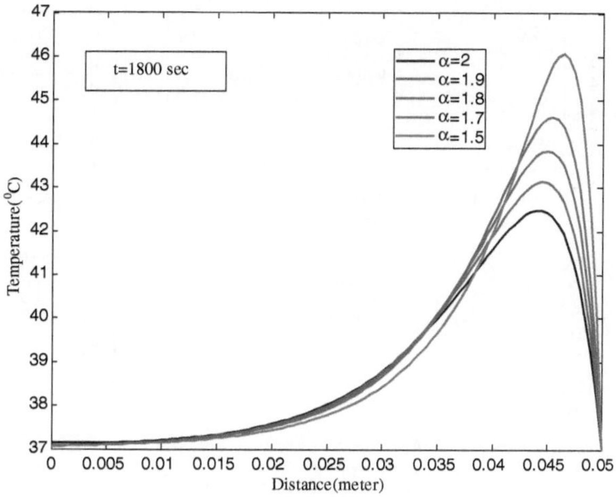

Fig. 1 Temperature distribution at t = 1,800 s

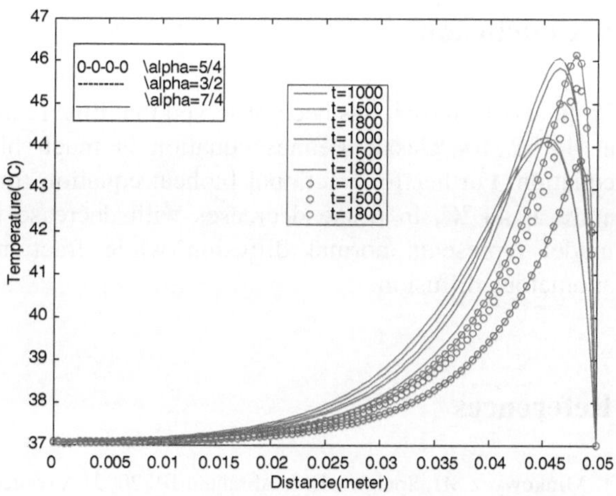

Fig. 2 Temperature distribution for $\alpha = 5/4$, 3/2 and 7/4

It observed that temperature 43 °C is obtained. Fig. 1 show the temperature distribution of classic Pennes bioheat equation ($\alpha = 2$) with respect to distance, at time $t = 300$, 600 and 1800 s here, time taken to reach at temperature 43 °C is 1,800 s.

Figure 2 indicates the temperature profile of fractional bioheat equation with $\alpha = 1.9$, 1.8 and 1.7 with respect to distance at time $t = 1,800$s in this case temperature is higher for decreasing values of α.

Figure 3 represents the temperature profile of fractional bioheat equation with $\alpha = 1.7$, 1.8, 1.9 and 2, with respect to distance at time $t = 300$, 600 and $1,800$ s in this case time taken to reach at temperature 44 °C is less for decreasing values of α.

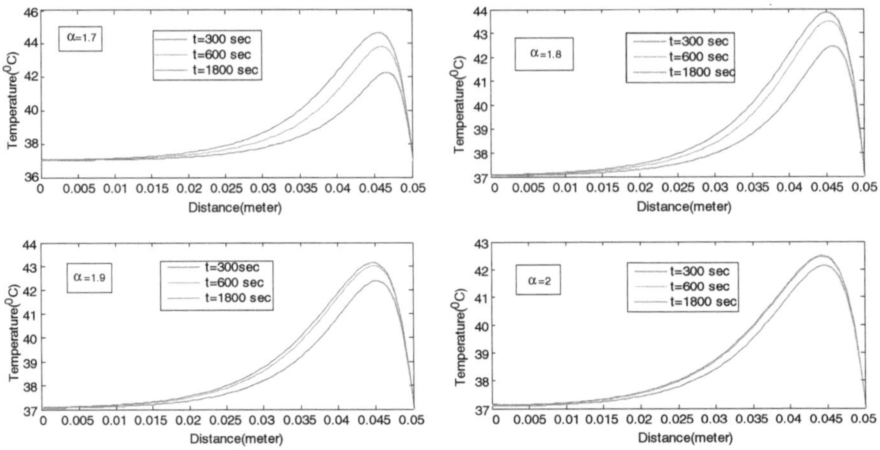

Fig. 3 Temperature distribution for different value of α

5 Conclusion

From above study it has been observed that time required reaching the temperature at 44 °C, for classic Pennes equation is much higher than fractional bioheat equation. Further for fractional bioheat equation time required to obtain temperature at 44 °C, in tissue decreases with decrease in α. Classic Pennes bioheat model represents normal diffusion while fractional bioheat model represent anomalous diffusion.

References

1. Minkowycz MJ, Sparoow EM, Abraham JP (2009) Advances in numerical heat transfer, vol 3. CRC press, New York
2. Meerschaert MM, Scheffler HP, Tadjeran C (2006) Finite difference methods for two-dimensional fractional dispersion equations. J Comput Phys 211:249–261
3. Murio DA (2008) Implicit finite difference approximation for time fractional diffusion equations. Comput Math Appl 56:1138–1145
4. Tunc M, Camdali U, Parmaksizoglu C, Cikrikci S (2006) The bioheat transfer equation and its applications in hyperthermia treatments. Eng Comput 23(4):451–463
5. Kumar S, Katiyar VK (2010) Mathematical modelling of freezing and thawing process in tissue a porous media approach. Int J Appl Mech 2(3):617–633
6. Singh J, Gupta PK, Rai KN (2011) Solution of fractional bioheat equations by finite difference method and HPM. Math Comput Model 54:2316–2325
7. Podlubny I (1999) Fractional differential equations. Academic Press, New York
8. Pennes HH (1948) Analysis of tissue and arterial blood temperature in the resting forearm. J Appl Physiol 1:93–122

About the Editors

Dr. Himanshu C. Patel is Associate Professor and Head of the Department of Mathematics at L. D. College of Engineering, Ahmedabad, India. He has been teaching for 13 years and has over 10 years of research experience. He has published several research papers in various international journals. He has also participated in several international conferences, including the ASME/STLE conference in 2012 at Denver, Colorado. He is Advisor—Research and Development in Science and Humanities, Gujarat Technological University.

Dr. Gunamani Deheri is Associate Professor of Mathematics at Sardar Patel University, Vallabh Vidhyanagar, Gujarat, India. He has 27 years of teaching/research experience. His research interests include tribology and functional analysis. He has over 100 research papers to his credit in various international journals. He has won the Hariom Ashram Award for the best research paper five times. He is also the winner of the Balkanji Bari Science Trust Award and the Excellence in Research Award, Saurashtra University.

Dr. H. S. Patel is Professor of Applied Mechanics at Government Engineering College, Patan, Gujarat, India. His areas of expertise include structural engineering, tsunami engineering and advanced computing in structural engineering. He has published more than 40 papers in international and national journals and three books in area of structural engineering. He has over 28 years of teaching experience and has conducted several training programs on computer aided design.

Dr. Shreya M. Mehta is Associate Professor of Mechanical Engineering at L. D. College of Engineering, Ahmedabad, Gujarat, India. She holds a doctorate from S. V. National Institute of Technology, Surat. She has been teaching for over 16 years. She has published several research papers at both the national and the international level. She had guided several students in their dissertation work.

H. C. Patel et al. (eds.), *Proceedings of International Conference on Advances in Tribology and Engineering Systems*, Lecture Notes in Mechanical Engineering, DOI: 10.1007/978-81-322-1656-8, © Springer India 2014

Author Index

H. C. Patel et al. (eds.), *Proceedings of International Conference on Advances in Tribology and Engineering Systems*, Lecture Notes in Mechanical Engineering, DOI: 10.1007/978-81-322-1656-8, © Springer India 2014